METHODS IN MOLECULAR BIOLOGY

Series Editor
John M. Walker
School of Life and Medical Sciences
University of Hertfordshire
Hatfield, Hertfordshire, UK

For further volumes:
http://www.springer.com/series/7651

For over 35 years, biological scientists have come to rely on the research protocols and methodologies in the critically acclaimed *Methods in Molecular Biology* series. The series was the first to introduce the step-by-step protocols approach that has become the standard in all biomedical protocol publishing. Each protocol is provided in readily-reproducible step-by-step fashion, opening with an introductory overview, a list of the materials and reagents needed to complete the experiment, and followed by a detailed procedure that is supported with a helpful notes section offering tips and tricks of the trade as well as troubleshooting advice. These hallmark features were introduced by series editor Dr. John Walker and constitute the key ingredient in each and every volume of the *Methods in Molecular Biology* series. Tested and trusted, comprehensive and reliable, all protocols from the series are indexed in PubMed.

Neutrophil

Methods and Protocols

Third Edition

Edited by

Mark T. Quinn

Department of Microbiology and Immunology, Montana State University, Bozeman, MT, USA

Frank R. DeLeo

Laboratory of Bacteriology, Rocky Mountain Laboratories, Division of Intramural Research, National Institute of Allergy and Infectious Diseases, National Institutes of Health, Hamilton, MT, USA

Humana Press

Editors
Mark T. Quinn
Department of Microbiology
and Immunology
Montana State University
Bozeman, MT, USA

Frank R. DeLeo
Laboratory of Bacteriology, Rocky Mountain Laboratories
Division of Intramural Research, National Institute of Allergy
and Infectious Diseases
National Institutes of Health
Hamilton, MT, USA

ISSN 1064-3745 ISSN 1940-6029 (electronic)
Methods in Molecular Biology
ISBN 978-1-0716-0153-2 ISBN 978-1-0716-0154-9 (eBook)
https://doi.org/10.1007/978-1-0716-0154-9

Cover Caption: Top row, from left to right: 1. Human neutrophil swarming against a cluster of Candida albicans. 2. Aspergillus fumigatus spores were patterned in a cluster and allowed to grow into hyphae. 3. Sytox green staining showing neutrophil extracellular trap release inside a human neutrophil swarm against C. albicans. 4. A patterned cluster of C. albicans yeast. Images taken by Alex Hopke and prepared for cover by Xiao Wang (Center for Engineering in Medicine, Massachusetts General Hospital, Harvard Medical School, Charlestown, MA).

This Humana imprint is published by the registered company Springer Science+Business Media, LLC, part of Springer Nature.
The registered company address is: 233 Spring Street, New York, NY 10013, U.S.A.

Dedication

This volume is dedicated to Dr. Niels Borregaard (1954–2017) in recognition of his extensive contributions to neutrophil cell biology, especially in the understanding of neutrophil granule formation, subcellular distribution, and function. Niels was a friend and contributor to the first two editions of this book. This volume is also dedicated to our families, who are tolerant of the time we spend researching neutrophils.

Preface

Neutrophils (also known as polymorphonuclear neutrophils (PMNs) or granulocytes) are the most abundant white cell in humans. Granulocytes and/or granulocyte precursors normally comprise ~60% of the nucleated cells in bone marrow and the bloodstream. Mature neutrophils have a typical circulating half-life of 6–8 h in the blood and then migrate through tissues for ~2–3 days. Their relatively short lifespan is devoted largely to surveillance for invading microorganisms. During infection, the neutrophil lifespan is extended, granulopoiesis increases, and large numbers of neutrophils are rapidly recruited to the site(s) of infection. Following recognition (binding) and phagocytosis of microorganisms, neutrophils utilize an extraordinary array of oxygen-dependent and oxygen-independent microbicidal weapons to destroy infectious agents. Oxygen-dependent mechanisms involve the production of reactive oxygen species (ROS), while oxygen-independent mechanisms include degranulation and release of lytic enzymes and bactericidal peptides. Inasmuch as these processes are highly effective at killing most ingested microbes, neutrophils serve as the primary cellular defense against infection.

The aim of *Neutrophils: Methods and Protocols, Third Edition,* is to provide (1) a set of protocols to assess basic neutrophil functions and (2) protocols for investigating specialized areas in neutrophil research. This volume is designed for the basic researcher involved in the study of neutrophil function. A wide variety of methods have been developed to assess neutrophil function, and these methods have been instrumental in advancing our understanding of the role of neutrophils in host defense and inflammatory disease. For those researchers and clinicians interested in the study of neutrophils, the availability of a comprehensive source of protocols describing the most modern methodological advances in neutrophil biology is invaluable, as many publications do not provide information on the finer details critical to success of a given method. As such, we have compiled a series of protocols written by leading researchers in the field that provide detailed guidelines for establishing and performing the most common neutrophil function assays. Hints of the best way to perform these methods as well as guidance in detecting associated problems are included, so novice investigators will also be able to effectively utilize these assays.

In the third edition of *Neutrophils: Methods and Protocols,* chapters retained from previous editions have been updated and include many new approaches. In addition, the *Third Edition* contains a number of new chapters that were not included in the first two editions. Part I contains overviews of neutrophil biology and function, and disorders of neutrophils. Part II describes commonly used methods to isolate neutrophils from humans and other animal species. This section also contains a chapter that describes use of a neutrophil transplant model with zebrafish larva. Part III details methods for investigating chemotaxis, transmigration, phagocytosis, and bactericidal activity. Three chapters provide methods used to assess neutrophil transmigration, chemotaxis, or swarming against live microbes. One of the chapters updated from the *Second Edition* covers a neutrophil microinjection approach for studying phagocytosis, and a new chapter details use of imaging flow cytometry to evaluate phagocytosis. Several of these chapters are new and contain innovative methods and approaches for studying neutrophils. Part IV includes protocols that measure neutrophil apoptosis, calcium signal transduction, degranulation and detection of cytoplasmic granules, gene expression, transcription factors, and apoptosis. Part V provides multiple assays for measuring production of intracellular and/or extracellular reactive oxygen species. In addition, there is a chapter that details the history and use of the cell-free NADPH oxidase assay, an iconic assay for studies of the neutrophil NADPH oxidase. Part VI provides chapters that describe how to analyze formation and function of neutrophil extracellular traps, including new chapters on visualization of NETs by intravital microscopy and detection of NETs in tissues. In addition to the step-by-step protocols, the **Notes** section of each chapter provides an outstanding depot of useful and interesting information not typically published in the Methods sections of standard journal articles.

We thank John M. Walker, Series Editor, and Springer Nature for the opportunity to assemble an outstanding collection of chapters and for help with the publication of the volume. We also thank the NIH IDeA Program (COBRE Grant GM110732) and the NIH Intramural Research Program, National Institutes of Allergy and Infectious Diseases, for support. Finally, we thank the authors for taking time to write outstanding chapters.

Bozeman, MT, USA *Mark T. Quinn*
Hamilton, MT, USA *Frank R. DeLeo*

Contents

Contributors

ULRIKE ABU-ABED • *Microscopy Core Facility, Max Planck Institute for Infection Biology, Berlin, Germany; Cellular Microbiology, Max Planck Institute for Infection Biology, Berlin, Germany*

SARAH L. ANZICK • *Genomics Unit, Rocky Mountain Laboratories, National Institute of Allergy and Infectious Diseases, National Institutes of Health, Hamilton, MT, USA*

LOUISA V. ASHBY • *Department of Pathology and Biomedical Science, Centre for Free Radical Research, University of Otago Christchurch, Christchurch, New Zealand*

JONATHAN W. ASTIN • *Department of Molecular Medicine and Pathology, Faculty of Medical and Health Sciences, University of Auckland, Auckland, New Zealand*

NICOLE D. BARTH • *Centre for Inflammation Research, Queen's Medical Research Institute, University of Edinburgh, Edinburgh, UK*

FLAVIA BAZZONI • *Department of Medicine, Section of General Pathology, University of Verona, Verona, Italy*

SAMIA BEDOUHÈNE • *Centre de Recherche sur l'Inflammation (CRI), INSERM-U1149, CNRS-ERL8252, Laboratoire d'Excellence Inflamex, Université Paris Diderot-Sorbonne Paris Cité, Faculté de Médecine, Site Xavier Bichat, Paris, France; Laboratoire de Biochimie Analytique et de Biotechnologie, Faculté des Sciences Biologiques et des Sciences Agronomiques, Université Mouloud Mammeri de Tizi-Ouzou, Tizi Ouzou, Algeria*

HALLA BJÖRNSDOTTIR • *Department of Oral Microbiology and Immunology, Institute of Odontology, Sahlgrenska Academy at University of Gothenburg, Gothenburg, Sweden*

KATJA BRANITZKI-HEINEMANN • *Department of Physiological Chemistry, University of Veterinary Medicine Hannover, Hannover, Germany; Research Center for Emerging Infections and Zoonoses (RIZ), University of Veterinary Medicine Hannover, Hannover, Germany*

VOLKER BRINKMANN • *Microscopy Core Facility, Max Planck Institute for Infection Biology, Berlin, Germany*

GRAHAM BROGDEN • *Department of Physiological Chemistry, University of Veterinary Medicine Hannover, Hannover, Germany*

JOHAN BYLUND • *Department of Oral Microbiology and Immunology, Institute of Odontology, Sahlgrenska Academy at University of Gothenburg, Gothenburg, Sweden*

JENNIE S. CAMPBELL • *Neutrophil Signalling Group, School of Dentistry, Cardiff University, Cardiff, UK*

MARCO A. CASSATELLA • *Department of Medicine, Section of General Pathology, University of Verona, Verona, Italy*

KARIN CHRISTENSON • *Department of Oral Microbiology and Immunology, Institute of Odontology, Sahlgrenska Academy at University of Gothenburg, Gothenburg, Sweden*

IWONA CICHON • *Department of Experimental Hematology, Institute of Zoology and Biomedical Research, Jagiellonian University, Kraków, Poland*

ALISON K. CRISS • *Department of Microbiology, Immunology, and Cancer Biology, University of Virginia, Charlottesville, VA, USA*

CLAES DAHLGREN • *Department of Rheumatology and Inflammation Research, Institute of Medicine, Sahlgrenska Academy at University of Gothenburg, Gothenburg, Sweden*

PHAM MY-CHAN DANG • *Centre de Recherche sur l'Inflammation (CRI), INSERM-U1149, CNRS-ERL8252, Laboratoire d'Excellence Inflamex, Université Paris Diderot-Sorbonne Paris Cité, Faculté de Médecine, Site Xavier Bichat, Paris, France*
HANNAH DARROCH • *Department of Molecular Medicine and Pathology, Faculty of Medical and Health Sciences, University of Auckland, Auckland, New Zealand*
NICOLE DE BUHR • *Department of Physiological Chemistry, University of Veterinary Medicine Hannover, Hannover, Germany; Research Center for Emerging Infections and Zoonoses (RIZ), University of Veterinary Medicine Hannover, Hannover, Germany*
FRANK R. DELEO • *Laboratory of Bacteriology, Rocky Mountain Laboratories, Division of Intramural Research, National Institute of Allergy and Infectious Diseases, National Institutes of Health, Hamilton, MT, USA*
SHARON DEWITT • *School of Biochemistry, University of Bristol, Bristol, UK; School of Medicine and School of Dentistry, Cardiff University, Cardiff, UK*
MARY C. DINAUER • *Department of Pediatrics (Hematology/Oncology), St. Louis Children's Hospital, Washington University School of Medicine in St. Louis, St. Louis, MO, USA; Department of Pathology and Immunology, St. Louis Children's Hospital, Washington University School of Medicine in St. Louis, St. Louis, MO, USA*
DAVID A. DORWARD • *Centre for Inflammation Research, Queen's Medical Research Institute, University of Edinburgh, Edinburgh, UK*
IAN DRANSFIELD • *Centre for Inflammation Research, Queen's Medical Research Institute, University of Edinburgh, Edinburgh, UK*
JAMEL EL-BENNA • *Centre de Recherche sur l'Inflammation (CRI), INSERM-U1149, CNRS-ERL8252, Laboratoire d'Excellence Inflamex, Université Paris Diderot-Sorbonne Paris Cité, Faculté de Médecine, Site Xavier Bichat, Paris, France*
NOAH FINE • *Faculty of Dentistry, University of Toronto, Toronto, ON, Canada*
PETER C. W. GAINES • *Department of Biological Sciences, University of Massachusetts Lowell, Lowell, MA, USA*
MICHAEL GLOGAUER • *Faculty of Dentistry, University of Toronto, Toronto, ON, Canada; Department of Dental Oncology, Maxillofacial and Ocular Prosthetics, Princess Margaret Cancer Centre, Toronto, ON, Canada; Centre for Advanced Dental Research and Care, Mount Sinai Hospital, Toronto, ON, Canada*
SERGIO GRINSTEIN • *Program in Cell Biology, Hospital for Sick Children, Toronto, ON, Canada; Department of Biochemistry, University of Toronto, Toronto, ON, Canada*
GAELEN GUZMAN • *Department of Molecular Microbiology and Immunology, Oregon Health and Science University, Portland, OR, USA*
CHRISTOPHER J. HALL • *Department of Molecular Medicine and Pathology, Faculty of Medical and Health Sciences, University of Auckland, Auckland, New Zealand*
MAURICE B. HALLETT • *Neutrophil Signalling Group, School of Medicine, Cardiff University, Cardiff, UK*
MARK B. HAMPTON • *Department of Pathology and Biomedical Science, Centre for Free Radical Research, University of Otago Christchurch, Christchurch, New Zealand*
ALEX HOPKE • *Center for Engineering in Medicine, Massachusetts General Hospital, Harvard Medical School, Shriners Hospital for Children, Boston, MA, USA*
MARGARITA HURTADO-NEDELEC • *Centre de Recherche sur l'Inflammation (CRI), INSERM-U1149, CNRS-ERL8252, Laboratoire d'Excellence Inflamex, Université Paris Diderot-Sorbonne Paris Cité, Faculté de Médecine, Site Xavier Bichat, Paris, France; AP-HP, Centre Hospitalier Universitaire Xavier Bichat, UF Dysfonctionnements Immunitaires, Paris, France*

Daniel Irimia • *Center for Engineering in Medicine, Massachusetts General Hospital, Harvard Medical School, Shriners Hospital for Children, Boston, MA, USA*
William Khoury • *Faculty of Dentistry, University of Toronto, Toronto, ON, Canada*
Liliya N. Kirpotina • *Department of Microbiology and Immunology, Montana State University, Bozeman, MT, USA*
Scott D. Kobayashi • *Laboratory of Bacteriology, Rocky Mountain Laboratories, National Institute of Allergy and Infectious Diseases, National Institutes of Health, Hamilton, MT, USA*
Elzbieta Kolaczkowska • *Department of Experimental Hematology, Institute of Zoology and Biomedical Research, Jagiellonian University, Kraków, Poland*
Douglas J. Kominsky • *Department of Microbiology and Immunology, Montana State University, Bozeman, MT, USA*
Silvie Kremserova • *Inflammation Program and Department of Medicine, Roy J. and Lucille A. Carver College of Medicine, University of Iowa, Iowa City, IA, USA*
Iraj Laffafian • *Neutrophil Signalling Group, School of Medicine, Cardiff University, Cardiff, UK*
Joanne Lannigan • *Department of Microbiology, Immunology, and Cancer Biology, University of Virginia, Charlottesville, VA, USA*
Benfang Lei • *Department of Microbiology and Immunology, Montana State University, Bozeman, MT, USA*
Nicholas J. Magon • *Department of Pathology and Biomedical Science, Centre for Free Radical Research, University of Otago Christchurch, Christchurch, New Zealand*
Natalia Malachowa • *Laboratory of Bacteriology, Rocky Mountain Laboratories, Division of Intramural Research, National Institute of Allergy and Infectious Diseases, National Institutes of Health, Hamilton, MT, USA*
Harry L. Malech • *Laboratory of Clinical Immunology and Microbiology, Division of Intramural Research, National Institute of Allergy and Infectious Diseases, National Institutes of Health, Bethesda, MD, USA*
Patrick P. McDonald • *Pulmonary Division, Medicine Faculty, Université de Sherbrooke, Sherbrooke, QC, Canada*
William M. Nauseef • *Inflammation Program and Department of Medicine, Roy J. and Lucille A. Carver College of Medicine, University of Iowa, Iowa City, IA, USA*
Weronika Ortmann • *Department of Experimental Hematology, Institute of Zoology and Biomedical Research, Jagiellonian University, Kraków, Poland*
Philip P. Ostrowski • *Program in Cell Biology, Hospital for Sick Children, Toronto, ON, Canada; Department of Biochemistry, University of Toronto, Toronto, ON, Canada*
Heather A. Parker • *Department of Pathology and Biomedical Science, Centre for Free Radical Research, University of Otago Christchurch, Christchurch, New Zealand*
Margery G. H. Pelletier • *Department of Biological Sciences, University of Massachusetts Lowell, Lowell, MA, USA*
Edgar Pick • *Department of Clinical Microbiology and Immunology, Sackler School of Medicine, Tel Aviv University, Tel Aviv, Israel*
Adeline R. Porter • *Laboratory of Bacteriology, Rocky Mountain Laboratories, Division of Intramural Research, National Institute of Allergy and Infectious Diseases, National Institutes of Health, Hamilton, MT, USA*
Mark T. Quinn • *Department of Microbiology and Immunology, Montana State University, Bozeman, MT, USA*
Rhiannon E. Roberts • *Neutrophil Signalling Group, Cardiff, UK*

Adriano G. Rossi • *Centre for Inflammation Research, Queen's Medical Research Institute, University of Edinburgh, Edinburgh, UK*

Ziv Roth • *Program in Cell Biology, Hospital for Sick Children, Toronto, ON, Canada*

Michal Santocki • *Department of Experimental Hematology, Institute of Zoology and Biomedical Research, Jagiellonian University, Kraków, Poland*

Igor A. Schepetkin • *Department of Microbiology and Immunology, Montana State University, Bozeman, MT, USA*

Daniel W. Siemsen • *Department of Microbiology and Immunology, Montana State University, Bozeman, MT, USA*

Asya Smirnov • *Department of Microbiology, Immunology, and Cancer Biology, University of Virginia, Charlottesville, VA, USA*

Michael D. Solga • *UVA Flow Cytometry Core, University of Virginia, Charlottesville, VA, USA*

Reuben J. Springer • *Department of Pathology and Biomedical Science, Centre for Free Radical Research, University of Otago Christchurch, Christchurch, New Zealand*

Dan E. Sturdevant • *Genomics Unit, Rocky Mountain Laboratories, National Institute of Allergy and Infectious Diseases, National Institutes of Health, Hamilton, MT, USA*

Martina Sundqvist • *Department of Rheumatology and Inflammation Research, Institute of Medicine, Sahlgrenska Academy at University of Gothenburg, Gothenburg, Sweden*

Steve D. Swain • *Department of Microbiology and Immunology, Montana State University, Bozeman, MT, USA*

Klaudia Szymczak • *Department of Biological Sciences, University of Massachusetts Lowell, Lowell, MA, USA*

Fikadu G. Tafesse • *Department of Molecular Microbiology and Immunology, Oregon Health and Science University, Portland, OR, USA*

Nicola Tamassia • *Department of Medicine, Section of General Pathology, University of Verona, Verona, Italy*

Marc Vendrell • *Centre for Inflammation Research, Queen's Medical Research Institute, University of Edinburgh, Edinburgh, UK*

Maren von Köckritz-Blickwede • *Department of Physiological Chemistry, University of Veterinary Medicine Hannover, Hannover, Germany; Research Center for Emerging Infections and Zoonoses (RIZ), University of Veterinary Medicine Hannover, Hannover, Germany*

Richard D. Ye • *Kobilka Institute of Innovative Drug Discovery, School of Life and Health Sciences, The Chinese University of Hong Kong, Shenzhen, China*

Part I

Neutrophils and Neutrophil Disorders: Overviews

Chapter 1

The Role of Neutrophils in the Immune System: An Overview

Harry L. Malech, Frank R. DeLeo, and Mark T. Quinn

Abstract

Neutrophils, also known as polymorphonuclear neutrophils (PMNs), have long been considered as the short-lived, nonspecific white cells that form pus—and also happen to kill invading microbes. Indeed, neutrophils were often neglected (and largely not considered) as immune cells. This historic view of neutrophils has changed considerably over the past several decades, and we now know that in addition to playing the predominant role in the clearance of bacteria and fungi, they have a major role in shaping the host response to infection and immune system homeostasis. The change in our view of the role of neutrophils in the immune system has been due in large part to the study of these cells in vitro. Such work has been made possible by new and/or improved methods and approaches used to investigate neutrophils. These methods are the focus of this volume.

Key words Polymorphonuclear leukocyte, Granulocyte, Neutrophil methods

1 Introduction

This valuable and unique book contains a compendium of methods and reviews that does much more than allow one to study the biology of neutrophils. What makes this collection of contributions so special is that it highlights and facilitates using the neutrophil as a simple, pure, single primary cell suspension model to study a remarkable array of generalized cellular functions (e.g., adhesion, chemotaxis and transmigration, phagocytosis, degranulation, oxygen radical production, apoptosis, and gene expression), as well as specialized functions (e.g., formation of extracellular traps) and molecules important to host defense against infection and the mediation and resolution of inflammation (*see* Fig. 1). Consideration of the array of chapter topics evokes some of the past history of inquiry into how neutrophils function and how we evolved into the current widespread use of the neutrophil as a convenient model system for studying so many types of cellular processes and biochemical pathways.

Mark T. Quinn and Frank R. DeLeo (eds.), *Neutrophil: Methods and Protocols*, Methods in Molecular Biology, vol. 2087, https://doi.org/10.1007/978-1-0716-0154-9_1, © Springer Science+Business Media, LLC, part of Springer Nature 2020

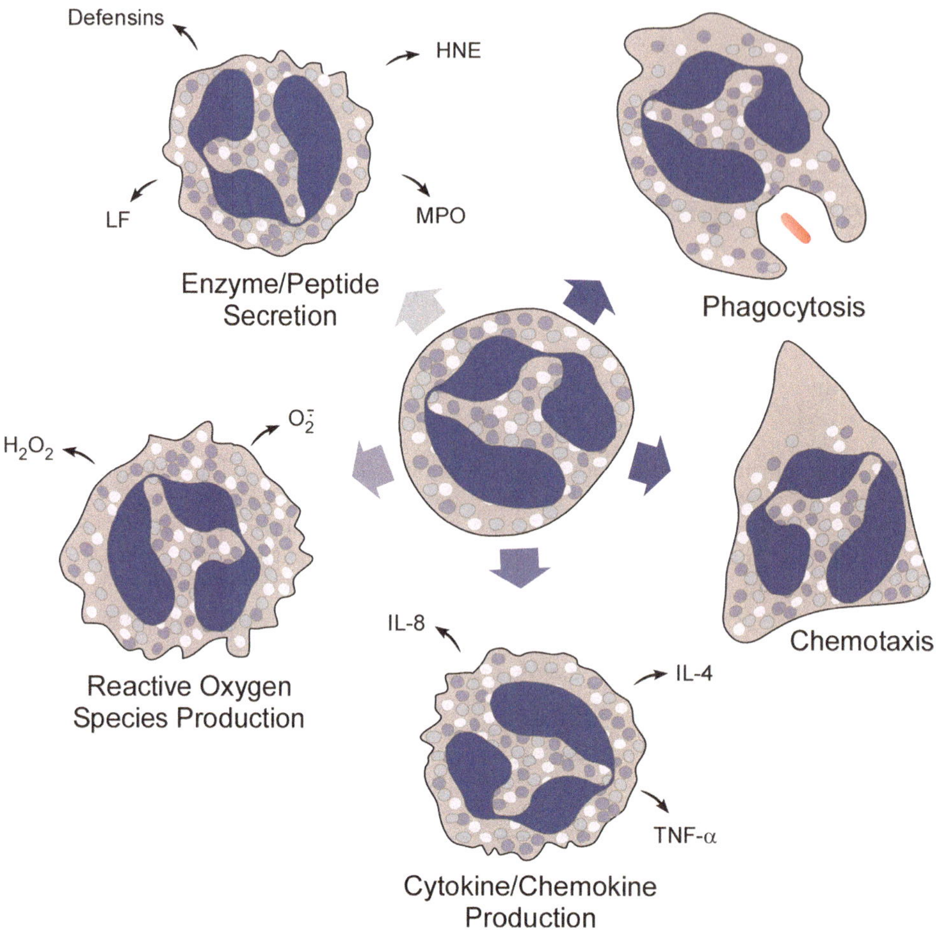

Fig. 1 Illustration of key neutrophil functions. Note that for production reactive oxygen species, secretion of granule components, and production of cytokines and chemokines, only a few representative molecules are shown. *HNE* human neutrophil elastase, *IL-8* interleukin-8, *IL-4* interleukin-4, *LF* lactoferrin, *MPO* myeloperoxidase, *TNF-α* tumor necrosis factor-α

2 Historical Overview

Only a few decades ago, in the 1970s and even the early 1980s, the biology and pathophysiology of the neutrophil was a boutique area of study involving a relatively small number of laboratories and investigators internationally. These investigators all tended to know each other and most of the active investigators in the field of neutrophil biology could easily meet together at the biannual Gordon Research Conference on Phagocytes. Even as recently as the early 1980s, "real" immunologists were investigators who delineated the subtypes and life cycle of lymphocytes, and within this scheme the only phagocytes of significance for lymphocyte immunologists were the monocytes. This was because only monocytes, which were long lived, and not neutrophils, which were short lived, were thought to be capable of antigen presentation, differentiation into tissue macrophages and other fixed tissue cell types, or capable of any significant protein synthesis, including production of potent immune modulating factors. The relatively recently coined phase, "innate immunity" encompasses in part the recent growing appreciation of the special role of neutrophils in host defense, immune regulation and regulation of inflammation, reflecting a vast body of new knowledge about how the neutrophil functions and affects the classic lymphocyte-oriented area of immunity encompassed by the term "acquired immunity" (reviewed in [1]).

Although the rapid amoeboid migration of neutrophils to sites of inflammation and their unique capacity to surround and engulf foreign bodies was known since the early twentieth century, it was only in the 1960s that it was generally appreciated that neutrophils produce microbicidal activated oxygen products or contained other nonoxidative potently microbicidal substances (e.g., *see* [2–4]). It was only in the late 1960s and early 1970s that a more detailed understanding of the different types of granules was delineated (e.g., *see* [5, 6]) and in the 1980s and 1990s that studies delineated the biochemistry of a large array of specialized cationic microbicidal proteins and a more complete understanding of the many proteolytic enzymes that were contained in those granules (reviewed in [7]). Only in the late 1980s and early 1990s were the biochemical details of the phagocyte oxidase delineated in fine detail (reviewed in [8–10]). Although investigators studying the biochemistry of nonmuscle actin in cell motility performed much of the critical early research in lower eukaryotic organisms, translation of this work to mammalian tissues was largely performed in the 1980s and 1990s in neutrophils and monocytes (e.g., *see* [11, 12]).

Since the human tritium tracer studies of the 1960s, it has been appreciated that when neutrophils emerge from bone marrow to peripheral blood, their half-life in blood is only 6 to 10 h and even shorter in infected patients, and that the lifespan in tissues is 3 d or less (e.g., *see* [5, 13, 14]). This provided a basis for considering neutrophils as end-stage cells only minimally more capable of anabolic processes than erythrocytes. This impression was further engendered by the observations that neutrophils, as compared to other cell types, produce energy for survival primarily through anaerobic metabolism, reserving most use of oxygen for production of superoxide in the context of the stimulated microbicidal respiratory burst [15]. There is a paucity of mitochondria and ribosomes in neutrophils compared, for example, to monocytes, and most investigators in the 1970s assumed that mature neutrophils in blood or tissues were devoid of significant protein synthetic capacity, functioning entirely on the store of enzymes and other proteins that were contained within their granules, membranes, and cytoplasm as these cells emerged from the bone marrow. It was not that investigators viewed neutrophils as inactive, since, after all, these are cells capable of remarkably rapid amoeboid migration, rapid engulfment of microorganisms, a prodigious respiratory burst-induced production of superoxide, and extremely rapid degranulation into phagosomes (reviewed in [16]). However, more recent studies have demonstrated unequivocally that neutrophils have significant stimulated biosynthetic capacity (e.g., *see* [17–19]). Of note is the production of a number of chemokines by neutrophils, in particular the production of large amounts of interleukin 8 [20]. This provides one important area of evidence for the important interface between the neutrophil component of innate immunity and the classic area of acquired immunity (e.g., *see* [1]).

One of the specialized motile properties of neutrophils is chemotaxis, and the delineation in the 1970s of bacteria-derived formyl peptides as chemotactic for neutrophils [21], in particular the discovery of the simple formylated tripeptide, formyl-methionyl-leucyl-phenyalanine (fMLF), as a potent chemoattractant, began the process of making the neutrophil, in the 1980s [22], a model of choice for investigators interested in delineating a large array of biochemical signaling pathways whose diverse enzymes and regulatory proteins were still being worked out (e.g., *see* [23]). Formylated peptides induced chemotaxis, but they also induced degranulation, which was associated metabolically with an ionized calcium transient, changes in electric potential similar to neural signaling, phosphorylation, metabolism of guanosine triphosphate (GTP), and metabolism of certain membrane phospholipids (e.g., *see* [24–26]). This was the beginning of the use of the neutrophil as the model system of choice for an increasing number of investigators delineating many types of newly identified biochemical signaling pathways, including the G-protein coupled signaling pathway and the large array of small GTPases of the Ras and Rho families that regulate so many cell functions (e.g., *see* [27–30]).

Although it was appreciated that neutrophils have a relatively short lifespan following release from the bone marrow, the final fate of "old" neutrophils remained a mystery until the emergence of the new paradigm of apoptosis defined the process of regulated cell death as a final stage in the differentiation of cells and delineated it from trauma, toxin-induced, or immune-induced cell necrosis [17, 31–33]. The recent application of the apoptosis paradigm to neutrophils has been used to explain how some processes lead to resolution of infection or inflammation without tissue damage by allowing neutrophil apoptosis to occur. The apoptotic process prevents release of the cytotoxic and proteolytic contents of the neutrophil by decreasing key cell functions and facilitating uptake and removal of apoptotic neutrophils by other phagocytic cells [32, 33]. Not only does this facilitate "cleanup" and resolution of infection and inflammation without tissue damage, but it also probably comprises an important interface of innate with acquired immunity in that contents of the phagocytosed, apoptotic neutrophils include components of killed microorganisms, which are processed by antigen presenting cells. Thus, it is appropriate that there is a chapter in this volume on methods to evaluate neutrophil apoptosis. Apoptosis of neutrophils can be replaced prematurely by necrosis and release of the potent proteolytic enzymatic contents of neutrophils by microorganisms that contain toxins that lyse the neutrophil (reviewed in [34]), by autoimmune processes such as antibodies to neutrophils or their contents that induce lysis or phagocytosis of neutrophils before normal apoptotic processes can occur, or by other causes of cytolysis. One such cytolytic process can result in extracellular extrusion of the nucleus to form web-like DNA structures called neutrophil extracellular traps (NETs), which are coated with histones and cationic granule proteins [35]. These interesting structures ensnare bacteria and fungi—and some reports suggest they kill bacteria—and can thereby contribute to host defense. The idea that DNA released from neutrophils can contribute to innate immunity was unheard of until NET discovery in 2004 [35]. Since that time, the number of studies published on the topic of NETs has increased dramatically [36]. Three chapters in this volume detail methods for analysis of NETs.

Neutrophils are primary mediators of the rapid innate host defense against most bacterial and fungal pathogens that occurs before the complex humoral and lymphocyte cellular processes of acquired immunity can be brought to bear on an infection. The importance of the neutrophil in this process is highlighted by the fact that iatrogenic neutropenia from cancer chemotherapy or reactions to cytotoxic drugs is the most common severe immune deficiency associated with significant morbidity encountered in medical practice [37–40]. Inherited forms of neutropenias are also associated with significant risk of infection from bacteria and fungi. There are a number of single gene disorders affecting primarily specific neutrophil functions, such as defects in the phagocyte oxidase responsible for the group of diseases with a common phenotype called chronic granulomatous disease (CGD), or defects in the CD18 β integrin responsible for leukocyte adhesion deficiency (LAD). The groups of bacterial and fungal organisms typically infecting patients with severe neutropenia, CGD, or LAD overlap a little but in general are different and distinct, suggesting strongly that there is an array of microbicidal defense systems and molecules used by the neutrophil in host defense and that different systems have evolved to provide specific defense against different organisms. Several chapters in this volume present methods for investigating key microbicidal defense systems employed by neutrophils. Also of note is that CGD patients have excessive inflammation and a tendency to develop a variety of autoimmune disorders including inflammatory bowel disease and poor wound healing, suggesting that the superoxide or hydrogen peroxide products of the respiratory burst may play an important role in the resolution of inflammation and in wound healing [41]. LAD patients have large nonhealing ulcers, also suggesting a role for neutrophils in wound healing, since LAD neutrophils have trouble migrating into tissues [42].

3 Future Prospects

Despite strong evidence in a variety of model systems of inflammation that superoxide dismutase, which catabolizes superoxide; catalase, which catabolizes hydrogen peroxide; or dimethylsulfoxide (DSMO), which scavenges hydroxyl radical, protect against tissue injury from neutrophil-mediated tissue damage, there remains considerable controversy regarding the role of neutrophil products of oxidative metabolism in mediating any of the tissue damage syndromes associated with autoimmune disease or septic shock (e.g., *see* [43–46]). Stronger evidence exists for actual clinical benefit accruing from blocking the activity a number of the proteolytic enzymes derived from neutrophils in a variety of autoimmune or other disease processes (reviewed in [47]). Because of this, discovery and development of antiproteolytic small molecules active against specific proteases found in neutrophils are part of the anti-inflammatory drug development programs of a number of pharmaceutical companies (e.g., *see* [48, 49]). Neutrophils are also a source of a number of small potent antimicrobial proteins and peptides that could be developed into new classes of antibiotics, and this is another area of interest to pharmaceutical development (reviewed in [50]). Thus, the chapters in this volume are likely to be of utility not only broadly to investigators interested in using the neutrophil as a model system, investigators specifically interested in neutrophil function, or investigators studying the interface between innate and acquired immunity but also to investigators developing new classes of natural biologicals as antibiotics or developing new classes of anti-inflammatory agents.

References

1. Hoebe K, Janssen E, Beutler B (2004) The interface between innate and adaptive immunity. Nat Immunol 5:971–974
2. Babior BM, Kipnes RS, Curnutte JT (1973) Biological defense mechanisms: production by leukocytes of superoxide,a potential bactericidal agent. J Clin Invest 52:741–744
3. Klebanoff SJ (1967) Iodination of bacteria: a bactericidal mechanism. J Exp Med 126:1063–1078
4. Lehrer RI, Hanifin J, Cline MJ (1969) Defective bactericidal activity in myeloperoxidase-deficient human neutrophils. Nature 223:78–79
5. Bainton DF, Ullyot JL, Farquhar MG (1971) The development of neutrophilic polymorphonuclear leukocytes in human bone marrow. J Exp Med 134:907–934
6. Bainton DF, Farquhar MG (1968) Differences in enzyme content of azurophil and specific

granules of polymorphonuclear leukocytes. I. Histochemical staining of bone marrow smears. J Cell Biol 39:286–298
7. Borregaard N, Cowland JB (1997) Granules of the human neutrophilic polymorphonuclear leukocyte. Blood 89:3503–3521
8. Segal AW, Abo A (1993) The biochemical basis of the NADPH oxidase of phagocytes. Trends Biochem Sci 18:43–47
9. Babior BM (1999) NADPH oxidase: An update. Blood 93:1464–1476
10. Clark RA (1990) The human neutrophil respiratory burst oxidase. J Infect Dis 161:1140–1147
11. Zigmond SH (1978) Chemotaxis by polymorphonuclear leukocytes. JCellBiol 77:269–287
12. Southwick FS, Stossel TP (1983) Contractile proteins in leukocyte function. Semin Hematol 20:305–321
13. Fliedner TM, Cronkite EP, Robertson JS (1964) Granulocytopoiesis. I. Senescence and random loss of neutrophilic granulocytes in human beings. Blood 24:402–414
14. Athens JW, Haab OP, Raab SO et al (1961) Leukokinetic studies. IV. The total blood, circulating and marginal granulocyte pools and the granulocyte turnover rate in normal subjects. J Clin Invest 40:989–995
15. Rossi F, Zatti M (1964) Changes in the metabolic pattern of polymorphonuclear leukocytes during phagocytosis. Br J Exp Pathol 45:548–559
16. DeLeo FR, Nauseef WM (2019) Granulocytic phagocytes. In: Bennett JE, Dolin R, Blaser M (eds) Mandell, Douglas, and Bennett's principles and practice of infectious diseases, 9th edn. Elsevier Limited, Oxford. In press
17. Kobayashi SD, Voyich JM, Buhl CL et al (2002) Global changes in gene expression by human polymorphonuclear leukocytes during receptor-mediated phagocytosis: cell fate is regulated at the level of gene expression. Proc Natl Acad Sci USA 99:6901–6906
18. Theilgaard-Mönch K, Knudsen S, Follin P et al (2004) The transcriptional activation program of human neutrophils in skin lesions supports their important role in wound healing. J Immunol 172:7684–7693
19. Zhang XQ, Kluger Y, Nakayama Y et al (2004) Gene expression in mature neutrophils: early responses to inflammatory stimuli. J Leukoc Biol 75:358–372
20. Strieter RM, Kasahara K, Allen RM et al (1992) Cytokine-induced neutrophil-derived interleukin-8. Am J Pathol 141:397–407
21. Schiffmann E, Corcoran BA, Wahl SM (1975) *N*-formylmethionyl peptides as chemoattractants for leucocytes. Proc Natl Acad Sci U S A 72:1059–1062
22. Snyderman R, Goetzl EJ (1981) Molecular and cellular mechanisms of leukocyte chemotaxis. Science 213:830–835
23. He HQ, Ye RD (2017) The formyl peptide receptors: diversity of ligands and mechanism for recognition. Molecules 22(3):455
24. O'Flaherty JT, Showell HJ, Ward PA (1977) Influence of extracellular Ca^{2+} and Mg^{2+} on chemotactic factor-induced neutrophil aggregation. Inflammation 2:265–276
25. Serhan CN, Broekman MJ, Korchak HM et al (1983) Changes in phosphatidylinositol and phosphatidic acid in stimulated human neutrophils. Relationship to calcium mobilization, aggregation and superoxide radical generation. Biochim Biophys Acta 762:420–428
26. McPhail LC, Clayton CC, Snyderman R (1984) A potential second messenger role for unsaturated fatty acids: activation of Ca+ +–dependant protein kinase. Science 224:622–625
27. Aharoni I, Pick E (1990) Activation of the superoxide-generating NADPH oxidase of macrophages by sodium dodecyl sulfate in a soluble cell-free system: evidence for involvement of a G protein. J Leukoc Biol 48:107–115
28. Quinn MT, Parkos CA, Walker L et al (1989) Association of a ras-related protein with cytochrome b of human neutrophils. Nature 342:198–200
29. Abo A, Pick E, Hall A et al (1991) Activation of the NADPH oxidase involves the small GTP-binding protein $p21^{rac1}$. Nature 353:668–670
30. Knaus UG, Heyworth PG, Evans T et al (1991) Regulation of phagocyte oxygen radical production by the GTP- binding protein Rac2. Science 254:1512–1515
31. Serhan CN, Savill J (2005) Resolution of inflammation: the beginning programs the end. Nat Immunol 6:1191–1197
32. Savill JS, Wyllie AH, Henson JE et al (1989) Macrophage phagocytosis of aging neutrophils in inflammation. Programmed cell death in the neutrophil leads to its recognition by macrophages. J Clin Invest 83:865–875
33. Whyte MK, Meagher LC, MacDermot J et al (1993) Impairment of function in aging neutrophils is associated with apoptosis. J Immunol 150:5124–5134

34. DeLeo FR (2004) Modulation of phagocyte apoptosis by bacterial pathogens. Apoptosis 9:399–413
35. Brinkmann V, Reichard U, Goosmann C et al (2004) Neutrophil extracellular traps kill bacteria. Science 303:1532–1535
36. Boeltz S, Amini P, Anders HJ et al (2019) To NET or not to NET: current opinions and state of the science regarding the formation of neutrophil extracellular traps. Cell Death Differ 26:395–408
37. Tobias JD, Schleien C (1991) Granulocyte transfusions--a review for the intensive care physician. Anaesth Intensive Care 19:512–520
38. Froland SS (1984) Bacterial infections in the compromised host. Scand J Infect Dis Suppl 43:7–16
39. Bodey GP, Buckley M, Sathe YS et al (1966) Quantitative relationships between circulating leukocytes and infection in patients with acute leukemia. Ann Intern Med 64:328–340
40. Dale DC, Guerry D, Wewerka JR et al (1979) Chronic neutropenia. Medicine (Baltimore) 58:128–144
41. Kobayashi SD, Voyich JM, Braughton KR et al (2004) Gene expression profiling provides insight into the pathophysiology of chronic granulomatous disease. J Immunol 172:636–643
42. Bunting M, Harris ES, McIntyre TM et al (2002) Leukocyte adhesion deficiency syndromes: adhesion and tethering defects involving beta 2 integrins and selectin ligands. Curr Opin Hematol 9:30–35
43. Weiss SJ (1989) Tissue destruction by neutrophils. N Engl J Med 320:365–376
44. Finkel T, Holbrook NJ (2000) Oxidants, oxidative stress and the biology of ageing. Nature 408:239–247
45. Temple MD, Perrone GG, Dawes IW (2005) Complex cellular responses to reactive oxygen species. Trends Cell Biol 15:319–326
46. Rahman I, Biswas SK, Kode A (2006) Oxidant and antioxidant balance in the airways and airway diseases. Eur J Pharmacol 533:222–239
47. Altieri DC (1995) Proteases and protease receptors in modulation of leukocyte effector functions. J Leukoc Biol 58:120–127
48. Zaidi SH, You XM, Ciura S et al (2002) Overexpression of the serine elastase inhibitor elafin protects transgenic mice from hypoxic pulmonary hypertension. Circulation 105:516–521
49. Zeiher BG, Matsuoka S, Kawabata K et al (2002) Neutrophil elastase and acute lung injury: prospects for sivelestat and other neutrophil elastase inhibitors as therapeutics. Crit Care Med 30:S281–S287
50. Ganz T (2004) Antimicrobial polypeptides. J Leukoc Biol 75:34–38

Chapter 2

Neutrophil Defects and Diagnosis Disorders of Neutrophil Function: An Overview

Mary C. Dinauer

Abstract

Primary disorders of neutrophil function result from impairment in neutrophil responses that are critical for host defense. This chapter summarizes inherited disorders of neutrophils that cause defects in neutrophil adhesion, migration, and oxidative killing. These include the leukocyte adhesion deficiencies, actin defects and other disorders of chemotaxis, hyperimmunoglobulin E syndrome, Chédiak–Higashi Syndrome, neutrophil specific granule deficiency, chronic granulomatous disease, and myeloperoxidase deficiency. Diagnostic tests and treatment approaches are also summarized for each neutrophil disorder.

Key words *Aspergillus* species, β_2 integrin, Chédiak–Higashi Syndrome, Chemotaxis, Chronic granulomatous disease, Hyperimmunoglobulin E, Leukocyte adhesion deficiency, Myeloperoxidase, NADPH oxidase, Neutrophil granule, *Staphylococcus aureus*

Abbreviations

CGD	Chronic granulomatous disease
DHR	Dihydrorhodamine 123
HIES	Hyperimmunoglobulin E
HSC	Hematopoietic stem cell
LAD	Leukocyte adhesion deficiency
LJP	Localized juvenile periodontitis
MPO	Myeloperoxidase
NBT	Nitroblue tetrazolium
phox	Phagocyte oxidase

1 Introduction

This chapter provides a brief overview of disorders of neutrophil function. Neutrophils play an essential role in the initial response to invading bacteria and fungi; thus, patients with defects in neutrophil function typically present in infancy or childhood with

Mark T. Quinn and Frank R. DeLeo (eds.), *Neutrophil: Methods and Protocols*, Methods in Molecular Biology, vol. 2087, https://doi.org/10.1007/978-1-0716-0154-9_2,

recurrent and/or difficult to treat bacterial infections [1–4]. Infections typically involve the skin, mucosa, gums, lung or draining lymph nodes, or cause deep tissue abscesses. The microorganisms causing these infections are often unusual or opportunistic pathogens. Many of the disorders have characteristic clinical and microbiological features that are related to the specific nature of the defect in neutrophil function. Note that congenital disorders affecting neutrophil function represent at most 20% of reported primary immune deficiencies. Thus, patients with suspected disorders of host defense should also be screened for defects in antibody production, T cell function, and the complement system. Figure 1 summarizes key steps in the response of neutrophils to invading microbes, which include initial adhesion to the endothelium, subsequent migration toward a site of infection, ingestion of the pathogen, and pathogen killing via oxidative metabolites, proteases and other toxic peptides in neutrophil granules. Disorders of neutrophil function can affect one or more of these pathways (Fig. 1).

In the diagnostic work-up of a patient with bacterial and fungal infections, a clinician must decide whether a complete evaluation of neutrophil function is warranted. Four key aspects of the patient history with infections should be considered in reaching this decision; namely, frequency, severity and location of infections, and the identity of the causal infectious agent. The presence of unusual features should trigger consideration of a neutrophil disorder. The patient's age and other medical conditions should also be taken into account. For example, one might suspect neutrophil dysfunction if an infectious agent that commonly affects children is observed in an older patient. Distinctive clinical findings can provide helpful guidelines in determining which patients merit further testing, and which tests are appropriate. A family history can also provide useful clues.

2 Disorders of Adhesion and Chemotaxis

The ability of neutrophils to adhere to the endothelium, tissue matrix, and to invading microbes is essential for their migration from the bloodstream to sites of infection, where they eliminate pathogens. Interactions between cell surface glycoproteins expressed on neutrophils and endothelial cells are fundamental to this process. The initial steps of adhesion require expression of E selectins on the surface of endothelial cells, which bind to fucosylated proteins on leukocytes. The next step involves induction and activation of cell surface integrins on leukocytes, which mediate tight adhesion between neutrophils and endothelial cells. Subsequent migration from capillaries into tissues uses additional cell-surface receptors. Defects in these interactions and/or chemotaxis impair recruitment of neutrophils into sites of infection or inflammation, often with relative neutrophilia in peripheral blood yet poor formation of pus.

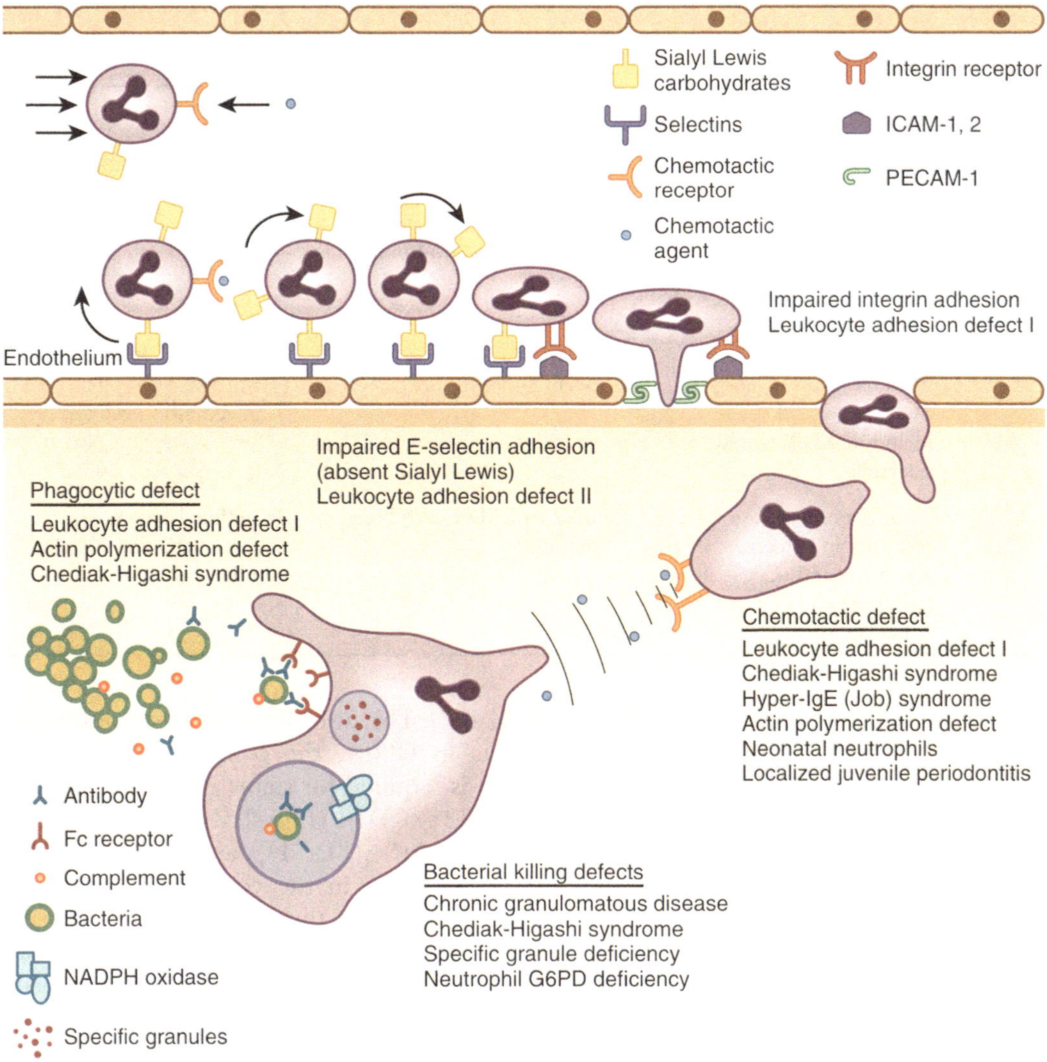

Fig. 1 Steps in the response of circulating neutrophils to infection or inflammation. The adhesion molecule E-selection is upregulated on endothelial cells in response to inflammatory mediators (IL-1, endotoxin, and TNF-α). Neutrophils interact with E-selectins on endothelial cells through sialyl Lewis carbohydrates, resulting in rolling attachment and margination. Chemoattractants, such as IL-8 upregulate neutrophil β2 integrins, which, in turn, mediate tight adhesion to ICAM-1 and PECAM-1 on endothelial cells. Activated neutrophils detect small changes in the chemoattractant gradient, which causes them to move toward the site of tissue infection. Neutrophils phagocytose bacteria opsonized by antibody and complement. Both oxidative and nonoxidative antimicrobial mechanisms mediate bacterial killing. Disorders of phagocyte function associated with each of these steps are listed. (Reprinted from *Hematology: Basic Principles and Practice*, sixth edition, Mary C. Dinauer and Thomas D. Coates, *Disorders of Phagocyte Function*, Chapter 48, Pages 655–673, Copyright Elsevier, 2012, with permission)

2.1 Leukocyte Adhesion Deficiency Type I

Leukocyte adhesion deficiency type I (LAD I) is an autosomal recessive disorder characterized by deficiency of three leukocyte glycoproteins that are members of the integrin superfamily of cell surface adhesion molecules (Table 1), and has been reported in

Table 1
Leukocyte adhesion deficiency

	Genetic defect	Clinical presentation	Diagnosis
LAD I	*ITGB2; encodes* CD18 subunit of β_2 integrins, resulting in impaired adhesion, chemotaxis, and neutrophil activation	Skin infection, soft tissue abscesses, delayed separation of umbilical cord and omphalitis, periodontal disease	Flow cytometry for CD11b/CD18 (Mac1)
LAD II	*SLC35C1*; encodes GDP-fucose transporter 1, resulting in impaired expression of fucosylated proteins, including SLeX ligand for selectins	Similar infections to LAD I but not as severe; developmental delay, short stature	Flow cytometry for leukocyteCD15s (SLeX) Bombay (hh) phenotype in red blood cell typing
LAD III	*FERMT3*; encodes kindlin-3, resulting in defective integrin activation and impaired leukocyte and platelet adhesion	Similar to LAD I; also bleeding tendency	Functional assays for neutrophil and platelet adhesion

hundreds of patients [1, 4–7]. This disorder results from autosomal recessive genetic defects in CD18, the common chain of the β_2 integrin family, which is required for stable expression of three distinct β_2 integrins; CD11a/CD18 (LFA-1; Lymphocyte Function Antigen), CD11b/CD18 (Mac-1), and CD11c/CD18 (p150,95). In LAD I, mutations in the common CD18 chain typically eliminate expression of these three leukocyte glycoprotein complexes. β_2 integrins interact with ICAM-1 proteins expressed on endothelial cells, which are upregulated in response to inflammatory cytokines. Thus, neutrophil adhesion to the endothelium is severely defective in LAD I. In addition, Mac-1 is the major receptor for the opsonic complement fragment C3bi, an important trigger for phagocytosis of complement-opsonized microbes. Mac-1 also mediates binding to fibrinogen. Finally, binding to Mac-1 provides an important costimulatory signal for activating other pathways important for adhesion, degranulation, and activation of reactive oxidant production [8]. Because of the multiple defects in adhesion-related functions, LAD I patients develop recurrent bacterial and fungal infections, typically with *Staphylococcus aureus* or gram-negative enteric microbes. Characteristic clinical features include frequent skin and periodontal infections, delayed separation of the umbilical cord and omphalitis, and deep tissue abscesses. Neutrophilia with paucity of neutrophils at inflamed or infected sites is characteristic. Recent studies suggest that impaired migration and clearance of neutrophils in periodontal tissues results in a dysregulated IL-23—IL-17 inflammatory axis, which drives the severe periodontal disease typical of LAD I rather than impaired control of gum bacteria [9].

Mutations in the CD18 gene in patients with LAD I are heterogeneous. Typically, cell surface expression of the CD11/CD18 heterodimers is absent, although variants with low level expression have been described; the latter phenotype is associated with reduced severity of disease.

Diagnosis of LAD I is straightforward, and can be made by flow cytometry of peripheral blood leukocytes, most often with a monoclonal antibody directed to the CD11b/CD18 heterodimer, Mac-1. Patients with LAD I also have diminished neutrophil migration in vivo. This can be studied with the Rebuck skin window; however, this test is no longer used for diagnostic purposes, due to the availability of flow cytometric methods.

Because of the severity of infectious complications, allogeneic hematopoietic stem cell (HSC) transplantation is generally recommended for patients with severe LAD I [1, 4]. Patients with partial expression of LAD I have a longer life expectancy without HSC transplantation, but require good supportive care including prophylactic antibiotics and scrupulous dental hygiene. Nevertheless, morbidity from periodontal disease, bacterial infection, and delayed healing are still problematic.

2.2 Leukocyte Adhesion Deficiency Types II and III

Two other autosomal recessive forms of leukocyte adhesion deficiency associated with distinct clinical and genetic defects have been described in a small number of patients (Table 1) [4, 5, 7]. Leukocyte adhesion deficiency type II (LAD II) is very rare (fewer than 10 patients) and caused by mutations in the membrane transporter for fucose, leading to loss of expression of fucosylated glycans on the cell surface. This disorder is also known as Congenital Disorder of Glycosylation Type IIc (CDG-IIc). Fucosylated proteins, such as sialyl-Lewis X (CD15s), are ligands for endothelial selectins and are important for the early phases of adhesion to endothelial cells. Patients with LAD II also have leukocytosis and form pus poorly, although infections tend to be less severe in LAD II patients compared to LAD I patients. In addition, LAD II patients have severe mental retardation and short stature, and the absence of fucosylated proteins on the surface of red blood cells results in the Bombay (hh) red cell phenotype. LAD II is readily diagnosed by flow cytometry for expression of leukocyte CD15s. Prolonged therapy with oral fucose has been beneficial in some but not all LAD II patients.

Leukocyte adhesion deficiency type III (LAD III) has been described in about 20 patients to date [4, 5, 7] LAD III is characterized by defects in the activation of multiple classes of integrins downstream of G-protein coupled receptors due to mutations in kindlin-3, a hematopoietic protein that regulates integrin activation. The phenotype of this disorder is similar to LAD I, but is also associated with defects in integrin-mediated platelet aggregation and excessive bleeding similar to Glanzmann thrombasthenia. The diagnosis of LAD III should be considered in patients with the clinical features of LAD I, but normal expression of β_2 integrins. Defective platelet aggregation can be evaluated using readily available clinical laboratory blood tests. Evaluation of leukocyte adhesion in such patients requires specialized functional assays that are usually carried out in research laboratories.

2.3 Defect in Adhesion and Chemotaxis Due to Mutation in Rac2 GTPase

A dominant-negative mutation in the Rac2 GTPase was described in an infant with impaired neutrophil adhesion and motility, along with decreased NADPH oxidase activation and degranulation in response to chemoattractants [10, 11]. Symptoms similar to LAD I developed in the first months of life. Leukocyte β_2 integrin expression was normal, and the phenotype presumably results from interference by the mutant Rac2 in Rac-regulated signaling pathways. This patient successfully underwent HSC transplantation. A similar patient with severe, recurrent skin and tissue abscesses, normal leukocyte β_2 integrin expression, and similar neutrophil function defects in response to chemoattractants died in infancy without determination of the genetic defect [12].

2.4 Other Genetic Disorders of Chemotaxis

Several rare human diseases characterized primarily by prominent defects in neutrophil chemotaxis have been described. In general, the neutrophils in these patients exhibit normal adhesive properties. Patients typically present with recurrent, severe skin and soft tissue infections in infancy. One patient exhibited markedly abnormal polymerization of neutrophil actin and higher than normal expression of an actin binding protein [13, 14]. A young female patient is reported to have a heterozygous point mutation in β-actin and impaired binding to actin-regulatory proteins [15]. The phenotype of this patient includes recurrent infections, mental retardation, and photosensitivity. Another rare autosomal recessive disorder involves mutations in the actin binding protein WDR1, which regulates cofilin-mediated actin depolymerization [16]. Neutrophils display abnormal chemotaxis, and herniation of nuclear lobes. Patients have poor wound healing and severe stomatitis with oral stenosis.

Localized Juvenile Periodontitis (LJP) is a rare disorder of unknown etiology that presents in children and adolescents as chronic and recurrent periodontal infections and severe alveolar bone loss [17, 18]. This disorder is heterogeneous and is associated with defective neutrophil chemotaxis in vitro. Some cases are sporadic, but others appear to be familial and thus may be linked to genetic defects. Children or teenagers who present with periodontal disease should be carefully evaluated for neutropenia or disorders of neutrophil function, since these are frequently associated with periodontal disease. LJP is diagnosed by its characteristic clinical features, the absence of nonperiodontal infections, normal neutrophil counts and morphology, normal expression of β_2 integrins, and normal oxidative burst. Recent studies suggest that dysregulated inflammatory signaling in LAP neutrophils contributes to the inflammatory gum disease [19, 20].

Hyperimmunoglobulin E (hyper-IgE or HIES) syndrome is characterized by elevated serum levels of IgE (i.e., $\geq$2000 IU-mL) and recurrent *Staphylococcal* infections of the skin and lung, pneumatoceles, chronic dermatitis, and skeletal and dental abnormalities [21–24]. Patients with this syndrome have variable and often profound defects in neutrophil chemotaxis, which are independent of fluctuations in serum IgE. Localized *Staphylococcal* infections are indolent and lack the usual characteristics of inflammation ("cold" abscesses). Chronic candidiasis of the mucosal surface and nail beds, hyperextensible joints and scoliosis are also frequently observed, and delay or failure to shed the primary teeth is not uncommon. Many patients have coarse facial features by early adolescence. Inherited or sporadic autosomal dominant mutations in *STAT3* account for the majority of cases of HIES. STAT3 is a Janus kinase (JAK)-activated transcription factor utilized downstream of many cytokines and growth factors. HIES-associated mutations in STAT3 involve DNA or protein binding domains, resulting in interference with the function of wild-type STAT3. Rare forms of HIES result from autosomal recessive mutations in either *DOCK8* or *TYK2,* which encode proteins regulating leukocyte signalling, and are additionally associated with profound lymphocyte function defects. The relationship between the immunologic abnormalities, neutrophil chemotactic defects, and the other features of HIES are only now emerging. STAT3-dependent responses to both proinflammatory and anti-inflammatory cytokines are altered. STAT3 may be important for neutrophil chemotaxis to certain chemokines [25]. Differentiation of T helper type 17 (Th17) lymphocytes, which are important for control of Candida, are impaired. The decreased numbers of Th17 lymphocytes are also associated with reduced epithelial production of neutrophil chemoattractants and antimicrobial peptides, which likely contribute to increased susceptibility to skin infections. High serum IgE levels may reflect a T-lymphocyte imbalance, which causes abnormally high production of IgE. Diagnostically, the hallmark laboratory finding in hyper-IgE syndrome is the marked elevation of serum IgE, especially in children or young adults. DNA testing should be used to confirm the diagnosis of HIES. Antibiotic prophylaxis for Staphylococcal infection is beneficial in patients with hyper-IgE syndrome. Atopic dermatitis is the major differential diagnosis, as it can also be associated with eosinophilia and elevated serum IgE.

2.5 Noninherited Disorders of Neutrophil Chemotaxis

A chemotactic abnormality, detected by in vitro assays, has been reported in neonatal neutrophils [26–28]. This abnormality may be due in part to defects in cellular adhesion and impaired mobilization and activation of intracellular adhesion proteins at the cell surface. Reduced neutrophil chemotaxis has also been described in burn victims and in patients with bacterial sepsis or diabetes. However, the chemotactic defects in these settings are variable, and their contribution to associated infectious complications is uncertain.

3 Disorders of Ingestion and Degranulation

Following phagocytosis, neutrophil granules fuse with phagosome membranes and release proteases, enzymes, and antibacterial proteins into the phagosome lumen. This process greatly facilitates microbial killing. In LAD I patients, C3bi-obsonized microbes fail to be ingested, because CD11b/CD18 (Mac-1), the receptor for C3bi, is not expressed. This receptor also provides an important costimulatory signal for Fcγ receptor–mediated phagocytosis. Deficiency of opsonization with complement is seen in serum complement deficiencies, and primary B cell deficiencies, such as Bruton's agammaglobulinemia, result in defective opsonization with specific immunoglobulins. Thus, these patients can present with recurrent infections with pyogenic bacteria, such as *S. aureus*, *Pneumococci*, and *H. influenzae*. In contrast, primary neutrophil defects causing disorders of ingestion and/or degranulation are very rare. Two such genetic disorders affecting neutrophil granules have been described (see below), both of which are associated with increased frequency of bacterial infections.

3.1 Chédiak–Higashi Syndrome

Chédiak–Higashi syndrome is a rare autosomal recessive disorder associated with widespread defects in granule morphogenesis and multiorgan pathology, including ineffective granulopoiesis, moderate neutropenia, and delayed and incomplete degranulation [29–31]. Neutrophils from patients with Chédiak–Higashi Syndrome have giant peroxidase-positive granules that appear to be a coalescence of azurophilic and specific granules. The giant granules are often more prominent in bone marrow neutrophils than in peripheral blood neutrophils. Giant granules are also seen in lymphocytes and natural killer cells from patients with Chédiak-Higashi Syndrome. Patients with Chédiak-Higashi Syndrome experience frequent *S. aureus* infections in lung and skin, and develop gingivitis and periodontitis. Patients also have partial oculocutaneous albinism, a mild tendency to bleed and neuropathies. These manifestations are all related to abnormal granule morphogenesis. Chédiak-Higashi Syndrome is caused by mutations in *CHS1*, which encodes a large protein thought to regulate lysosomal and granule trafficking. Supportive care with prophylactic antibiotics is a mainstay of treatment. Of importance, Chédiak–Higashi Syndrome is a cause of inherited hemophagocytic histiocytosis (HLH), which results from the granule defects in natural killer cells and other lymphocytes and can be precipitated by viral infections including Epstein–Barr virus [31]. The majority of Chédiak-Higashi Syndrome patients that survive childhood develop this HLH. In this phase of the disease, which was often referred to as the "accelerated phase," patients experience fever, lymphadenopathy, lymphohistiocytic infiltration, and progressive pancytopenia, which is fatal unless treated with allogeneic HSC transplantation.

3.2 Neutrophil Specific Granule Deficiency

Neutrophil Specific Granule Deficiency (SGD) is a very rare disorder characterized by the absence of specific or secondary granules in developing neutrophils [32, 33]. Neutrophils from SGD patients also typically have morphologically abnormal bilobed nuclei. SGD neutrophils are markedly deficient in many important microbicidal granule proteins, including lactoferrin and the defensins. SGD neutrophils also demonstrate relatively severe chemotactic defects, which are thought to result from a decrease in the pool of intracellular leukocyte adhesion molecules normally mobilized to the cell surface in response to inflammatory stimuli. SGD patients present with recurrent and difficult to treat bacterial and fungal infections, primarily involving the skin and lungs. *S. aureus*, enteric gram-negative bacteria, *P. aeruginosa*, and *Candida albicans* are the major pathogens. This disorder appears to be inherited in an autosomal recessive manor. The molecular defect responsible for some cases of SGD involves the myeloid transcription factor C/EBPε [33, 34]. C/EBPε plays an important role as a gene-specific transcriptional regulator during late promyelocyte and early myelocyte development. The diagnosis of SGD is made if microscopic examination of neutrophils reveals the absence of specific granules. The diagnosis can be confirmed by assessing expression of granule-specific proteins (i.e., lactoferrin or gelatinase) by staining or other assays. Treatment of SGD is supportive, with prophylactic antibiotics, and prompt and prolonged treatment of infections.

4 Disorders of Oxidative Metabolism

Reactive oxygen species play an important role in killing microbial pathogens. The major source of reactive oxygen species is the phagocyte respiratory burst pathway, which is activated in response to phagocytosis or soluble inflammatory stimuli. The initial reaction in this pathway is catalyzed by an NADPH oxidase, which is found in plasma and phagosome membranes of neutrophils, monocytes/macrophages, dendritic cells, eosinophils, and B lymphocytes [35, 36]. NADPH oxidase produces the superoxide free radical by catalyzing the transfer of electrons from NADPH to molecular oxygen. Superoxide is converted to hydrogen peroxide, and, in the presence of neutrophil myeloperoxidase, HOCl, along with numerous other microbicidal oxidants that synergize with granule proteins to kill microbes in the phagosome. Thus, patients lacking key steps in the respiratory burst pathway are deficient in microbial killing and can develop recurrent bacterial and fungal infections.

4.1 Chronic Granulomatous Disease

Chronic granulomatous disease (CGD) is a group of inherited disorders caused by defects in the phagocyte NADPH oxidase complex (Table 2). Genetic defects in four core subunits of NADPH oxidase cause "classic" CGD, and these mutations lead to complete or, less frequently, markedly reduced NADPH oxidase activity [1, 36–44]. In addition, recessive mutations in a fifth NADPH oxidase subunit or in a chaperone protein important for assembly of the oxidase flavocytochrome lead to partial loss of oxidase activity [39, 45–47]. The estimated incidence of CGD is approximately 1 in 200,000 live births, and it is the most common clinically significant inherited disorder of neutrophil function. Because respiratory burst oxidants are an important component of microbial killing mechanisms, CGD patients suffer recurrent, often life-threatening bacterial and fungal infections. The infections typically involve microorganisms for which oxidant-mediated killing is particularly critical for effective host defense. An additional and distinctive hallmark of CGD is the formation of inflammatory granulomas and other inflammatory complications. Dysregulated inflammatory responses, which are not always related to infection, reflect a broad impact of NADPH oxidase-derived reactive oxygen species on cellular pathways involved in innate and adaptive immunity [36]. Interestingly, common hypomorphic variants in NADPH oxidase genes are linked to autoimmune disorders [36].

The NADPH subunits are referred to by their apparent molecular mass (kDa) and the designation *phox* (phagocyte oxidase) (Table 2). Approximately 70 percent of CGD cases result from X-linked recessive defects in the gene encoding the gp91^{phox} subunit of flavocytochrome b_{558}, a membrane heterodimer that is the redox center of NADPH oxidase. This subunit contains both flavoprotein and heme-binding domains, and is also sometime referred to as NOX2 (NOX, NADPH oxidase; 2, referring to its number in a series of homologous flavocytochromes) [48]. An uncommon autosomal recessive form of CGD is associated with mutations in the gene encoding p22^{phox}, the other component of the flavocytochrome b_{558} heterodimer. This subunit mediates translocation of two regulatory subunits of NADPH oxidase, p47^{phox} and p67^{phox}. Mutations in genes encoding p47^{phox} and p67^{phox} are affected in two other autosomal recessive subgroups of "classic" CGD. p47^{phox} binds to p22^{phox} and acts as an adapter protein for recruitment and positioning of the p67^{phox} subunit, which contains an activation

Table 2
Genetic defects in NADPH oxidase in chronic granulomatous disease

Subunit	Inheritance	Gene locus and affected gene	Function	Mutations	Frequency in CGD
gp91^{phox} (NOX2)	X	Xp21.1 *CYBB*	Integral membrane glycoprotein; contains flavoprotein domain and heme groups for electron transport	Heterogeneous, most with absent flavocytochrome b_{558}	~70%
gp22^{phox}	AR	16p24 *CYBA*	Integral membrane protein, flavocytochrome subunit; contains docking site for p47^{phox}	Heterogeneous, most with absent flavocytochrome b_{558}	~5%
p47^{phox}	AR	7q11.23 *NCF1*	Cytosolic protein, activated by phosphorylation, mediates translocation of p67^{phox} to flavocytochrome b_{558}	Majority with GT deletion in exon 2; absent expression of p47^{phox}	~25%
p67^{phox}	AR	1q25 *NCF2*	Cytosolic protein, activates electron transport after translocation	Heterogeneous, most with absent expression of p67^{phox}	~5%
p40^{phox}	AR	22q13.1 *NCF4*	Cytosolic protein, regulates phagosome membrane NAPDH oxidase through binding to phosphatidylinositol 3-phosphate, with biggest impact on intracellular oxidase activity	24 patients, heterogeneous mutations with absent expression of p40^{phox}	<1%
CYBC1 (EROS)	AR	17q25.3 *CYBC1*	Chaperone protein in the endoplasmic reticulum, helps mediate assembly of flavocytochrome b_{558} heterodimer; oxidase activity may be impacted more in macrophages compared to neutrophils	9 patients, 8 with same mutation due to founder effect	<1%

domain that stimulates electron transport through flavocytochrome *b*. Mutations in the genes encoding the flavocytochrome *b* subunits and p67^{phox} are heterogeneous and cause deficient expression of the affected subunit. In patients with p47^{phox} mutations, a deletion of a GT dinucleotide at the start of Exon 2 in the p47^{phox} gene accounts for almost all mutations, which results from recombination with a closely linked pseudogene. This deletion causes a frameshift and premature stop codon that terminates translation of the p47^{phox} mRNA transcript [49]. Finally, two rare genetic subgroups of CGD with residual oxidase activity were recently described, and affected patients have less severe clinical symptoms than "classic" CGD. One subgroup involves heterogeneous autosomal recessive defects in the gene encoding p40^{phox} [39, 45]. This subunit plays a selective role in stimulating NADPH oxidase activity on phagosome membranes via a domain that binds to the membrane lipid phosphatidylinositol 3-phosphate. Another uncommon form of CGD results from autosomal recessive mutations in the endoplasmic reticulum-resident protein EROS (essential for ROS), also known as CYBC1, which helps to mediate assembly of the flavocytochrome b_{558} heterodimer [46, 47]. CYBC1 defects appear to affect monocyte and macrophage oxidase flavocytochrome b_{558} expression more substantially than neutrophils.

NADPH oxidase activity also requires activated Rac, which binds to and activates p67^{phox}. As described above, a dominant-negative mutation in the Rac2 GTPase has been reported, which inhibits NADPH oxidase activity in response to certain agonists. However, the clinical phenotype resembled leukocyte adhesion deficiency rather than CGD, reflecting the important role for Rac2 in regulating the neutrophil cytoskeleton, adhesion, and motility.

Patients with "classic" CGD typically present in infancy or early childhood with recurrent infections involving the skin, lung and draining lymph nodes; liver abscesses and osteomyelitis also develop due to hematogenous spread of microorganisms [1, 50]. *S. aureus* is the most common microbe involved, but CGD patients are also prone to infection by opportunistic pathogens such as *Burkholderia cepacia*, *Serratia marcescens*, *Aspergillus* species, and *Nocardia*. Many of these organisms express catalase, which removes microbe-generated hydrogen peroxide and thus inhibits ROS-mediated killing in the phagosomes of CGD cells. The lymph nodes, liver, and spleen of CGD patients can become enlarged. CGD patients can also develop increased neutrophilic inflammation, granulomas and other inflammatory manifestations, which are a hallmark of this subgroup of neutrophil disorders [36, 51]. Chronic inflammatory granulomas can obstruct the gastric outlet or urinary tract or cause granulomatous colitis and the symptoms of inflammatory bowel disease. Aberrant inflammation with autoimmune features, including discoid lupus-like skin lesions, can also occur. Interestingly, women who are carriers of X-linked mutations in *CYBB* can also develop CGD-associated inflammatory complications, although CGD-like infections occur only in women with less than 10% NADPH oxidase-positive leukocytes [52, 53]. These chronic inflammatory complications are thought to reflect an important role for oxidants generated from the NADPH oxidase in down-regulating the inflammatory response through a variety of redox-mediated effects [36].

Patients with X-linked CGD or autosomal recessive CGD involving the $p22^{phox}$ or $p67^{phox}$ subunits have a more severe clinical course than individuals with $p47^{phox}$-deficient CGD [41]. In the latter subgroup, very low levels of superoxide activity are often still detectable [54], which may explain the typically milder clinical course. Polymorphisms in oxygen-independent antimicrobial defense mechanisms or other components of the innate immune response are also thought to modify the severity of CGD in some patients [55].

In contrast to "classic" CGD, the clinical spectrum of disease in patients with mutations involving $p40^{phox}$ or CYBC1 resembles an atypical form of CGD [39, 45–47]. Inflammatory disorders are prominent, including granulomatous GI disease. Several patients with CYBC1 defects developed mycobacterial disease, but invasive bacterial and fungal infections have not yet otherwise been reported.

The diagnosis of CGD is made by measuring respiratory burst activity in neutrophils. The nitroblue tetrazolium test (NBT) using peripheral blood neutrophils is the classic method, in which the soluble yellow dye NBT is reduced by activated neutrophils, forming insoluble dark purple formazan deposits. However, a flow cytometry assay based on oxidation of dihydrorhodamine 123 (DHR) by activated neutrophils has become the standard in clinical laboratories for diagnosis of CGD [54, 56]. Both the NBT and DHR tests identify two populations of neutrophils in female carriers of X-linked CGD. Neutrophils that lack functional $p47^{phox}$ can make a small amount of superoxide, and therefore NBT and DHR assays can be weakly positive in this form of CGD. Chemiluminescence-based assays for respiratory burst activity are also occasionally used for clinical diagnosis. Family history and carrier testing will frequently discriminate between X-linked and autosomal recessive forms of CGD. However, approximately 10 percent of patients with X-linked CGD carry new mutations. Molecular genetic testing is commercially available for all subgroups of CGD, and is recommended unless the subgroup is known based on prior family testing. Note that because the most common gene defect involving $p47^{phox}$ results from recombination with a pseudogene, a PCR-based method specific to the recombination event must be used to identify this mutation.

All forms of "classic" CGD are treated similarly [1, 57]. Trimethoprim–sulfamethoxazole (or dicloxacillin in sulfa-allergic patients) and itraconazole as antibiotic prophylaxis to reduce the incidence of bacterial and fungal infections are mainstays of treatment [58]. Prophylactic interferon gamma, which is thought to enhance nonoxidative phagocyte functions, is also recommended for routine use, based on reports that it substantially decreases infectious complications in CGD patients [59, 60]. With optimal medical management, including aggressive treatment of acute infections, the mortality for CGD is now estimated to be approximately 2 percent per patient per year in the United States [59, 60]. Allogeneic HSC transplantation has also been used to treat selected cases of CGD with frequent serious or refractory infections or gastrointestinal disease [40]. Despite improved success rates and reduced complications of graft versus host disease and other transplant-associated toxicities, use of this procedure remains an individualized decision. However, marrow transplantation is becoming more widely used with the advent of improved reduced intensity condition regimens [61, 62]. Gene replacement therapy targeted at hematopoietic stem cells has been under development, and early phase clinical trials for gene therapy of X-linked CGD are underway [63, 64].

4.2 Myeloperoxidase Deficiency

Myeloperoxidase (MPO) deficiency is the most common inherited disorder of phagocytes, with complete deficiency occurring in approximately 1 in 4000 individuals [65]. However, MPO deficiency is rarely associated with clinical symptoms. MPO is inherited in an autosomal recessive manner, and is caused by mutations in the MPO gene on chromosome 17. Mutations in the MPO gene are frequently point mutations that cause defective posttranslational processing of the MPO precursor protein. Deficiency in myeloperoxidase inhibits formation of hypochlorous acid from chloride and hydrogen peroxide. There is a remarkable lack of clinical symptoms in most individuals with MPO deficiency, despite in vitro defects in the ability to kill *Candida albicans* and *Aspergillus fumigatus* hyphae. Bacterial killing in vitro is also slower than normal. Nevertheless, patients with MPO deficiency rarely develop symptoms unless they also suffer from diabetes mellitus, which leads to disseminated candidiases and other fungal infections. Diagnosis of MPO deficiency is straightforward using histochemical analysis of peroxidase activity. Because patients with MPO deficiency are typically asymptomatic, no prophylactic antibiotics are prescribed. Individuals with MPO deficiency and diabetes are usually treated aggressively to prevent fungal infections.

4.3 Neutrophil Glucose-6-Phosphate Dehydrogenase Deficiency

NADPH, the primary substrate for the superoxide-generating NADPH oxidase, is generated by the first two reactions of the hexose monophosphate shunt pathway, catalyzed by glucose-6-phosphate dehydrogenase (G6PD) and 6-phosphogluconate dehydrogenase (6PGD). The leukocyte and erythrocyte G6PD are encoded by the same gene. However, the only symptom of vast majority of individuals with inherited G6PD deficiency is red cell hemolysis triggered by oxidative stress [66]. Most G6PD mutations cause a gradual decay in G6PD, which has little or no impact in short-lived neutrophils; in contrast, longer-lived erythrocytes are severely affected by loss of G6PD activity. Even most of the more unstable G6PD variants that cause chronic nonspherocytic hemolytic anemia even in the absence of oxidative stress do not result in critically low levels of NADPH in neutrophils. A few rare G6PD mutations lead to extremely low levels of G6PD in both neutrophils and erythrocytes, resulting in chronic, severe hemolytic anemia and CGD-like symptoms; these combined features clearly distinguish this very rare form of NADPH oxidase deficiency from CGD.

References

1. Dinauer M, Coates T (2018) Disorders of phagocyte function. In: Hoffman (ed) Hematology: basic principles and practice, 7th edn. Elsevier Inc, Philadelphia, PA, pp 691–709
2. Lekstrom-Himes JA, Gallin JI (2000) Immunodeficiency diseases caused by defects in phagocytes. N Engl J Med 343(23):1703–1714
3. Nauseef WM, Borregaard N (2014) Neutrophils at work. Nat Immunol 15(7):602–611. https://doi.org/10.1038/ni.2921
4. Bouma G, Ancliff PJ, Thrasher AJ, Burns SO (2010) Recent advances in the understanding of genetic defects of neutrophil number and function. Br J Haematol 151(4):312–326. https://doi.org/10.1111/j.1365-2141.2010.08361.x
5. Hanna S, Etzioni A (2012) Leukocyte adhesion deficiencies. Ann N Y Acad Sci 1250:50–55. https://doi.org/10.1111/j.1749-6632.2011.06389.x
6. van de Vijver E, Maddalena A, Sanal O, Holland SM, Uzel G, Madkaikar M, de Boer M, van Leeuwen K, Koker MY, Parvaneh N, Fischer A, Law SK, Klein N, Tezcan FI, Unal E, Patiroglu T, Belohradsky BH, Schwartz K, Somech R, Kuijpers TW, Roos D (2012) Hematologically important mutations: leukocyte adhesion deficiency (first update). Blood Cells Mol Dis 48(1):53–61. https://doi.org/10.1016/j.bcmd.2011.10.004
7. Harris ES, Weyrich AS, Zimmerman GA (2013) Lessons from rare maladies: leukocyte adhesion deficiency syndromes. Curr Opin Hematol 20(1):16–25. https://doi.org/10.1097/MOH.0b013e32835a0091
8. Schymeinsky J, Mocsai A, Walzog B (2007) Neutrophil activation via beta2 integrins (CD11/CD18): molecular mechanisms and clinical implications. Thromb Haemost 98(2):262–273
9. Moutsopoulos NM, Konkel J, Sarmadi M, Eskan MA, Wild T, Dutzan N, Abusleme L, Zenobia C, Hosur KB, Abe T, Uzel G, Chen W, Chavakis T, Holland SM, Hajishengallis G (2014) Defective neutrophil recruitment in leukocyte adhesion deficiency type I disease causes local IL-17-driven inflammatory bone loss. Sci Transl Med 6(229):229ra240. https://doi.org/10.1126/scitranslmed.3007696
10. Ambruso DR, Knall C, Abell AN, Panepinto J, Kurkchubasche A, Thurman G, Gonzalez-Aller C, Hiester A, deBoer M, Harbeck RJ, Oyer R, Johnson GL, Roos D (2000) Human neutrophil immunodeficiency syndrome is associated with an inhibitory Rac2 mutation. Proc Natl Acad Sci U S A 97(9):4654–4659
11. Williams DA, Tao W, Yang F, Kim C, Gu Y, Mansfield P, Levine JE, Petryniak B, Derrow CW, Harris C, Jia B, Zheng Y, Ambruso DR, Lowe JB, Atkinson SJ, Dinauer MC, Boxer L (2000) Dominant negative mutation of the hematopoietic-specific rho GTPase, Rac2, is associated with a human phagocyte immunodeficiency. Blood 96(5):1646–1654
12. Roos D, Kuijpers TW, Mascart-Lemone F, Koenderman L, de Boer M, van Zwieten R, Verhoeven AJ (1993) A novel syndrome of severe neutrophil dysfunction: unresponsiveness confined to chemotaxin-induced functions. Blood 81(10):2735–2743
13. Coates TD, Torkildson JC, Torres M, Church JA, Howard TH (1991) An inherited defect of neutrophil motility and microfilamentous cytoskeleton associated with abnormalities in 47-Kd and 89-Kd proteins. Blood 78(5):1338–1346
14. Howard T, Li Y, Torres M, Guerrero A, Coates T (1994) The 47-kD protein increased in neutrophil actin dysfunction with 47- and 89-kD protein abnormalities is lymphocyte-specific protein. Blood 83(1):231–241
15. Nunoi H, Yamazaki T, Tsuchiya H, Kato S, Malech HL, Matsuda I, Kanegasaki S (1999) A heterozygous mutation of beta-actin associated with neutrophil dysfunction and

recurrent infection. Proc Natl Acad Sci U S A 96(15):8693–8698

16. Kuhns DB, Fink DL, Choi U, Sweeney C, Lau K, Priel DL, Riva D, Mendez L, Uzel G, Freeman AF, Olivier KN, Anderson VL, Currens R, Mackley V, Kang A, Al-Adeli M, Mace E, Orange JS, Kang E, Lockett SJ, Chen SPJ, Hsu AP, Zarember KA, Malech HL, Gallin JI, Holland SM (2016) Cytoskeletal abnormalities and neutrophil dysfunction in WDR1 deficiency. Blood 128 (17):2135–2143. https://doi.org/10.1182/blood-2016-03-706028
17. Van Dyke TE, Vaikuntam J (1994) Neutrophil function and dysfunction in periodontal disease. Curr Opin Periodontol:19–27
18. Oh TJ, Eber R, Wang HL (2002) Periodontal diseases in the child and adolescent. J Clin Periodontol 29(5):400–410
19. Nibali L, O'Dea M, Bouma G, Parkar M, Thrasher AJ, Burns S, Donos N (2010) Genetic variants associated with neutrophil function in aggressive periodontitis and healthy controls. J Periodontol 81(4):527–534. https://doi.org/10.1902/jop.2010.090543
20. Shaddox L, Wiedey J, Bimstein E, Magnuson I, Clare-Salzler M, Aukhil I, Wallet SM (2010) Hyper-responsive phenotype in localized aggressive periodontitis. J Dent Res 89 (2):143–148. https://doi.org/10.1177/0022034509353397
21. Grimbacher B, Holland SM, Gallin JI, Greenberg F, Hill SC, Malech HL, Miller JA, O'Connell AC, Puck JM (1999) Hyper-IgE syndrome with recurrent infections--an autosomal dominant multisystem disorder. N Engl J Med 340(9):692–702
22. Sowerwine KJ, Holland SM, Freeman AF (2012) Hyper-IgE syndrome update. Ann N Y Acad Sci 1250:25–32. https://doi.org/10.1111/j.1749-6632.2011.06387.x
23. Zhang Q, Su HC (2011) Hyperimmunoglobulin E syndromes in pediatrics. Curr Opin Pediatr 23(6):653–658. https://doi.org/10.1097/MOP.0b013e32834c7f65
24. Farmand S, Sundin M (2015) Hyper-IgE syndromes: recent advances in pathogenesis, diagnostics and clinical care. Curr Opin Hematol 22(1):12–22. https://doi.org/10.1097/MOH.0000000000000104
25. Nguyen-Jackson H, Panopoulos AD, Zhang H, Li HS, Watowich SS (2010) STAT3 controls the neutrophil migratory response to CXCR2 ligands by direct activation of G-CSF-induced CXCR2 expression and via modulation of CXCR2 signal transduction. Blood 115(16):3354–3363. https://doi.org/10.1182/blood-2009-08-240317
26. Koenig JM, Yoder MC (2004) Neonatal neutrophils: the good, the bad, and the ugly. Clin Perinatol 31(1):39–51. https://doi.org/10.1016/j.clp.2004.03.013
27. Hill HR (1987) Biochemical, structural, and functional abnormalities of polymorphonuclear leukocytes in the neonate. Pediatr Res 22(4):375–382. https://doi.org/10.1203/00006450-198710000-00001
28. Lawrence SM, Corriden R, Nizet V (2017) Age-appropriate functions and dysfunctions of the neonatal neutrophil. Front Pediatr 5:23. https://doi.org/10.3389/fped.2017.00023
29. Kaplan J, De Domenico I, Ward DM (2008) Chediak-Higashi syndrome. Curr Opin Hematol 15(1):22–29. https://doi.org/10.1097/MOH.0b013e3282f2bcce
30. Manoli I, Golas G, Westbroek W, Vilboux T, Markello TC, Introne W, Maynard D, Pederson B, Tsilou E, Jordan MB, Hart PS, White JG, Gahl WA, Huizing M (2010) Chediak-Higashi syndrome with early developmental delay resulting from paternal heterodisomy of chromosome 1. Am J Med Genet A 152A(6):1474–1483. https://doi.org/10.1002/ajmg.a.33389
31. Janka GE (2012) Familial and acquired hemophagocytic lymphohistiocytosis. Annu Rev Med 63:233–246. https://doi.org/10.1146/annurev-med-041610-134208
32. Gallin JI (1985) Neutrophil specific granule deficiency. Annu Rev Med 36:263–274. https://doi.org/10.1146/annurev.me.36.020185.001403
33. Gombart AF, Koeffler HP (2002) Neutrophil specific granule deficiency and mutations in the gene encoding transcription factor C/EBP (epsilon). Curr Opin Hematol 9(1):36–42
34. Lekstrom-Himes JA, Dorman SE, Kopar P, Holland SM, Gallin JI (1999) Neutrophil-specific granule deficiency results from a novel mutation with loss of function of the transcription factor CCAAT/enhancer binding protein epsilon. J Exp Med 189(11):1847–1852
35. Nunes P, Demaurex N, Dinauer MC (2013) Regulation of the NADPH oxidase and associated ion fluxes during phagocytosis. Traffic 14(11):1118–1131. https://doi.org/10.1111/tra.12115
36. Dinauer MC (2019) Inflammatory consequences of inherited disorders affecting neutrophil function. Blood (Epub ahead of print) 133(20):2130–2139. https://doi.org/10.1182/blood-2018-11-844563

37. Dinauer MC (2016) Primary immune deficiencies with defects in neutrophil function. Hematology Am Soc Hematol Educ Program 2016 (1):43–50. https://doi.org/10.1182/asheducation-2016.1.43
38. Seger RA (2008) Modern management of chronic granulomatous disease. Br J Haematol 140(3):255–266. https://doi.org/10.1111/j.1365-2141.2007.06880.x
39. Matute JD, Arias AA, Wright NA, Wrobel I, Waterhouse CC, Li XJ, Marchal CC, Stull ND, Lewis DB, Steele M, Kellner JD, Yu W, Merouch SO, Nauseef WM, Dinauer MC (2009) A new genetic subgroup of chronic granulomatous disease with autosomal recessive mutations in p40 phox and selective defects in neutrophil NADPH oxidase activity. Blood 114(15):3309–3315. https://doi.org/10.1182/blood-2009-07-231498
40. Kang EM, Marciano BE, DeRavin S, Zarember KA, Holland SM, Malech HL (2011) Chronic granulomatous disease: overview and hematopoietic stem cell transplantation. J Allergy Clin Immunol 127(6):1319–1326.; quiz 1327-1318. https://doi.org/10.1016/j.jaci.2011.03.028
41. Kuhns DB, Alvord WG, Heller T, Feld JJ, Pike KM, Marciano BE, Uzel G, DeRavin SS, Priel DA, Soule BP, Zarember KA, Malech HL, Holland SM, Gallin JI (2010) Residual NADPH oxidase and survival in chronic granulomatous disease. N Engl J Med 363(27):2600–2610. https://doi.org/10.1056/NEJMoa1007097
42. Roos D, Kuhns DB, Maddalena A, Bustamante J, Kannengiesser C, de Boer M, van Leeuwen K, Koker MY, Wolach B, Roesler J, Malech HL, Holland SM, Gallin JI, Stasia MJ (2010) Hematologically important mutations: the autosomal recessive forms of chronic granulomatous disease (second update). Blood Cells Mol Dis 44(4):291–299. https://doi.org/10.1016/j.bcmd.2010.01.009
43. Roos D, Kuhns DB, Maddalena A, Roesler J, Lopez JA, Ariga T, Avcin T, de Boer M, Bustamante J, Condino-Neto A, Di Matteo G, He J, Hill HR, Holland SM, Kannengiesser C, Koker MY, Kondratenko I, van Leeuwen K, Malech HL, Marodi L, Nunoi H, Stasia MJ, Ventura AM, Witwer CT, Wolach B, Gallin JI (2010) Hematologically important mutations: X-linked chronic granulomatous disease (third update). Blood Cells Mol Dis 45(3):246–265. https://doi.org/10.1016/j.bcmd.2010.07.012
44. Bustamante J, Arias AA, Vogt G, Picard C, Galicia LB, Prando C, Grant AV, Marchal CC, Hubeau M, Chapgier A, de Beaucoudrey L, Puel A, Feinberg J, Valinetz E, Janniere L, Besse C, Boland A, Brisseau JM, Blanche S, Lortholary O, Fieschi C, Emile JF, Boisson-Dupuis S, Al-Muhsen S, Woda B, Newburger PE, Condino-Neto A, Dinauer MC, Abel L, Casanova JL (2011) Germline CYBB mutations that selectively affect macrophages in kindreds with X-linked predisposition to tuberculous mycobacterial disease. Nat Immunol 12(3):213–221. https://doi.org/10.1038/ni.1992
45. van de Geer A, Nieto-Patlan A, Kuhns DB, Tool AT, Arias AA, Bouaziz M, de Boer M, Franco JL, Gazendam RP, van Hamme JL, van Houdt M, van Leeuwen K, Verkuijlen PJ, van den Berg TK, Alzate JF, Arango-Franco CA, Batura V, Bernasconi AR, Boardman B, Booth C, Burns SO, Cabarcas F, Bensussan NC, Charbit-Henrion F, Corveleyn A, Deswarte C, Azcoiti ME, Foell D, Gallin JI, Garces C, Guedes M, Hinze CH, Holland SM, Hughes SM, Ibanez P, Malech HL, Meyts I, Moncada-Velez M, Moriya K, Neves E, Oleastro M, Perez L, Rattina V, Oleaga-Quintas C, Warner N, Muise AM, Lopez JS, Trindade E, Vasconcelos J, Vermeire S, Wittkowski H, Worth A, Abel L, Dinauer MC, Arkwright PD, Roos D, Casanova JL, Kuijpers TW, Bustamante J (2018) Inherited p40phox deficiency differs from classic chronic granulomatous disease. J Clin Invest 128(9):3957–3975. https://doi.org/10.1172/JCI97116
46. Thomas DC, Charbonnier LM, Schejtman A, Aldhekri H, Coomber EL, Dufficy ER, Beenken AE, Lee JC, Clare S, Speak AO, Thrasher AJ, Santilli G, Al-Mousa H, Alkuraya FS, Chatila TA, Smith KGC (2019) EROS/CYBC1 mutations: decreased NADPH oxidase function and chronic granulomatous disease. J Allergy Clin Immunol 143(2):782–785 e781. https://doi.org/10.1016/j.jaci.2018.09.019
47. Arnadottir G (2018) A homozygous loss-of-function mutation leading to CYBC1 deficiency causes chronic granulomatous disease. Nat Commun 9(1):4447. https://doi.org/10.1038/s41467-018-06964-x
48. Al Ghouleh I, Khoo NK, Knaus UG, Griendling KK, Touyz RM, Thannickal VJ, Barchowsky A, Nauseef WM, Kelley EE, Bauer PM, Darley-Usmar V, Shiva S, Cifuentes-Pagano E, Freeman BA, Gladwin MT, Pagano PJ (2011) Oxidases and peroxidases in cardiovascular and lung disease: new concepts in reactive oxygen species signaling. Free Radic Biol Med 51(7):1271–1288. https://doi.org/10.1016/j.freeradbiomed.2011.06.011

49. Roesler J, Curnutte JT, Rae J, Barrett D, Patino P, Chanock SJ, Goerlach A (2000) Recombination events between the p47-phox gene and its highly homologous pseudogenes are the main cause of autosomal recessive chronic granulomatous disease. Blood 95 (6):2150–2156
50. Marciano BE, Spalding C, Fitzgerald A, Mann D, Brown T, Osgood S, Yockey L, Darnell DN, Barnhart L, Daub J, Boris L, Rump AP, Anderson VL, Haney C, Kuhns DB, Rosenzweig SD, Kelly C, Zelazny A, Mason T, DeRavin SS, Kang E, Gallin JI, Malech HL, Olivier KN, Uzel G, Freeman AF, Heller T, Zerbe CS, Holland SM (2015) Common severe infections in chronic granulomatous disease. Clin Infect Dis 60 (8):1176–1183. https://doi.org/10.1093/cid/ciu1154
51. Henrickson SE, Jongco AM, Thomsen KF, Garabedian EK, Thomsen IP (2018) Noninfectious manifestations and complications of chronic granulomatous disease. J Pediatric Infect Dis Soc 7(suppl_1):S18–S24. https://doi.org/10.1093/jpids/piy014
52. Marciano B, Zerbe C, Falcone E, Ding L, DeRavin S, Daub J, Kreuzburg S, Yockey L, Hunsberger S, Foruraghi L, Barnhart L, Matharu K, Anderson V, Darnell D, Frein C, Fink D, Lau K, Long Priel D, Gallin J, Malech H, Uzel G, Freeman A, Kuhns D, Rosenzweig S, Holland S (2018) X-linked carriers of chronic granulomatous disease: illness, lyonization, and stability. J Allergy Clin Immunol 141(1):365–371
53. Battersby AC, Cale AM, Goldblatt D, Gennery AR (2013) Clinical manifestations of disease in X-linked carriers of chronic granulomatous disease. J Clin Immunol 33(8):1276–1284. https://doi.org/10.1007/s10875-013-9939-5
54. Vowells SJ, Fleisher TA, Sekhsaria S, Alling DW, Maguire TE, Malech HL (1996) Genotype-dependent variability in flow cytometric evaluation of reduced nicotinamide adenine dinucleotide phosphate oxidase function in patients with chronic granulomatous disease. J Pediatr 128(1):104–107
55. Foster CB, Lehrnbecher T, Mol F, Steinberg SM, Venzon DJ, Walsh TJ, Noack D, Rae J, Winkelstein JA, Curnutte JT, Chanock SJ (1998) Host defense molecule polymorphisms influence the risk for immune-mediated complications in chronic granulomatous disease. J Clin Invest 102(12):2146–2155
56. Vowells SJ, Sekhsaria S, Malech HL, Shalit M, Fleisher TA (1995) Flow cytometric analysis of the granulocyte respiratory burst: a comparison study of fluorescent probes. J Immunol Methods 178(1):89–97
57. Thomsen IP, Smith MA, Holland SM, Creech CB (2016) A comprehensive approach to the management of children and adults with chronic granulomatous disease. J Allergy Clin Immunol Pract 4(6):1082–1088. https://doi.org/10.1016/j.jaip.2016.03.021
58. Gallin JI, Alling DW, Malech HL, Wesley R, Koziol D, Marciano B, Eisenstein EM, Turner ML, DeCarlo ES, Starling JM, Holland SM (2003) Itraconazole to prevent fungal infections in chronic granulomatous disease. N Engl J Med 348(24):2416–2422
59. Marciano BE, Wesley R, De Carlo ES, Anderson VL, Barnhart LA, Darnell D, Malech HL, Gallin JI, Holland SM (2004) Long-term interferon-gamma therapy for patients with chronic granulomatous disease. Clin Infect Dis 39(5):692–699
60. The International Chronic Granulomatous Disease Cooperative Study Group (1991) A controlled trial of interferon gamma to prevent infection in chronic granulomatous disease. N Engl J Med 324(8):509–516. https://doi.org/10.1056/NEJM199102213240801
61. Gungor T, Teira P, Slatter M, Stussi G, Stepensky P, Moshous D, Vermont C, Ahmad I, Shaw PJ, Telles da Cunha JM, Schlegel PG, Hough R, Fasth A, Kentouche K, Gruhn B, Fernandes JF, Lachance S, Bredius R, Resnick IB, Belohradsky BH, Gennery A, Fischer A, Gaspar HB, Schanz U, Seger R, Rentsch K, Veys P, Haddad E, Albert MH, Hassan M, Inborn Errors Working Party of the European Society for B, Marrow T (2014) Reduced-intensity conditioning and HLA-matched haemopoietic stem-cell transplantation in patients with chronic granulomatous disease: a prospective multicentre study. Lancet 383(9915):436–448. https://doi.org/10.1016/S0140-6736(13)62069-3
62. Morillo-Gutierrez B, Beier R, Rao K, Burroughs L, Schulz A, Ewins AM, Gibson B, Sedlacek P, Krol L, Strahm B, Zaidman I, Kalwak K, Talano JA, Woolfrey A, Fraser C, Meyts I, Muller I, Wachowiak J, Bernardo ME, Veys P, Sykora KW, Gennery AR, Slatter M (2016) Treosulfan based conditioning for allogeneic HSCT in children with chronic granulomatous disease: a multicentre experience. Blood 128(3):440–448. https://doi.org/10.1182/blood-2016-03-704015
63. Booth C, Gaspar HB, Thrasher AJ (2016) Treating immunodeficiency through HSC gene therapy. Trends Mol Med 22 (4):317–327. https://doi.org/10.1016/j.molmed.2016.02.002

64. Kuo CY, Kohn DB (2016) Gene therapy for the treatment of primary immune deficiencies. Curr Allergy Asthma Rep 16(5):39. https://doi.org/10.1007/s11882-016-0615-8
65. Klebanoff SJ, Kettle AJ, Rosen H, Winterbourn CC, Nauseef WM (2013) Myeloperoxidase: a front-line defender against phagocytosed microorganisms. J Leukoc Biol 93(2):185–198. https://doi.org/10.1189/jlb.0712349
66. Beutler E (1994) G6PD deficiency. Blood 84 (11):3613–3636

Part II

Neutrophil Isolation

Chapter 3

Isolation of Human Neutrophils from Venous Blood

Silvie Kremserova and William M. Nauseef

Abstract

Venous blood provides a ready source of large numbers of unstimulated granulocytes and mononuclear cells. Exploiting the differences in the relative densities of the leukocytes circulating in venous blood, one can separate leukocytes from erythrocytes as well as isolate the individual leukocyte populations in high purity for use in ex vivo studies. For selected functional studies, such as transcriptional analysis or cytokine quantitation, addition of an immunomagnetic negative selection step to the standard isolation protocol can yield highly purified human neutrophils.

Key words Granulocytes, Mononuclear cells, Ficoll-Hypaque, Dextran sedimentation, Ultrapure neutrophils

1 Introduction

Under normal conditions, there are ~7.4 K/mm^3 (4.5–11.0) white blood cells in circulation, approximately 59% are polymorphonuclear neutrophils (PMN), 2.7% eosinophils, and 4% monocytes. The different densities of the circulating hematopoietic cells are exploited in order to separate erythrocytes from leukocytes and then to isolate the individual leukocyte populations. Although a variety of methods for rapid, one-step isolation of PMN have been developed recently and used successfully, we routinely use one of two variants of a method first described by Böyum in 1968 [1]. PMN isolated in this way do not consume oxygen and maintain their secretory vesicles intracellularly, two features that suggest that the PMN are minimally, if at all, activated by the isolation procedure.

Two major steps are involved in the isolation, each using specific conditions to separate cells based on their intrinsic density: sedimentation in dextran at 1 × *g* and differential sedimentation in a discontinuous density gradient of Ficoll-Hypaque. In dextran, erythrocytes form rouleaux and thus sediment more rapidly than do granulocytes in suspension. In the Ficoll-Hypaque, granulocytes

Mark T. Quinn and Frank R. DeLeo (eds.), *Neutrophil: Methods and Protocols*, Methods in Molecular Biology, vol. 2087, https://doi.org/10.1007/978-1-0716-0154-9_3,

and erythrocytes pellet to the bottom, whereas mononuclear cells (i.e., lymphocytes and monocytes), basophils, and platelets remain at the interphase between plasma/buffer and the Ficoll-Hypaque. Because we routinely process relatively large quantities of blood (e.g., ≥200 mL) and are primarily interested in recovering PMN (not monocytes), we generally perform dextran sedimentation first, followed by Ficoll-Hypaque sedimentation. When smaller volumes of blood are processed or when monocytes are the targeted cell of interest, the sedimentation in Ficoll-Hypaque can be performed first. Both approaches are described below.

However, PMN isolated by centrifugation through density gradients are contaminated with other blood cells, mainly monocytes and eosinophils, which compromise accurate assessment of a subset of PMN activities, including transcriptional responses and cytokine production. The study of transcriptional events by PMN is demanding because of their relative paucity of RNA. On a per cell basis, PMN have 10 to 20-fold less RNA than do other leukocytes [2], and 1% contamination of cell preparations with mononuclear cells produces as much as 30% RNA contamination [3]. Consequently, *it is essential* to use ultrapure PMN preparations when measuring transcription of genes that may be expressed by the contaminating cells. By the same token, the robust relative cytokine production by contaminating mononuclear cells or eosinophils in a routine PMN isolation can be misleading, as illustrated by assertions that human PMN produce IL-10. In fact, when studies are performed using ultrapure PMN, it is clear that the IL-10 locus in human PMN is inactive [4].

When an immunomagnetic negative selection step [5] (Stem Cell Technologies) is added to our standard PMN isolation protocol (serial dextran sedimentation and Ficoll-Hypaque gradient centrifugation [6]), we obtain preparations that are 99.7 ± 0.06% PMN ($n = 30$), as judged by flow cytometry ($CD15^{+}CD16^{+}$) and the inability to secrete IL-6 in response to lipopolysaccharide (LPS), since human PMN produce IL-6 only when stimulated overnight and exposed to TLR8 agonists that increase IFNα [7, 8]. The procedure to recover ultrapure PMN is described below in Subheading 3.4.

2 Materials

1. Anticoagulant: The preferred anticoagulant is preservative-free 1000 U/mL sodium heparin, although lithium heparin, EDTA, or citrate are acceptable.
2. 3% (w/v) Dextran-500 (average molecular weight 200,000–500,000) solution: dissolve 30 g/L of dextran in endotoxin-free, sterile 0.9% NaCl (*see* **Note 1**). Heat the

solution if necessary to promote dissolution of the dextran, and handle the solution using sterile technique.

3. Ficoll-Hypaque solution: originally prepared by mixing Ficoll 400 and isopaque to create a solution with a density of 1.077 g/mL [1]. Currently premixed, sterilely prepared Ficoll-Hypaque Plus can be purchased. Each 100 mL contains 5.7 g Ficoll 400 and 9.0 g diatrizoate solution with .0231 g of disodium calcium EDTA in endotoxin-free H_2O. If protected from light, the solution is stable for 3 years when stored at 4–30 °C. The appearance of yellow color or particulate material indicates deterioration.
4. Endotoxin-free, sterile H_2O.
5. 1.8% (w/v) NaCl solution.
6. Sterile, endotoxin-free, Hanks' balanced salt solution (HBSS) without Ca^{2+} or Mg^{2+}.
7. Ultrapure PMN purification buffer: sterile, endotoxin-free, phosphate buffered saline (PBS) without Ca^{2+} or Mg^{2+} containing 2% fetal bovine serum (FBS) and 1 mM EDTA.
8. FACS buffer: sterile, endotoxin-free HBSS with Ca^{2+} and Mg^{2+} containing 20 mM HEPES, 0.2% human serum albumin (HSA), and 0.2% NaN_3.
9. Neutrophil Enrichment EasySep kit (Stemcell).
10. DynaMag magnet, holds 15 mL conical tubes.
11. Anti-human CD15 conjugated with allophycocyanin (APC); anti-human CD16 conjugated with phycoerythrin (PE).

3 Methods (*See* Note 2)

3.1 Dextran Sedimentation Followed by Ficoll-Hypaque Density Centrifugation

1. Draw blood into a syringe containing sufficient preservative-free heparin to have a final concentration of 20 U/mL in the blood sample.
2. In a 50 mL conical tube containing the blood, add an equal volume of 3% dextran.
3. Mix tubes by repeated inversion (ten times), and set tubes upright for 18–20 min at room temperature (*see* **Note 3**).
4. Aspirate the straw-colored, leukocyte-rich, erythrocyte-poor upper layer with a sterile plastic pipette or a 10–15 mL sterile syringe and transfer the aspirate to a sterile 50 mL conical tube (*see* **Note 4**).
5. Pellet leukocytes by centrifugation at 500 × *g* for 10 min at 4 °C; aspirate and discard the supernatant.
6. Resuspend pellets in 10 mL, shake well but *do not* vortex. Three resuspended pellets can be pooled into a single conical

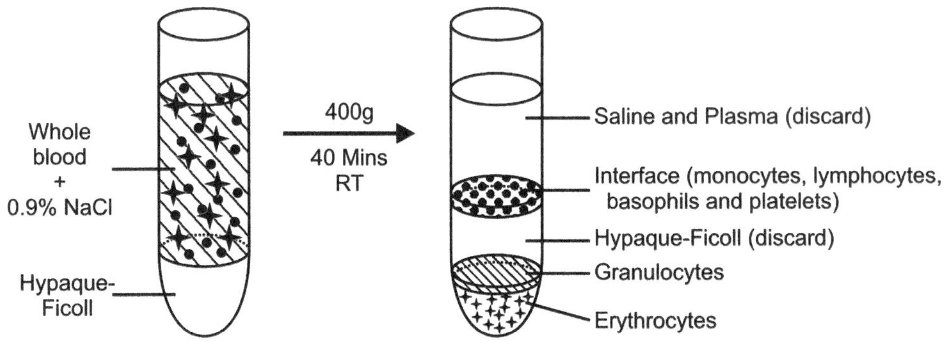

Fig. 1 Ficoll-Hypaque gradient separation of granulocytes and peripheral blood mononuclear cells. Ficoll-Hypaque gradient before (left side) and after (right side) centrifugation. (Reproduced from [6] by permission of Humana Press©2007)

tube to achieve a final volume of 35 mL. Top off with sterile saline if needed to reach 35 mL.

7. Carefully underlay the leukocyte suspension with 10 mL of Ficoll-Hypaque, using a sterile, plastic 10 mL pipette (Fig. 1).
8. Centrifuge at 400 × *g* for 40 min at room temperature (*see* **Note 5**).
9. After centrifugation, two bands should be apparent in the conical tube (Fig. 1). The lighter band contains mononuclear cells, whereas the denser band has both granulocytes and erythrocytes. If monocytes are desired, aspirate the less dense band using a sterile plastic pipette and mix with an equal volume of cold sterile saline in a separate conical tube.
10. Aspirate and discard supernatant above the PMN-erythrocyte layer, taking care not to lose any of the PMN-rich pellet.
11. Resuspend each pellet in sterile water and mix well (but *do not* vortex) for 28 s. Promptly restore tonicity by adding an equal volume of 1.8% saline and mixing (*see* **Note 6**).
12. Centrifuge 500 × *g* for 5 min at 4 °C. Discard supernatant and repeat **step 11** *if* more lysis is needed, but repeat **step 11** only once (*see* **Note 6**).
13. Resuspend cells in HBSS without Ca^{2+} and Mg^{2+} at $\leq 3 \times 10^7$/mL.
14. Determine the cell concentration by manually counting using a hemacytometer. The differential of leukocytes can be assessed by examining a stained slide microscopically (*see* **Note 7**).

3.2 Ficoll-Hypaque Density Centrifugation-Mononuclear Cell Recovery

1. Draw blood into a syringe as in Subheading 3.1 and dilute with 1 volume of 0.9% NaCl at room temperature to a total volume of 40 mL in a sterile 50 mL conical plastic tube (*see* **Note 8**).
2. Carefully underlay the diluted blood with 10 mL of Ficoll-Hypaque.

3. Centrifuge at 400 × *g* for 40 min at room temperature.
4. After centrifugation, remove the HBSS and plasma above the mononuclear band at the interface (*see* Fig. 1) and discard.
5. Recover the fraction containing mononuclear cells, combining bands from no more than two gradients into a separate 50 mL conical tube.
6. Bring each of the pooled fractions to 50 mL with sterile HBSS without Ca^{2+} and Mg^{2+} to dilute the Ficoll-Hypaque (*see* **Note 9**).
7. Pellet mononuclear cells at 600 × *g* for 10 min at 4 °C.
8. Discard supernatant, pool, and wash pellets twice more with HBSS without Ca^{2+} and Mg^{2+}.
9. Resuspend final pellet in 5 mL of HBSS without Ca^{2+} and Mg^{2+} and count.

3.3 Ficoll-Hypaque Density Centrifugation First-PMN Recovery

1. After **step 5** in Subheading 3.2, when the supernatant and the mononuclear band have been removed, aspirate and discard the remaining gradient above the PMN-erythrocyte pellet (*see* Fig. 1).
2. Resuspend the pellet in a volume up to 25 mL with HBSS without Ca^{2+} and Mg^{2+}.
3. Add 25 mL of 3% dextran, mix by inverting the tube ten times, and let stand upright at room temperature for 18–20 min.
4. Aspirate the straw-colored, leukocyte-rich, erythrocyte-poor upper layer with a sterile plastic pipette or a 10–15 mL sterile syringe and transfer the aspirate to a sterile 50 mL conical tube (*see* **Note 4**).
5. Pellet leukocytes by centrifugation at 500 × *g* for 10 min at 4 °C; aspirate and discard the supernatant.
6. Resuspend each pellet in sterile water and mix well (but *do not* vortex) for 28 s. Promptly restore tonicity by adding an equal volume of 1.8% saline and mixing (*see* **Note 6**).
7. Resuspend cells in HBSS without Ca^{2+} and Mg^{2+} at $\leq 3 \times 10^7$/mL.
8. Determine the cell concentration by manually counting using a hemacytometer. The differential of leukocytes can be assessed by examining a stained slide microscopically.

3.4 Purification of Ultrapure PMN

1. After hypotonic lysis, resuspend PMN pellet in HBSS without Ca^{2+} and Mg^{2+}. Determine the cell concentration by manually counting using a hemocytometer. Use maximum of 3×10^8 cells per tube (*see* **Notes 10** and **11**).

2. Pellet PMN by centrifugation at 250 × *g* for 5 min at 4 °C in polystyrene 15 mL conical tube. Resuspend PMN pellet to 5×10^7 cells/mL in Ultrapure PMN purification buffer.
3. To the cell suspension add Human Neutrophil Enrichment cocktail at 50 μL/mL cell suspension. Mix well by inverting tube or gently pipet up and down more than five times and incubate for 10 min at 4 °C. *Do not vortex!!*
4. Mix Magnetic particles from the kit to make sure they are uniformly suspended; vigorously pipet up and down more than five times. *Do not vortex!!* Add 100 μL of Magnetic particles per 1 mL of the cell suspension. Mix well and incubate for 10 min at 4 °C.
5. Bring volume to 10 mL with Ultrapure PMN purification buffer (3×10^7 PMN/mL). Invert tube up and down three times.
6. Place the tube without cap into the magnet and incubate for 10 min at room temperature.
7. In one continuous motion, invert the magnet with the tube to pour the enriched cell suspension into a new polystyrene 15 mL conical tube. Do not shake or blot off any drops that may remain hanging from the mouth of the tube.
8. Remove the empty tube from the magnet.
9. Place the new tube containing the supernatant without a cap into the magnet and incubate for 10 min at room temperature.
10. In one continuous motion, invert the magnet with tube to pour the enriched cell suspension into another new polystyrene 15 mL conical tube. Do not shake or blot off any drops that may remain hanging from the mouth of the tube.
11. Remove the empty tube from the magnet.
12. Determine the cell concentration by manually counting using a hemocytometer.
13. Assess purity by flow cytometry and staining cells with anti-human CD16-PE and anti-human CD15-APC antibodies (*see* **Note 12**).
14. Take 1×10^6 cells per each condition to examine by flow cytometry (*see* **Note 13**).
15. Pellet cells by centrifugation at 250 × *g* for 5 min at 4 °C and resuspend in 100 μL FACS buffer.
16. Add 5 μL of each antibody and incubate 30 min at 4 °C in dark (flick tubes every 10 min).
17. After staining, wash the cells with 500 μL FACS buffer. Pellet at 250 × *g* for 5 min at 4 °C, resuspend pellet in 500 μL FACS buffer and read on flow cytometer.

4 Notes

1. The concentration and the molecular weight range of the dextran are critical determinants of the speed with which cells sediment [9]. The preparation of 3% in the 200,000–500,000 weight range provides excellent sedimentation of erythrocytes within 18–20 min, with rates slower using either the lower molecular weight dextrans (<20,000) or the high molecular weight (>7,000,000) preparations. Given the range in size of the components in the dextran being used, one can anticipate corresponding variation in the rate of sedimentation (reflected as differences in the number of cells recovered). For example, we currently allow sedimentation to occur for 18 min at room temperature to obtain maximal number of granulocytes. Be prepared to modify the time allotted for sedimentation accordingly. The volume of the dextran-blood mixture has no effect on the efficiency of sedimentation. In the event that sedimentation has proceeded for longer than the intended time, the sample can be mixed again and the sedimentation repeated with no adverse effect on yield or purity. Whenever alternative agents are used for sedimentation, it is important to determine if the recovered PMN are primed, as occurs with gelatin [10]
2. In general, always use polypropylene tubes, as glass and polyethylene may activate the PMN during isolation. Pipettes should be sterile and buffers should be prepared with sterile, endotoxin-free reagents, as endotoxin will prime PMN. Cell suspensions should never be vortexed, since vigorous mechanical agitation will stimulate PMN. Resuspend by gentle pipetting.
3. Adequate mixing of the sample is essential for reproducible sedimentation, presumably to distribute the erythrocytes throughout the suspension and provide maximal opportunities for rouleaux formation. When directly examined, ten mixing inversions were sufficient to guarantee reproducible and maximum sedimentation; fewer inversions compromised the reproducibility in yield.
4. Once the leukocyte-rich supernatant from the dextran sedimentation has been centrifuged, proceed as quickly as possible with the isolation procedure. Do not leave cell pellets at intermediate steps. Aspirates and cell pellets are biohazard waste and should be handled accordingly.
5. The centrifugation of the Ficoll-Hypaque gradient should be at 20 °C and with no brake on the centrifuge. When performed at 4 °C, there is more PMN contamination of the mononuclear band than seen when centrifugation is at 20 °C (2.3 vs 0.1%, respectively), and more lymphocyte and mononuclear

contamination of the PMN pellet (6.9 and 0.6% vs. 1.4 and 0.1%, respectively). It is best to have a dedicated room-temperature centrifuge for this use; alternatively, have the refrigeration in the centrifuge turned off overnight before running the gradient the following morning.

6. The hypotonic lysis of erythrocytes exploits the relative resistance of leukocytes to osmotic stress. This difference is relative and prolonged or repeated hypotonic lysis will damage PMN. Attention to limiting the time of exposure to hypotonicity to 18–20 s should be strictly maintained. More than 2 cycles of hypotonic lysis must be avoided, as they will not lyse additional erythrocytes but will begin to damage PMN.
7. When counting the PMN suspension, one can directly count the PMN by diluting the cell suspension 1:20 in 3% acetic acid and observe under 40× objective. In the acetic acid solution the nuclear morphology of PMN is clearly identified, allowing direct counting of the PMN in suspension. Alternatively, PMN suspension can be diluted 1:20 in HBSS and counted to determine the leukocyte concentration. A separate sample (e.g., 10 μL) can be diluted 1:5 in HBSS or saline and placed on a slide (either using a cytospin or simply maneuver the slide to obtain a thin layer) and allowed to air-dry. The slide can then be stained with Wright stain (or an equivalent) and a differential performed. With the total number of leukocytes and the percent of PMN on differential staining, the number of PMN isolated can be calculated. In general, the yield should be 2–four million PMN/mL blood drawn and the suspension should be ~94–96% PMN with a few contaminating eosinophils.
8. Efficient separation is achieved by 1:1 dilution of blood with saline prior to centrifugation in the Ficoll-Hypaque gradient. The tendency of lymphocytes to be trapped in erythrocyte-granulocyte aggregates decreases the yield of lymphocytes while simultaneously contaminating the granulocyte pellet with lymphocytes. Dilution of whole blood at the start of the isolation will decrease both problems.
9. Neither mononuclear cells nor granulocytes tolerate long incubation in Ficoll-Hypaque, so it is best to dilute the gradient matrix in the isolated cell fractions as soon as feasible.
10. Keep in mind that you will lose ~50% of cells during ultrapurification process; that is, your yield will be approximately half of the initial number of PMN.
11. Use no more than 3×10^8 cells per tube resuspended in 6 mL of Ultrapure PMN purification buffer at this step. In **step 5**, added Ultrapure PMN purification buffer will bring the volume to 10 mL, a volume of cell suspension that is below the

level of the magnet. A larger volume will exceed the height of the magnet.

12. Gating for the populations of different leukocytes is based on the size (FCS) and granularity (SSC) of the cells. Figure 2 shows distribution of leukocyte populations in traditionally purified PMN (left) and ultrapure PMN (right). Granulocyte populations include eosinophils and neutrophils, which are distinguished by staining for CD15 and CD16. Whereas both PMN and eosinophils express surface CD15, only PMN stain for CD16 (Fig. 3).

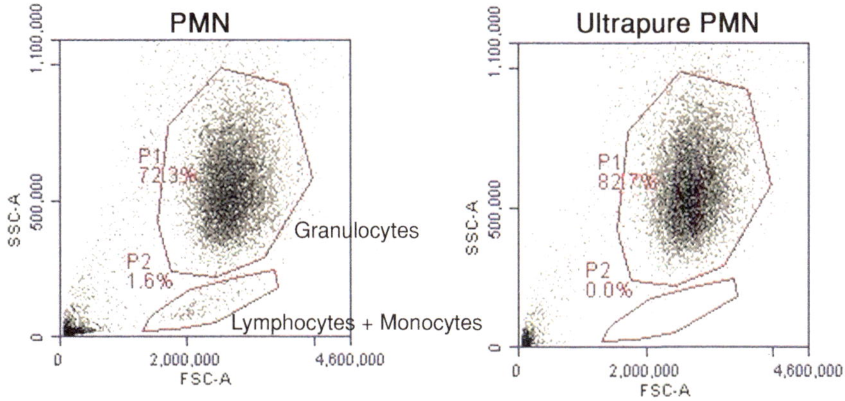

Fig. 2 Gating strategy of PMN. Distribution of leukocyte populations after isolation by dextran sedimentation followed by Ficoll-Hypaque density centrifugation (left) and in PMN after ultrapurification step (right)

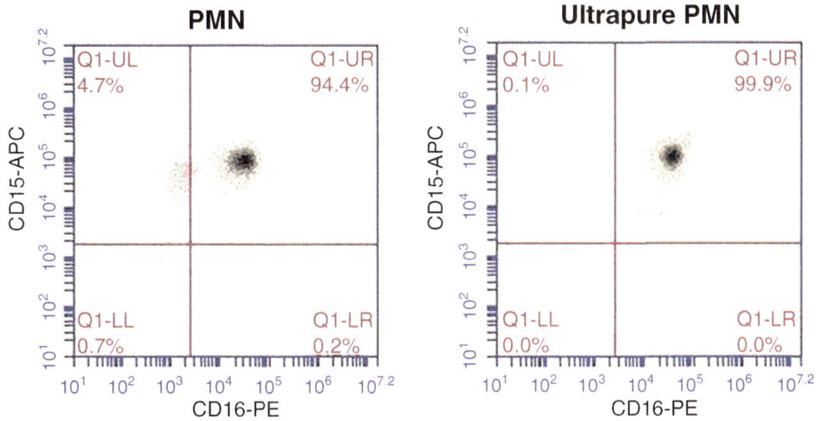

Fig. 3 Distribution of granulocyte population in PMN isolated by dextran sedimentation followed by Ficoll-Hypaque density centrifugation (left) and ultrapure PMN (right). Eosinophils are $CD15^+$ (upper left quadrant), whereas PMN are both—$CD15^+CD16^+$ (upper right quadrant)

13. Using 1×10^6 cells for each, five different conditions need to be examined: (a) unstained cells; (b + c) cells stained only with anti-CD15 or with anti-CD16; and (d + e) cells stained with both anti-CD15 and anti-CD16 for PMN before and after ultrapurification.

References

1. Böyum A (1968) Isolation of mononuclear cells and granulocytes from human blood: isolation of mononuclear cells by one centrifugation, and of granulocytes by combining centrifugation and sedimentation at 1 g. Scand J Clin Lab Invest Suppl 97:77–89
2. Scapini P, Calzetti F, Cassatella MA (1999) On the detection of neutrophil-derived vascular endothelial growth factor (VEGF). J Immunol Methods 232:121–129
3. Tamassia N, Cassatella MA, Bazzoni F (2014) Fast and accurate quantitative analysis of cytokine gene expression in human neutrophils. Methods Mol Biol 1124:451–467
4. Tamassia N, Zimmermann M, Castellucci M et al (2013) Cutting edge: an inactive chromatin configuration at the IL-10 locus in human neutrophils. J Immunol 190:1921–1925
5. Tecchio C, Micheletti A, Cassatella MA (2014) Neutrophil-derived cytokines: facts beyond expression. Front Immunol 5:508
6. Nauseef WM (2014) Isolation of human neutrophils from venous blood. Methods Mol Biol 1124:13–18
7. Zimmermann M, Aguilera FB, Castellucci M et al (2015) Chromatin remodelling and autocrine TNFα are required for optimal interleukin-6 expression in activated human neutrophils. Nat Commun 6:6061
8. Zimmermann M, Arruda-Silva F, Bianchetto-Aguilera F et al (2016) IFNalpha enhances the production of IL-6 by human neutrophils activated via TLR8. Sci Rep 6:19674
9. Skoog WA, Beck WS (1956) Studies on the fibrinogen, dextran, and phytohemagglutinin methods of isolating leukocytes. Blood 11:436–454
10. Stie J, Jesaitis AJ (2007) Reorganization of the human neutrophil plasma membrane is associated with functional priming: implications for neutrophil preparations. J Leuk Biol 81:672–685

Chapter 4

Isolation of Neutrophils from Nonhuman Species

Daniel W. Siemsen, Liliya N. Kirpotina, Natalia Malachowa, Igor A. Schepetkin, Adeline R. Porter, Benfang Lei, Frank R. DeLeo, and Mark T. Quinn

Abstract

The development of new advances in understanding the role of neutrophils in inflammation requires effective procedures for isolating and purifying neutrophils. Methods for isolating human neutrophils are fairly standard, and some are covered in other chapters of this volume and previous editions. However, procedures for isolating neutrophils from nonhuman species used to model human diseases vary from those used in isolating human neutrophils and are not as well developed. Since neutrophils are highly reactive and sensitive to small perturbations, the methods of isolation are important to avoid isolation technique-induced alterations in cell function. We present methods here for reproducibly isolating highly purified neutrophils from large animal models (bovine, equine, ovine), small animal models (murine and rabbit), and nonhuman primates (cynomolgus macaques) and describe optimized details for obtaining the highest cell purity, yield, and viability.

Key words Inflammation, Large animal model, Granulocyte, Polymorphonuclear leukocyte, Cell isolation, Flow cytometry, Blood, Bone marrow

1 Introduction

Over the years, various animal models have been developed for investigation of the pathogenesis of human inflammation and infectious diseases (reviewed in [1, 2]). Although small animal models, such as rodents, are easier to handle, breed easily, require much less in the way of housing facilities, and are generally less expensive, they often do not provide an accurate reflection of human physiology [1, 3]. Thus, nonhuman primates and larger animal models, such as pigs, cattle, and sheep, are often desirable as models for human disease pathogenesis [3, 4]. In these models, it is important to characterize neutrophil function, which requires efficient methods for purification of these phagocytic cells.

Currently, much of our understanding of neutrophil biology is based on studies using human cells; whereas, much less is known

Mark T. Quinn and Frank R. DeLeo (eds.), *Neutrophil: Methods and Protocols*, Methods in Molecular Biology, vol. 2087, https://doi.org/10.1007/978-1-0716-0154-9_4, © Springer Science+Business Media, LLC, part of Springer Nature 2020

regarding the biology of these cells in nonhuman species. It is clear, however, that neutrophils from other species differ from their human counterpart in a number of important functional characteristics [2, 5], and one cannot assume that the basic features of human neutrophil biochemistry and function are representative of neutrophils in all other species. Many species are used as models to understand human pathophysiology; thus, it is essential that we gain a thorough understanding of neutrophil functions in these models. It is also important to characterize neutrophil biology in various nonhuman species to determine immune status and host defense mechanisms in these organisms. Additionally, comparison of conserved features between neutrophils from different species can contribute to our understanding of neutrophil biochemistry in general.

One of the major challenges associated with neutrophil studies is the isolation of highly purified cell preparations that are morphologically and functionally similar to cells found in the blood in vivo. Neutrophils are temperamental and can be easily altered by improper handling [6]. Thus, the method of cell isolation is extremely important to avoid isolation technique-induced alterations in neutrophil function [6]. For example, neutrophils can be primed during isolation, resulting in altered neutrophil responses to subsequent stimuli and changes in surface antigen expression [7–9]. Furthermore, some neutrophil functions, such as chemotaxis, can actually be inhibited by isolation procedures [6]. To completely eliminate any isolation artifacts, methods have been developed for analysis of neutrophils in whole blood (e.g., *see* [7, 10]). Conversely, the use of whole blood can be impractical for many biochemical studies, which require purified cells in the absence of other contaminating cells and serum proteins, or in some small animal models (e.g., mice), where the volume of blood available is very small.

We present methods here for neutrophil isolation from a range of large (bovine, equine, nonhuman primate, ovine) and small (murine, rabbit) animal models and describe optimized details for obtaining the highest cell purity, yield, and viability. Importantly, all of these isolation methods are relatively inexpensive, utilize commonly available reagents, and do not require the acquisition of specialized equipment. Thus, they can be easily implemented in any lab.

2 Materials

1. 15-mL Vacutainer tubes (Becton Dickinson) containing 150 μL of 500 mM disodium ethylenediamine tetraacetic acid (EDTA), pH 7.4, prepared in sterile H_2O and filtered (*see* **Note 1**).

2. Sterile endotoxin-free disposable plastic pipettes, polypropylene centrifuge tubes, and disposable polyethylene transfer pipettes (*see* **Note 2**).
3. CD66abce MicroBead kit (Miltenyi Biotec).
4. MACS LS separation columns and MACS magnetic stand (Miltenyi Biotec) (*see* **Note 3**).
5. Sodium heparin 1000 U/mL.
6. Sterile 6″ cotton-tipped applicators.
7. 2% acetic acid solution.
8. 0.4% trypan blue solution.

2.1 Buffers

1. Sterile injection-grade H_2O and sterile 0.9% NaCl solution (*see* **Note 4**).
2. Phosphate-buffered saline (PBS): 140 mM NaCl, 2.7 mM KCl, 8 mM Na_2HPO_4, 1.5 mM KH_2PO_4 dissolved in sterile water. Adjust pH to 7.2 and sterile filtered. Store at 4 °C.
3. Dulbecco's Modified Eagle Medium (DMEM), Dulbecco's Phosphate Buffered Saline (DPBS), 10× Hank's balanced-salt solution (10× HBSS) (without Ca^{2+}, Mg^{2+}, and phenol red), RPMI Medium 1640 without phenol red (RPMI). For 1× HBSS: Dilute 10× HBSS in sterile H_2O, adjust pH to 7.4, and sterile filter. Store at 4 °C (*see* **Notes 4** and **5**).
4. RPMI/Hepes: RPMI supplemented with 10 mM HEPES buffer.
5. Hetastarch solution: 6% hetastarch in 0.9% NaCl.
6. Solutions of 9% (w/v) and 10% (w/v) NaCl in sterile H_2O. Prepare fresh and sterile filter (*see* **Notes 4** and **5**).
7. Acid citrate dextrose (ACD) solution: 65 mM citric acid, 85 mM sodium citrate, and 2% dextrose dissolved in sterile H_2O and sterile filtered. Store at 4 °C (*see* **Notes 4** and **5**).
8. Murine neutrophil buffer: HBSS containing 0.1% (w/v) bovine serum albumin, 1% (w/v) glucose. Prepare fresh and sterile filter (*see* **Notes 4** and **5**).
9. Rabbit neutrophil buffer: 138 mM NaCl, 27 mM KCl, 8.1 mM Na_2HPO_4, 1.5 mM KH_2PO_4, and 5.5 mM glucose in sterile H_2O and sterile filtered. Store at 4 °C (*see* **Notes 4** and **5**).
10. 6% dextran solution: mix 4.5 mL of 20% sterile dextran solution (average MW 500,000; Sigma), 540 μL of 25% NaCl solution prepared in endotoxin-free H_2O, and 9.96 mL of endotoxin free H_2O. Sterile filter and store at 4 °C (*see* **Notes 4–6**).
11. 10× PBS/EDTA buffer: Dissolve 100 mM KH_2PO_4, 9% (w/v) NaCl, 2 mg/L EDTA in sterile H_2O, adjust pH to

7.4, and sterile filter (*see* **Notes 4** and **5**). For 1× PBS/EDTA, dilute 10× PBS/EDTA 1:10 in sterile H_2O, adjust pH to 7.4 if needed, and sterile filter (*see* **Notes 4** and **5**).

12. MACS erythrocyte lysis buffer: 155 mM NH_4Cl, 10 mM $KHCO_3$ and 0.1 mM EDTA dissolved in sterile water. Adjust pH to 7.3 and sterile filtered. Store at 4 °C. (NHP protocol).
13. MACS neutrophil buffer: PBS (pH 7.2), 0.5% human serum albumin, 2 mM EDTA. Sterile filtered and stored at 4 °C (*see* **Note 7**).

2.2 Density Gradient Solutions

1. Percoll, Histopaque 1077, and Histopaque 1119.
2. Histopaque 1077/1119 solution: Mix equal volumes of Histopaque 1077 and Histopaque 1119.
3. Percoll stock solution (100% Percoll): mix Percoll and 10× HBSS, pH 7.4 at a ratio of 9:1 (v/v) (*see* **Note 8**).
4. Percoll solutions: Mix 100% Percoll stock with the appropriate volumes of 1× HBSS to obtain 85%, 81%, 70%, 65%, 62%, 55%, 50%, and 45% (v/v) Percoll solutions (*see* **Note 8**).
5. Percoll/EDTA stock solution (100% Percoll/EDTA): Percoll and 10× PBS/EDTA, pH 7.4 mixed at a ratio of 9:1 (v/v) (*see* **Note 8**).
6. Percoll/EDTA solution: Mix 100% Percoll/EDTA stock with the appropriate volume of 1× PBS/EDTA to obtain 65% (v/v) Percoll/EDTA solution (*see* **Note 8**).

3 Methods

In the methods described below, we outline the steps to obtain highly purified and functionally active neutrophils from bovine, equine, nonhuman primates, ovine, and rabbit blood, as well as murine bone marrow (*see* **Note 9**). Note that some of the methods detailed here have been adapted with modifications from previously published methods for isolation of equine [11], murine [12], ovine [13], and rabbit neutrophils [14, 15]. Additionally, the method for isolation of neutrophils from nonhuman primates utilizes positive selection, which requires labeling of neutrophils with primary and secondary antibodies conjugated to Microbeads and was adapted from the Miltenyi Microbead Kit protocol (Miltenyi Biotec Inc.). Importantly, the presence of these antibodies on the surface of nonhuman primate neutrophils should be considered in the experimental design.

In addition to the basic neutrophil isolation procedures, we also provide details for quantifying cell yield, evaluating cell purity, and viability. These parameters are compared with those of human neutrophils purified by standard methods. Although neutrophil isolation from all species except nonhuman primates is based on density gradient separation techniques, the gradient media and composition vary because of slight differences in neutrophil density between species. In addition, differences in red blood cell reactivity to aggregating reagents between species are reflected in the methods described below. Overall, these methods are efficient, easy to perform, and reproducibly generate high-quality neutrophil populations for biochemical and functional studies.

3.1 Bovine Neutrophil Isolation

1. Collect bovine blood into Vacutainer tubes containing EDTA (*see* **Note 9**). For the method outlined here, we collected 50 mL of blood. If different volumes of blood are required, adjust the indicated volumes proportionally.
2. Pool 50 mL of blood into a conical 50-mL polypropylene centrifuge tube and centrifuge at 740 × *g* for 10 min at room temperature with low brake.
3. Remove the upper plasma layer and buffy coat found at the plasma-red blood cell interface with a plastic transfer pipette. Transfer the remaining red blood cell layer into a conical 250-mL polypropylene tube.
4. Lyse red blood cells by adding 50 mL of sterile H_2O. Mix by gently inverting the tube for 20 s at room temperature (*see* **Note 10**).
5. Immediately add 5 mL of 10% NaCl solution and mix well by gently inverting the tube.
6. Centrifuge at 585 × *g* for 10 min at room temperature with low brake.
7. Remove the supernatant using a plastic transfer pipette and resuspend the cell pellet in 50 mL of HBSS.
8. Lyse any remaining red blood cells by repeating **steps 4** through **6**.
9. Resuspend the leukocyte pellet in 10 mL of HBSS.
10. Prepare Histopaque gradients by first pipetting 15-mL of Histopaque 1077 into the bottom of a conical 50-mL centrifuge tube. Place a borosilicate glass Pasteur pipette into the tube so the pipette tip rests on the bottom of the tube. Use this pipette as a funnel to carefully underlay 15 mL of Histopaque 1077/1119 solution.
11. Layer the 10-mL leukocyte suspension on top of the Histopaque gradient using a plastic transfer pipette. This must be done carefully to avoid mixing the cell suspension with the Histopaque.
12. Centrifuge the gradient at 440 × *g* for 25 min at room temperature with no brake.

13. Remove the supernatant with a plastic transfer pipette and discard (*see* **Note 11**).
14. Wash the neutrophil pellet by resuspending the cells in 50 mL of HBSS and centrifuging at 585 × *g* for 10 min at room temperature.
15. Resuspend purified cells in the desired assay buffer.

3.2 Equine Neutrophil Isolation

1. Collect equine blood into Vacutainer tubes containing EDTA (*see* **Note 9**). For the method outlined here, we collected 24 mL of blood. If different volumes of blood are required, adjust the indicated volumes proportionally.
2. Prepare Percoll gradients by underlaying 2.5-mL of 85% Percoll solution below 2.5 mL of 70% Percoll solution in a conical 15-mL polypropylene centrifuge tube. Use a borosilicate glass Pasteur pipette as a funnel to underlay the Percoll solution (*see* **step 11** under Subheading 3.1.).
3. Carefully layer 3 mL of blood on top of each gradient using a plastic transfer pipette.
4. Centrifuge the gradients for 20 min at 400 × *g* with no brake at room temperature.
5. The neutrophil band sediments at the interface between 70% and 85% Percoll solutions. Carefully remove all supernatant above the neutrophil band with a plastic transfer pipette and discard (*see* **Note 11**).
6. Collect the neutrophil band with a clean plastic transfer pipette.
7. Wash the cells by resuspending them in 50 mL of HBSS and centrifuging at 200 × *g* for 10 min at room temperature.
8. Wash the cells twice more by repeating **step** 7 above.
9. Resuspend purified cells in the desired assay buffer.

3.3 Human Neutrophil Isolation

1. Collect human blood into Vacutainer tubes containing EDTA (*see* **Note 9**). For the method outlined here, we collected 30 mL of blood. If different volumes of blood are required, adjust the indicated volumes proportionally.
2. Combine 6.7 mL of dextran solution with 30 mL of blood in a conical 50-mL polypropylene centrifuge tube. Mix by gently inverting the tube.
3. Allow the blood–dextran mixture to sediment for 45 min at room temperature. Dextran causes the red blood cells to form aggregates, which sediment to the bottom of the tube. This leaves a clear, red blood cell-depleted layer above the red blood cell-rich lower layer.
4. Transfer the upper cell layer to a clean conical 50-mL polypropylene centrifuge tube with a plastic transfer pipette.

5. Centrifuge the tube at 740 × *g* for 10 min with low brake at room temperature.
6. Remove the supernatant using a plastic transfer pipette and discard.
7. Resuspend the white blood cell pellet in 7 mL of sterile 0.9% NaCl solution.
8. Place 7 mL of Histopaque 1077 into a conical 50-mL polypropylene centrifuge, and carefully layer the white blood cell suspension on top of the Histopaque. This must be done carefully to avoid mixing the cell suspension with the Histopaque.
9. Centrifuge at 700 × *g* for 15 min with no brake at room temperature.
10. Remove the supernatant using a plastic transfer pipette and discard (*see* **Note 11**).
11. Resuspend the neutrophil pellet in 6 mL of sterile 0.9% NaCl solution.
12. Lyse contaminating red blood cells by adding 20 mL of sterile H_2O. Mix by gently inverting tubes for 20 s at room temperature (*see* **Note 10**).
13. Immediately add 1.8 mL of 10% NaCl solution and mix well by gently inverting tubes.
14. Centrifuge at 740 × *g* for 10 min at room temperature with low brake.
15. Remove the supernatant using a plastic transfer pipette and discard.
16. Resuspend the neutrophil pellet in 6 mL of sterile 0.9% NaCl solution.
17. Lyse any remaining red blood cells by repeating **steps 12** through **15** above.
18. Remove the supernatant with a plastic transfer pipette.
19. Wash the neutrophil pellet by resuspending the cells in 50 mL of 0.9% NaCl solution and centrifuging at 740 × *g* for 10 min at room temperature.
20. Resuspend purified cells in the desired assay buffer.

3.4 Murine Neutrophil Isolation

Unlike with the other species described in this Chapter, mice have a very small blood volume. Thus, the number of neutrophils that can be isolated from murine blood is limited and usually not sufficient for biochemical and pharmacological studies or adoptive transfer experiments [16]. To overcome this issue, bone marrow is commonly used to isolate resting murine neutrophils, which have been well characterized and shown to be functionally competent [17].

1. Dissect femurs and tibias from 8–12 week old mice (*see* **Note 9**). BALB/c mice were used here, but this procedure should also work for other strains of mice.
2. Clip the ends of each tibia and femur with dissecting scissors to expose the marrow.
3. Flush bone marrow cells from the tibias and femurs with murine neutrophil buffer using a syringe with 27-G needle. Use two 1-mL volumes of buffer for tibias and three 1-mL volumes of buffer for femurs.
4. Resuspend the pooled bone marrow eluates by gentle pipetting, followed by filtration through a 70-μm nylon cell strainer to remove cell clumps and bone particles.
5. Centrifuge pooled bone marrow cells at 600 × *g* for 10 min at 4 °C with low brake.
6. Remove the supernatant with a plastic transfer pipette and discard.
7. Resuspend the cell pellet in 3 mL of 45% Percoll solution.
8. Prepare Percoll gradients by layering 2 mL each of the 62%, 55%, and 50% Percoll solutions successively on top of 3 mL of 81% Percoll solution in a conical 15-mL polypropylene tube.
9. Carefully layer the bone marrow cell suspension on top of the gradient.
10. Centrifuge at 1600 × *g* for 30 min with no brake at 10 °C.
11. Remove the supernatant down through the 62% Percoll layer using a plastic transfer pipette and discard. Be careful not to disturb the neutrophil band (*see* **Note 11**).
12. Collect the neutrophil band that is located just at the interface of the 62% and 81% Percoll layers using a plastic transfer pipette.
13. Wash the collected cells by resuspending them in 10 mL of murine neutrophil buffer and centrifuging at 600 × *g* for 10 min at 10 °C.
14. Wash the cells again by repeating **step 13** above and resuspend the final pellet in 3 mL of murine neutrophil buffer.
15. Carefully layer the cell suspension on top of 3 mL of Histopaque 1119 in conical 15-mL polypropylene tubes.
16. Centrifuge the gradients at 1600 × *g* for 30 min at 10 °C and no brake to remove contaminating red blood cells.
17. Remove the supernatant using a plastic transfer pipette and discard (*see* **Note 11**).
18. Collect the cell layer between the Histopaque and buffer layers with a plastic transfer pipette.

19. Wash the cells by resuspending them in 10 mL of murine neutrophil buffer and centrifuging at 600 × *g* for 10 min at 10 °C.
20. Wash the cells again by repeating **step 19** above.
21. Resuspend purified cells in the desired assay buffer.

3.5 Nonhuman Primate Neutrophil Isolation

1. Collect blood from cynomolgus macaques (*Macaca fascicularis*) into syringes containing 100–150 μL of heparin (1000 U/mL) per 10 mL of blood (*see* **Notes 9** and **12**).
2. Transfer blood to a fresh tube and mix gently with 5–10 volumes of MACS erythrocyte lysis buffer (*see* **Note 13**).
3. Rotate tube continuously at low settings on a rotation mixer for slow and gentle rotation of cells for 10 min at room temperature or rotate tubes by hand several times during the incubation.
4. Pellet cells by centrifugation at 300 × *g* for 10 min at room temperature.
5. Aspirate supernatant and wash cells once with 10 mL of MACS neutrophil buffer (*see* **Note 14**).
6. Pellet cells by centrifugation at 200 × *g* for 10 min at room temperature.
7. Remove supernatant and resuspend cell in 0.5 mL of MACS neutrophil buffer.

3.5.1 Magnetic Labeling

1. Count cells using a hemocytometer or estimate based on the known number of neutrophils per mL of blood.
2. If it is necessary to concentrate cells, centrifuge the cell suspension at 200 × *g* for 10 min at room temperature.
3. Aspirate supernatant and resuspend cell pellet in 40 μL MACS neutrophil buffer per 10^7 cells.
4. Add 10 μL of anti-CD66abce biotin-labeled antibody to per 10^7 cells. Mix and incubate for 10 min at 2–8 °C.
5. Add 30 μL of MACS neutrophil buffer per 10^7 cells.
6. Add 20 μL of anti-biotin MicroBeads to the cell suspension. Mix and incubate for 15 min at 2–8 °C.
7. Wash cells by adding 1–2 mL of MACS neutrophil buffer per 10^7 cells and centrifuge at 300 × *g* for 10 min at 4 °C. Aspirate supernatant.
8. Resuspend cells in 500 μL MACS neutrophil buffer (up to 10^8 cells for every 500 μL of buffer).

3.5.2 Magnetic Separation

1. Place the LS column in the magnetic field of the MACS Separator (*see* **Note 3**).
2. Rinse the LS column with 3 mL of MACS neutrophil buffer.
3. Transfer cell suspension to the LS column.
4. Collect cells that pass through the column (these will be unlabeled cells), and wash the column three times with 3 mL of MACS neutrophil buffer. The total effluent should be collected (i.e., the unlabeled cell fraction) (*see* **Note 15**).
5. Transfer the column from the MACS Separator to the top of an appropriate collection tube.
6. Pipet 5 mL of MACS buffer onto the LS column. Flush out the magnetically labeled cells by depressing the plunger into the column.
7. Pellet cells by centrifugation at 300 × *g* for 10 min at 4 °C.
8. Resuspend cell pellet in 500 μL of R PMI/Hepes.
9. Count cells and dilute to desired concentration.

3.6 Ovine Neutrophil Isolation

1. Collect ovine blood into Vacutainer tubes containing EDTA (*see* **Note 9**). For the method outlined here, we collected 50 mL of blood. If different volumes of blood are required, adjust the indicated volumes proportionally.
2. Transfer 50 mL of blood into a conical 50-mL polypropylene tube and centrifuge at 400 × *g* for 20 min with low brake at room temperature.
3. Remove the upper plasma layer and buffy coat found at the plasma-red blood cell interface with a plastic transfer pipette.
4. Dilute the red blood cell layer up to the starting blood volume (50 mL in this case) with PBS/EDTA buffer.
5. Pipet 25 mL of the diluted cells into each of two conical 250-mL polypropylene tubes.
6. Lyse red blood cells by adding 150 mL of sterile H_2O into each tube. Mix by gently inverting tubes for 20 s at room temperature (*see* **Note 10**).
7. Immediately add 15 mL of 9% NaCl solution and mix well by gently inverting tubes.
8. Centrifuge at 250 × *g* for 5 min at room temperature with low brake.
9. Remove the supernatant using a plastic transfer pipette and resuspend the cell pellet in 50 mL of PBS/EDTA buffer.
10. Centrifuge at 250 × *g* for 5 min at room temperature.
11. Resuspend the leukocyte pellet in 9 mL of PBS/EDTA buffer.
12. Carefully layer 3 mL of the white blood cell suspension on top of 5 mL of 65% Percoll/EDTA solution using a plastic transfer pipette.

13. Centrifuge the gradients at 400 × *g* for 20 min at room temperature with no brake.
14. Remove supernatant using a plastic transfer pipette and discard (*see* **Note 11**).
15. Wash the cells by resuspending them in 50 mL of PBS/EDTA buffer and centrifuging at 400 × *g* for 10 min at room temperature.
16. Wash the cells again by repeating **step 15**.
17. Resuspend purified cells in the desired assay buffer.

3.7 Rabbit Neutrophil Isolation

1. Collect rabbit blood into a conical 50-mL polypropylene tube containing ACD solution so that a 4:1 (v/v) ratio of blood–ACD solution is achieved (*see* **Note 9**). For the method outlined here, we collected 24 mL of blood into a tube containing 6 mL of ACD solution. If different volumes of blood are required, adjust the indicated volumes proportionally.
2. Transfer 30 mL of blood into a conical 250-mL conical centrifuge tube and add 5 volumes of Hetastarch to each tube (150 mL in this case). Mix by gently inverting the tube.
3. Allow the blood–hetastarch mixture to sediment for 40 min at room temperature. Hetastarch causes the rabbit red blood cells to form aggregates, which sediment to the bottom of the tube. This leaves a clear, red blood cell-depleted layer above the red blood cell-rich lower layer (*see* **Note 16**).
4. Transfer the upper red blood cell-depleted layer to a clean conical 250-mL polypropylene tube with a plastic transfer pipette.
5. Centrifuge the solutions at 585 × *g* for 10 min with low brake at room temperature.
6. Remove the supernatant using a plastic transfer pipette and discard.
7. Resuspend the white blood cell pellet in 10 mL of rabbit neutrophil buffer.
8. Lyse red blood cells by adding 100 mL of sterile H_2O. Mix by gently inverting tubes for 20 s at room temperature (*see* **Note 10**).
9. Immediately add 10 mL of the 10% NaCl solution and mix well by gently inverting tubes.
10. Centrifuge at 585 × *g* for 10 min with low brake at room temperature.
11. Remove the supernatant using a plastic transfer pipette and discard.
12. Resuspend the cell pellet in 10 mL of rabbit neutrophil buffer.

13. Lyse any remaining red blood cells by repeating **steps 8–10** above.
14. Resuspend the leukocyte pellet in 5 mL of rabbit neutrophil buffer.
15. Carefully layer cell suspension on top of 7 mL of Histopaque 1077 in a conical 50-mL polypropylene tube.
16. Centrifuge the gradients at 475 × *g* for 25 min with no brake at room temperature.
17. Remove the supernatant using a plastic transfer pipette and discard (*see* **Note 11**).
18. Wash the neutrophil pellet by resuspending the cells in 50 mL of rabbit neutrophil buffer and centrifuging at 585 × *g* for 10 min at room temperature.
19. Resuspend purified cells in the desired assay buffer.

3.8 Quantifying Cell Number and Viability

1. Resuspend the final neutrophil pellet into the desired volume of assay buffer to achieve the appropriate cell concentration (usually 2 to 5 mL) and remove an aliquot for counting.
2. To quantify cell number, dilute 10 μL of the final cell suspension in 190 μL of 2% acetic acid. Pipet a few microliters onto a hemocytometer, and count the cells contained in the 25 squares inside the central double lines. Count only neutrophils, which are easily identified by their characteristic multilobed nuclei. Divide the neutrophil count by 25 to obtain the average per square. Multiply the average per square by 5×10^6 and then by the volume (in mL) of the final cell suspension to determine the total number of isolated neutrophils. A summary of the neutrophil recovery data determined for all species is shown in Table 1.

Table 1
Average neutrophil purity, yield, and cell viability using the described methods

Species	Total neutrophils[a]	Neutrophil purity (%)	Yield (per mL blood)	Viability (%)
Bovine	9.79×10^7	93.7	6.94×10^5	>99
Equine	3.9×10^7	97.8	1.63×10^6	>99
Human	2.53×10^7	99.1	8.42×10^5	>99
Murine	5.12×10^6	85.9	–	>99
Nonhuman primate[b]	1.74×10^7	97.4	2.4×10^6	≥98
Ovine	1.8×10^7	93.6	4.89×10^5	>99
Rabbit	1.83×10^6	90.7	7.62×10^4	>99

[a]Number of neutrophils obtained from the described method and volumes of blood; murine bone marrow neutrophil yield is presented as average number of neutrophils per mouse
[b]Cynomolgus macaques. The data represent the average from at least three separate neutrophil preparations per species

3. Determine cell viability by mixing equal aliquots of neutrophil suspension and trypan blue, pipetting the mixture onto microscope slides, and viewing the cells under a microscope. Cells that exclude the trypan blue and appear transparent are counted as viable, whereas cells that turn blue are counted as dead cells. A summary of the cell viability data determined for all species is shown in Table 1.

3.9 Analysis of Cell Purity

1. Purity can be evaluated with the hemocytometer (*see* **step 2** under Subheading 3.8.) by differential counting of neutrophils versus non-neutrophils.
2. Analysis of cell purity can also be performed by flow cytometry, which provides an effective approach to evaluate the cells present and their level of activation.
3. Collect 10,000 events for each sample using a flow cytometer with linear amplification of forward and side scatter channels.
4. Create a forward-scatter versus side-scatter dot plot and gate out any cellular debris. Set a gate around the neutrophil population to obtain gate statistics, such as percent of total events (a measure of purity) and relative size and granularity (*see* **Note 17**). As an example, Fig. 1a shows a representative dot plot from an equine blood neutrophil preparation, where the neutrophils form a relatively uniform profile, indicating a high level of purity. Figure 1b shows a representative dot plot from a murine bone marrow neutrophil preparation, where there is slightly lower purity (*see* Table 1). The greater variability in cell granularity and size is also likely due to the presence of some less mature neutrophils, which is a characteristic of murine bone marrow neutrophil preparations [17]. Likewise, Fig. 2 shows representative dot plots of highly purified neutrophils isolated from human and nonhuman primate blood using positive selection on Microbeads.
5. A summary of the neutrophil purity data obtained for all species is shown in Table 1.

4 Notes

1. EDTA was added to Vacutainer tubes by injection with a 1-mL syringe and 27-G needle.
2. It is essential that the blood and subsequently isolated neutrophils do not ever come into contact with glass, which leads to cell activation. Thus, plasticware should be used throughout all procedures, with exception of the Vacutainer tubes, which are silicone-coated.

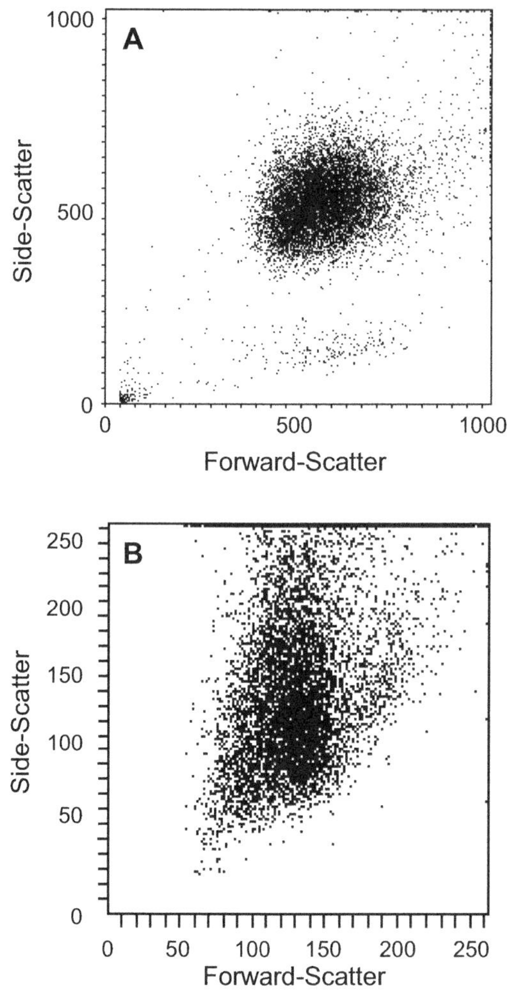

Fig. 1 Analysis of neutrophil purity by flow cytometry. Equine blood (*Panel A*) and murine bone marrow (*Panel B*) neutrophils were purified, and the isolated cells were analyzed by flow cytometry, as described in this chapter. Forward-scatter versus side-scatter dot plots are shown. Neutrophils from all species showed similar forward-scatter versus side-scatter profiles

3. An adequate sized column should be selected based on its maximum cell capacity. The volumes provided here are for LS columns. Check the Miltenyi instructions on the package insert for appropriate volumes for use with other columns.
4. Neutrophils are highly susceptible to priming and/or activation by endotoxin or lipopolysaccharide (LPS) (e.g., *see* [19]), which is often a contaminant in biological reagents. Thus, all plasticware must be endotoxin-free. In addition, all buffers and reagents are prepared in sterile H_2O or saline and sterile filtered to avoid endotoxin contamination.

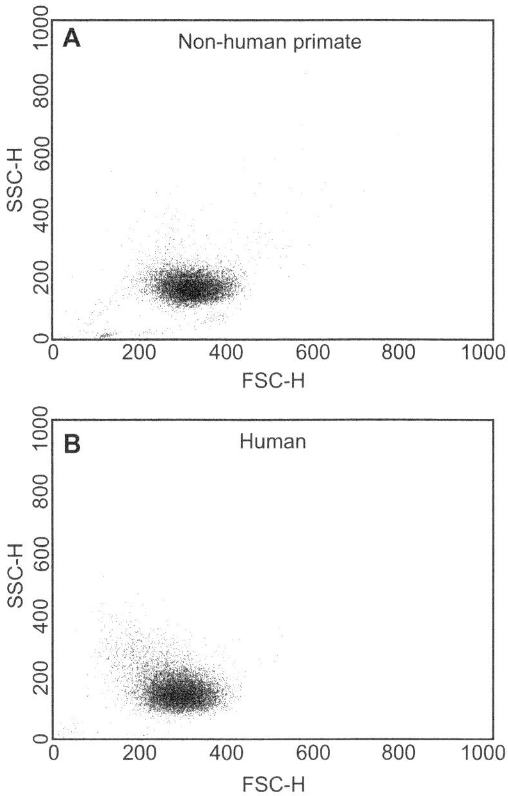

Fig. 2 Analysis of neutrophils from nonhuman primates (*Macaca fascicularis*) and humans by flow cytometry. Neutrophils were obtained from venous blood of nonhuman primates (*Panel A*) or humans (*Panel B*) using the positive selection methods described in this chapter and analyzed by flow cytometry. Forward (FSC-H) and side (SSC-H) angle light scatter plots are shown. (Reproduced by permission of Springer Nature©2014 [18])

5. All buffers should be sterile filtered through 0.2-μm filter units (Fisher Scientific).
6. To avoid possible contamination, which is a common problem with dextran, 30 g of Dextran 500 is weighed directly into a sterile plastic 500-mL Nalgene container and dissolved in sterile 0.9% NaCl solution, followed by sterile filtering.
7. As suggested by manufacturer, human serum albumin (HSA) can be substituted with bovine serum albumin, fetal bovine serum, or human serum.
8. Use extreme care and accuracy when preparing Percoll mixtures, as small variations in the final density of Percoll mixtures affects the purity and yield of neutrophil preparation.
9. Before use of humans or animals in any research project, appropriate approvals must be obtained from the Institution Review

Board (for human blood) or Institutional Animal Care and Use Committee (for all animal use).

10. Do not extend this incubation longer than 20 s, as longer incubation in hypotonic solution can alter and/or damage the neutrophils.
11. After the supernatant has been removed from the gradients, cotton applicators may be used to wipe the walls of the centrifuge tube to remove any adherent debris, which may contaminate the preparation. Be sure to avoid touching the neutrophil band or pellet with the applicator.
12. Sodium heparin can be replaced by other anticoagulants such as EDTA or sodium citrate. Alternatively, blood can be collected into Vacutainer tubes (Becton Dickinson) as described for the other neutrophil isolation methods.
13. Typically we use 35 mL of lysis buffer per 5 mL of heparinized blood (7:1 ratio).
14. Handle neutrophils as gently as possible (e.g., have a pipette on the lowest settings).
15. Both purified neutrophil fraction and effluent (unlabeled) fraction should be analyzed to determine purification efficiency.
16. As an alternative, rabbit red blood cells can also be aggregated with 6% dextran (100,000–200,000 molecular weight) for 30–40 min [15]; however, hetastarch seems to be more efficient. For some reason, rabbit red blood cells do not lyse as readily as those from other species. Even after two rounds of H_2O lysis, some red blood cells may still be present. If this is the case, remaining red blood cells may be removed by very gently washing the surface of the neutrophil pellet.
17. Note that neutrophil priming or activation causes an increase in cell size and granularity, which can also be evaluated with flow cytometry dot plots.

Acknowledgments

This work was supported in part by National Institutes of Health IDeA Program COBRE Grant GM110732; the Intramural Research Program of the National Institute of Allergy and Infectious Diseases, National Institutes of Health; USDA National Institute of Food and Agriculture Hatch project 1009546; and the Montana State University Agricultural Experiment Station.

References

1. Wiles S, Hanage WP, Frankel G et al (2006) Modelling infectious disease - time to think outside the box? Nat Rev Microbiol 4:307–312
2. Webb DR (2014) Animal models of human disease: inflammation. Biochem Pharmacol 87:121–130
3. Casal M, Haskins M (2006) Large animal models and gene therapy. Europ J Hum Genet 14:266–272
4. Ziegler A, Gonzalez L, Blikslager A (2016) Large animal models: the key to translational discovery in digestive disease research. Cell Mol Gastroenterol Hepatol 2:716–724
5. Styrt B (1989) Species variation in neutrophil biochemistry and function. J Leukoc Biol 46:63–74
6. Glasser L, Fiederlein RL (1990) The effect of various cell separation procedures on assays of neutrophil function. A critical appraisal. Am J Clin Pathol 93:662–669
7. Watson F, Robinson JJ, Edwards SW (1992) Neutrophil function in whole blood and after purification - changes in receptor expression, oxidase activity and responsiveness to cytokines. Biosci Rep 12:123–133
8. Forsyth KD, Levinsky RJ (1990) Preparative procedures of cooling and re-warming increase leukocyte integrin expression and function on neutrophils. J Immunol Methods 128:159–163
9. Macey MG, Jiang XP, Veys P et al (1992) Expression of functional antigens on neutrophils. Effects of preparation. J Immunol Methods 149:37–42
10. Alvarez-Larrán A, Toll T, Rives S et al (2005) Assessment of neutrophil activation in whole blood by flow cytometry. Clin Lab Haematol 27:41–46
11. Pycock JF, Allen WE, Morris TH (1987) Rapid, single-step isolation of equine neutrophils on a discontinuous percoll density gradient. Res Vet Sci 42:411–412
12. Lowell CA, Fumagalli L, Berton G (1996) Deficiency of Src family kinases p59/61hck and p58c-fgr results in defective adhesion-dependent neutrophil functions. J Cell Biol 133:895–910
13. Woldehiwet Z, Scaife H, Hart CA et al (2003) Purification of ovine neutrophils and eosinophils: anaplasma phagocytophilum affects neutrophil density. J Comp Pathol 128:277–282
14. White-Owen C, Alexander JW, Sramkoski RM et al (1992) Rapid whole-blood microassay using flow cytometry for measuring neutrophil phagocytosis. J Clin Microbiol 30:2071–2076
15. Doerschuk CM, Allard MF, Martin BA et al (1987) Marginated pool of neutrophils in rabbit lungs. J Appl Physiol 63:1806–1815
16. Swamydas M, Lionakis MS (2013) Isolation, purification and labeling of mouse bone marrow neutrophils for functional studies and adoptive transfer experiments. J Vis Exp (77): e50586
17. Boxio R, Bossenmeyer-Pourié C, Steinckwich N et al (2004) Mouse bone marrow contains large numbers of functionally competent neutrophils. J Leukoc Biol 75:604–611
18. Siemsen DW, Malachowa N, Schepetkin IA et al (2014) Neutrophil isolation from nonhuman species. Methods Mol Biol 1124:19–37
19. DeLeo FR, Renee J, Mccormick S et al (1998) Neutrophils exposed to bacterial lipopolysaccharide upregulate NADPH oxidase assembly. J Clin Invest 101:455–463

Chapter 5

Isolation of Neutrophils from Larval Zebrafish and Their Transplantation into Recipient Larvae for Functional Studies

Hannah Darroch, Jonathan W. Astin, and Christopher J. Hall

Abstract

Live imaging of neutrophils within optically transparent larval zebrafish has proved a powerful technique to investigate how specific gene products control neutrophil function. To resolve whether a gene contributes to neutrophil function in a cell-autonomous manner necessitates a way to examine gene-deficient neutrophils in an otherwise wild type background. To this end, here we describe methods to harvest fluorescent neutrophils from larval donor zebrafish and transplant them into age-matched recipients. We show that transplanted neutrophils can survive in recipient larvae for at least 3 days providing ample opportunity for functional studies. Focusing on bactericidal activity, we show that transplanted neutrophils phagocytose and kill live bacteria with similar kinetics to nontransplanted neutrophils, indicating that the transplantation process does not influence these neutrophil effector functions. Following the methods described here to transplant neutrophils between gene-deficient and wild type larval zebrafish will enable investigations into whether a gene's contribution to neutrophil function is cell-autonomous.

Key words Neutrophil, Transplantation, Live cell imaging, Phagocytosis, Zebrafish, Cell autonomous, Bactericidal activity

1 Introduction

Neutrophils are a critical component of the innate immune system and are regarded as the "front-line" defenders against pathogens. They are highly mobile cells armed with a variety of potent antimicrobial functions including phagocytosis, NETosis, and degranulation [1]. To move toward a complete understanding of their function, neutrophils need to be studied in vivo, within live animal models. Zebrafish (*Danio rerio*) are a well-established vertebrate model that, like all vertebrates, possess a complete repertoire of innate and adaptive blood cell lineages [2, 3]. Zebrafish boast high fecundity, short generation times, and are also largely amenable to high resolution, noninvasive live imaging due to their optical transparency at larval stages. Live imaging of fluorescent neutrophils within transgenic larval zebrafish has been exploited extensively to

Mark T. Quinn and Frank R. DeLeo (eds.), *Neutrophil: Methods and Protocols*, Methods in Molecular Biology, vol. 2087, https://doi.org/10.1007/978-1-0716-0154-9_5,

reveal novel neutrophil behaviors such as reverse migration from sites of inflammation, as well uncovering new roles during development, wound healing, inflammation, and infection [4–8]. In addition, forward and reverse genetic techniques including ENU mutagenesis and CRISPR-Cas9 gene editing have been used to uncover genes that regulate specific neutrophil functions [9, 10].

Due to a relative paucity of genetic techniques available in zebrafish to knock out gene function in a tissue-specific manner, it has been difficult to explore if specific genes/mutations contribute to neutrophil function in a cell autonomous manner. This is particularly relevant when a gene's expression is not restricted to neutrophils. Recently, this has been overcome by utilizing cell-specific CRISPR-Cas9 gene editing to create mutations only in the neutrophil compartment [9]. However, this technique generates an array of indels (not all of which will be inactivating) that differ from cell to cell meaning that any resulting phenotype is unlikely to be as strong as that resulting from a null allele shared by all cells. Transplantation of neutrophils from animals with known mutations into wild type recipients would provide an opportunity to examine the contribution of specific mutations to neutrophil function. Larval zebrafish are particularly amenable to such an approach given they do not yet have a functional adaptive immune system, meaning that transplant rejection is not an issue [11].

Transgenic zebrafish reporter lines that specifically label neutrophils have been previously generated [4, 6, 12]. Utilization of these reporter lines greatly facilitates the isolation of pure populations of neutrophils through fluorescence-activated cell sorting (FACS). In this chapter, we describe protocols for FACS-isolation and transplantation of functional neutrophils into recipient larvae. Transplanted neutrophils are retained for at least 3 days providing ample opportunity to explore how genes/mutations of interest contribute to neutrophil function in a cell autonomous manner. As an example of a phenotypic readout, and to demonstrate that this transplantation procedure does not have a negative impact on neutrophil function, we also describe protocols for live imaging and quantification of neutrophil bactericidal activity.

2 Materials

2.1 Dissociating Larvae into Single Cells

1. E3 medium: 5 mM NaCl, 0.17 mM KCl, 0.33 mM $CaCl_2$, and 0.33 mM $MgSO_4$ in sterile H_2O.
2. Pronase solution: 1 mg/mL pronase in E3 medium, as previously described [13].
3. Tricaine solution: 0.168 mg/mL tricaine in E3 medium.

4. 10× Ringer's solution: 20 mM KCl, 15 mM K_3PO_4 (dibasic), 10 mM $MgSO_4$, 100 mM HEPES, and 1.4 M NaCl in sterile H_2O. Dilute 1:10 for 1× Ringer's solution.
5. Ringer's solution: 5 mL 10× Ringer's, 500 μL of 1 M D-glucose, and 100 μL of 1 M $MgCl_2$ made up to 50 mL with sterile H_2O. Keep at 4 °C and use ice-cold.
6. 0.5% trypsin-EDTA solution (e.g., Gibco).
7. Sterile PBS.
8. 0.25% trypsin-EDTA/PBS solution: dilute 0.5% trypsin-EDTA 1:1 with PBS.
9. Fetal Bovine Serum (FBS).
10. Inhibition solution: Add 8 μL of 500 mM $CaCl_2$ and 200 μL of FBS into 792 μL of sterile H_2O. Requires 1 mL per sample.
11. 1.5 and 2.0 mL microcentrifuge tubes.
12. Round bottom flow cytometry tube.
13. 40 μm nylon cell strainers.
14. 3 mL plastic transfer pipettes.
15. Centrifuge (rotor appropriate for 1.5–2 mL microcentrifuge tubes).
16. Dumont fine tip forceps.

2.2 FACS Isolation of Fluorescent Neutrophils

1. Round bottom flow cytometry tube.
2. 1.5 mL microcentrifuge tube.
3. Resuspension solution: mix 9 parts 1× Ringer's solution with 1 part PBS, resulting in 0.9× Ringer's solution in 0.1× PBS.
4. Dissociated whole larval single cell suspension.
5. Centrifuge (rotor appropriate for 1.5–2 mL microcentrifuge tubes).

2.3 Transplantation of FACS-Isolated Neutrophils

1. E3 medium: 5 mM NaCl, 0.17 mM KCl, 0.33 mM $CaCl_2$, and 0.33 mM $MgSO_4$ in sterile H_2O.
2. Methyl cellulose solution: 3% (w/v) methyl cellulose in E3 medium.
3. Tricaine solution: 0.168 mg/mL tricaine in E3 medium.
4. Microinjection needles: thin wall borosilicate capillary tubes (e.g., Warner Instruments; 1 mm O.D. × 0.78 mm I.D. × 10 cm length). These are pulled with a micropipette puller (Sutter Instruments Co., flaming/brown puller set to: heat 680, pull 75, velocity 40, time 55, pressure 530, to produce tapered needles).
5. Microloader pipette tip (Eppendorf).
6. CellTram vario (Eppendorf), with mineral oil to fill.

7. Microscope (e.g., Nikon SMZ1500 with fluorescence via Nikon Intensilight C-HGF1).
8. 3 mL plastic transfer pipette.
9. Dumont fine tip forceps.

2.4 Live Imaging Bactericidal Activity of Transplanted Neutrophils

1. E3 medium: 5 mM NaCl, 0.17 mM KCl, 0.33 mM $CaCl_2$, and 0.33 mM $MgSO_4$ in sterile H_2O.
2. Tricaine solution: 0.168 mg/mL tricaine in E3 medium.
3. Methyl cellulose solution: 3% (w/v) methyl cellulose in E3 medium.
4. 1-phenyl-2-thiourea (PTU) solution: 0.003% PTU in E3 medium. Use from 24 h postfertilization on to inhibit pigmentation.
5. GFP-tagged *Salmonella enterica* serovar *Typhimurium*.
6. Phenol red solution: 0.5% phenol red in sterile PBS, pH 7.
7. Dulbecco's Modified Eagle's Medium (DMEM).
8. Microinjection needles: thin wall borosilicate capillary tubes (e.g., Warner Instruments; 1 mm O.D. × 0.78 mm I.D. × 10 cm length). These are pulled with a micropipette puller (Sutter Instruments Co., flaming/brown puller set to: heat 680, pull 75, velocity 40, time 55, pressure 530, to produce tapered needles).
9. Luria Broth (LB).
10. LB agar.
11. Low melting point (LMP) agarose: mix 1% w/v agarose in E3 medium supplemented with 0.003% PTU and 0.168 mg/mL tricaine. Microwave until completely dissolved.
12. 35 mm small culture dish.
13. 3 mL plastic transfer pipette.
14. Dumont fine tip forceps.
15. Air pressure microinjector.
16. Confocal laser scanning microscope (e.g., Olympus FV1000).

3 Methods

3.1 Dissociation of Larvae into Single Cells

Neutrophil-specific transgenic reporter lines enable the isolation of pure populations of neutrophils for transplantation. Larvae must be dissociated into a single-cell suspension before undergoing FACS. Here we use the *Tg(lyz:DsRED2)* reporter line [12] that will facilitate the live imaging of GFP-tagged *Salmonella* (hereafter referred to as Sal-GFP) within red fluorescent neutrophils following transplantation (Fig. 1). We routinely dissociate larvae aged between 2 to 5 days postfertilization (dpf) for transplantation experiments into age-matched recipients.

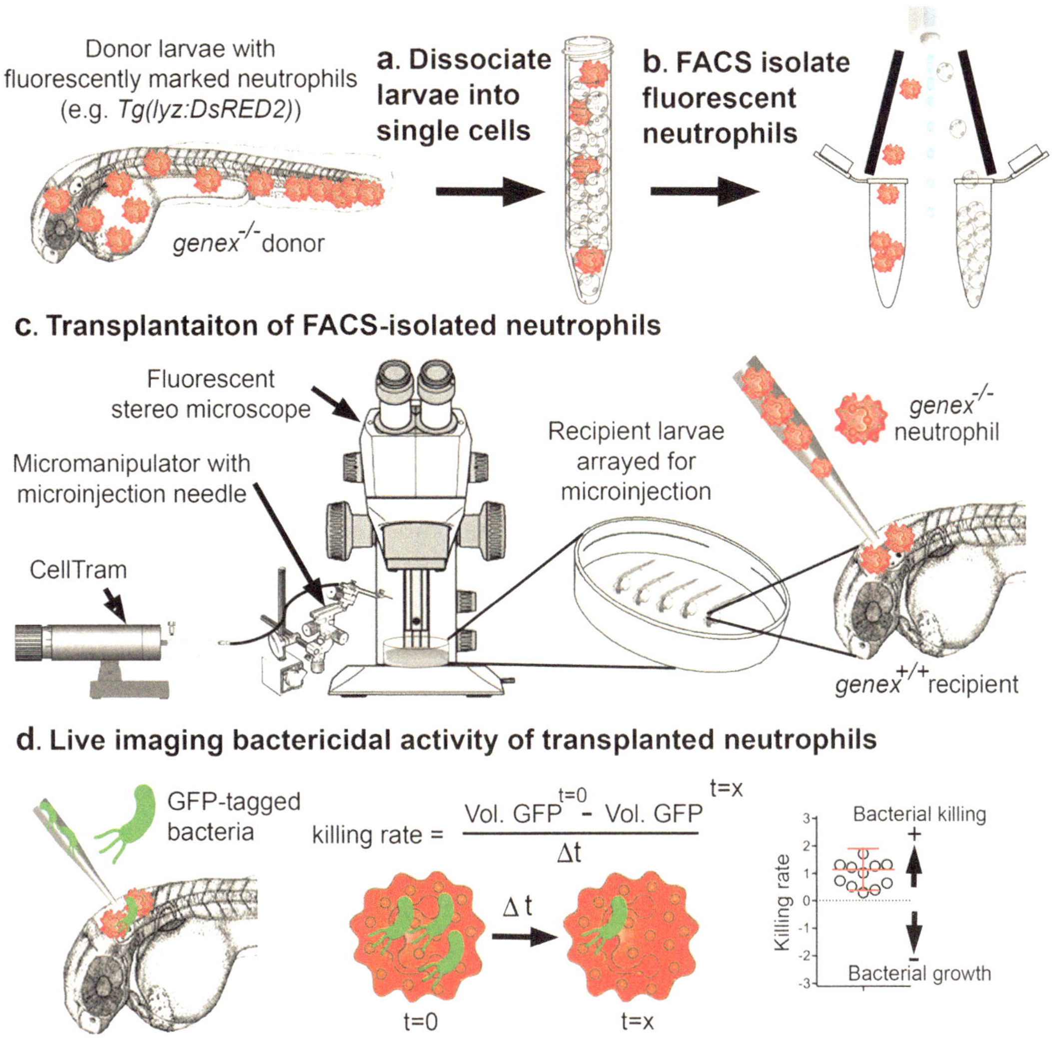

Fig. 1 Schematic illustrating the work flow for methods to isolate and transplant neutrophils between larval zebrafish for the purpose of examining the cell-autonomous contribution of *genex* to neutrophil function. (**a**) Larvae possessing fluorescent neutrophils are dissociated to form a single cell suspension. In this example *Tg (lyz:DsRED2)* larvae are homozygous for a null allele of *genex*. (**b**) A pure population of *genex*-deficient neutrophils is generated by FACS. (**c**) Neutrophils are transplanted into wild type larvae utilizing a dissecting microscope, CellTram injector, and micromanipulator with microinjection needle. (**d**) As an example functional readout for neutrophils, the bacterial killing capacity of transplanted *genex*-deficient neutrophils is determined by live confocal imaging. The volume of fluorescent bacteria is measured at the start and end of a time-lapse imaging experiment, and a killing rate can be measured by dividing the change in bacterial volume over change in time

1. Raise transgenic reporter lines until age of interest. If unhatched, manually dechorionate the larvae using fine tip forceps or pronase digestion (*see* **Note 1**).
2. Pool approximately 50–100 larvae and anesthetize in tricaine solution.

3. Transfer anesthetized larvae into a 15 mL tube using a plastic transfer pipette. Let the larvae settle to the bottom of the tube and then remove the majority of the E3 medium. Add 5 mL of ice-cold 1× Ringer's solution. Store on ice for 15 min to euthanize the larvae.
4. While waiting, prepare the resuspension solution (1 mL/sample) plus an additional 100 μL to prepare 1.5 mL microcentrifuge tubes to collect the sorted cells. Place on ice until ready for **step 12**.
5. Deyolk the larvae by repeated passage through a 200 μL pipette tip for 5 min. Time may vary depending on age and number of larvae. Solution will go cloudy, and you will see a separation between the fatty yolks (risen to top) and deyolked larvae (sunk to bottom). Remove all of the Ringer's solution and consequently all of the yolk material, taking care not to discard any larvae. Place the 15 mL tube at room temperature.
6. Add 3 mL of 0.25% trypsin-EDTA/PBS solution to the larvae. Keep at 28 °C.
7. Incubate at 28 °C for 1–2 h while manually dissociating the larvae by repeatedly pipetting through a 1 mL pipette tip, every 10–15 min. Solution should go cloudy, with no large larval debris (*see* **Note 2**).
8. Once dissociated, add 1 mL of inhibition solution. Sample will now be 4 mL in total.
9. Divide the sample evenly between 2 mL microcentrifuge tubes.
10. Centrifuge for 10 min at 260 × *g* and 4 °C.
11. Remove and discard the supernatant carefully by decanting (*see* **Note 3**).
12. Resuspend the pellet of one 2 mL tube in 800 μL of ice-cold resuspension solution (*see* **Note 4**). Transfer this same 800 μL of resuspended material into the second 2 mL tube to resuspend the second pellet. Keep spare resuspension buffer on ice.
13. Pass the material through a 40 μm cell strainer into a 50 mL conical polypropylene tube (*see* **Note 5**).
14. Transfer strained sample into a 1.5 mL microcentrifuge tube or round bottom flow cytometry tube. Store sample on ice. Proceed to FACS isolation immediately.

3.2 FACS Isolation of Fluorescent Neutrophils

Dissociating approximately 100 4 dpf *Tg(lyz:DsRED2)* larvae will typically yield around 10,000–15,000 neutrophils following FACS (*see* **Note 6**).

1. Take the following, on ice, when FACS-isolating the neutrophils:

(a) Single cell suspension in FACS tube or 1.5 mL microcentrifuge tube.
(b) 1–3 1.5 mL microcentrifuge tubes with 20 μL resuspension buffer from additional buffer made in Subheading 3.1, **step 4**.
(c) Spare resuspension buffer (*see* **Note 7**).

2. In a quick presort run, ensure that a pure population of neutrophils can be isolated from your single cell suspension. Be sure to appropriately gate for neutrophils based on their fluorescence marker and forward/side scatter properties. Neutrophils will typically constitute approximately 0.5% of a whole-larval suspension with this dissociation method.
3. After appropriate gating, collect the neutrophils into a 1.5 mL microcentrifuge tube that contains approximately 20 μL of resuspension solution to avoid the cell droplets from drying out. Sort no more than 10,000 cells into a single microcentrifuge tube. If you have more cells, then sort these into additional 1.5 mL tubes.
4. Sort the maximum number of neutrophils from your sample as you can, but no more than 30,000 is usually necessary. Sorting will typically take 0.5–1 h (*see* **Note 8**).
5. Centrifuge FACS-isolated neutrophils, 10 min at 260 × *g* and 4 °C (*see* **Note 9**).
6. Carefully remove as much of the supernatant as possible with a pipette positioned toward the opposite wall of the tube containing the cell pellet so as not to disturb the pellet. Leave 20–30 μL of resuspension solution behind. Place the supernatant into a separate 1.5 mL microcentrifuge tube in case the pellet is accidentally disturbed (*see* **Note 10**).
7. Resuspend the neutrophils by gently pipetting up and down in the remaining resuspension solution. Focus the pipette on the side of the tube you expect the cells to be pelleted against. Keep on ice and proceed to transplantation.

3.3 Transplantation of FACS-Isolated Neutrophils

This protocol was developed for transplanting between 10 and 50 neutrophils into the hindbrain ventricle of recipient larvae. The hindbrain ventricle was chosen as it is a large fluid-filled cavity that can accommodate the necessary liquid volume that accompanies the transplanted cells.

1. Set up injection station as shown in Fig. 2a. Ensure that the CellTram injector is full of mineral oil before you begin (*see* **Note 11**).
2. Anesthetize recipient larvae in tricaine solution.

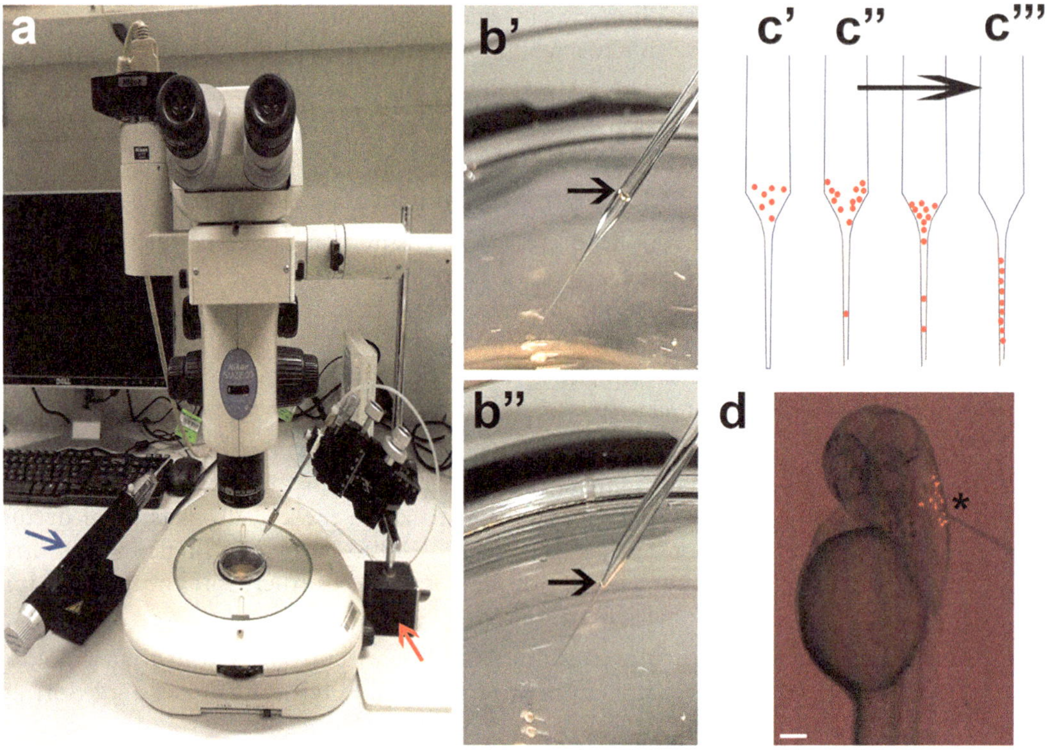

Fig. 2 Setup for transplantation. (**a**) Injector setup. Larvae are mounted in a small petri dish in 3% methyl cellulose and viewed on an SMZ1500 (dissecting) fluorescent stereo microscope. A microinjection needle is held in place with a micromanipulator (red arrow) and connected to a CellTram (blue arrow). (**b** and **c**) Actual and illustrated views of cut microinjection needle immediately after cutting (**b**′ and **c**″) and when ready for transplanting cells (**b**″ and **c**″). Black arrows indicate fluid level in **b**. (**c**′) Uncut needle following loading with neutrophils (red dots) suspended in FACS resuspension buffer (orange shading). Following cutting of the cell-loaded microinjection needle, over time the neutrophils become progressively concentrated at which point the density of neutrophils is sufficient for transplantation. (**d**) Live image of red fluorescent transplanted neutrophils in the hindbrain ventricle of a 2 dpf wild type recipient. Asterisk marks microinjection site. Scale bar, 100 μm in **d**. Abbreviations: *dpf* days postfertilization

3. Coat the bottom of a 60 mm × 15 mm petri dish with a thin layer of 3% methyl cellulose solution.
4. Deposit recipient larvae into the petri dish with minimal E3 medium transfer to avoid diluting the methyl cellulose (*see* **Note 12**).
5. Position the larvae laterally with the dorsal side facing toward the microinjection needle to target the hindbrain ventricle (Fig. 2d). Move larvae through the methyl cellulose using a shortened microloader pipette tip (or eyelash manipulator) taking care not to damage the larvae.
6. Load injection needle with 3 μL of resuspended neutrophils.

7. Cut a microinjection needle with a pair of fine tip Dumont forceps (*see* **Note 13**).
8. Turn the dial of the Celltram to start the flow of fluid/cells. Turn on the fluorescence of the microscope to enable detection of the fluorescently labeled neutrophils as they flow out of the needle.
9. Figure 2 shows diagrammatically how the cells will typically flow out of the microinjection needle. Neutrophils will typically flow immediately after cutting the microinjection needle (Fig. 2b′ and c″). The flow rate is controlled by CellTram operation. Initially, there will be a large amount of fluid passing from the needle containing a small number of neutrophils. At this stage the volume associated with these sparse neutrophils is too great for transplantation and would result in significant damage to the recipient larvae.
10. Over time the concentration of neutrophils increases as the liquid is expelled. When a large number of neutrophils begin to descend down the needle, place the microinjection needle into the hindbrain ventricle and microinject the neutrophils (Fig. 2b″ and c‴) (*see* **Note 14**).
11. Return successfully transplanted larvae into fresh E3 medium and incubate at 28 °C. It is easiest to move larvae from the methyl cellulose by flooding the plate with E3 to dislodge the larvae and pouring this into a fresh Petri dish. Transplanted neutrophils can be detected in recipient larvae for at least 3 days posttransplant (dpt), with a high abundance at 1 dpt (Fig. 3a, b).

3.4 Live Imaging Bactericidal Activity of Transplanted Neutrophils

The hindbrain ventricle is a well-established site in larval zebrafish for microinjection of microbial challenges and its superficial location facilitates live imaging of leukocyte behavior at single cell resolution [14]. Given the primary use of this transplantation protocol is to examine the cell-autonomous contribution of specific genes/mutations to neutrophil function, here we describe our protocol for live imaging and quantifying bactericidal activity of transplanted neutrophils, that is modified from that previously described in Yang et al. [15]. In this assay, fluorescently labeled bacteria are injected into the hindbrain ventricle and the bacterial volume within individual neutrophils is quantified over time to determine a killing rate (Figs. 1 and 4) [15, 16]. These measurements are taken from time-lapse confocal imaging of individual neutrophils for approximately 10 min. To allow recipient larvae to recover from the transplantation procedure we typically measure the killing rate of transplanted neutrophils at 1 dpt.

1. An overnight culture of Sal-GFP in LB medium is diluted 1:10 with a 50:50 mix of DMEM and LB, up to 40 mL. This is

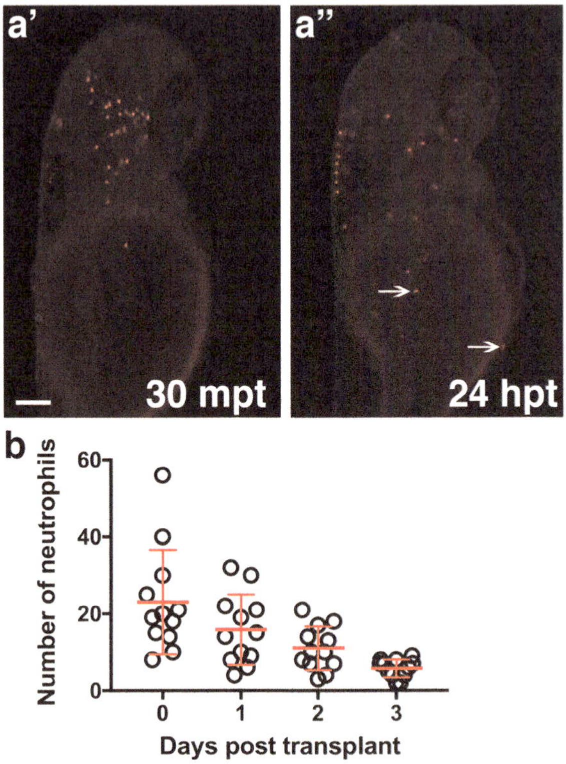

Fig. 3 Survival of wild type neutrophils transplanted into 2 dpf recipient larvae. (**a**) Live images of 2 dpf recipient larva 30 min (**a**′) and 24 h (**a**″) posttransplantation. Many cells are still present in the hindbrain of recipient larvae 24 h following transplantation while some have migrated away over the yolk (white arrows). (**b**) Quantification of transplanted neutrophils in recipient larvae from the day of transplant to 3 days posttransplantation. Larvae were initially transplanted at 2 dpf for this experiment and neutrophils were counted using fluorescence microscopy. Error bars show mean ± SD. Scale bar, 100 μm in **a**′. Abbreviations: *dpf* days postfertilization, *hpt* hours posttransplantation, *mpt* minutes posttransplantation

incubated for 45 min at 37 °C in a shaking incubator at 200 rpm. Make an injection mix of 1500 colony-forming units (CFU)/nL.

2. Screen larvae for those that have retained transplanted neutrophils. Anesthetize larvae in E3 medium supplemented with tricaine solution and array in 3% methyl cellulose, as previously described (Subheading 3.3, **steps 3–5**).
3. Load injection mix into a microinjection needle and cut with fine tip Dumont forceps.
4. Inject 1 nL of the injection mix into the hindbrain ventricle of larvae (*see* **Note 15**).

5. To check dose, plate an injection bolus onto appropriate growth medium. Inject 1 nL of the injection mix into 100 μL of PBS. Dilute this 1:100 and then plate 10 μL in triplicate on appropriate agar (e.g., LB). Incubate overnight at 28 °C.
6. Remove larvae from methyl cellulose and place into E3 medium, as described in (Subheading 3.3, **step 11**).
7. Immediately mount larvae for live imaging.
8. Microwave the LMP agarose to liquefy and use a transfer pipette to put a thin layer (~5 mm) of molten 1% LMP agarose into a small culture dish (6 mm × 10 mm). Cool to harden. Keep stock LMP agarose in a 50 °C water bath to keep it molten.
9. When the LMP agarose has hardened, dig a small trench in the surface of the agarose bed to accommodate the yolk sac.
10. Anesthetize larvae in tricaine solution. Use about 8 larvae per mounting dish. Concentrate the anesthetized larvae in the middle of dish so that they can be transferred into the mounting media with minimal E3 medium.
11. Collect approximately 2 mL of the LMP agarose solution using a plastic transfer pipette (*see* **Note 16**). Expel approximately 0.5 mL from the pipette and use the pipette to collect the anesthetized larvae. Transfer the larvae into the small culture dish with the preset agarose bed.
12. Arrange larvae ventral side down with their yolk sacs positioned into the small trenches. Use a shortened microloader pipette tip or eyelash manipulator to gently move the larvae. Maintain larvae in this position until the agarose has set, then proceed to live time-lapse imaging using a laser-scanning confocal microscope.
13. Typical imaging settings we use for measuring the killing rates of *DsRED2*-expressing neutrophils at 1 dpt, following Sal-GFP injection, are as follows:
    (a) 60× water immersion lens.
    (b) Multiline Argon (488 nm) laser set to 4% and green Helium Neon (543 nm) laser set to 20%.
    (c) Scan speed of 2 μs/pixel.
    (d) Image format 512 × 512.
    (e) 15–20 z sections no more than 1.5 μm apart.
    (f) Time-lapses were taken for 10–12 min per neutrophil.
    (g) 2× zoom.
14. Select neutrophils that have phagocytosed bacteria and begin time-lapse imaging.

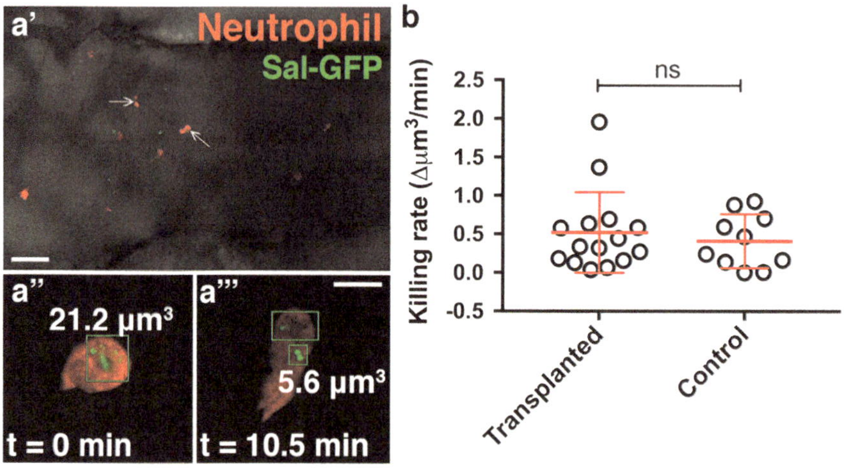

Fig. 4 Transplantation procedure does not impact neutrophil bacterial killing capacity. (**a**′) Live confocal image showing dorsal view of the hindbrain ventricle of a 5 dpf (1 dpt) recipient larva with transplanted wild type neutrophils following microinjection of Sal-GFP. White arrows mark Sal-GFP-laden neutrophils. (**a**″) Confocal image of individual transplanted neutrophil containing intracellular Sal-GFP at the beginning of a time-lapse experiment ($t = 0$ min). (**a**‴) Confocal image of transplanted neutrophil (same cell as in **a**″) at the end of the time-lapse experiment ($t = 10.5$ min). The volume of internalized Sal-GFP is measured within each neutrophil (green boxes). (**b**) Quantification of killing rates of wild type transplanted neutrophils compared to control neutrophils (not transplanted). Error bars show mean ± SD, $n = 3$ separate experiments, Student's *t*-test. Scale bars, 50 μm in **a**′ and 10 μm in **a**″. Abbreviations: *dpf* days postfertilization, *dpt* days posttransplantation, *ns* not significant, *t* time

15. Analyze the images to determine their suitability for measuring bacterial killing rates. We employ the following criteria that must be met to accept a time lapse experiment:
    (a) Neutrophils must be imaged for at least 10 min to ensure similar light exposure for all samples.
    (b) Neutrophils must not phagocytose additional bacteria during the imaging period.
    (c) Neutrophils must remain within the X, Y and Z boundaries of the imaging parameters.
16. 3D image analysis software (such as Volocity, PerkinElmer) is then used to measure the volume of green fluorescent bacteria inside individual neutrophils at the beginning and end of the time-lapse movies. Killing rates are determined by measuring the change in bacterial volume over time ($\Delta\mu m^3$/min) (Figs. 1 and 4a). Using this method, we have shown that transplanting wild type neutrophils does not impact the bactericidal activity of neutrophils, when compared to age-matched nontransplanted control neutrophils (Fig. 4b).

4 Notes

1. Before you start dissociating it will save you time later on if you cool down a centrifuge to 4 °C, defrost trypsin at room temperature, and have a 28 °C incubator available to incubate the dissociation mixture.
2. The time required to dissociate the larvae will depend on how many larvae you have as well as their age. 2 dpf larvae will typically take no more than 1 h to dissociate, whereas a tube of 100 4 dpf larvae will take almost 2 h to fully dissociate. A well-digested solution will look cloudy and you should not be able to see any large larval debris when you pipette the mixture.
3. Do not remove the supernatant by pipette. The suction of the pipette can dislodge the pellet and you may throw away a large proportion of your cells.
4. This volume should be adjusted depending on how many larvae are being dissociated, but making this volume too small will mean that your sample is too concentrated to be efficiently run through the FACS machine. Smaller volumes will reduce FACS time, so you can trial volumes as low as 400 μL to see what works for your specific dissociation mix.
5. Be sure to use a fresh pipette tip when you collect the strained material to avoid reintroducing larger debris into your cell suspension after straining. Larger debris will block the FACS machine. A single strain is usually sufficient.
6. This yield occurs following 1.5 h in 0.25% trypsin-EDTA solution. Longer or shorter incubations are likely to affect this yield.
7. Take some spare resuspension buffer with you during FACS sorting in case your sample is too concentrated to be run efficiently on the machine, especially if it is the first time running this type of sample.
8. We typically continue on to transplant the cells if >5000 neutrophils are sorted. We usually cap the number of cells sorted to about 30,000 because this is plenty to transplant and helps to lower the time/cost of FACS.
9. You will not see a visible pellet of cells so be mindful about the orientation of the tube in the centrifuge.
10. If you see no cells when you try to transplant it is likely you disturbed the pellet by accident when removing the supernatant. In that event, spin down the supernatant for 10 min at 260 *g* and 4 °C, and try again from Subheading 3.2, **step 5**.
11. The CellTram is used for these transplants instead of an air-pressure injector because it enables fine control over

pressure and cell movement. The CellTram enables you to drain most of the resuspension fluid away from the cells before they are ejected, allowing a high concentration of cells to be ready for injection.

12. To transfer larvae with minimal E3 medium, suck all of the larvae into a 3 mL transfer pipette. Allow the larvae to settle at the tip of the transfer pipette by gravity and deposit all of the larvae onto the methyl cellulose in a single drop of E3 medium.
13. Consideration to the size of the microinjection needle is important because if the needle bore is too small the cells will get stuck/damaged, and if it is too big the hindbrain of the recipient larvae will be damaged and any transplanted cells will leak out the large hole formed by the injection.
14. The amount of liquid that you inject alongside the neutrophils needs to be limited to avoid damaging the larvae, as well as forcing the transplanted cells back out of the hindbrain ventricle. The best way to do this is to wait until the neutrophils are sufficiently concentrated before placing the microinjection needle into the recipient. This highlights the importance of using a fluorescence microscope so that the flow rate of the fluorescent neutrophils can be visualized and the correct time to place the needle into the recipient can be determined.
15. Look at the larvae under fluorescence to see where the transplanted neutrophils are in the head of the recipients. It is common for the cells to migrate a slight distance after transplantation (Fig. 3a), so it is beneficial to target the bacterial injection to where the cells are located, that is, if most have moved to the forebrain or otic vesicle, inject there.
16. It is important to get the temperature of the LMP agarose correct. If it is too hot it will kill the larvae, but if it is too cold it will set too quickly to get the larvae into the correct orientation for imaging. Keep the LMP agarose at 50 °C so that it does not harden over time. Just before mounting, place the LMP agarose at room temperature. Use when the LMP agarose solution is warm to the touch.

References

1. Amulic B, Cazalet C, Hayes GL et al (2012) Neutrophil function: from mechanisms to disease. Annu Rev Immunol 30(1):459–489
2. Davidson AJ, Zon LI (2004) The 'definitive' (and 'primitive') guide to zebrafish hematopoiesis. Oncogene 23(43):7233–7246
3. Trede NS, Langenau DM, Traver D et al (2004) The use of zebrafish to understand immunity. Immunity 20(4):367–379
4. Mathias JR, Perrin BJ, Liu T-X et al (2006) Resolution of inflammation by retrograde chemotaxis of neutrophils in transgenic zebrafish. J Leukoc Biol 80(6):1281–1288
5. Mathias JR, Dodd ME, Walters KB et al (2007) Live imaging of chronic inflammation caused by mutation of zebrafish Hai1. J Cell Sci 120 (19):3372–3383
6. Renshaw SA, Loynes CA, Trushell DMI et al (2006) A transgenic zebrafish model of neutrophilic inflammation. Blood 108 (13):3976–3978
7. Davis JM, Clay H, Lewis JL et al (2002) Real-time visualization of mycobacterium-macrophage interactions leading to initiation of granuloma formation in zebrafish embryos. Immunity 17(6):693–702
8. Henry KM, Loynes CA, Whyte MKB et al (2013) Zebrafish as a model for the study of neutrophil biology. J Leukoc Biol 94 (4):633–642
9. Zhou W, Cao L, Jeffries J et al (2018) Neutrophil-specific knockout demonstrates a role for mitochondria in regulating neutrophil motility in zebrafish. Dis Model Mech 11(3)
10. Pase L, Layton JE, Wittmann C et al (2012) Neutrophil-delivered myeloperoxidase dampens the hydrogen peroxide burst after tissue wounding in zebrafish. Curr Biol 22 (19):1818–1824
11. Lam SH, Chua HL, Gong Z et al (2004) Development and maturation of the immune system in zebrafish, Danio rerio: a gene expression profiling, in situ hybridization and immunological study. Dev Comp Immunol 28 (1):9–28
12. Hall C, Flores MV, Storm T et al (2007) The zebrafish lysozyme C promoter drives myeloid-specific expression in transgenic fish. BMC Dev Biol 4(7):42
13. Westerfield M (2000) The zebrafish book, 4th edn. University of Oregon Press, Eugene
14. Hall CJ, Boyle RH, Astin JW et al (2013) Immunoresponsive gene 1 augments bactericidal activity of macrophage-lineage cells by regulating β-oxidation-dependent mitochondrial ROS production. Cell Metab 18 (2):265–278
15. Yang C-T, Cambier CJ, Davis JM et al (2012) Neutrophils exert protection in the early tuberculous granuloma by oxidative killing of mycobacteria phagocytosed from infected macrophages. Cell Host Microbe 12 (3):301–312
16. Astin JW, Keerthisinghe P, Du L et al (2017) Chapter 2 - Innate immune cells and bacterial infection in zebrafish. In: Detrich HW, Westerfield M, Zon LI (eds) Methods in cell biology, vol 138. Academic Press, Cambridge, MA, USA.

Part III

Neutrophil Chemotaxis, Phagocytosis, and Bactericidal Activity

Chapter 6

Analysis of Neutrophil Transmigration Through Epithelial Cell Monolayers

Liliya N. Kirpotina, Douglas J. Kominsky, Mark T. Quinn, and Steve D. Swain

Abstract

Transmigration of neutrophils through an epithelial layer, such as in the intestine or lung, is a necessary response to a perceived attack at the mucosal surface of that tissue. This process is dynamically regulated by a number of interactive events between the neutrophil and other cell types and allows for an effective and localized neutrophil response. However, in certain inflammatory diseases, including inflammatory bowel disease and chronic obstructive pulmonary disease (COPD), persistent neutrophil accumulation can contribute to disease pathology. Elucidating the mechanisms of this aberrant neutrophil accumulation is crucial for understanding and ameliorating these disease processes. The method we describe here is a controlled model system that allows for the investigation of the interactive signals involved in neutrophil transmigration through epithelial barriers, and possible mechanisms of deregulation of this process.

Key words Neutrophil, Epithelium, Transmigration, Inflammation

1 Introduction

Epithelial cells of mucosal tissues, such as lung and intestine, are crucial for the maintenance of physiological barriers that restrict entry of environmental pathogens into the body. At locations where these barriers break down, there is often a strong inflammatory response, and in some cases widespread infection can result. At the same time, these epithelial cells constitute a functional barrier to the egress of inflammatory cells from the bloodstream side of the epithelium (basolateral) to the luminal (apical) side. However, under the appropriate stimuli, inflammatory cells, typically neutrophils, can be recruited from their sequestered state in the vasculature and actively transmigrate through the epithelium [1]. In most cases, this is a necessary and transient process, where neutrophils are recruited to help eliminate a potential microbial invader and aid in the wound healing process. However, in some cases, this process persists beyond what is necessary, and long-term accumulation of

Mark T. Quinn and Frank R. DeLeo (eds.), *Neutrophil: Methods and Protocols*, Methods in Molecular Biology, vol. 2087,
https://doi.org/10.1007/978-1-0716-0154-9_6,

neutrophils acts to propagate a pathological inflammatory process, such as in inflammatory bowel disease [2, 3].

Transmigration of neutrophils across the epithelium is a dynamic process that involves active cellular events in both cell types. Before recruitment, neutrophils circulate freely in the bloodstream or are found marginated along capillary and venue walls [4]. To eventually reach the epithelial mucosal surface, they must first adhere to the local vascular endothelium and then traverse sequentially the endothelial and epithelial barriers [5]. Epithelial cells can be an important driver of this process. Chemokines released by epithelial cells (including CXCL1, CXCL2, and CXCL8), often in response to stimulation by specific microbial components through Toll-like receptors, are important in attracting neutrophils to the basolateral aspect of the epithelium [6]. Early phase cytokines, such as tumor necrosis factor α (TNF-α) and interleukin (IL)-1α can prime neutrophils during this process so that they can achieve their maximal functional capabilities [7]. Other epithelial derived factors, such as the bioactive lipids leukotriene B_4 (LTB_4) and lipoxin A_4 (LXA_4), play important roles in the final process of neutrophil transmigration to the apical side of the epithelial mucosa [6]. Neutrophils themselves are not simply passive responders to this process: even before transmigration, they can signal to the basolateral epithelial receptors Par-1 and Par-2, resulting in a decrease of barrier integrity [8]. During transmigration, neutrophils also interact with selected tight junction proteins, in particular the junctional adhesion molecules (JAMs), further facilitating neutrophil movement across the epithelial layer [8].

There is clear evidence of dysregulation of recruitment in some chronic inflammatory diseases. While Crohn's disease and ulcerative colitis are both inflammatory bowel diseases, the role of neutrophils in these diseases is complex and sometimes contradictory, with evidence existing for involvement of both excessive activity and functional deficiency of neutrophils and innate immune responses [2]. The method described here can be useful to study the interaction of neutrophils with epithelial and endothelial cells. This method involves analysis of the in vitro transmigration of neutrophils through defined endothelial and/or epithelial cell layers. This basic system can be modified in many ways, including the types of soluble activators or chemoattractants used, the use of functionaly modified neutrophils or epithelial cells, or as the basis of more complex coculture models [9, 10].

2 Materials

2.1 Epithelial Cell Culture

1. Epithelial cell line: T84 colorectal carcinoma (ATCC CCL-248). These cells will form a strong polarized barrier in culture and are typically the cell line of choice for intestinal cell culture experiments where a measure of barrier integrity is required [11]. Other possible cell types are discussed in **Note 1**.
2. Cell culture medium: Dulbecco's Modified Eagle's Medium and Hams F12 Medium at 1:1 mix (DMEM/F12). Select a formulation that comes with HEPES buffer in addition to sodium bicarbonate. Supplement the medium at the time of use with 10% fetal bovine serum, 2 mM glutamine, 100 U/mL penicillin, and 100 μg/mL streptomycin.
3. Tissue culture plasticware: Stock T84 cultures can be kept in normal T75 flasks. Transmigration experiments must be set up in inserts with membrane permeable supports placed in tissue culture wells. Insert composition, size, and membrane pore size are important variables. For neutrophil transmigration experiments, pore size should be 3.0 μm or greater (*see* **Note 2**).
4. Coating medium: An important variable in these experiments can be extracellular matrix proteins used for coating the inserts before cells are applied. These can be omitted in some cases, but it is common to coat with a collagen solution (*see* **Note 3**).
5. Trypsin–ethylenediamine tetraacetic acid (EDTA) solution: 0.25% Trypsin with 1 mM EDTA.
6. Barrier measurement instrument: Epithelial barrier integrity is measured as the trans-epithelial electrical resistance (TEER). The most commonly used instrument is the EVOM2 epithelial voltmeter (World Precision Instruments) with "chopstick" electrodes (*see* **Note 4**).

2.2 Neutrophil Isolation

1. Blood collection tubes: 9 mL Vacutainer tubes with dry EDTA or desired anticoagulant.
2. 6% dextran solution: Mix 4.5 mL of dextran stock solution (Sigma D8802) with 540 μL of sterile filtered 25% NaCl solution in endotoxin-free H_2O and add endotoxin-free H_2O to bring the final volume up to 15 mL (*see* **Note 5**).
3. Endotoxin-free H_2O.
4. Hank's balanced salt solution without Ca^{2+} and Mg^{2+} ($HBSS^-$). Commercially available HBSS supplemented with 10 mM HEPES to improve buffering capacity.
5. Hank's balanced salt solution with Ca^{2+} and Mg^{2+} ($HBSS^+$): $HBSS^-$ supplemented with 1.3 mM $CaCl_2$ and 1.0 mM $MgSO_4$.

6. 0.9% NaCl solution: Dissolve 0.9 g NaCl in 90 mL endotoxin free H_2O and then adjust volume to 100 mL with endotoxin-free H_2O. Sterile filter.
7. 25% NaCl solution: Dissolve 25 g NaCl in 90 mL endotoxin free H_2O and then adjust volume to 100 mL with endotoxin free H_2O. Sterile filter.
8. Density gradient media: Ficoll-Paque, sterile solution, density 1.077 g/mL.
9. Sterile plasticware, including 50 mL tubes and pipettes.

2.3 Transepithelial Migration Assay (TEM)

1. Polystyrene culture plates (6, 24, and 96 well).
2. Chemoattractant: *N*-formyl methionyl-leucyl-phenylalanine (*f*MLF). Prepare a 10 mM stock in DMSO and store aliquots at −20 °C.
3. 0.5% Triton X-100 solution: Mix 0.5 mL of Triton X-100 with 99.5 mL of $HBSS^+$.
4. 10% Triton X-100 solution: Mix 10 mL of Triton X-100 with 90 mL of $HBSS^+$.
5. 1 M citrate buffer: Dissolve 129.24 g of citric acid and 113.53 g of sodium citrate in 800 mL of deionized/distilled H_2O, and adjust the solution to pH 4.2 using HCl or NaOH. Add H_2O up to 1 L.
6. 100 mM citrate buffer: dilute 1 M citrate buffer 1:10 with deionized/distilled H_2O.
7. Peroxidase substrate solution: dissolve 8.25 mg 2,2′-azino-bis (3-ethylbenzothiazoline-6-sulfonic acid) diammonium salt (ABTS) in 15 mL of 100 mM citrate buffer and add 15 μL of H_2O_2. Prepare fresh for each experiment.
8. Microplate spectrophotometer capable of reading at 405 nm.

3 Methods

Measuring neutrophil transmigration in a defined model system such as we describe here provides a system in which experimental parameters can be modified to elucidate the importance of specific factors. Here, we use the colonic epithelial T84 cell line, although other cells lines can be used as well. Although not without the drawbacks of any cell lines, these defined cells allow for gene knockdowns and other manipulations, facilitating study of the mechanistic involvement of specific factors. T84 cells in particular are useful because they form a polarized monolayer with basolateral and apical surfaces, and we describe here the method that allows for studying neutrophil transmigration in the physiologically relevant basolateral to apical direction. Because of the short lifespan of neutrophils, those types of genetic manipulations are not possible for isolated neutrophils. However, neutrophils from donors with different diseases and/or pharmacologically modified neutrophils can be used (*see* **Note 6**). In all cases, procedures for the isolation of neutrophils for transmigration experiments should follow certain criteria for optimal reproducibility, and we describe those here. This model system is also optimal for the application of defined chemoattractants. Although we used the classic chemoattractant *f*MLF, other chemokines or activators could be easily used. Finally, the simplicity of this model allows for modifications depending on the desired experimental goals and procedures, including the types of inserts used (and the possibility of collecting cell layers for transcriptional or protein expression analysis) and adoption of cellular quantification methods already in use.

3.1 Epithelial Monolayer Preparation

1. If inserts are to be coated, they should be prepared beforehand. Apply sterile collagen solution to the surface and allow adherence for at least 2 h. Many collagen solutions are slightly acidic, so rinse twice with sterile PBS or medium before applying cells.
2. Dissociate cells from stock cell culture by rinsing the cells once with sterile PBS (or serum-free medium) and then apply room temperature trypsin–EDTA solution (2 mL for a T75). Incubate in a 37 °C tissue culture incubator for about 10 min or until all cells have detached from the plastic. Add cell culture medium supplemented with serum to the flask (8 mL for a T75), and pipet repeatedly to break up the cells. Count the cells using a hemocytometer and adjust cell concentration to 2–5 × 10^5 cells/mL.
3. Since the physiological direction of neutrophil transmigration is normally basal to apical, the T84 cells must be applied to the bottom of the inserts ("inverts," *see* Fig. 1a). Using sterile forceps, remove the inserts from their wells, and place upside down in a sterile 150 mm petri dish. Carefully apply 100 μL of T84 cells to the bottoms of the inserts (for the 6.5 mm inserts) and place them in the tissue culture incubator; make sure they are not disturbed for 24 h (*see* **Note 7**).
4. After 24 h, take sterile forceps and flip each 12 mm insert into 1 mL of prewarmed medium in the wells of a 24-well plate. Add 200 μL of medium to the upper chamber of each insert (basal compartment).
5. After 3–4 days, TEER can be monitored, but it may take seven or more days before a consistent strong barrier forms. TEER measurements will vary as the temperature changes when cultures are removed from the 37 °C incubator, so it is important to be consistent. For example, let the plates sit at room temperature for 10–15 min in the tissue culture hood, rinse the electrode tips in a well of the plate with medium only, then

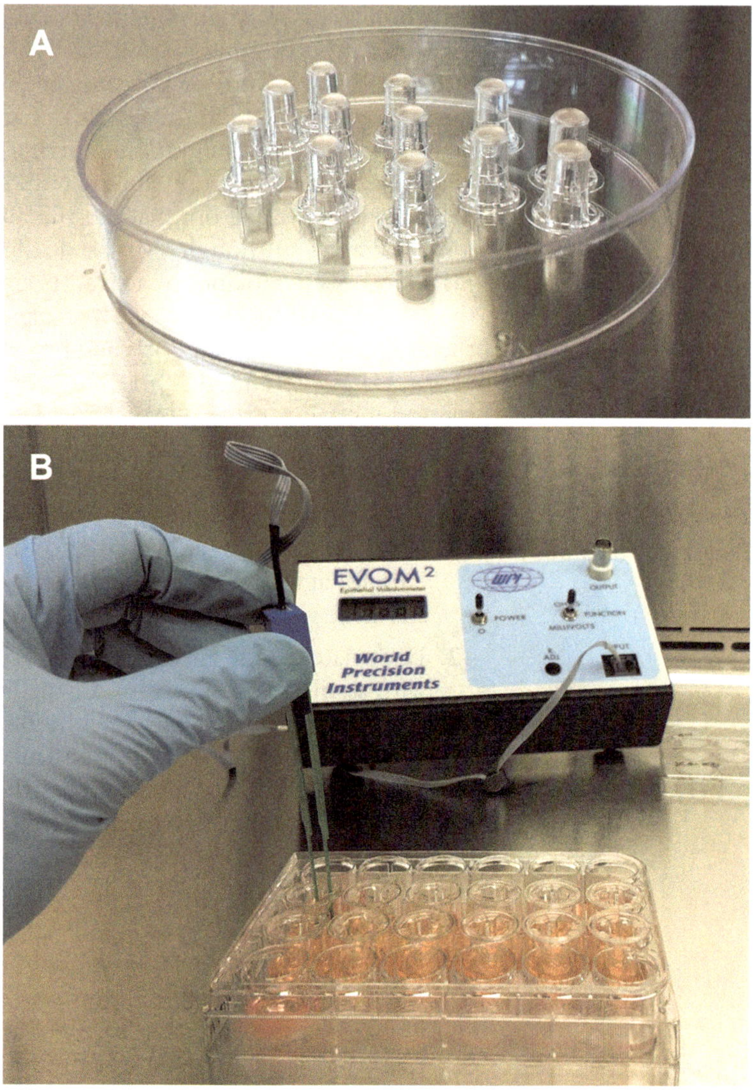

Fig. 1 Establishment of epithelial inverted monolayers. **Panel A:** Suspended epithelial cells are applied to the bottom of transwell inserts for 24 h before the inserts are placed into the wells. **Panel B:** Barrier integrity (TEER) is measured using a paddle electrode with an electrode tip in both the inner chamber (basal compartment in the inverted configuration) and bottom well (apical compartment)

insert the probe tips into the samples with one tip each in the basal and apical compartments (Fig. 1b). Readings will briefly fluctuate, so allow the measurement to stabilize for several seconds before recording the readings. TEER is usually reported as $\Omega \times$ area, so for a 12 mm insert, this is (measured Ω) $\times$ 0.3 cm^2. It is also good practice to have an empty insert in each plate, and the measured background resistance in that well can be subtracted from the measurements in inserts with cells (*see* **Note 8**).

6. Reported stable TEER for T84 monolayers is variable but is typically between 800–3000 Ω cm^2. When your measurements are stable in this range for 1–2 days, the monolayers are ready for use in transmigration experiments (*see* **Note 9**).

3.2 Human Neutrophil Isolation

Human neutrophils are isolated from venous blood using dextran sedimentation followed by Ficoll-Paque density gradient centrifugation.

1. Collect venous blood samples in Vacutainer tubes containing anticoagulant. 40 mL of blood is typically enough to generate sufficient neutrophils for an experiment with 24 6.5 mm inserts (*see* **Note 10**).
2. Mix blood with 6% dextran solution in a 50 mL tube at a ratio of 2.21 mL of dextran for each 10 mL of blood (*see* **Note 11**). After mixing by gentle inversion, let the erythrocytes sediment for 45 min at room temperature. Two layers will form: a yellow, leukocyte-rich upper layer and a red erythrocyte-rich lower layer.
3. Transfer the entire upper layer into a new 50 mL tube and pellet the leukocytes by centrifugation at 500 × *g* for 10 min, then remove the supernatant without disturbing the cell pellet.
4. Gently resuspend the pellet in 6 mL of HBSS$^-$ solution. Endotoxin free 0.9% NaCl solution can also be used if desired.
5. Pipet 7 mL of Ficoll-Paque solution into a clean 50 mL tube. Carefully layer the resuspended leukocytes on top of the Ficoll-Paque. To produce a distinct layer, it can be useful to slightly incline the tube and slowly release the cell solution from the pipet onto the Ficoll-Paque.
6. Centrifuge gradients at 400 × *g* for a minimum of 15 min (up to 40 min). This should result in two visible bands: mononuclear cells (monocytes and lymphocytes) at the interface of the HBSS and Ficoll-Paque layers and a white film of granulocytes on the top of the pelleted erythrocytes. Carefully aspirate the layers above the pellet (the mononuclear cells can be used for other experiments, if desired). Gently resuspend the pellet in 5.3 mL of HBSS$^-$ or 0.9% NaCl solution.
7. Lyse residual erythrocytes by adding 18.6 mL of endotoxin free water for 45 s. Immediately stop this process by rapidly mixing in 690 μL of 25% NaCl to restore isotonicity.
8. Centrifuge at 500 × *g* for 5–10 min. Repeat **step 8** if necessary. Resuspend cells in HBSS$^-$. Count the granulocytes using a hemocytometer and adjust the concentration to 2.5 × 10^7 cells/mL (*see* **Note 12**).

3.3 Transepithelial Migration Assay (TEM)

1. Prewarm sterile filtered HBSS$^+$ to 37 °C. Keep all reagents in a 37 °C water bath.
2. Prepare desired dilutions of *f*MLF in DMSO so that the final DMSO concentration in the well is not >1%, since the DMSO itself can affect cells at higher concentrations. We typically prepare a set of tenfold dilutions in DMSO from 1 mM down to 100 nM *f*MLF so that a concentration range can be evaluated, and DMSO alone is used as a negative control.
3. Prepare a 24-well plate with 1 mL of HBSS$^+$ in each well. Add 10 μL per well of desired *f*MLF dilutions in duplicates (or more replicates if sufficient inserts and neutrophils are available). Reserve 12 wells with 1 mL of HBSS$^+$ per well to perform serial dilutions of neutrophils to generate a standard curve for neutrophil quantification. These wells will not receive inserts.
4. Rinse inverts five times with HBSS$^+$ to remove all residual serum factors, and place them immediately into the plate containing the fMLF dilutions (*see* **Note 13**).
5. Centrifuge 1 mL of the neutrophil solution (2.5×10^7 cells/mL) for 5 min at 600 × *g* to pellet the cells. Discard the supernatant, and gently resuspend the cell pellet in 5 mL of HBSS$^+$ to achieve a final neutrophil concentration of 5×10^6 cells/mL (*see* **Note 14**).
6. Pipet 200 μL of the cells (1×10^6 cells) into each insert. Avoid touching the bottom of the insert.
7. Prepare neutrophil serial dilutions in the reserved wells to form a standard curve, starting with 1×10^6 neutrophils in the first well. To prepare two-fold serial dilutions of the neutrophils for the standard curve, add 600 μL of HBSS$^+$ and 400 μL of neutrophils from the 5×10^6 cells/mL stock to the first of the 12 reserved wells already containing 1 mL of HBSS$^+$. This results in 2 mL final volume containing 2×10^6 neutrophils in the first well. Gently mix and transfer 1 mL from the first well to the second well and so on 10 more times, changing the pipette tips at every transfer. Discard 1 mL from well #11. Well #12 will have no cells and serves as a medium-only background.
8. Cover the plates and incubate for 90–120 min at 37 °C and 5% CO_2 on a level surface without disturbance (*see* **Notes 15**). Neutrophils will migrate through the monolayers toward the chemoattractant in the lower well, which represents migration from the basolateral to the apical side of the cells.
9. During the incubation, precool 10% and 0.5% Triton X-100 solutions to 4 °C and prepare the peroxidase substrate solution.
10. Stop the assay by placing the plate on ice.

11. Quantify the number of migrated neutrophils. An easy, inexpensive, and neutrophil-specific quantification method is the colorimetric assay for myeloperoxidase (MPO) activity [12], which is described here. However, other methods of cell quantification can be used (*see* **Note 16**).
12. Carefully remove inserts from the plate. Be careful to keep all of the apical buffer in the lower wells, since the plate wells contain the migrated neutrophils. If you want to quantify the number of neutrophils stuck to the monolayers, keep the inserts and process as described below.
13. To quantify the number of migrated neutrophils, as well as the neutrophil dilutions used to generate the standard curve, add 50 μL of 10% Triton X-100 solution per mL of liquid in the plate wells to achieve a final Triton X-100 concentration of 0.5%.
14. Shake the plates for 15 min at 4 °C to solubilize the neutrophils.
15. Acidify the samples by adding 100 μL of 1.0 M citrate buffer to each well. Mix by swirling gently.
16. Pipet 100 μL of the solubilized neutrophil mixtures (preferably in duplicates or triplicates) into a 96-well plate.
17. Pipet 100 μL of peroxidase substrate solution into each well.
18. Incubate at room temperature until the desired color develops, and read absorbance at 405 nm using a microplate spectrophotometer. The best results are obtained by letting development proceed until the reading for the well with the highest activity is approximately 2.5 O.D.
19. If you wish to quantify cells associated with the epithelial monolayer, save the inserts and wash them four times with 200 μL of $HBSS^+$.
20. Place the washed inserts into a new 24-well plate containing 1 mL of 0.5% Triton X-100 solution in each well. Puncture the inserts with a 200 μL pipette tip to dislodge the cells. Remove the inserts and quantify the number of neutrophils present, as described above.
21. Create a standard curve from the serial dilutions of neutrophils (Fig. 2A). This neutrophil standard curve will then allow you to quantify the number of neutrophils migrating in the experimental samples. As an example of typical results, Fig. 2B shows the number of neutrophils migrating from the basolateral to apical side of T84 monolayers in response to various *f*MLF concentrations.

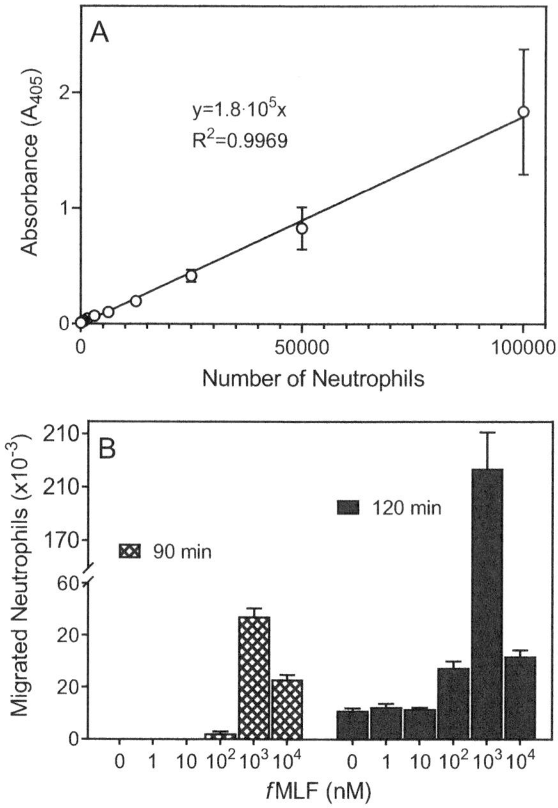

Fig. 2 Neutrophil transmigration through T84 cell monolayers. Transmigration of human neutrophils through T84 cell monolayers (basal to apical) toward *f*MLF was evaluated in 24-well plates containing T84 monolayer inserts, as described. **Panel A:** To quantify the number of transmigrated cells, serial dilutions of purified neutrophils were aliquoted in duplicate into wells containing 1 mL of HBSS$^+$. These wells did not receive monolayer inserts but were incubated for the same times as the wells containing inserts. Following the incubations, Triton X-100 solution was added to the wells (0.5% final concentration), and MPO in the lysates was measured to obtain the standard curve shown (mean ± S.D. of duplicate samples). **Panel B** Purified human neutrophils were added to the upper wells (basal side) of transwell inserts containing T84 cell monolayers. The lower wells contained HBSS$^+$ supplemented with the indicated concentrations of *f*MLF. After incubation for the indicated times at 37 °C and 5% CO_2, the plates were placed on ice, the inserts were removed, Triton X-100 solution was added to the wells, and the number of migrated cells was determined by measuring MPO in the cell lysates and extrapolating actual migrated neutrophil numbers from the standard curve (mean ± S.D. of duplicate samples)

4 Notes

1. Another intestinal cell line commonly used is Caco-2 (ATCC HTB-37), although it does not have as strong a barrier (in terms of TEER) as T84 cells [11]. For respiratory epithelial cell lines, the commonly used A549 cells (ATCC CCL-185) do not form a tight barrier, but the less well known cell lines NCI-H441 or Calu-3 do form tight barriers, although not with as high a TEER as T84 cells [13]. When respiratory cells are used, it is common practice to have an air–liquid interface, such that the apical aspect of the cells are exposed to air and not medium. Primary epithelial cell lines are not commonly used, although there are some examples [14]. Primary cell cultures are commonly used to analyze barrier in endothelial cells. The most well-known of these is human umbilical vein endothelial cells (HUVECs). Under the correct culture conditions, these cells exhibit consistent TEER, although it is much less than that seen in T84 cells.
2. Besides pore size, another consideration when choosing an insert is membrane composition. Polycarbonate is the most common, but polyester membranes provide more visibility for microscopic assessment of monolayers, and other membranes are available precoated with collagen. Insert size (diameter) can be chosen to optimize the numbers of cells needed for each experimental treatment. If the monolayer cells will be harvested for RNA or protein, larger inserts (e.g., 24 mm) are often useful.
3. Type I collagen is most commonly used. It can be prepared from rat tail [15], but premade concentrated solutions are commercially available. Type IV collagen can also be used and is more representative of collagen in epithelial and endothelial basement membranes. Other materials used to coat inserts include laminin, fibronectin, poly-L-lysine, or gelatin. Cell behavior can be significantly affected by the type of substrate, so it is important to be consistent.
4. The EVOM2 system is used to manually measure a single well at a time, but World Precision also produces a system to automatically measure TEER over an entire plate over time. Other manufacturers of TEER measurement devices are Millipore and Flocel.
5. Use of sterile techniques and sterile reagents is important for avoiding neutrophil activation, as is the use of plastic tubes and pipettes. Any glass used during neutrophil preparation will activate the cells and ruin the experiment.
6. An alternative for primary cells is HL-60 cells, which can be differentiated into neutrophil-like cells that migrate in response

to chemoattractants [16]. These cells can be manipulated genetically and, thus, may be useful to manipulate various granulocyte signaling pathways activated during transmigration [17].

7. The number of inserts to be seeded depends on the experimental design and desired number of replicates. If possible, it can be worthwhile to seed extra inserts, since in spite of uniform seeding conditions, there is sometimes variability in the level of the stable TEER.
8. A very good discussion of the variables inherent in the measurement of TEER can be found in a review by Srinivasan et al. [18].
9. Because there is sometimes variability between insert TEER values in the same experiment, extra seeded inserts might be used to help reduce variability in the transmigration assay. For example, inserts could be grouped according to range of TEERs (800–1200, 1200–1800, or >1800 Ω cm^2), and inserts not in the desired range could be omitted.
10. Blood must be collected from donors in accordance with a protocol approved by the Institutional Review Board that includes the informed consent of the donors. Always follow the general precautious working with the biohazardous materials, especially with potential blood borne pathogens.
11. All steps of neutrophil isolation are performed at room temperature. Refrigerated reagents should be brought to room temperature prior to the procedure.
12. After **step 8**, the neutrophils should be kept at 4 °C (on ice) until use in the transmigration assay.
13. Use a hemostat or tweezers to handle the inserts containing monolayers. Grasp the insert with the hemostat, lift it out of the well carefully, and dip it five times in the five different wells of a 6-well plate with prewarmed HBSS$^+$. Take care not to touch the inverted monolayer to the bottom and the walls of the well. Gently invert the insert after each wash to remove HBSS$^+$. Lightly touch the upper edge of the insert with filter paper to wick away remaining liquid from the insert. All inserts should be processed in the same way. To verify that the washing procedure does not affect the TEER, you can place the inserts in the wells with HBSS$^+$ and measure the TEER until a stable level is reached. Otherwise, immediately place washed inserts into the 24-well plate with prepared *f*MLF dilutions and allow to equilibrate for 5–15 min.
14. When resuspending the neutrophils in HBSS$^+$, first add 0.5–1.0 mL of HBSS$^+$ and gently resuspend cells. If the cells resuspend easily, add an additional 4.0–4.5 mL of HBSS$^+$ and gently invert the 15 mL tube a few times. If the cells clump,

they probably were activated during the isolation procedure. In this case, these cells will not be useful in the experiment, and you will need to repeat the isolation procedure with new blood. You may also want to investigate possible sources of endotoxin contamination in the reagents. Make sure to use only plastic (never glass) labware.

15. The optimal time should be determined empirically. We found that 120 min of incubation consistently provides reproducible numbers of transmigrated neutrophils.
16. For example, an alternative method of cell quantification is measurement of total ATP released after cell lysis (CellTiter-Glo Luminescent Cell Viability Assay, Promega); however, this assay is more expensive and not specific to neutrophils.

References

1. Bekkering A, Torensma R (2013) Another look at the life of a neutrophil. World J Hematol 2:44–58
2. Wéra O, Lancellotti P, Oury C (2016) The dual role of neutrophils in inflammatory bowel diseases. J Clin Med 5:118
3. Linden A, Laan M, Anderson GP (2005) Neutrophils, interleukin-17A and lung disease. Eur Respir J 25:159–172
4. Bekkering A, Torensma R (2013) Another look at the life of a neutrophil. World J Hematol 2:44-58.
5. Liu Y, Shaw SK, Ma S et al (2004) Regulation of leukocyte transmigration: cell surface interactions and signaling events. J Immunol 172:7–13
6. Szabady RL, McCormick BA (2013) Control of neutrophil inflammation at mucosal surfaces by secreted epithelial products. Front Immunol 4:220
7. Vogt KL, Summers C, Chilvers ER et al (2018) Priming and de-priming of neutrophil responses in vitro and in vivo. Eur J Clin Investig 48(Suppl 2):e12967
8. Parkos CA (2016) Neutrophil-epithelial interactions: a double-edged sword. Am J Pathol 186:1404–1416
9. Kusek ME, Pazos MA, Pirzai W et al (2014) In vitro coculture assay to assess pathogen induced neutrophil trans-epithelial migration. J Vis Exp:e50823
10. Colgan SP, Parkos CA, Delp C et al (1993) Neutrophil migration across cultured intestinal epithelial monolayers is modulated by epithelial exposure to IFN-gamma in a highly polarized fashion. J Cell Biol 120:785–798
11. Devriese S, Van den Bossche L, Van Welden S et al (2017) T84 monolayers are superior to Caco-2 as a model system of colonocytes. Histochem Cell Biol 148:85–93
12. Parkos CA, Delp C, Arnaout MA et al (1991) Neutrophil migration across a cultured intestinal epithelium. Dependence on a CD11b/CD18-mediated event and enhanced efficiency in physiological direction. J Clin Invest 88:1605–1612
13. Ren H, Birch NP, Suresh V (2016) An optimised human cell culture model for alveolar epithelial transport. PLoS One 11:e0165225
14. Yamaura Y, Chapron BD, Wang Z et al (2016) Functional comparison of human colonic carcinoma cell lines and primary small intestinal epithelial cells for investigations of intestinal drug permeability and first-pass metabolism. Drug Metab Dispos 44:329–335
15. Rajan N, Habermehl J, Coté M-F et al (2006) Preparation of ready-to-use, storable and reconstituted type I collagen from rat tail tendon for tissue engineering applications. Nat Protoc 1:2753–2758
16. Millius A, Weiner OD (2009) Chemotaxis in neutrophil-like HL-60 cells. Methods Mol Biol 571:167–177
17. Carrigan SO, Weppler AL, Issekutz AC et al (2005) Neutrophil differentiated HL-60 cells model Mac-1 (CD11b/CD18)-independent neutrophil transepithelial migration. Immunology 115:108–117
18. Srinivasan B, Kolli AR, Esch MB et al (2015) TEER measurement techniques for in vitro barrier model systems. J Lab Autom 20:107–126

Chapter 7

Quantification of Chemotaxis or Respiratory Burst Using Ex Vivo Culture-Derived Murine Neutrophils

Klaudia Szymczak, Margery G. H. Pelletier, and Peter C. W. Gaines

Abstract

Two critical functional responses of neutrophils are chemotaxis, a response driven by concentration gradients of chemokines released by infected or inflamed tissues, and production of reactive oxygen species (ROS), molecules essential to their capacity to kill pathogens. Assays to accurately test each response have been important to assess efficacies of pharmaceuticals predicted to block recruitment of neutrophils or attenuate their ROS production. Identified antagonists to neutrophil functions may help to reduce tissue damage following inflammation. Described are detailed assays to test these functions, along with steps to generate neutrophils from ex vivo-cultured murine bone marrow that produce robust responses in either assay. The first function protocol details a quantitative assay for chemotaxis that involves culture plates with dual chamber wells that separate cells from a chemokine with small pore-sized membranes. Quantitative measurements of cell numbers in the chemokine-containing chamber are performed with either fluorescence or luminescence detection reagents, which provide signals directly proportional to the numbers of migrated cells. Multiwell plates are used for rapidly testing a variety of conditions and/or chemoattractants. Described in the second function protocol is an assay to measure ROS produced by stimulated neutrophils, again using a multiwell platform for rapid, quantitative measurements of several conditions simultaneously.

Key words Murine neutrophils, Bone marrow, Chemotaxis, Keratinocyte-derived chemokine, Reactive oxygen species, Phorbol myristate acetate, Opsonized zymosan

1 Introduction

Upon maturation in the bone marrow, neutrophils can be stimulated to elicit multiple functional responses vital to their capacity to eliminate invading pathogens. Two of these functional responses are chemotaxis and production of reactive oxygen species (ROS), each indispensable to innate immunity. Chemotaxis is a migration process whereby neutrophils follow concentration gradients of chemoattractants, in particular the C-X-C motif chemokines GRO-α (CXCL1) and IL-8 (CXCL8) in humans, or the homologs keratinocyte-derived chemokine (KC, also a CXCL1 chemokine) and macrophage inflammatory protein 2 (MIP-2, a CXCL2 chemokine) in mice [1–5]. These chemokines activate multiple CXC

Mark T. Quinn and Frank R. DeLeo (eds.), *Neutrophil: Methods and Protocols*, Methods in Molecular Biology, vol. 2087, https://doi.org/10.1007/978-1-0716-0154-9_7,

receptor family members on the surface of neutrophils. Once activated, the G protein-coupled receptors stimulate signaling pathways that promote cytoskeletal reorganization important to shape changes during migration and subsequent phagocytosis of pathogens [3, 6]. Efficient assays to analyze the neutrophil chemotaxis response have allowed for the evaluation of therapeutics that influence neutrophil migration. Identified therapeutics are now being used as interventions for the destructive actions of recruited neutrophils in both acute and chronic disorders, including cystic fibrosis, chronic obstructive pulmonary disease, asthma, bronchiectasis, and rheumatoid arthritis [7–12]. The first neutrophil function protocol provided here details a method for rapidly and reliably measuring neutrophil chemotaxis using a microplate-based design. This protocol allows for a multiwell platform for testing the effects of multiple concentrations or different types of chemokines, with or without effector molecules that may influence neutrophil chemotaxis.

Once neutrophils engulf pathogens, they release ROS that include superoxide anion (O_2^-) and hydrogen peroxide (H_2O_2), each generated from assembled NADPH oxidase (NOX) complexes. NOX assembly begins with association of the integral membrane-bound factors gp91^{phox} (also known as Nox2) plus p22^{phox}, followed by recruitment of the cytosolic proteins p40^{phox}, p47^{phox}, and p67^{phox}, among other factors. The multiprotein complex assembles both on the plasma membrane and within membranes of the phagosome. Upon catalytic activation, NOX reduces molecular oxygen, the consumption of which causes an oxidative or respiratory burst, the common terms for this function. Loss of respiratory burst results in deficient neutrophil-mediated pathogen killing, as clearly demonstrated by studies of chronic granulomatous diseases that result from genetic mutations of NOX subunits [13–15]. Aberrant overproduction of ROS is also problematic, and is associated with chronic inflammatory diseases that include rheumatoid arthritis, psoriasis, inflammatory bowel disease, atherosclerosis, and pulmonary fibrosis [16]. The second neutrophil function protocol detailed here provides quantitative measurements of the respiratory burst stimulated in mature neutrophils by two commonly used agents, phorbol myristate acetate (PMA) or opsonized zymosan (OZ). PMA directly activates protein kinase C, a signaling factor normally activated by cell surface proteins on neutrophils, including the formyl-peptide receptor, Fc receptors (e.g., FcγRII and FcγRIII in humans, or FcγIII and FcγIV in mice), and complement/integrin receptors (CD11b/CD18 or Mac-1) [17–19]. Activation of these receptors typically requires stimuli that initiate phagocytosis (e.g., bacterial peptide formyl-methionyl-leucyl-phenylalanine (fMLF), immunoglobulins, or serum opsonins), but PMA bypasses this requirement and therefore elicits a rapid (e.g., within seconds) and robust response.

By comparison, zymosan (typically Zymosan A, a polysaccharide prepared from *Saccharomyces cerevisiae*) that has been opsonized triggers phagocytosis of the particles and then activation of NOX; the resulting respiratory burst is prolonged and generates lower levels of ROS compared to that stimulated by PMA. However, this response is considered to be more physiologically representative of the respiratory burst in vivo, and therefore may be more appropriate for certain experimental questions. The protocols that utilize either stimulating agent are set up in multiwell platforms along with addition of an enhanced chemiluminescent compound that detects superoxide radicals, providing for quantitative and rapid tests of the respiratory burst produced by neutrophils under a variety of conditions. Importantly, the chemiluminescence detection system provides negligible background signals from unstimulated neutrophils and will not interfere with compounds that produce autofluorescence.

2 Materials

Prepare all solutions using purified, deionized water and cell culture-grade reagents when possible, unless indicated otherwise. Most cytokines, chemokines, or stimulating reagents are stored at −20 °C until the day of experiment. Repeated freeze–thaw cycles should be minimized for all inducing and stimulatory reagents. All cell culture reagents are stored at 4 °C; fetal bovine serum (FBS) is stored at −80 °C and thawed only one time. Note that several inducing agents and diluents require strict regulations for waste disposal. Do not add sodium azide to any of the reagents.

2.1 Cell Preparation

1. Iscove's Modified Dulbecco's Medium (IMDM) containing 20% horse serum (IMDM + HS).
2. Dulbecco's phosphate buffered saline (DPBS).
3. Stem cell factor (SCF), interleukin-3 (IL-3), interleukin-6 (IL-6), granulocyte-colony stimulating factor (G-CSF).
4. IMDM + HS supplemented with 50 ng/mL SCF and 50 ng/mL IL-3.
5. IMDM + HS supplemented with 50 ng/mL G-CSF.
6. IMDM + HS supplemented with 50 ng/mL each of SCF, IL-3, and G-CSF.
7. 6-well Tissue Culture (TC)-Treated Plates (Corning).

2.2 Chemotaxis Chambers and Reagents

1. Chemoattractant: Dissolve 100 μg/mL keratinocyte-derived chemokine (KC, also known as CXCL1; Peprotech) in DPBS containing 0.1% FBS and store at −80 °C in single-use aliquots (e.g., 50 μL). On the day of analysis, thaw the aliquot(s) and store at 4 °C until ready for use (*see* **Note 1**).

2. Cell quantitation system (*see* **Note 2**): For CellTiter-Glo (chemiluminescence, Promega), prepare according to manufacturer's protocol and store the resuspended reagents in aliquots at −20 °C. On the day prior to the experiment, thaw the required aliquot(s) overnight at room temperature (RT). For CyQUANT (fluorescence, ThermoFisher Scientific), prepare fresh according to manufacturer's protocol and use immediately or store for up to 24 h at room temperature.
3. 96-well transwell plates with 3 μm pore size, and TC-treated polycarbonate (HTS Transwell-96, Corning).
4. Phenol red-free IMDM.
5. Dulbecco's phosphate buffered saline (DPBS).
6. Fetal bovine serum (FBS), Certified (*see* **Note 3**).
7. Plate reader suitable for measuring luminescence or fluorescence (e.g., Synergy HT, BioTek).

2.3 ROS Detection

1. Diogenes (for detection of superoxide radicals, National Diagnostics): Prepare reagents according to manufacturer's protocol and store at 4 °C (shelf life of nonreconstituted reagent is 1 year and reconstituted Diogenes can be stored at 4 °C for at least 1 month).
2. Phorbol 12-myristate 13-acetate (PMA) solution: 1 mg/mL PMA in dimethyl sulfoxide (DMSO) as a stock concentration, store at −20 °C. Working concentration is then made by diluting the stock solution 1:1000 in HBSS with 0.1% glucose (final concentration of PMA is 1 μg/mL or 1.6 μM) (*see* **Note 4**).
3. Zymosan A solution: 10 mg/mL Zymosan A in HBSS, store at 4 °C.
4. Hank's Balanced Salt Solution (HBSS).
5. HBSS containing 0.1% (wt/vol) glucose (HBSS + glucose).
6. Opsonized zymosan (OZ): Transfer an appropriate volume of 10 mg/mL zymosan solution to a microcentrifuge tube, planning for 10 μL of opsonized zymosan per reaction, and then add twice this volume of fresh or freshly thawed mouse serum (*see* **Note 5**). Incubate the combined reagents in a heat block at 37 °C for 1 h with occasional flicking of the tube to mix the zymosan particles in the serum. Centrifuge the tube at 1400 × *g* for 15 min, and carefully aspirate and discard supernatant. Resuspend the pellet in 750 μL HBSS and repeat the centrifugation step. Aspirate and discard the supernatant. Resuspend the pellet in HBSS + glucose, using the same total volume of prepared zymosan as used in the first step of the procedure, to yield a final concentration of 10 mg/mL OZ (or 100 μg/reaction).

7. 96-well white polystyrene TC-treated microplate(s) with clear flat bottom (Corning).
8. Plate reader suitable for measuring luminescence (e.g., Synergy HT, BioTek).

3 Methods

Culturing of lineage-depleted cells from the bone marrow toward terminally differentiated neutrophils is accomplished in three steps, each yielding populations of cells at distinct stages of neutrophil maturation. The process begins with a 3-day culture in two cytokines, SCF plus IL-3, known to support the survival and proliferation of common myeloid progenitors (CMP, including myeloblasts). The cells are then cultured in media with these cytokines plus G-CSF, which will initiate differentiation toward promyelocytes as neutrophil maturation begins. Following 2 more days of culture (5 days total), the population is comprised of multiple cell types: residual myeloblasts, promyelocytes, and myelocytes, plus a small number of late-stage cells with characteristic lobulated nuclei (here termed propolymorphonuclear neutrophils, or pro-PMN). Despite the presence of morphologically mature cells, in general the population lacks typical functional responses of fully mature neutrophils (*see* Ref [20]). Finally, the population is cultured for 2 more days in G-CSF alone to yield primarily polymorphonuclear neutrophils (PMN). Prior to performing any of the functional tests, the cells derived from each phase of ex vivo culture should be inspected for characteristics of their developmental stage. Morphologic maturation can be easily examined by staining the cells after cytocentrifugation with Wright-Giemsa stains, as previously described (Fig. 1a and Ref [20]). However, it is recommended to also examine the ex vivo-cultured cells for typical cell surface marker expression profiles as the cells mature from early CMP (e.g., Day 3 culture after SCF/IL-3) toward pro-PMN (Day 5 after G-CSF addition), and finally PMN (Day 7 in G-CSF alone) (*see* Fig. 1b, c). Analyses of cell surface markers should include examining levels of Mac-1 (CD11b/CD18) and either Gr-1 or Ly6G (antibodies to Gr-1 actually detect either Ly6G or Ly6C, but anti-Ly6G antibodies will more definitively indicate terminal neutrophil maturation as Ly6C is also expressed by monocytes and macrophages) [21–23]. An abbreviated step-by-step procedure for culturing the bone marrow cells ex vivo is presented in the first protocol.

The second and third protocols allow for quantitative measurements of chemotaxis and the respiratory burst, each induced with different stimuli. The assays are typically processed in 96-well microplates, which allows for multiple experimental conditions to

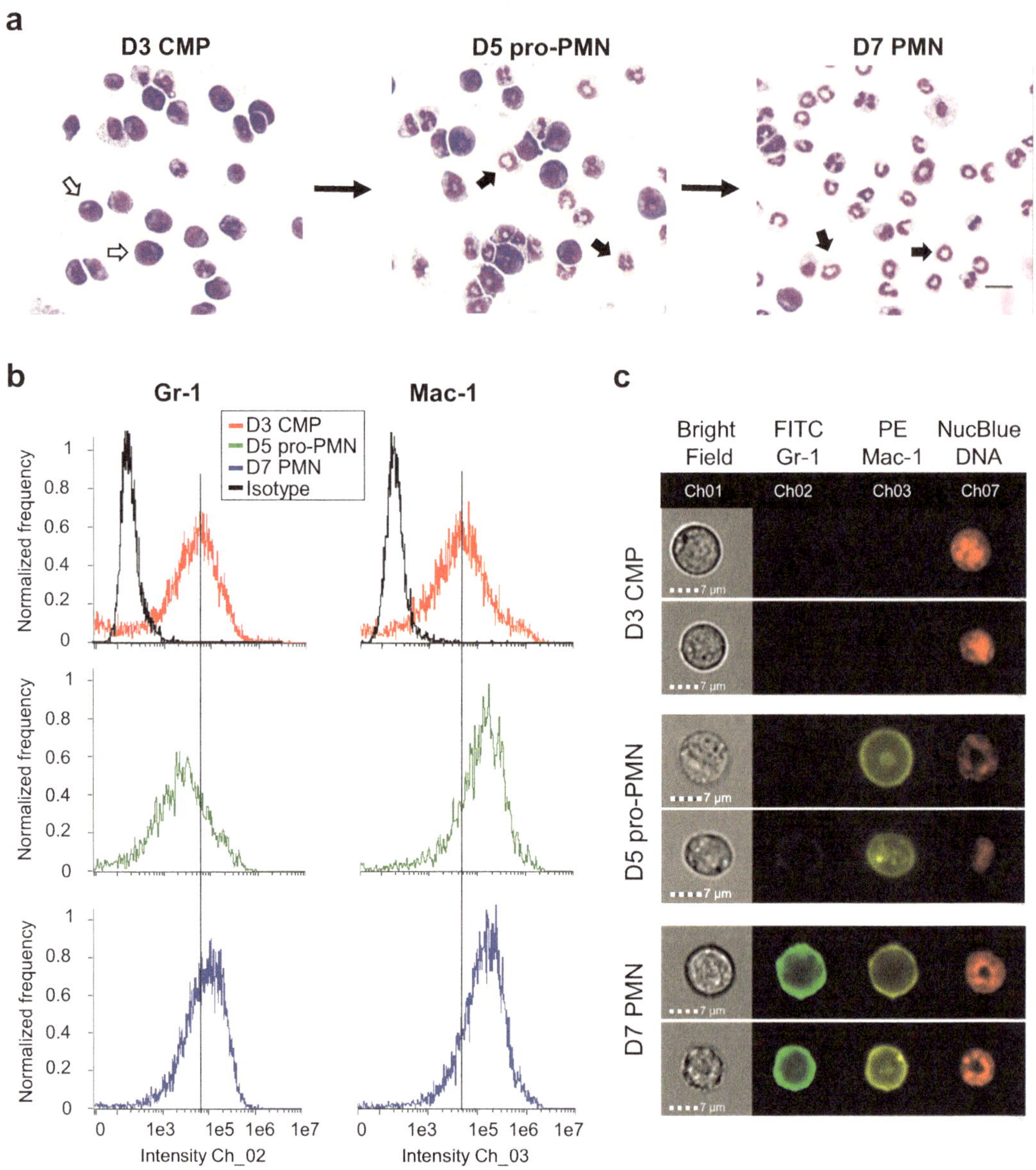

Fig. 1 Morphologic characteristics and cell surface marker expression profiles of differentiating neutrophils derived from ex vivo-culture of mouse bone marrow. (**a**) Representative photomicrographs of cells stained with Wright and Giemsa dyes reveal changes in nuclear shape during the neutrophil differentiation process. Common myeloid progenitors that exhibit large, circular nuclei occupying most of the cell volume (D3 CMP; left panel, open arrows) transform into mature neutrophils with condensed, lobulated nuclei (black arrows in D5 pro-PMN and D7 PMN photomicrographs). The scale bar in D7 PMN indicates 20 μm. (**b**) Detection of immunolabeled cell surface proteins with imaging flow cytometry (IFC, ImageStream, Luminex/Amnis) demonstrates upregulation of neutrophil markers Gr-1 and Mac-1 as the D3 progenitors mature into D7 PMN. (**c**) Representative images from IFC of differentiating neutrophils at each stage of differentiation show characteristic changes in Gr-1 and Mac-1 cell surface expression (using FITC-conjugated anti-Gr-1 or PE-conjugated anti-Mac-1 antibodies, respectively) as well as nuclear lobulation (via NucBlue staining of DNA)

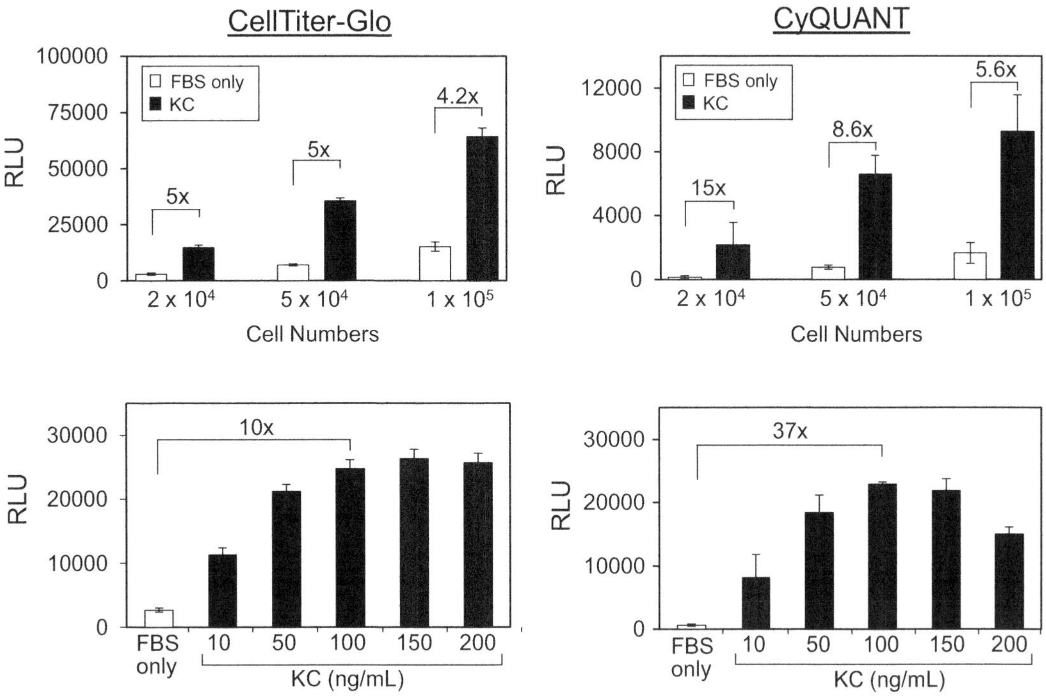

Fig. 2 Chemotaxis of neutrophils from ex vivo-cultured bone marrow in response to KC. Shown are typical relative light units (RLU) detected from migrated neutrophils after 2 h of exposure to KC in chemotaxis chambers, or FBS alone as a control. Signals are produced with either CellTiter-Glo (left panels) or CyQUANT (right panels), each indicating numbers of migrated cells that are dependent on the numbers of cells added to the chambers (upper panels) or concentrations of KC (lower panels). All data shown are averages ± standard deviations of RLU from five replicate chambers for each condition. Fold differences between responses to KC vs. FBS are also indicated for multiple conditions

be simultaneously analyzed. For chemotaxis, KC is routinely used as this provides the most robust response (Fig. 2), however MIP-2 can also be used. Analyses of cell numbers can be performed with either luminescence or fluorescence measurements, which are then adjusted to relative light units (RLU). Each detection reagent provides linear relationships between RLU vs. predetermined cell concentrations, with R^2 values >0.99 (data not shown). For detecting ROS, the use of a microplate reader not only allows for many assay conditions but also provides for temporal analyses of the cells as the responses progress. This time component is particularly important for analyzing responses caused by OZ, as this stimulant requires phagocytosis of the particles prior to assembly and activation of NOX. Typically, PMA is used as the stimulus as this provides for robust responses; however, OZ also provides for a significant release of ROS and may be viewed as more physiologically relevant (Fig. 3).

Bone marrow can be used directly from euthanized mice or after freezing in liquid nitrogen, as detailed in Refs [20] and [24]. Once

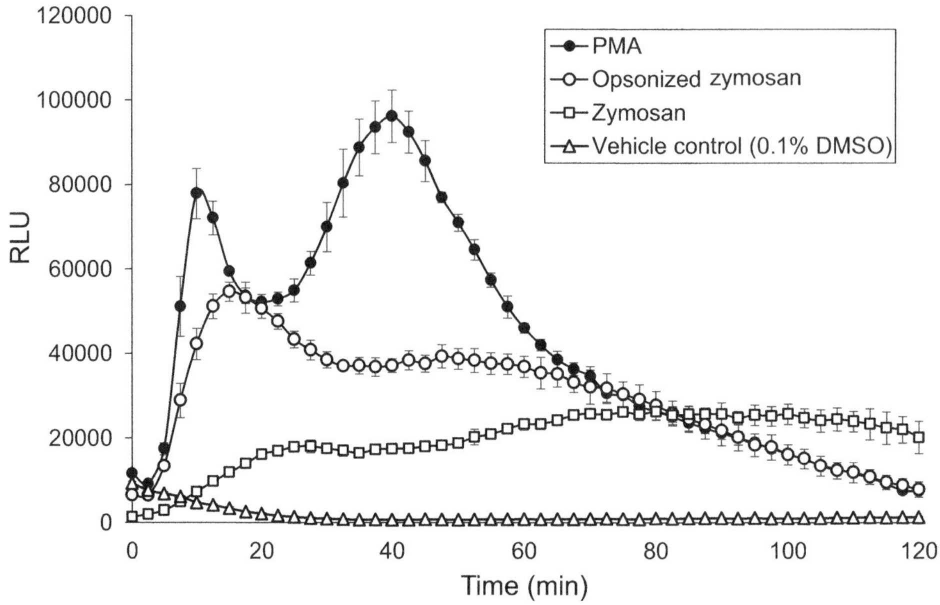

Fig. 3 Respiratory bursts produced by neutrophils from ex vivo-cultured bone marrow in response to PMA or zymosan. Shown are RLU generated from neutrophils stimulated with PMA, opsonized zymosan, untreated zymosan, or the diluent for PMA (vehicle control). All responses shown are average values of RLU ± standard deviations from five replicates; data shown is a representative response profile from more than three independent assays of each stimulant

3.1 Ex Vivo-Culture of Lineage-Depleted Bone Marrow Toward Mature Neutrophils

the cells are prepared, they can be expanded in 5 mL cultures using 6-well plates and then differentiated into mature neutrophils as described.

1. Deplete the isolated bone marrow of differentiated cells using the BD IMag Cell Separation System, typically using 10^7 cells from whole bone marrow according to the manufacturer's protocol (*see* **Note 6**).
2. After lineage depletion steps, combine and centrifuge all cells at $250 \times g$ for 5 min, carefully decant the supernatant and resuspend the pellet in IMDM + HS. Adjust the final cell concentration to $1–2 \times 10^5$ cells /mL and transfer into 6 well plates, each with 5 mL of total medium plus cells.
3. Add 50 ng/mL of SCF and IL-3 to each well and incubate the cells for 2 days at 37 °C and 5% CO_2.
4. Perform a cell count and further dilute the cells by adding fresh IMDM + HS supplemented with 50 ng/mL of SCF and IL-3 at a final concentration of no more than 3×10^5 cells/mL. Incubate the cells for 1 day at 37 °C with 5% CO_2.
5. Perform a cell count to determine concentration and then collect all cells in a conical tube and centrifuge at $250 \times g$ for 5 min. Decant the supernatant and wash the pelleted cells by

resuspending them in 5 mL of DPBS, followed by centrifugation at 250 × *g* for 5 min. Decant the supernatant and resuspend the cells in IMDM + HS supplemented with 50 ng/mL each of fresh SCF, IL-3, and G-CSF. Transfer all cells to new wells in a 6-well plate and dilute to a final concentration of 1–2 × 10^5 cells/mL. Incubate the cells at 37 °C with 5% CO_2 for 2 days (*see* **Note 7**).

6. Perform a cell count and collect all cells in a conical tube and centrifuge at 250 × *g* for 5 min. Decant the supernatant and wash the pelleted cells by resuspending cells in 5 mL of DPBS, followed by centrifugation at 250 × *g* for 5 min. Decant the supernatant and resuspend the pellet in IMDM + HS supplemented with 50 ng/mL of G-CSF. Transfer all cells to new wells in a 6-well plate adjusting the final concentration to 3 × 10^5 cells/mL. Incubate all plates at 37 °C with 5% CO_2.
7. Monitor cell numbers over the next 2 days, maintaining final concentrations between 5–10 × 10^5 cells per mL (*see* **Note 8**). After 2 days of culture, use small aliquots of the cells to perform analyses for morphologic maturation and/or cell surface marker expression, as previously described (*see* Ref [20]; typical results are depicted in Fig. 1). Once maturation is confirmed, proceed with functional assays as detailed.

3.2 Quantitation of Neutrophil Chemotaxis

1. Harvest 1 × 10^5 neutrophils per reaction by centrifugation at 250 × *g* for 5 min. Prepare sufficient numbers of cells for 3–5 replicates per cell type and/or condition, along with unstimulated cells to serve as controls.
2. Carefully aspirate and discard the supernatant, then resuspend cells in 5 mL DPBS to wash, and repeat centrifugation.
3. Carefully aspirate and discard the supernatant, and resuspend cells in 80 μL phenol red-free IMDM per reaction (based on the numbers of reactions/cells established at **step 1**).
4. Add 230 μL media to bottom chambers of a transwell plate (*see* **Note 9**). Below is an example:
   (a) Phenol red-free IMDM + 1% certified FBS (background).
   (b) Phenol red-free IMDM + 1% certified FBS (negative control, with added cells in **step 5**).
   (c) Phenol red-free IMDM + 1% certified FBS + 100 ng/mL KC.
5. Apply upper chambers on the bottom plate and carefully add 80 μL of cells into each appropriate well or 80 μL of phenol red-free IMDM only in background wells (i.e., for background signals) (*see* **Note 10**).
6. Cover the plate and incubate at 37 °C in a humidified incubator with 5% CO_2 for 2 h.

7. After incubation is complete, carefully remove the upper chambers with gentle agitation to make sure all the liquid from the bottom part of the membrane stays in wells.
8. Add detection reagents as follows (*see* **Note 11**).

   CellTiter-Glo: Add 230 μL of prepared CellTiter-Glo reagent to the bottom wells of all reaction chambers and incubate protected from light at RT for 10 min.

   CyQuant: Add 230 μL of prepared CyQUANT reagent to bottom wells of all reaction chambers and incubate protected from light at 37 °C and 5% CO_2 for 1 h.

9. Read luminescent/fluorescence signals using a plate reader (*see* **Note 12**).
10. Assemble all readings and calculate RLU for each condition as follows:
    (a) Determine the average levels generated by the wells with media + serum only (background).
    (b) Subtract this background average level from each value produced from either negative control or test wells.
    (c) Average the remaining light units for each condition to yield RLU plus/minus standard deviations (*see* **Note 13**).

3.3 Quantitation of Neutrophil Respiratory Bursts

1. Harvest 1×10^6 neutrophils per reaction by centrifugation at $250 \times g$ for 5 min. Plan for at least 3 (optimally 5) replicates and control wells with no stimulant (DMSO for PMA or HBSS + glucose for OZ).
2. Carefully aspirate and discard supernatant, resuspend cells with 5 mL HBSS and repeat centrifugation.
3. Carefully aspirate and discard supernatant, then resuspend cells in 160 μL of HBSS + glucose per reaction.
4. Plate cells in designated wells in a white 96-well microplate (with clear flat bottom) at 160 μL per well (use a repeat pipettor if necessary).
5. Read background luminescence signal using an appropriate plate reader.
6. Remove the plate and add 40 μL Diogenes reagent into each well and incubate for 3 min at RT, either in the plate reader or on the benchtop protected from light.
7. Read control luminescence with Diogenes added using the plate reader.
8. Add 10 μL of 1 μg/mL PMA (i.e., 10 ng per reaction) or 10 mg/mL OZ (i.e., 100 μg per reaction) (*see* **Note 14**) to the appropriate wells. Use 10 μL of diluted DMSO (1:1000 in HBSS + glucose) or HBSS + glucose for negative controls (PMA or OZ, respectively).

9. Read luminescence signal every 2 min (or desired intervals) for 2 h at 37 °C using the plate reader in kinetic mode (*see* **Note 15**).
10. Assemble all readings and calculate RLU for each condition as described in Subheading 3.2, **step 10**.

4 Notes

1. The protocol and figures showing responses all refer to the use of KC as the chemoattractant, however additional inducers of chemotaxis can also be used, including MIP-2 (*see* Ref [25]). However, KC provides the most robust response (typically maximal migration of cells toward MIP-2 is approximately 50% that of KC). The cells will also respond to FBS, but the percentage of migrating cells is considerably lower as compared to the aforementioned chemokines.
2. The choice of detection reagent for quantifying the number of cells that have migrated into the bottom chamber of the trans-well will depend on the types of conditions within which the cells are manipulated, and/or the types of monitoring equipment available to the researcher. For reagents that exhibit substantial autofluorescence, the CellTiterGlo system may be preferred since detection does not depend on fluorescence. However, the reagent is substantially more expensive than CyQUANT, and our results indicate that either exhibits strong, highly reproducible responses (*see* Fig. 2).
3. FBS for all reactions should be certified, not qualified. We recommend using highest grade FBS, such as that available from Gibco. While FBS alone will induce migration, rates of migration and actual numbers of cells that migrate into the bottom chamber will be significantly less than FBS combined with a chemokine (e.g., ~40-fold increase in light units or fluorescence, depending on the detection method, *see* Fig. 2). Amounts of FBS alone will also determine the background levels of migration; while 1% FBS causes minimal migration, 5% FBS will induce ~5–6-fold increased migration (data not shown).
4. PMA can be purchased from a variety of manufacturers, although we have typically used Millipore Sigma. Certain cell lines might require the molecular biology grade reagent, but this is not required for most applications (and is more expensive).
5. Opsonization of the zymosan particles is easily performed with fresh mouse serum, but frozen/thawed (one time only) serum also works well. Serum is collected and processed as described

(*see* Ref [20]); briefly blood is incubated at RT for at least 15 min, the samples are vortexed for 30 s, then centrifuged for 10 min at 2000 × *g* at 4 °C, and the serum carefully decanted from the sample.

6. Details of the cell separation system and steps can be found in Gupta et al. (Ref [24]). All culture prior to lineage depletion is performed with certified FBS. Once depleted of differentiated cells, the collected cells are centrifuged and then resuspended in IMDM + HS, preparing the cells for expansion toward myeloblasts.
7. We typically find that cells exhibit a burst of proliferation upon exposure to the added G-CSF, characteristic of promyelocytes proliferating in the bone marrow. Therefore the cells must be monitored over the 2 day culture period and expanded if necessary (typically two-fold dilution), with the addition of fresh media and cytokines.
8. Cell numbers do not significantly increase after transfer into G-CSF alone, since most cells begin to reach terminal differentiation over the first 24 h.
9. Use of a repeat pipettor is recommended to aliquot media into the chosen wells of the plate.
10. A p200 pipettor is recommended for adding the cells into the top chamber for better control of speed and accurate additions.
11. Incubation times are different for each detection reagent, but all samples should be stored in the dark during incubation (covering the plates with aluminum foil works well). For CyQUANT measurements, fluorescence is detected using excitation/emission wavelengths of 508/527, respectively. Either reagent identifies significant increases in migrated cell numbers in the bottom chamber as compared to FBS alone (*see* Fig. 2, top panels), and can sensitively measure chemotaxis even at low KC concentrations (Fig. 2, bottom panels).
12. Some plate reader settings such as gain must be tested empirically. For the Synergy HT (BioTek) used to obtain results shown in Fig. 2, the gain was set to 150.
13. Approximate numbers of migrated cells can be determined by generating standard curves using known concentrations of the cells being analyzed and RLU generated by those concentrations with either detection reagent, and then extrapolating cell numbers from the curve.
14. Zymosan itself (without opsonization) can also cause ROS production, however levels are significantly less than those caused by opsonized zymosan (typically <50%, *see* Fig. 3). Since pathogens that invade any tissue or circulatory system rapidly become coated with opsonins, this step is most often used when testing for the respiratory burst caused by these yeast particles.

15. Some plate reader settings such as gain must be tested empirically. For this assay we have been using gains within the 150–225 range. The whole protocol starting from measuring background luminescence can be programmed using designated software so that the plate is automatically ejected for adding the reagents and inserted for the reads. Measurements must be adjusted with background signals, and then averages will yield RLU, as depicted in Fig. 3.

Acknowledgments

This work was supported in part by the National Heart, Lung, and Blood Institute at the National Institutes of Health (Academic Research Enhancement Award grant No. 1R15HL104593 to P.G.).

References

1. De Filippo K, Henderson RB, Laschinger M, Hogg N (2008) Neutrophil chemokines KC and macrophage-inflammatory protein-2 are newly synthesized by tissue macrophages using distinct TLR signaling pathways. J Immunol 180:4308–4315
2. Tecchio C, Cassatella MA (2016) Neutrophil-derived chemokines on the road to immunity. Semin Immunol 28:119–128
3. Rajarathnam K, Schnoor M, Richardson RM, Rajagopal S (2019) How do chemokines navigate neutrophils to the target site: dissecting the structural mechanisms and signaling pathways. Cell Signal 54:69–80
4. Lee J, Cacalano G, Camerato T, Toy K, Moore MW, Wood WI (1995) Chemokine binding and activities mediated by the mouse IL-8 receptor. J Immunol 155:2158–2164
5. Matzer SP, Zombou J, Sarau HM, Röllinghoff M, Beuscher HU (2004) A synthetic, non-peptide CXCR2 antagonist blocks MIP-2-induced neutrophil migration in mice. Immunobiology 209:225–233
6. Sadik CD, Kim ND, Luster AD (2011) Neutrophils cascading their way to inflammation. Trends Immunol 32:452–460
7. McColl SR, Clark-Lewis I (1999) Inhibition of murine neutrophil recruitment in vivo by CXC chemokine receptor antagonists. J Immunol 163:2829–2835
8. Durham AL, Caramori G, Chung KF, Adcock IM (2016) Targeted anti-inflammatory therapeutics in asthma and chronic obstructive lung disease. Transl Res 167:192–203
9. Jackson PL, Noerager BD, Jablonsky MJ, Hardison MT, Cox BD, Patterson JC, Dhanapal B, Blalock JE, Muccio DD (2011) A CXCL8 receptor antagonist based on the structure of N-acetyl-proline-glycine-proline. Eur J Pharmacol 668:435–442
10. Planagumà A, Domènech T, Pont M, Calama E, García-González V, López R, Aulí M, López M, Fonquerna S, Ramos I, de Alba J, Nueda A, Prats N, Segarra V, Miralpeix M, Lehner MD (2015) Combined anti CXC receptors 1 and 2 therapy is a promising anti-inflammatory treatment for respiratory diseases by reducing neutrophil migration and activation. Pulm Pharmacol Ther 34:37–45
11. Todd CM, Salter BM, Murphy DM, Watson RM, Howie KJ, Milot J, Sadeh J, Boulet LP, O'Byrne PM, Gauvreau GM (2016) The effects of a CXCR1/CXCR2 antagonist on neutrophil migration in mild atopic asthmatic subjects. Pulm Pharmacol Ther 41:34–39
12. Podolin PL, Bolognese BJ, Foley JJ, Schmidt DB, Buckley PT, Widdowson KL, Jin Q, White JR, Lee JM, Goodman RB, Hagen TR, Kajikawa O, Marshall LA, Hay DW, Sarau HM (2002) A potent and selective nonpeptide antagonist of CXCR2 inhibits acute and chronic models of arthritis in the rabbit. J Immunol 169:6435–6444
13. Quie PG, White JG, Holmes B, Good RA (1967) In vitro bactericidal capacity of human polymorphonuclear leukocytes: diminished activity in chronic granulomatous disease of childhood. J Clin Invest 46:668–679

14. Good RA, Quie PG, Windhorst DB, Page AR, Rodey GE, White J, Wolfson JJ, Holmes BH (1968) Fatal (chronic) granulomatous disease of childhood: a hereditary defect of leukocyte function. Semin Hematol 5:215–254
15. Panday A, Sahoo MK, Osorio D, Batra S (2015) NADPH oxidases: an overview from structure to innate immunity-associated pathologies. Cell Mol Immunol 12:5–23
16. Mittal M, Siddiqui MR, Tran K, Reddy SP, Malik AB (2014) Reactive oxygen species in inflammation and tissue injury. Antioxid Redox Signal 20:1126–1167
17. Selvatici R, Falzarano S, Mollica A, Spisani S (2006) Signal transduction pathways triggered by selective formylpeptide analogues in human neutrophils. Eur J Pharmacol 534:1–11
18. Lim JJ, Grinstein S, Roth Z (2017) Diversity and versatility of phagocytosis: roles in innate immunity, tissue remodeling, and homeostasis. Front Cell Infect Microbiol 7:191
19. Bertram A, Ley K (2011) Protein kinase C isoforms in neutrophil adhesion and activation. Arch Immunol Ther Exp 59:79–87
20. Pelletier MG, Szymczak K, Barbeau AM, Prata GN, O'Fallon KS, Gaines P (2017) Characterization of neutrophils and macrophages from ex vivo-cultured murine bone marrow for morphologic maturation and functional responses by imaging flow cytometry. Methods 112:124–146
21. Jutila MA, Kroese FG, Jutila KL, Stall AM, Fiering S, Herzenberg LA, Berg EL, Butcher EC (1988) Ly-6C is a monocyte/macrophage and endothelial cell differentiation antigen regulated by interferon-gamma. Eur J Immunol 18:1819–1826
22. Lagasse E, Weissman IL (1996) Flow cytometric identification of murine neutrophils and monocytes. J Immunol Methods 197:139–150
23. Fleming TJ, Fleming ML, Malek TR (1993) Selective expression of Ly-6G on myeloid lineage cells in mouse bone marrow. RB6-8C5 mAb to granulocyte-differentiation antigen (Gr-1) detects members of the Ly-6 family. J Immunol 151(5):2399–2408
24. Gupta D, Shah HP, Malu K, Berliner N, Gaines P (2014) Differentiation and characterization of myeloid cells. Curr Protoc Immunol. 104: Unit 22F.5
25. Gaines P, Chi J, Berliner N (2005) Heterogeneity of functional responses in differentiated myeloid cell lines reveals EPRO cells as a valid model of murine neutrophil functional activation. J Leukoc Biol 77:669–679

Chapter 8

Ex Vivo Human Neutrophil Swarming Against Live Microbial Targets

Alex Hopke and Daniel Irimia

Abstract

Neutrophils often communicate with each other and coordinate their actions to seal off large sites of injury and infection that individual neutrophils could not cover. The concerted actions of neutrophils are essential for the expeditious protection of healthy tissues from wounds and microbes. These processes, collectively known as swarming, are typically studied in vivo in mice. However, these studies are low throughput and their relevance to human disease is limited. Recently, new tools have been developed for the study of human neutrophil swarming ex vivo. The emergent microscale swarming assays are providing significant insights into the molecular mediators of swarming. By enabling the direct study of human cells, these assays also shed new light on human diseases and host responses against infections. Here, we describe a robust technique for manufacturing microscale swarming arrays with live microbial targets (e.g., clusters of *Candida albicans*). These arrays allow for the direct, precise, and efficient interrogation of the antimicrobial functions of human swarming against a variety of targets.

Key words Human, Array, Micropatterning, Swarming, Candida, Time-lapse, Migration, Phagocytosis, Wound healing, Infections

1 Introduction

Recent characterization of neutrophil behavior has revealed that neutrophils communicate with each other, coordinating their actions to accomplish functions that independent neutrophils could not achieve [1–4]. This process, termed "neutrophil swarming," helps seal off sites of injury and infections, separating these from healthy tissues. The process also allows neutrophils to restrict and contain microbes and microbe clusters that would be too big to phagocytose by individual neutrophils. Swarming is characterized by exponential neutrophil recruitment and driven by feedback loops that the neutrophils themselves perpetuate [2–4]. Swarming is distinct from phagocytosis and is only triggered against targets above a certain size threshold [4]. While important insights have been provided by in vivo mouse models, these systems present

Mark T. Quinn and Frank R. DeLeo (eds.), *Neutrophil: Methods and Protocols*, Methods in Molecular Biology, vol. 2087, https://doi.org/10.1007/978-1-0716-0154-9_8,

numerous challenges including low throughput, poor access to molecular signals for analysis and difficulty in translating protocols for the study of human neutrophil swarming [2, 3]. Recently, microscale and microfluidic technologies have been used to probe swarming biology. These technologies are particularly beneficial in elucidating the constellation of signaling molecules coordinating neutrophil swarming and interactions with other immune and nonimmune cells. The microscale neutrophil swarming arrays enable the monitoring of hundreds of swarms at once. The arrays are particularly useful for enriching the supernatant in the molecules mediating neutrophil communication through the synchronized release from hundreds of swarms [4].

Here we describe recent work that has extended the swarming array technology to directly interrogate the ability of human neutrophils to restrict the growth of live fungi. We print microarrays of poly-L-lysine/ZETAG directly onto glass slides in predetermined patterns. We use commercially-available multiwell attachment to create individual wells above each printed array. We then add a solution of live microbes to each well, allow them to adhere to the arrays, and then wash off the excess. This creates patterned arrays of the desired microbial targets, ready for use. As each array is isolated from one another in individual wells, many different microbial targets can be tested at once. Alternatively, the same target can be run in multiple wells with neutrophils from multiple sources or in multiple treatment conditions. Neutrophil swarming responses can be directly observed against live microbes by time-lapse microscopy. As this is an open system, the supernatants or cells can be readily recovered and subject to molecular analysis.

2 Materials

2.1 Array Printing

1. Poly-L-lysine solution: 0.1% poly-L-lysine (w/v) in sterile H_2O.
2. ZETAG solution: 1.6 mg/mL ZETAG 8185 (BASF) in sterile H_2O.
3. Printing Solution: Mix 1 mL of the poly-L-lysine solution with 12 μL of the ZETAG solution. Add 0.1 mg/mL FITC or another fluorophore to allow for visualization of the printed arrays. Vortex vigorously for 20 s.
4. PolyPico microspotter.
5. PolyPico cartridges.
6. Ultraclean glass slides.
7. Heat block or 37 °C incubator.

2.2 Microorganism Patterning

1. Appropriate media to culture desired microorganism (Yeast Extract-Peptone-Dextrose (YPD) broth for *C. albicans*).
2. Proplate multiarray 16 well attachment.
3. Spray bottle with sterile phosphate-buffered saline (PBS) solution.

2.3 Isolation of Human Neutrophils from Blood

1. EasySep Direct Human Neutrophil Isolation kit and EasySep Buffer.
2. Appropriate magnet for the selected blood volume and isolation kit.
3. Iscove's Modified Dulbecco's Medium (IMDM) + 20% Fetal bovine serum (FBS).

2.4 Neutrophil Swarming Assay

1. Microscope with fluorescent imaging capabilities and temperature control chamber (e.g., Nikon BioStation IM-Q or higher throughput), automated Nikon Eclipse Ti2 series.

3 Methods

Carry out all procedures at room temperature unless otherwise noted.

3.1 Printing of Arrays

1. Open a disposable PolyPico cartridge and place the suction cup attachment on the end of a P200 pipet. Set the pipet to 100 μL. Aliquot printing solution in the well of a 24-well plate. Align the holding rack over the well and place your cartridge in the well. Place the pipet with the suction cup over the top of cartridge and press down to create a seal. Draw up on the pipet to load solution into the cartridge (*see* **Note 1**).
2. Transfer the loaded cartridge into the head of the PolyPico microspotting machine (*see* **Note 2**). Place an ultraclean glass slide on the platform of the machine, with its edge aligned at the "A1" corner.
3. Open the PolyPico Software and turn on the PolyPico microspotting machine. Select the appropriate array design in the software. Go to the "Calibrate" panel and select "find tip." The machine will automatically find the tip of the loaded cartridge. Next, click the "dispense on" box to view the droplet being produced. Adjust the amplitude, width, and frequency settings as necessary to see a well formed (spherical) droplet. Move to the "Dispense" panel. Select the "use current parameters" option and then hit the play button to dispense droplets in the selected array pattern (*see* **Note 3**).
4. Screen the arrays by microscopy to ensure accuracy. For small volume arrays, the addition of a fluorophore like FITC in the printing solution is very helpful to visualize the arrays (*see* Fig. 1).

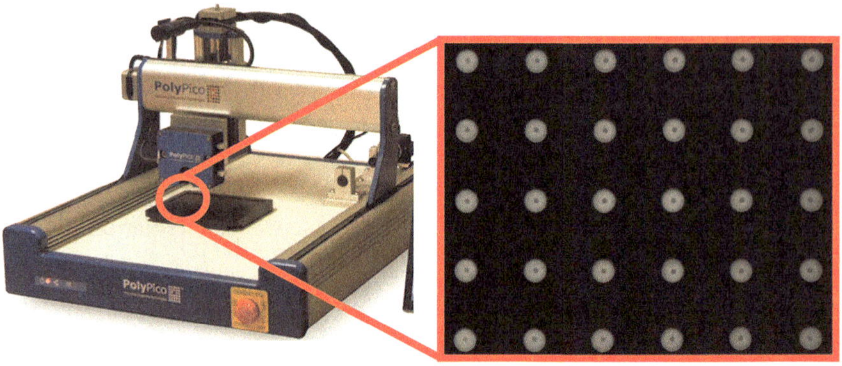

Fig. 1 Printing of microscale arrays. Swarming arrays were printed on glass slides using a PolyPico microspotter and a solution of poly-L-lysine/ZETAG. FITC was spiked in the solution to allow visualization of the spots in the arrays (shown on the right). The printed spots shown are 100 μm in diameter

5. Dry the slides on a heat block at 37–40 °C for 2 h. Slides can then be left at room temperature until use.

3.2 Patterning of Microorganisms

1. Grow the desired target microorganism according to standard culture methods. Methods described here will be for *C. albicans*. Spin down an overnight culture in a centrifuge at max speed for 30 s. Remove the media and discard it. Resuspend the pellet in an equivalent amount of sterile water (*see* **Note 4**).
2. Place the 16-well attachment and gasket system onto the glass slide with the printed arrays. Slide on the clips on each side to hold the attachment in place (Fig. 2a).
3. Load 50 μL of the microorganism solution into each well. Incubate on a rocker for 5–10 min.
4. Wash the wells clean with PBS, using a spray bottle. Vigorous and repeated washing is required to ensure nonbound microorganism is removed from the glass. Check periodically with microscopy to determine when arrays are clean (Fig. 2b).
5. Dispense 60 μL of PBS into each well to ensure that the microorganism arrays do not dry out before running the assay.

3.3 Isolation of Neutrophils from Human Blood

1. Many methods are available for isolating neutrophils from human peripheral blood including density gradient centrifugation and both positive and negative magnetic selection (e.g., *see* Chapters 3 and 4 of this volume). For our swarming assays, we isolate neutrophils from peripheral blood using negative selection with Direct EasySep Neutrophil Isolation kits. Isolations are done according to the manufacturer's protocols.
2. Validating the functionality of neutrophils could be performed by checking their migration signatures in response to

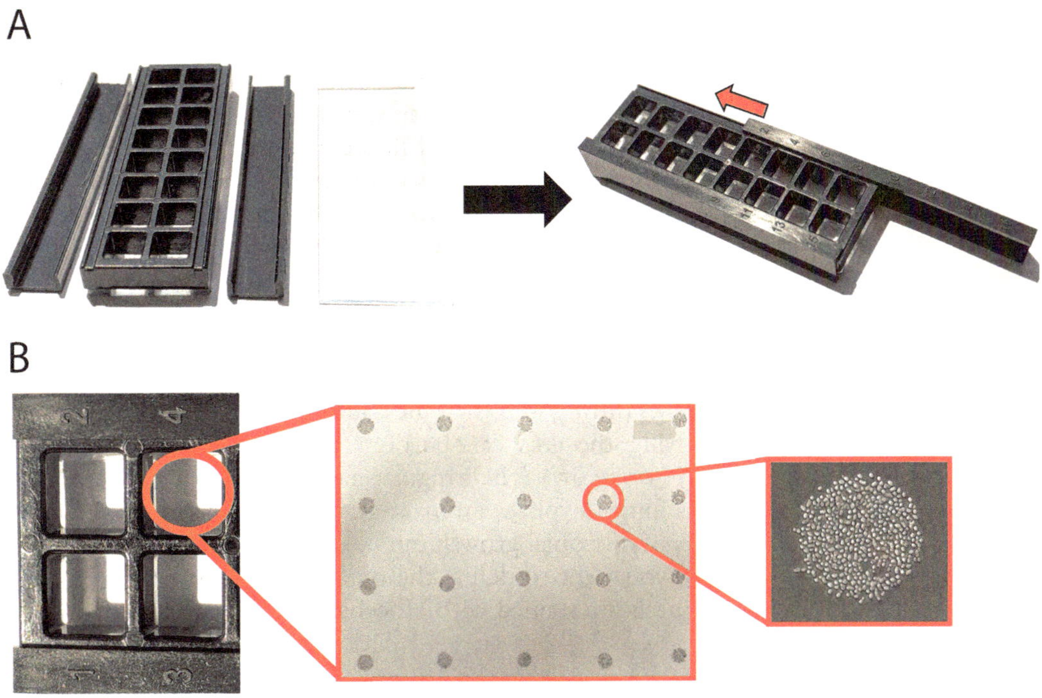

Fig. 2 Assembly of swarming chambers and microbe patterning. After printing and drying, slides are ready for use. Grace Bio-labs multiwell chambers are attached on top of the glass slide. It is secured in place using the slide on clips on each side (**a**). A solution containing the desired target is added to each well and incubated with rocking. Following incubation, each well is thoroughly washed to remove excess target, leaving behind only those adhered to the printed arrays. An image of an array of the live fungus *C. albicans* is shown as an example (**b**)

traditional chemoattractants (fMLF, LTB_4, etc.) [5]. If migration defects are observed, the neutrophil separation protocol can be validated by comparing the migration of isolated neutrophils with that of neutrophils in whole blood, using microfluidic devices designed to accommodate both whole blood and isolated neutrophils.

3.4 Neutrophil Swarming Assay

1. Place the slides with patterned microorganism arrays on an appropriate microscope for imaging. If multipoint functionality is available, select all points prior to the loading of neutrophils into the wells (*see* **Note 5**).
2. Add 5×10^5 neutrophils per well. Typically, this is delivered as 200 μL of a cell suspension at a concentration of 2.5×10^6 cells/mL.
3. Start the time-lapse imaging within seconds after adding the neutrophil suspension. Incubate the arrays at 37 °C for the duration of time-lapse imaging.

4. To track the dynamics of swarming over time, the area of individual swarms can be determined using the analysis software appropriate for the microscope used, or by using ImageJ. The addition of certain stains, such as Hoechst or SYTOX, also enables measurements of fluorescent intensity to be useful in examining swarming dynamics and function. We use Nikon TiE microscopes and therefore NIS-elements or FIJI for our area and fluorescent intensity analysis.

3.5 Swarming Area Quantification

For analysis of area, there are two main outputs: "area of the swarm" and "area of microbial growth." For area of microbial growth, you will want to outline the area covered by the microbe as it grows throughout the time course (Fig. 3a–c). Identifying and quantifying the area can be done automatically or manually in ImageJ (*see* below). The irregular nature of microbial components such as fungal hyphae can make automated analysis inaccurate, so analysis of microbial growth can be done manually. For area of the swarm, you want to define the outlines of the swarm (Fig. 4). If neutrophils are stained with Hoechst, this fluorescent channel can

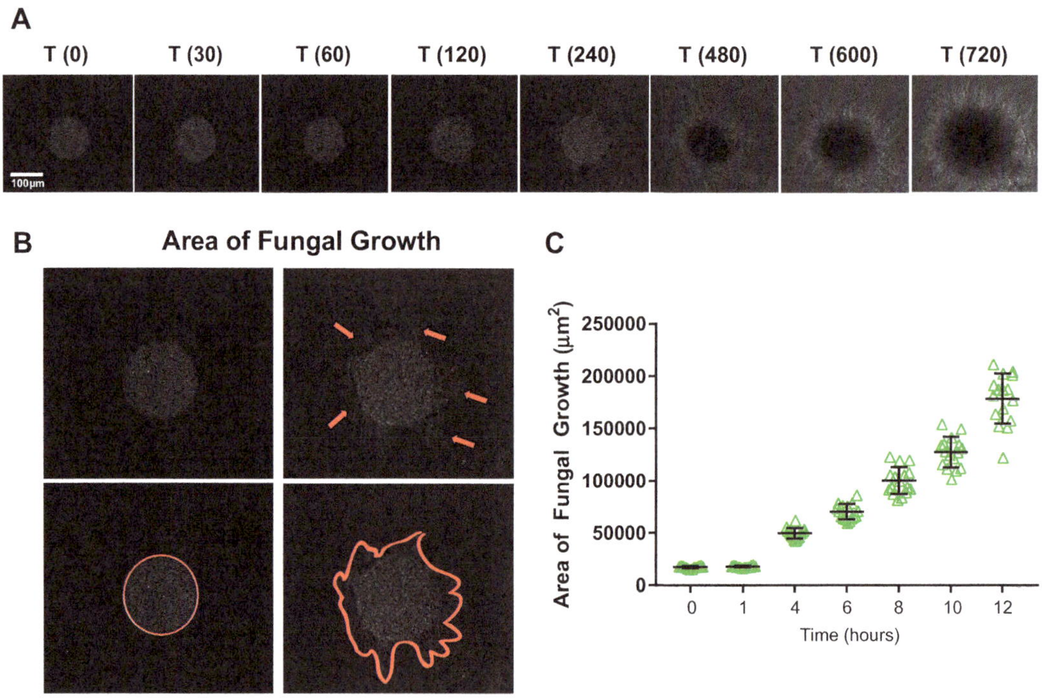

Fig. 3 Analysis of microbial growth. Individual spots in the arrays can be followed over time to determine the dynamics of microbial growth. Panels from a time course show the germination and growth of live *C. albicans* over time (**a**). To quantify this growth, the area can be determined (**b**, **c**). For filamentous fungi like *C. albicans*, irregular growths like hyphae should be included in the area of growth quantitation (**b**, right panels). Quantitation of *C. albicans* growth over time is shown. This quantitation follows 16 individual *C. albicans* spots over a 12 h time course. Error bars represent standard deviation

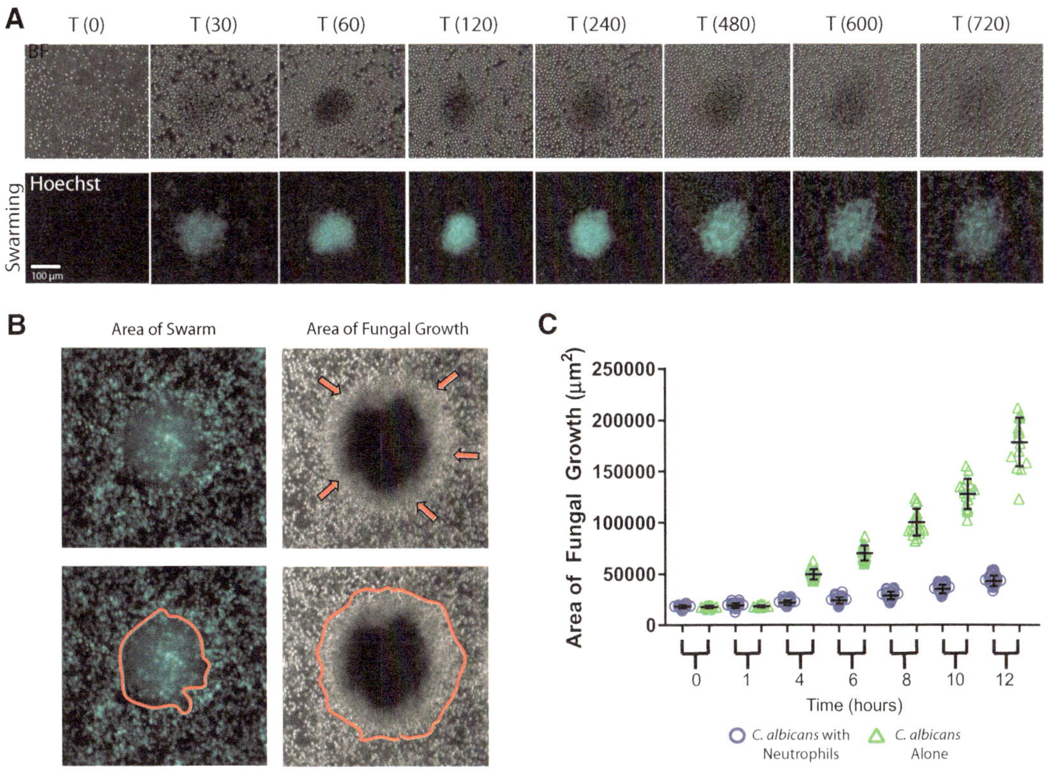

Fig. 4 Analysis of swarming area. By adding neutrophils to the arrays of live microbes, the impact of swarming on microbial growth can be examined. Human neutrophils isolated from peripheral blood were added to arrays of live *C. albicans* to look at swarming. A representative sequences of images showing neutrophils (stained with Hoechst) swarming to *C. albicans* is shown (**a**). The dynamics of swarming can be tracked by determining the area of the swarm, using Hoechst fluorescence to identify just the neutrophils in the swarm (**b**, left panels). By examining microbial growth in the context of swarming and comparing to microbial growth without neutrophils, you can determine the impact of swarming on the microbe. To determine microbial growth, use the brightfield channel to identify any elements of microbial growth that have escaped beyond the area of the swarm and determine area as outlined previously (Fig. 3, Fig. 4**b** left panels). If the microbe expresses fluorescence, this can be combined with the brightfield images during quantitation to provide greater accuracy (*see* also Fig. 5). Quantitation showing the ability of neutrophils to restrict *C. albicans* growth is shown. The area of fungal growth seen during neutrophil swarming is overlaid on the data of *C. albicans* growing alone to illustrate the significant restriction mediated by neutrophils (**c**). $N \geq 16$ individual swarms per time point. Error bars represent standard deviation

be used to identify and outline the boundaries of area covered by neutrophils in the swarm (Fig. 4a, b). This can often be done automatically or manually in ImageJ.

1. Automated area analysis:
   (a) Open your image file in ImageJ. Convert the image to 8-bit by opening the Image tab, going to "Type" and selecting 8-bit.
   (b) Create a thresholded binary image. Go to the "Process" tab, go to "Binary" and select "Make Binary."

(c) Analyze Area. Open the "Analysis" tab and select "Analyze Particles." Enter the upper and lower size limits for what should be considered a swarm in your image. Check the "show outlines" and "display results" boxes before clicking OK. Data measurements will be in a new window that you can export as an Excel file and will include the area. A copy of your image with the swarms outlined will also appear. Be sure to check this image closely to ensure that swarms were accurately outlined.

The automated method outlined above works well for getting just the area of the swarm, but if swarms are irregular in shape or if you want to capture irregular microbial growth that is escaping the swarm, you may need to use manual outlining for analysis (Fig. 3).

2. Manual area analysis:

(a) Open your image in FIJI. Zoom in on the swarms to an appropriate level so you can accurately see all the details you wish to include in the analysis.

(b) Select the "Freehand Tool." Click and hold to manually outline the area to be analyzed. Release once finished.

(c) Analyze area. Use the "Control" and "M" keys on the keyboard to measure the outlined area. Results will appear in a new window and can be exported or copy/pasted into excel.

3. Fluorescent Intensity Analysis

By including different fluorescent probes, you can measure many aspects of neutrophil function within a swarm (Fig. 5). Inclusion of SYTOX Green in the assay, for example, can allow the measurement of the release of neutrophil extracellular traps over time within the swarm (Fig. 5a, b). Furthermore, the use of microbial strains which constitutively express fluorescent proteins can allow you to use fluorescent intensity as a measure of microbial viability and growth (or killing) within the swarm (Fig. 5c, d).

(a) Open your image in ImageJ. Select the appropriate fluorescent channel. You will now need to define the area over which you want to measure the fluorescent intensity. This can be done using the "rectangle," "oval," "polygon," or "freehand" tools (*see* **Note 6**).

(b) Use the "Control" and "M" on the keyboard to measure the intensity in the defined area. Results will appear in a new window and can be exported to Excel.

4 Conclusion

Fully understanding neutrophil behaviors during infection represents a key barrier to harnessing these critical cells in the design of novel therapies against infectious disease. Neutrophil swarming

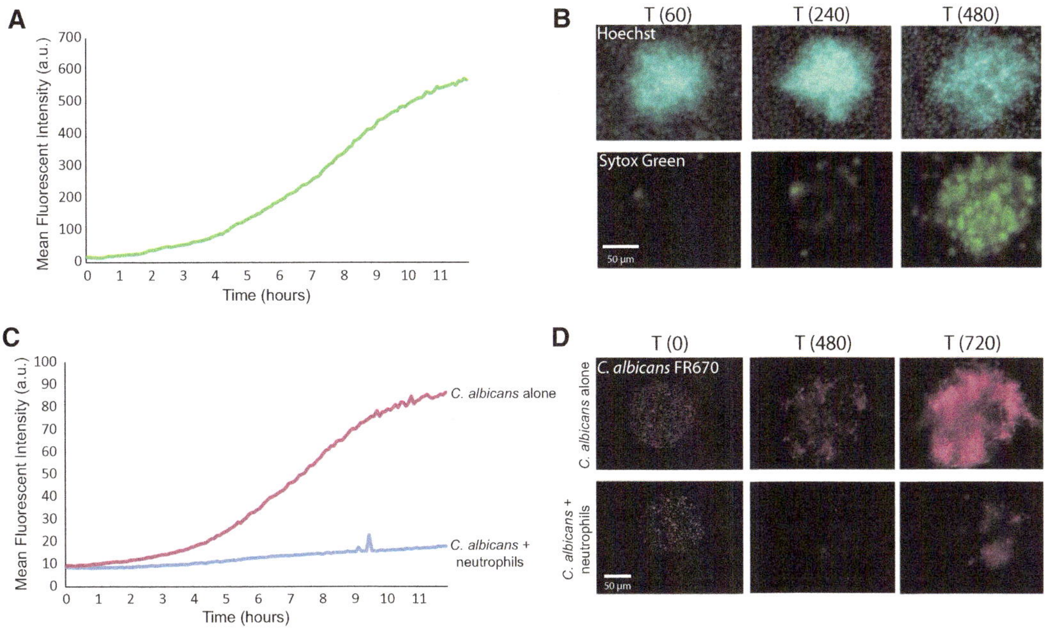

Fig. 5 Analysis of fluorescent intensity during swarming. In addition to area measurements, examination of fluorescent intensity can provide useful information of events during swarming. By incorporating fluorescent probes into the assay, such as SYTOX Green, you can use fluorescent intensity measurements to quantify dynamics within the swarm. In the case of SYTOX Green, you can monitor the release of DNA and NETs within the swarm. The fluorescent intensity and a selected panel of images of SYTOX Green staining throughout a time course are shown for a representative swarm (**a**). Using microbes that express fluorescent proteins allows the use of fluorescent intensity to track microbial growth (or killing) during swarming. Using *C. albicans* that expresses a far red fluorescent protein, fluorescent intensity was followed over time to examine microbial growth [6]. Fluorescent intensity of *C. albicans* was also followed after the addition of neutrophils to the swarming array, showing the restriction of fungal growth (**b**)

represents a novel neutrophil function that is in the early stages of being characterized [2–4]. Neutrophil swarming may play a critical role in how neutrophils attack and contain the invasion of large pathogens like fungal hyphae and microbe clusters [1, 3]. Microscale arrays present an important tool by which to interrogate neutrophil swarming in a high-throughput format, in molecular detail, in human cells [4]. Here, we provide a detailed protocol for the creation and use of microscale arrays of microbial targets, for use in neutrophil swarming experiments with human cells. While described here in the context of examining swarming in restricting infection, these arrays are well positioned to be leveraged to study neutrophil swarming in other situations, particularly in the context of inflammatory diseases where exuberant neutrophil responses could contribute to tissue damage and disease symptoms. Taken together, this technique can be utilized to provide significant insight into the process of neutrophil swarming in the context of infection and inflammation.

5 Notes

1. Break the seal between the pipet and the cartridge before lifting the cartridge out of the solution. Failure to do so could result in air bubbles being drawn into the cartridge, which will interfere with droplet dispensing.
2. During transfer, check the cartridge to ensure that it is free of any air bubbles, as these will impair droplet dispensing. If air bubbles are seen, place the cartridge back in the well, push the solution out and then reload the cartridge as in **step 2**.
3. To ensure that the arrays are printing correctly, one can print a single test array and screen it before committing to printing large matrices of arrays.
4. Each microorganism may have an optimal concentration for patterning. For initial testing, pattern with a series of dilutions to find this optimal range.
5. It is critical to set all experimental parameters and multipoint selections prior to adding neutrophils. The initial neutrophil/microbe interactions that lead to swarming can begin in under 5 min, so early data will be lost if neutrophils are added before parameters are set.
6. To appropriately compare fluorescent intensities between images, the area over which you measure the intensity should the same. We therefore suggest that you define the area with a tool such as "oval," as you easily recreate a circle of a defined area across multiple images.

Acknowledgments

This work was supported by funding from National Institute of Health (awards EB002503 and GM092804). Alex Hopke is supported by a fellowship from Shriners Hospital for Children.

References

1. Warnatsch A, Tsourouktsoglou TD, Branzk N et al (2017) Reactive oxygen species localization programs inflammation to clear microbes of different size. Immunity 46:421–432
2. Lammermann T, Afonso PV, Angermann BR et al (2013) Neutrophil swarms require LTB4 and integrins at sites of cell death in vivo. Nature 498:371–375
3. Kienle K, Lämmermann T (2016) Neutrophil swarming: an essential process of the neutrophil tissue response. Immunol Rev 273:76–93
4. Reategui E, Jalali F, Khankhel AH et al (2017) Microscale arrays for the profiling of start and stop signals coordinating human-neutrophil swarming. Nat Biomed Eng 1:0094
5. Boneschansker L, Yan J, Wong E et al (2014) Microfluidic platform for the quantitative analysis of leukocyte migration signatures. Nat Commun 5:4787
6. Hopke A, Nicke N, Hidu EE et al (2016) Neutrophil attack triggers extracellular trap-dependent Candida cell wall remodeling and altered immune recognition. PLoS Pathog 12:e1005644

Chapter 9

Microinjection and Micropipette-Controlled Phagocytosis Methods for Neutrophils

Maurice B. Hallett, Jennie S. Campbell, Iraj Laffafian, and Sharon Dewitt

Abstract

The ability to microinject substances into the cytosol of living neutrophils opens the possibility of manipulating the chemistry within the cell and also of monitoring changes using indicators which otherwise cannot be introduced into the cell. However, neutrophils cannot be microinjected by the conventional glass pipette insertion method. Here we outline two techniques which work well with neutrophils, namely, SLAM (Simple Lipid-Assisted Microinjection) and electromicroinjection. As these methods utilize micropipettes, we also include a simple method which uses a micropipette to deliver a phagocytic stimulus to a specific cell at a defined time, enable detailed study of the phagocytic process from particle contact to particle internalization.

Key words Microinjection, SLAM, Electroinjection, Phagocytic delivery, Cell signaling

1 Introduction

Microinjection of molecules into the cytosol of living cells is a powerful technique in cell biology. However, microinjection of neutrophils has always been difficult to achieve without damaging the cell. In conventional microinjection, penetration of the plasma membrane is achieved by a rapid entry and exit "stab." The period of time for ejection of material when the pipette tip is within the cell is thus short (on the order of 100 ms) and so requires high pressure (100–200 mbar) to introduce sufficient material from the micropipette into the cell during that time. This must be carefully controlled, as insufficient pressure results in too little material being injected, and excessive pressure causes cell damage or even cell rupture. Also, in the absence of full 3D control of the micropipette tip, its location within the cell is uncontrolled and is liable to cause damage to intracellular organelles. Whereas this may not be a problem for larger cells, the percentage of the cell cytoplasm occupied by organelles in neutrophils, such as the nucleus and granules, is high. Since the velocity at which the micropipette tip enters the

Mark T. Quinn and Frank R. DeLeo (eds.), *Neutrophil: Methods and Protocols*, Methods in Molecular Biology, vol. 2087,
https://doi.org/10.1007/978-1-0716-0154-9_9,

cell may be on the order of 700 μm/s [1], the tip is likely to displace, damage, or enter the nucleus rather than directly enter the cytosol.

1.1 Principles of Microinjection for Neutrophils

We have found only two approaches which work well for microinjection of neutrophils, namely SLAM (Simple Lipid-Assisted Microinjection) and electroinjection. The principles underlying each of these are first discussed before the methodology protocol is given.

1.1.1 Simple Lipid-Assisted Microinjection (SLAM)

An older approach that does not involve glass micropipettes employs lipid fusion, either liposomes [2, 3] or erythrocyte ghost fusion [4, 5] as a means of introducing material into small cells. However, this approach has two limitations. First, the amount of material injected per cell is limited by the space within the liposome or erythrocyte ghost and can be very small. Thus, the technique is useful for introducing nucleic acid into cells, as with lipofection, where amplification of the effect of the injected molecule can occur but may be insufficient for experiments with no amplifying effect. The SLAM technique is a "hybrid" technique involving both the glass micropipette and lipid fusion approaches [6]. The micropipette tip is coated with a lipid bilayer so that contact with the neutrophil plasma membrane results in fusion between the lipid at the micropipette tip and the cell membrane (Fig. 1). This results in

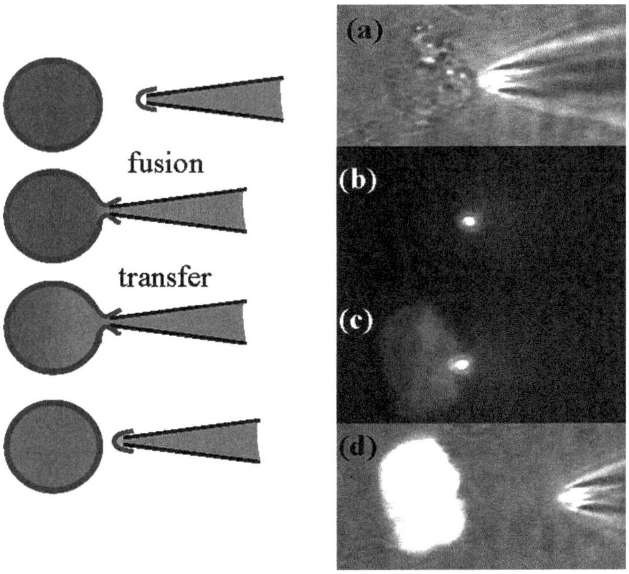

Fig. 1 SLAM transfer of Lucifer yellow into a human neutrophil. The diagrams on the left and an example of an experimental transfer are shown. (**a**) Approach of the lipid-coated micropipette. (**b**) Fusion of the lipid on the micropipette with the cell membrane. (**c**) Transfer of fluorescent material. (**d**) Withdrawal of the micropipette. Images (**a**) and (**d**) are phase contrast and fluorescence combined; and images (**b**) and (**c**) show fluorescent fields only

a channel into the cell cytosol and the transfer of molecules into the neutrophil cytosol at low pressure [6, 7]. In a similar manner to internal perfusion, the pipette tip does not enter the cytosol and so avoids the possibility of organelle damage. However, unlike internal perfusion, the micropipette can be withdrawn, and cell survival is good. Furthermore, the low pressure in the micropipette ensures that the amount of material injected is controlled and does not unduly damage the cell [6, 8]. The technique has been used successfully to introduce large molecular weight inhibitors of IP_3 receptor [9] and high molecular weight Ca^{2+} indicators (e.g., dextran fura2) [10]. The cells retain low cytosolic Ca^{2+} and the ability to undergo chemotaxis and phagocytosis [10].

1.1.2 Electroinjection

Although many problems are overcome by SLAM, fusion between the lipid coating on the micropipette and the plasma membrane of the neutrophil requires a close contact between the two bilayers for several (often tens of) seconds, making it difficult to use on neutrophils, which are moving or undergoing other dynamic shape changes. Electroinjection is a "no touch" approach that we found was surprisingly benign and can be used on neutrophils engaged in chemotaxis without affecting their motile behavior.

The technique involves passing controllable electrical voltage pulses from the open tip of a small-bore micropipette (containing the molecules to be injected) through the cell to be injected. Provided that the pipette is close enough to (but not necessarily touching) the cell, the voltage pulses will cause a localized and transient electroporation of the cell membrane, during which time molecules diffusing out of the micropipette will enter the cell (Fig. 2). Since many molecules also carry a charge, provided

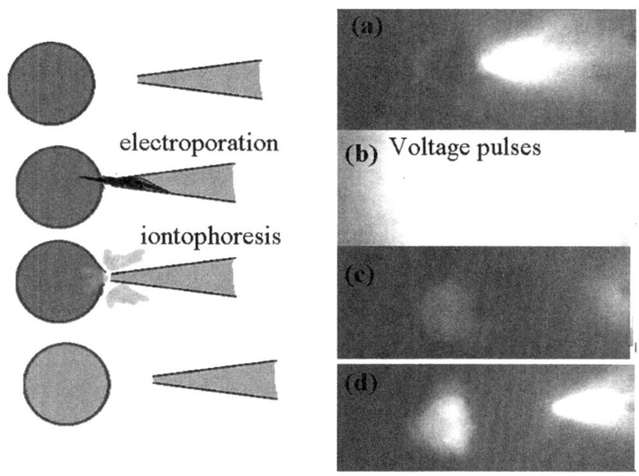

Fig. 2 Electroinjection of Lucifer yellow into a human neutrophil. The diagrams on the left and an example of an experimental transfer are shown. (**a**) Approach of the micropipette. (**b**) Voltage pluses causing electroporation and ejection of fluorescent material. (**c**) Transfer of fluorescent material. (**d**) Ability of the cell to undergo chemotaxis after electroinjection

the electrical polarity is in the appropriate direction, the voltage pulses will also have an iontophoretic effect, forcing molecules out of the pipette synchronously with the opening of the electroporation pore. Since the electroporation effect is dependent on the membrane curvature [11], it is selective for the larger radius of curvature of the cell membrane over the smaller curvatures of intracellular organelles. Haas and Cline have used this approach for transfecting neurones in vivo with GFP–expressing plasmids [12, 13] and for other molecules [14]. We have also shown that this is useful for neutrophils, introducing membrane impermeant molecules into the neutrophil cytosol [15], such as substrates for calpain activity [16]. The approach is surprisingly gentle and simple, and is a "no-touch" (point and shoot) method for introducing material into the neutrophil cytosol with minimal impact on cell shape change dynamics.

1.2 Directed Phagocytosis

Perhaps the main function of neutrophils is phagocytosis. However, the dynamics of the process are surprisingly difficult to observe in real-time and in detail. One way to overcome this problem is to deliver the particle to the cell using a micropipette. In this way, the time of contact between the particle and the neutrophil can be determined precisely; and the subsequent process of phagocytosis followed in real time. The technique is also extremely useful if a membrane impermeant indicator has been introduced into a single neutrophil or the chemistry of a single neutrophil has been modified by microinjection. The phagocytic stimulus can be delivered specifically to that injected neutrophil and, perhaps, compared to a nearby uninjected cell. The technique employs slight suction in a micropipette to hold a particle in its mouth for delivery to the cell (Fig. 3). The technique is simple and works well after electromicroinjection, as he same micropipette can be used to inject an identified cell and then deliver a particle to the same cell to observe phagocytosis (*see* examples in [10, 17, 18]).

2 Materials

1. Purified neutrophils (e.g., *see* chapters in this volume).
2. Micropipettes: can either be pulled using a micropipette puller to give the required tip diameter of 0.5–1 μm or premade (e.g., Eppendorf).
3. Micromanipulator: a good micromanipulator is required to achieve the fine control required for neutrophils (e.g., Eppendorf micromanipulator).
4. SLAM micropipettes (Warner Instruments).

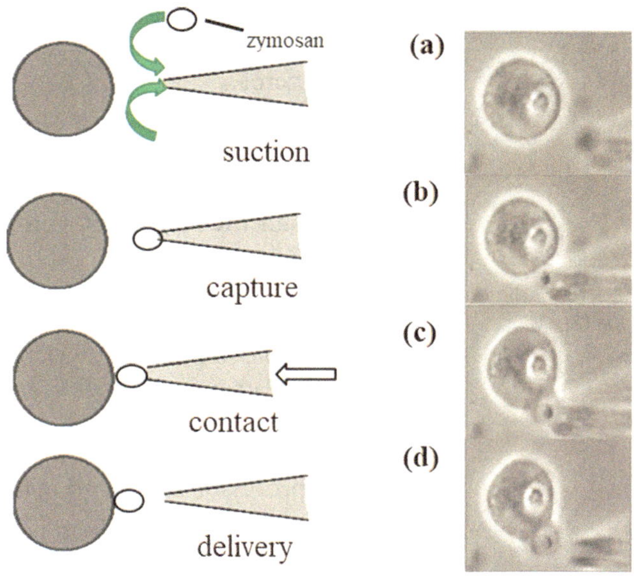

Fig. 3 Directed phagocytosis of a zymosan particle into a human neutrophil. The diagrams on the left and an example of an experimental transfer are shown. (**a**) Capture of a zymosan particle by suction within the micropipette. (**b**) Contact between the particle and the neutrophil. (**c**) Delivery of the phagocytic stimulus and (**d**) the withdrawal of the micropipette

5. Lipid coating for SLAM can be achieved using 20 mg/ml oleoyl-palmitoyl phosphatidylcholine (POPC) dissolved in chloroform and stored below 0 °C. Dilute aliquots of this solution with chloroform before use (final POPC concentration 1 mM).
6. DiIC18 (3)(1,1-dioctadecyl-3,3,3′,3′-tetramethylindocarbocyanine perchlorate) (Molecular Probes) and Lucifer yellow CH (Sigma).

3 Methods

3.1 SLAM

3.1.1 Lipid Coating of Micropipette

1. Back-load the micropipette with sufficient volume of the injection medium, containing the material to be injected (*see* **Notes 1** and **2**) to exert a pressure to offset capillary pressure.
2. Dip the tip of the loaded micropipette in the 1 mM POPC lipid solution that is kept on ice, or a drop (~10 μl) of the lipid solution may be applied to the tip of the micropipette. The dipping method is simpler but requires a larger volume of lipid solution.
3. Evaporation of the chloroform after the micropipette is withdrawn from the lipid solution results in a dry thin coating of lipid on the glass.

4. The micropipette can then be connected to a control pressure device (e.g., Eppendorf micromanipulator) with the pressure set to zero.
5. The micropipette is then carefully lowered into the aqueous medium bathing the neutrophils on the microscope stage. This results in swelling of the dried lipid on the tip of the micropipette and the formation of a bilayer across the open tip (*see* **Note 3**).

3.1.2 The "SLAM" Procedure

1. Bring the loaded lipid-coated micropipette into the field of view of the neutrophils, which have been allowed to sediment onto a glass coverslip (*see* **Note 4**). As neutrophils are microscopically "small," a high magnification objective (e.g., oil immersion objective 100× or 63× and a motorized, microprocessor-controlled micromanipulator are required.
2. The tip of the SLAM pipette is now in gentle contact with the surface of the chosen neutrophil.
3. A short delay of 1–5 s is often required before transfer of lipid, and the aqueous contents of the micropipette to the cell are observed. The pressure within the micropipette is not increased during the microinjection but is held constant at 5–10 mbar (*see* **Note 5**).
4. When the required amount of material has been transferred (monitored by the fluorescent signal), the tip of the pipette is removed from the neutrophil, and the effect of the molecule on the neutrophil behavior/responsiveness can be observed (*see* **Note 6**).

3.2 Electroinjection

1. Back-fill the micropipette with the appropriate loading solution (~ 1–2 μl) (*see* **Notes 1** and 7). Pass a silver wire (0.25 mm diameter) through the micropipette holder into the micropipette and into the loading solution. A second silver wire is fixed in place in the cell-containing medium on a microscope slide. The two wires are then connected to a voltage stimulator terminal. We use a Grass SD9 stimulator, but any stimulator capable of delivering low voltage pulses could be used.
2. The micropipette is positioned next to the target neutrophil (preferably within 1 μm) using a micromanipulator, and electroporation is initiated by a 1 s train of voltage pulses. We use 1 ms square pulses; 10–50 V volts; 200 Hz (*see* **Note** 7).
3. This will result in the expulsion of fluorescent material from the pipette tip and its incorporation into the neutrophil (Fig. 2).
4. EGTA should be included in the solution within the micropipette, as localized electroporation can result in a transient elevation of cytosolic free Ca^{2+} within the neutrophil [16]. When EGTA is expelled from the micropipette

synchronously with electroporation, the effect on cytosolic Ca^{2+} is reduced or eliminated, depending on the EGTA concentration [16].

3.3 Directed Phagocytosis

1. Particles are sedimented on to a field of neutrophils. Zymosan particles (~ 1–3 μm in length) are useful, as they are large enough to be retained at the micropipette mouth opening (~ 1 μm); but small enough to be phagocytosed (and not too small to be sucked into the interior of the micropipette). Zymosan is also useful, as it can be easily opsonized with C3bi and can be fluorescently labeled if needed.
2. Position the micropipette within the microscopic field, and apply a small positive pressure. The micropipette is then placed near a zymosan particle, and the positive pressure released. The fluid within the micropipette is then drawn back and fluid around the pipette is sucked into the pipette. If the pipette is empty, suction into the pipetted is caused by capillary attraction and can be difficult to control. The movement of fluid into the micropipette draws the nearby particle to the mouth of the pipette.
3. The micropipette with the "captured" zymosan particle is then be positioned next to (but not touching) the target neutrophil using a micromanipulator. When desired, the micropipette is advanced gently so that the zymosan particle touches the neutrophil (time of contact). The micropipette can now be withdrawn, leaving the particle adherent to the cell (Fig. 3).

4 Notes

1. It is advisable to include a fluorescent marker along with the material to be injected. We routinely use 10–50 mg/ml Lucifer yellow. Often, however, the injectate is itself fluorescent (e.g., GFP-tagged protein or FITC labeled).
2. It is important to remember that there will be a large dilution of the material within the micropipette when it enters the cell. This can be quantified by monitoring fluorescence transfer and quantified from the intensity of the injected cell. However, the dilution may be of the order of 1:1000 (i.e., 0.1%) to 1:100 (1% transfer). This may limit the use of some proteins which cannot be solubilized at high concentrations. In this case, a prolonged transfer time can be tried (similar to internal perfusion, where a far smaller dilution occurs). However, it must be remembered that diffusion of molecules out of the cytosol will also occur. Therefore, it is important to include in the micropipette molecules that might be important for the survival of the microinjected cell and for its responsiveness (e.g., ATP).

3. The formation of a lipid bilayer at the micropipette tip can be checked by increasing the pressure in the micropipette to 5–10 mbar. Lack of ejection or leakage of the dye from the micropipette provides evidence that an effective lipid seal had formed at the micropipette tip.
4. For good cell response, it is, of course, important to maintain the neutrophils at 37 °C using a microscope stage heater or similar device.
5. This low pressure was important to prevent rupture of the lipid seal at the micropipette tip and to prevent subsequent damage to the cell undergoing microinjection.
6. The lipid-coated micropipette can be used more than once, but the number of possible successful injections depends on the amount of lipid in the coating and the amount of cell "debris" picked up as the micropipette is raised and lowered into the cell bathing medium. This can result in blockage of the pipette tip by particulate material. Debris can often be dislodged by applying a high pressure to the micropipette (cleaning pressure), and the lipid seal may reform.
7. We have found a range of parameters work, but the stated parameters may not be optimal for the molecules of your particular interest. The duration and voltage should be experimentally varied to establish the optimum delivery of your material into the neutrophil. It is important to remember that using Lucifer yellow as a surrogate marker may be misleading with this technique (which relies on molecular charge) and that ideally the material being injected should itself be fluorescent.

References

1. Guse AH, Berg I, da Silva CP et al (1997) Ca^{2+} entry by cyclic ADP-ribose in intact T-lymphocytes. J Biol Chem 272:8546–8550
2. Hallett MB, Campbell AK (1980) Uptake of liposomes containing the photoprotein obelin by rat isolated adipocytes; adhesion, endocytosis or fusion? Biochem J 192:587–596
3. Gao X, Huang L (1995) Cationic liposome-mediated gene transfer. Gene Ther 2:710–722
4. Hallett MB, Campbell AK (1982) Measurement of changes in cytoplasmic free calcium in fused cell hybrids. Nature 294:155–158
5. Campbell AK, Hallett MB (1983) Measurement of intracellular calcium ions and oxygen radicals in polymorphonuclear leucocyte-erythrocyte "ghost" hybrids. J Physiol 338:537–550
6. Laffafian I, Hallett MB (1998) Lipid-assisted microinjection: introducing material into the cytosol and membranes of small cells. Biophys J 75:2558–2563
7. Peters R, Sikorski R (1998) Gentle slam. Science 282:2213–2214
8. Laffafian I, Hallett MB (2000) Gentle microinjection for myeloid cells using SLAM. Blood 95:3270–3271
9. Davies-Cox EV, Laffafian I, Hallett MB (2001) Control of Ca^{2+} influx in human neutrophils by IP_3 binding: differential effects of microinjected IP_3 receptor antagonists. Biochem J 355:139–143
10. Dewitt S, Laffafian I, Hallett MB (2003) Phagosomal oxidative activity during β_2 integrin (CR3)-mediated phagocytosis by neutrophils is triggered by a non-restricted Ca^{2+} signal: Ca^{2+} controls time not space. J Cell Sci 116:2857–2865

11. Zimmermann U (1986) Electrical breakdown, electropermeabilization and electrofusion. Ann Rev Physiol Biochem 105:175–256
12. Haas K, Sin WC, Javaherian A et al (2001) Single-cell electroporation for gene transfer in vivo. Neuron 29:583–591
13. Haas K, Jensen K, Sin WC et al (2002) Targeted electroporation in Xenopus tadpoles in vivo - from single cells to the entire brain. Differentiation 70:148–154
14. Bestman JE, Ewald RC, Chiu SL et al (2006) In vivo single-cell electroporation for transfer of DNA and macromolecules. Nat Protoc 1:1267–1272
15. Lewis KJ, Masternam B, Laffafian I (2014) Minimal impact electro-injection of cells undergoing dynamic shape change reveals calpain activation. Biochem Biophys Acta (Mol Cell Biol) 1843:1182–1187
16. Campbell JS, Hallett MB (2015) Active calpain in phagocytically competent human neutrophils: electroinjection of fluorogenic calpain substrate. Biochem Biophys Res Commun 457:341–346
17. Dewitt S, Hallett MB (2002) Cytosolic free Ca^{2+} changes and calpain activation are required for beta integrin-accelerated phagocytosis by human neutrophils. J Cell Biol 159:181–189
18. Dewitt S, Tian W, Hallett MB (2006) Localised PtdIns(3,4,5)P-3 or PtdIns(3,4)P-2 at the phagocytic cup is required for both phagosome closure and Ca^{2+} signalling in HL60 neutrophils. J Cell Sci 119:443–451

Chapter 10

Using Imaging Flow Cytometry to Quantify Neutrophil Phagocytosis

Asya Smirnov, Michael D. Solga, Joanne Lannigan, and Alison K. Criss

Abstract

Neutrophils are professional phagocytes that are important for innate host defenses against pathogens and resolution of inflammation. Traditionally, the phagocytic capacity of neutrophils was quantified by enumeration of cells containing either internalized or bound bacteria or other cargo from a series of microscopic images. Here we describe an imaging flow cytometry-based protocol and analysis method for quantifying the binding and uptake of *Neisseria gonorrhoeae* by primary adherent human neutrophils. Imaging flow cytometry combines the capacity for quantitative, high-throughput analysis of tens of thousands of cells per condition, with the imaging power of fluorescence microscopy. Here, all bacteria are labeled with Tag-it Violet™ and bound bacteria are differentially stained with a DyLight™ 650-conjugated antibody. Images are analyzed using spot count and other algorithms. Outputs include the percent of neutrophils associated with bacteria, the percent of neutrophils with internalized bacteria, and the percent of internalized bacteria. This basic protocol can be adapted to a variety of particle types and can be used for multiplex analysis in combination with staining for different neutrophil surface and intracellular markers.

Key words Imaging flow cytometry, Neutrophils, Bacteria, Phagocytosis, Binding, Internalization, Fluorescence

1 Introduction

Neutrophil phagocytic activity plays an important role in clearing microbial pathogens, resolving inflammation, and regulating immune responses [1–4]. Phagocytosis of a target particle involves the coordinated steps of recognition, binding to phagocytic receptors, and internalization, which if successful leads to its elimination. Measurement of neutrophil phagocytic capacity has historically been performed in one of two ways. One, fluorescent target particles such as bacteria are incubated with neutrophils, and fluorescence-positive neutrophils are quantified by flow cytometry. However, this does not discriminate between bound and internalized targets [5–7]. Two, neutrophils are incubated with fluorescent targets, and surface-bound targets are recognized with an antibody or other reagent of

Mark T. Quinn and Frank R. DeLeo (eds.), *Neutrophil: Methods and Protocols*, Methods in Molecular Biology, vol. 2087,
https://doi.org/10.1007/978-1-0716-0154-9_10,

a different fluorophore. Differentially labeled bacteria can be visualized by immunofluorescence microscopy and quantified, but this approach is laborious, time consuming, and low throughput [8, 9]. To overcome the limitations of each approach, we and others have employed imaging flow cytometry to enable simultaneous fluorescent and morphological analysis of phagocytes and their targets in a large cell population [10–14]. Different methods have been used to quantify the binding and internalization of targets by phagocytes, including masks to delineate the outside of cells, bright detail similarity score between single- and dual-labeled targets, and spot counting [11, 14, 15]. We have found spot counting to be the most accurate approach for protrusive cells like neutrophils, especially in conditions where a sizable percentage of target particles are bound and not internalized [14].

Here we describe a protocol that was developed to quantify *Neisseria gonorrhoeae* bacteria that are bound to and/or internalized by adherent, interleukin 8 (IL-8)-treated, primary human neutrophils [16]. Neutrophils purified from peripheral blood were infected with Tag-it Violet™ labeled bacteria. Neutrophils were fixed, lifted, and stained with an anti-*Neisseria gonorrhoeae* antibody coupled to DyLight™ 650. In the absence of permeabilization, intracellular bacteria are inaccessible to the antibody and remain single stained (Tag-it Violet™), while bound bacteria are double stained (Tag-it Violet™ and DyLight™ 650). Use of these fluorophores is an improvement over earlier versions of this protocol that used carboxyfluorescein succinimidyl ester, which reserves fluorophores with emission wavelengths in the 500–550 nm range for additional staining procedures if desired by the user [14, 16]. Fluorescent and imaging data were acquired using ImageStream^X Mk II with INSPIRE® acquisition software. Data analysis was performed using IDEAS® v. 6.2, and the spot count algorithm was applied to count the number of Tag-it Violet™ and DyLight™ 650 positive bacteria in individual neutrophils in a population of thousands of cells. These two measurements were used to calculate: (1) percent of neutrophils with cell-associated bacteria (either bound or internalized); (2) percent of cell-associated bacteria that are internalized; and (3) percent of neutrophils with internalized bacteria (neutrophils with at least one internalized bacterium).

2 Materials

All the solutions for working with primary human neutrophils are prepared in endotoxin-free water using aseptic techniques.

2.1 Primary Human Neutrophils

Neutrophils are purified from peripheral human blood using dextran sedimentation of erythrocytes, followed by separation on a Ficoll-Paque PLUS gradient (GE Healthcare) and hypotonic lysis of residual erythrocytes [17, 18] (also *see* Chapters 3 and 4 of this volume). Purified neutrophils are resuspended in Dulbecco's modified PBS, pH 7.4 without calcium and magnesium (Thermo Scientific), kept on ice, and used within 2 h from purification (*see* **Note 1**).

2.2 Tissue Culture

25 mm tissue culture plastic coverslips (Sarstedt).

6-well tissue culture dish.

Interleukin 8 (IL-8) (e.g., R&D Systems).

RPMI without glutamine supplemented with 10% of heat-inactivated fetal bovine serum (FBS).

2.3 Bacterial Culture

Neisseria gonorrhoeae culture grown to mid-logarithmic phase in liquid gonococcal medium base medium containing Kellogg's supplements I and II [19] (*see* **Note 2**).

2.4 Bacteria Labeling

Tag-it Violet™ Proliferation and Cell Tracking Dye (TIV) (Biolegend).

5 mM $MgSO_4$ in phosphate-buffered saline (PBS), pH 7.4.

2.5 Fixation and Staining

16% paraformaldehyde solution.

Blocking solution: 10% normal goat serum in PBS, pH 7.4.

Goat anti-*Neisseria gonorrhoeae* antibody (Biosource) labeled with DyLight™ 650 (DL650) according to the manufacturer's protocol (Thermo).

2.6 Instrument and Software

ImageStream^X Mk II imaging flow cytometer with INSPIRE® and IDEAS® v. 6.2 Software packages (Amnis Luminex Corporation).

3 Methods

3.1 Short-Term Culture of IL-8 Treated Primary Adherent Human Neutrophils

1. Add 25 mm tissue culture plastic coverslips to 6-well tissue culture dishes (*see* **Note 3**).
2. Dilute the neutrophil suspension into RPMI + 10% FBS containing 10 nM human interleukin 8 (IL-8) to a final concentration of 2.5×10^6 neutrophils/ml (*see* **Note 4**).
3. Gently pipet 400 μl of the cell suspension onto the middle of the coverslips, allowing it to form "a dome."
4. Incubate neutrophils for 30 min in a humidified incubator at 37 °C, 5% CO_2.

3.2 Labeling of Bacteria

1. Measure optical density of the bacterial culture. Using a standard curve of enumerated bacterial colony-forming units per optical density reading, set by the researcher's laboratory, pellet 2×10^8 bacteria by centrifugation at 10,000 × *g* for 3 min at room temperature.
2. Resuspend bacteria in 1 ml of PBS containing 5 mM $MgSO_4$.
3. Add 3 μl of TIV solution (50 μM TIV in DMSO).
4. Incubate for 20 min in a water bath at 37 °C.
5. Pellet bacteria by centrifugation at 10,000 × *g* for 3 min at room temperature.
6. Remove and discard supernatant by aspiration.
7. Resuspend pellet in 1 ml PBS containing 5 mM $MgSO_4$.
8. Pellet bacteria by centrifugation at 10,000 × *g* for 3 min at room temperature.
9. Resuspend pellet in 0.5 ml RPMI + 10% FBS.
10. Dilute the bacteria to a final concentration of 1×10^7 per ml in RPMI + 10% FBS (*see* **Note 5**).

3.3 Infection of Neutrophils

1. Remove 6-well plates with adherent neutrophils from the incubator and place them on ice or shipping packs that are prechilled at 4 °C.
2. Add 700 μl of RPMI + 10% FBS per well and incubate in the cold for 5 min.
3. Add 100 μl of the TIV-labeled bacterial suspension per well. Gently swirl the plate to mix.
4. Centrifuge the plates in a refrigerated tabletop centrifuge with swinging bucket rotor and microplate carriers at 600 × *g* for 4 min at 4 °C.
5. Return the plates to the humidified 37 °C, 5% CO_2 incubator and incubate for desired times (*see* **Note 6**).

3.4 Sample Fixation and Collection

1. Add 400 μl of 16% paraformaldehyde solution (final concentration of 4%) and incubate for 10 min at room temperature protected from light (*see* **Note 7**).
2. Remove neutrophils from the coverslips using a cell scraper, and collect them from replicate wells into one 15 ml conical tube per experimental condition.
3. Pellet neutrophils by centrifugation in a tabletop centrifuge with a swinging bucket rotor and conical tube carriers at 600 × *g* for 5 min at room temperature (*see* **Note 8**).
4. Remove and discard supernatant.

5. Resuspend neutrophils in 0.5 ml of PBS and pour through a 70 μm pore size cell strainer to remove any clumps of neutrophils.
6. Add 2.5 ml of PBS, vortex gently, and pellet neutrophils by centrifugation in a tabletop centrifuge with a swinging bucket rotor and conical tube carriers at 600 × *g* for 5 min at room temperature.
7. Remove and discard supernatant.
8. Add 3 ml of PBS, vortex gently, and pellet neutrophils by centrifugation in a tabletop centrifuge with a swinging bucket rotor and conical tube carriers at 600 × *g* for 5 min at room temperature.
9. Remove and discard supernatant.
10. Repeat **steps 8** and **9**.

3.5 Staining Cell-Bound Bacteria

1. Resuspend pelleted neutrophils in 200 μl of 10% normal goat serum in PBS and vortex gently (*see* **Note 9**).
2. Incubate for 15 min at room temperature, protected from light.
3. Pellet neutrophils by centrifugation in a tabletop centrifuge with a swinging bucket rotor and conical tube carriers at 600 × *g* for 5 min at room temperature.
4. Resuspend neutrophils in 100 μl of PBS containing 10% normal goat serum and 1 μg/ml of DL650-conjugated goat anti-*Neisseria gonorrhoeae* antibody (*see* **Note 10**).
5. Incubate for 1 h at room temperature, protected from light.
6. Add 2 ml of PBS, vortex gently, and pellet neutrophils by centrifugation in a tabletop centrifuge with a swinging bucket rotor and conical tube carriers at 600 × *g* for 5 min at room temperature.
7. Remove and discard supernatant.
8. Repeat **steps 6** and 7 (*see* **Note 11**).
9. Resuspend neutrophils in 40 μl of PBS and transfer to 1.5 ml microfuge tubes (*see* **Note 12**).

3.6 Data Acquisition

Analyze samples on the ImageStream^X Mk II imaging flow cytometer with INSPIRE® software. Collect fluorescence data for each fluorophore and brightfield imaging data using the following parameters:

Magnification 60×.

Channel 1 (Ch1) Camera 1 Brightfield 1; LED intensity 38.5 mW; emission collected with 420–480 nm filter.

Channel 6 (Ch6) Camera 1; Darkfield (side scatter, SSC); excitation with 4 mW 785 nm laser; emission collected with 740–880 nm filter.

Channel 7 (Ch7) Camera 2; Excitation with 405 nm laser at 100 mW; emission collected with 420–550 nm filter.

Channel 11 (Ch11) Camera 2; excitation with 642 nm laser at 45 mW; emission collected with 660–740 nm filter.

Channel 9 (Ch9) Brightfield 2; Camera; 2 LED intensity 52.75 mW; emission collected with 570–595 nm filter.

In order to define the compensation matrix, prepare and collect data from single color samples with Brightfield and Darkfield illumination turned off: (1) TIV single color control, containing neutrophils infected with TIV-labeled bacteria and no DL650-conjugated antibody; (2) DL650 single color control, containing neutrophils infected with unlabeled bacteria and stained with DL650-conjugated anti-*Neisseria gonorrhoeae* antibody.

3.7 Data Analysis

1. For data analysis, use IDEAS® v6.2 software (*see* **Note 13**).
2. Create a new compensation matrix.
3. Open one data file and apply the compensation matrix to it.
4. Create a data analysis file by defining populations as outlined in Fig. 1:
   (a) *Singlet Gate*: Generate a scatter plot with Area_M01 on the *X*-axis vs. Aspect Ratio_M01 on the *Y*-axis, and identify single neutrophils by gating on the population of neutrophils with Area_M01 between 85 and 270 and Aspect ratio_M01 between 0.5 and 1 (*see* Fig. 2a).

   (b) *Focus gate*: From the Singlet gate generate a histogram of Brightfield Gradient Root Mean Square (RMS). Identify the focused neutrophil population as neutrophils with RMS value ≥ 50 (*see* **Note 14**) (*see* Fig. 2b).

   (c) From the Singlet and Focus gated population generate a scatter plot Intensity of TIV on the *X*-axis vs. Intensity of DL650 on the *Y*-axis, and gate on the neutrophils with lower DL650 intensity, excluding cells with high DL650 intensity, which are not intact (*see* **Note 15**, Fig. 2c).

   (d) Use the Spot Count Wizard to create a TIV spot count feature (Spot_count_TIV) (*see* **Note 16**).

   (e) Create a histogram for the TIV spot count and gate on (1) the population of neutrophils with 0 TIV spots;

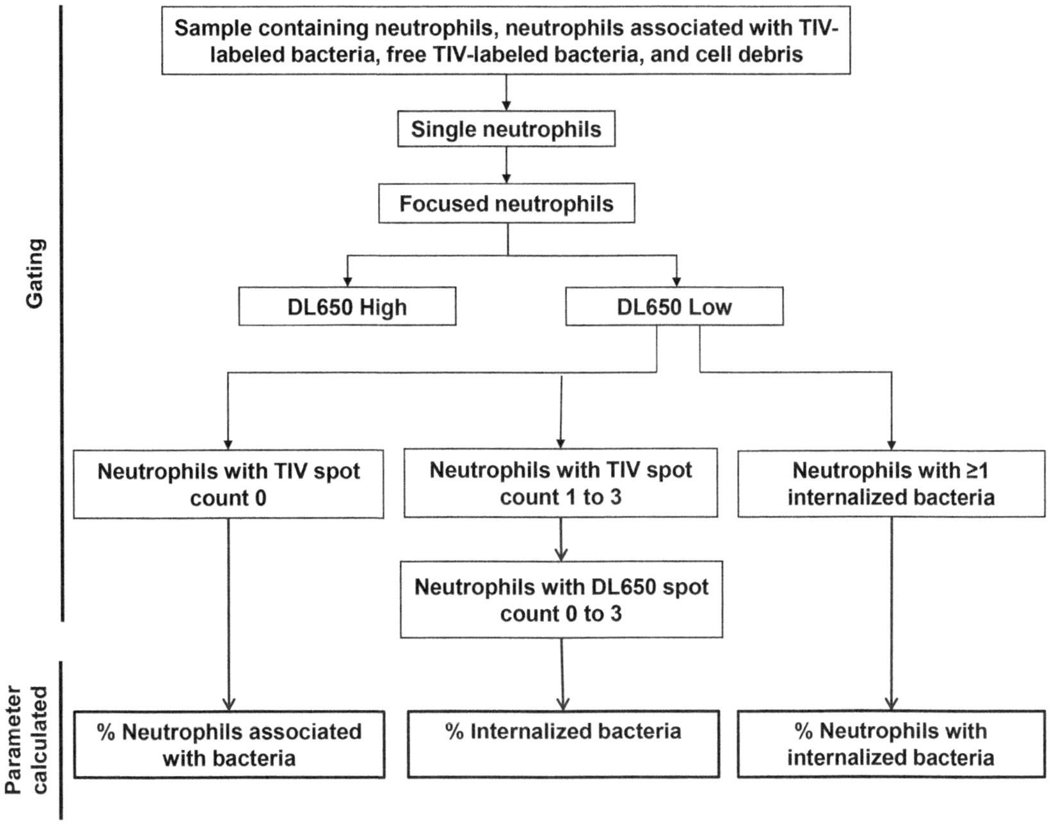

Fig. 1 Gating strategy and data analysis workflow

(2) the population of neutrophils with 1 to 3 TIV spots (*see* **Note 17**) (Figs. 3a and 2d).

(f) Use the Spot Count Wizard to create a DL650 spot count feature (Spot_count_DL650).

(g) Create a histogram for the DL650 spot count for the population of neutrophils with 1–3 TIV spots and gate on the population of neutrophils with 0 to 3 DL650 spots (*see* **Note 18**) (Figs. 3b and 2e).

(h) From the DL650 low neutrophil population (Subheading 3.7, **step 4c**), create an additional plot with Spot_count_-TIV on the *X* axis and Spot_count_DL650 on the *Y* axis.

(i) Identify the population of neutrophils with $\geq$1 internalized bacterium, the population of neutrophils with no TIV spots (0 TIV), and the population of neutrophils with number of DL650 spots higher than TIV spots using the coordinates indicated in Table 1 and Figs. 4a and 2f (*see* **Note 19**).

(j) Save the file as an analysis template.

(k) Apply compensation matrix and this analysis template to the rest of the files for the experiment.

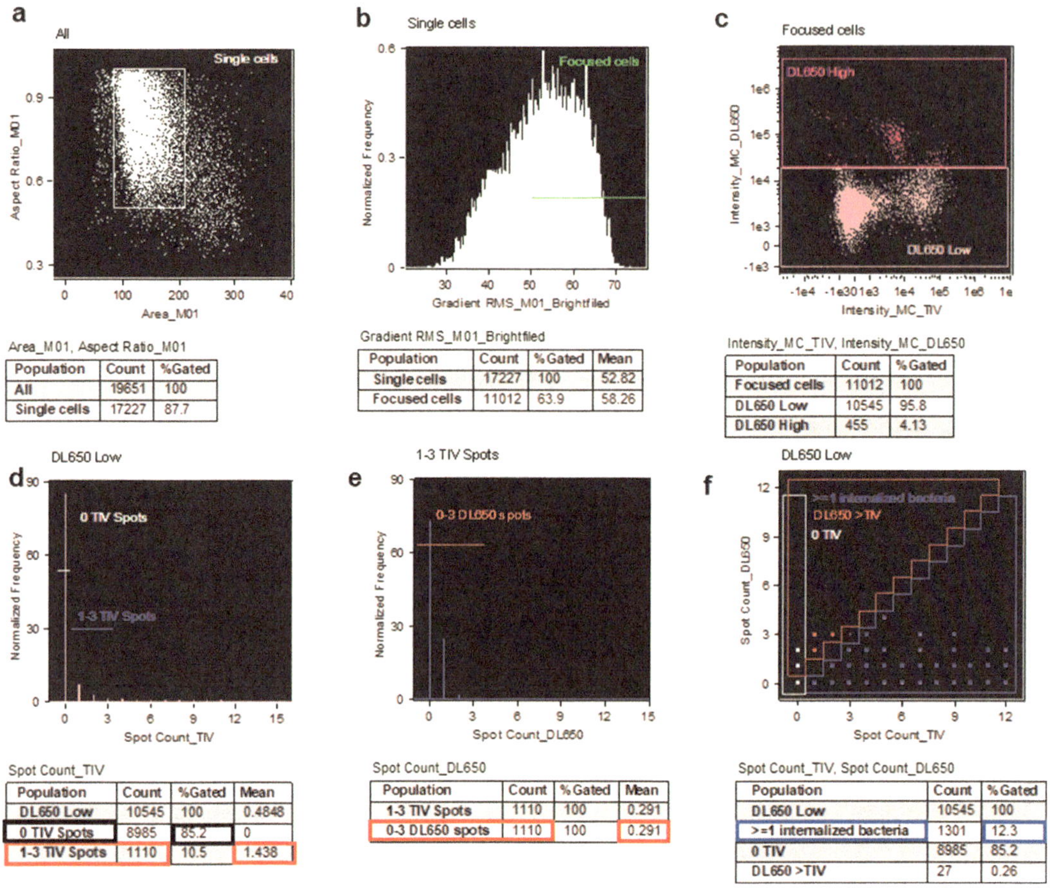

Fig. 2 Experimental data analysis. Data were collected from IL-8 primed, adherent human neutrophils that were infected with Tag-it Violet™ -labeled *Neisseria gonorrhoeae* for 1h, fixed, blocked and stained with anti-*Neisseria gonorrhoeae* specific antibody labeled with DyLight™ 650. Gating was performed as described in Subheading 3.7. TIV spot count is performed on single (a), focused (b), and intact DyLight™ 650 low (c) population of neutrophils. The percent of neutrophils with associated bacteria is 14.8% (d; data used for calculation outlined in black); the percent of internalized bacteria is 79.8% (d, e data used for calculation outlined in red); the percent of neutrophils with internalized bacteria is 12.3% (f; data used for calculation outlined in blue)

3.8 Parameter Calculations

1. Calculate the percent of neutrophils associated with bacteria (this population includes neutrophils with both internalized and bound bacteria) (Fig. 3a and 2d outlined in black):

   % neutrophils associated with bacteria = 100%–% Gated in "0 TIV" gate

2. Calculate the percent of cell-associated bacteria that are internalized within 1–3 TIV gate (Figs. 3 and 2d, e):

   % internalized bacteria = 100 × (number of internalized bacteria/number of total bacteria).

   Number of total bacteria = Count $_{1\text{-}3TIV}$ X Mean$_{1\text{-}3\ TIV}$.

   Count$_{1\text{-}3\ TIV}$ is the number of neutrophils in the 1-3 TIV Spot gate.

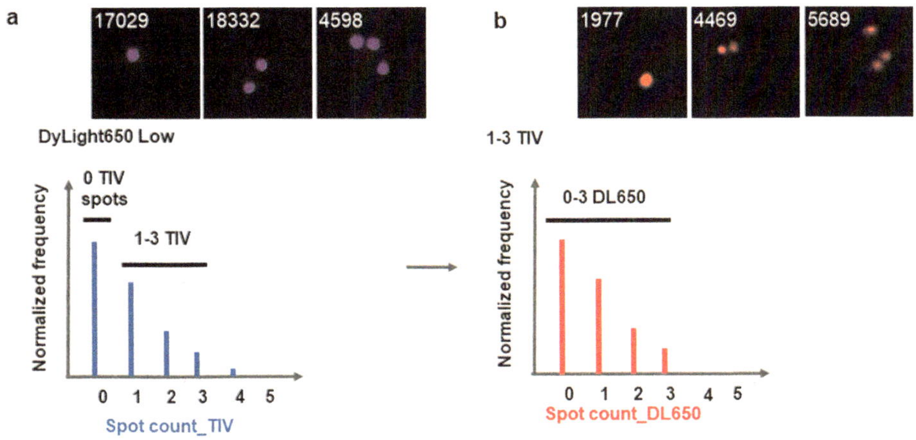

Percent neutrophils associated with bacteria = 100 - Percent gated 0 TIV spots

Number of total bacteria (TIV+ bacteria) = $\text{Count}_{1\text{-}3\ \text{TIV}}$ X $\text{Mean}_{1\text{-}3\ \text{TIV}}$ in 1-3 TIV gate

Number of bound bacteria (DL650+ bacteria) = $\text{Count}_{0\text{-}3\ \text{DL650}}$ X $\text{Mean}_{0\text{-}3\ \text{DL650}}$ in 0-3 DL650 gate

Number of internalized bacteria = Number of total bacteria – Number of bound bacteria

Percent of internalized bacteria = 100 X (Number of internalized bacteria/ Number of total bacteria)

Fig. 3 Gating strategy, representative images, and calculations of the percent of neutrophils with associated bacteria and the percent of internalized bacteria using TIV (a) and DL650 (b) Spot count features. Images are examples of neutrophils with 1, 2, and 3 TIV(a) or DL650 (b) spots

Table 1
Coordinates for gates described in Fig. 4

Neutrophils with no bacteria (0TIV)												
X	−0.5	0.5	−0.5	0.5								
Y	−0.5	−0.5	5.5	5.5								
Neutrophils with ≥1 internalized bacteria												
X	0.5	0.5	1.5	1.5	2.5	2.5	3.5	3.5	4.5	4.5	5.5	5.5
Y	−0.5	0.5	0.5	1.5	1.5	2.5	2.5	3.5	3.5	4.5	4.5	−0.5
Neutrophils with number of DL650 spots higher than number of TIV spots(DL650 > TIV)												
X	−0.5	0.5	0.5	1.5	1.5	2.5	2.5	3.5	3.5	4.5	4.5	−0.5
Y	0.5	0.5	1.5	1.5	2.5	2.5	3.5	3.5	4.5	4.5	5.5	5.5

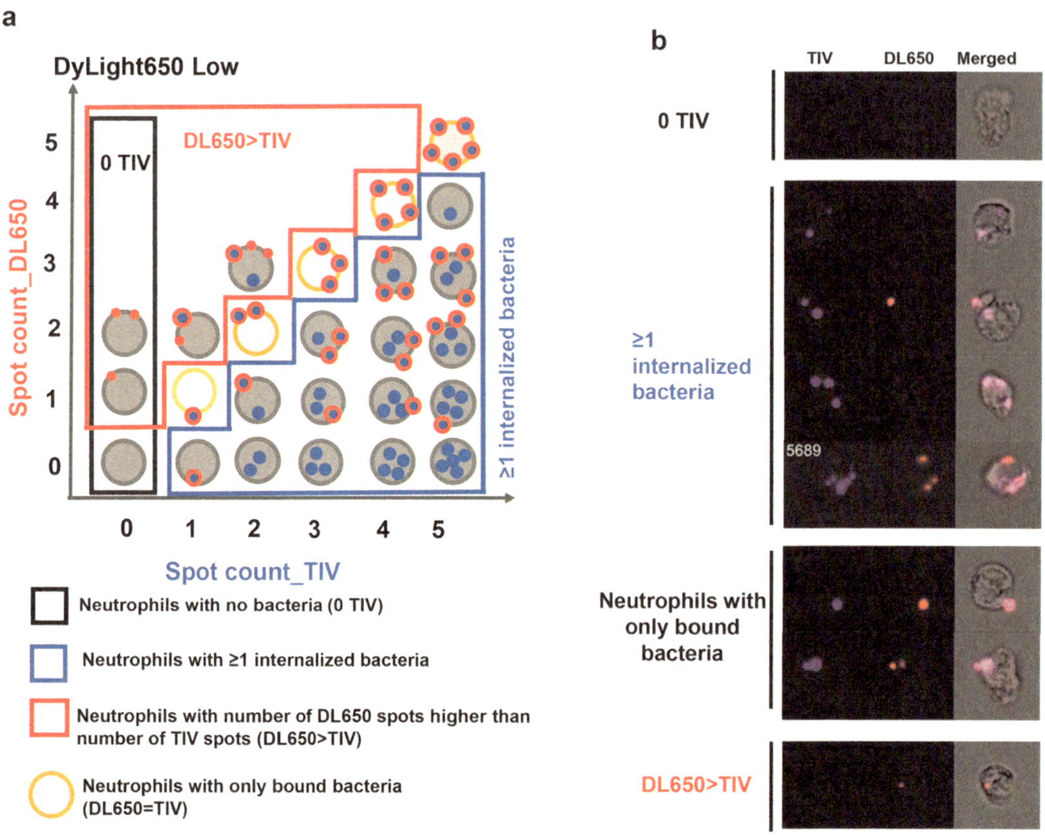

Fig. 4 Gating strategy for calculating the percent of neutrophils with internalized bacteria. (**a**) Schematic representation and (**b**) representative images of cells in the four outlined populations: neutrophils with 0 TIV spots; neutrophils with ≥1 internalized bacteria; neutrophils with the number of TIV spots equal to the number of DL650 spots, and neutrophils with the number of DL650 spots higher than the number of TIV spots

$Mean_{1\text{-}3\ TIV}$ is the mean number of spots per neutrophil in the 1-3 TIV gate (Figs. 3a and 2d outlined in red).

Number of bound bacteria = $Count_{0\text{-}3\ DL650}$ X $Mean_{0\text{-}3\ DL650}$.

$Count_{0\text{-}3\ DL650}$ is the number of neutrophils in the 0-3 DL650 gate.

$Mean_{0\text{-}3\ DL650}$ is the mean number of spots per neutrophil in the 0-3 DL650 gate (Figs. 3b and 2e outlined in red).

Number of internalized bacteria = number of total bacteria–number of bound bacteria

3. Calculate the percent of neutrophils with internalized bacteria in neutrophils within the DL650 Low population (Fig. 2f outlined in blue):

   % neutrophils with internalized bacteria = % in the ≥1 internalized bacteria gate (*see* **Note 20**).

4 Notes

1. We purify neutrophils using the protocol as described by Stohl et al. [18]. The neutrophil purity is ≥90%. Purified neutrophils are resuspended in DPBS pH 7.4 and kept on ice to prevent activation. Any other protocol for purification of intact quiescent human neutrophils, including those published in this book, will be suitable for the procedure.
2. In this protocol we used *Neisseria gonorrhoeae* as a model organism. The bacteria were grown in rich liquid media with sequential dilution to ensure maximum bacterial viability at the time of infection, as described previously [20].
3. We observe better adherence of neutrophils to plastic coverslips compared to tissue culture plastic.
4. This system seeks to model the state of extravasated and transmigrated neutrophils in vivo, where they are exposed to the neutrophil chemoattractant and priming agent IL-8 [21]. We found that adding 1 to 3 × 10^6 neutrophils per 25-mm diameter coverslip results in an optimal density of neutrophils. Neutrophils plated in higher density tend to clump during the course of the experiment. Normally 2 to 3 coverslips per condition are required to ensure there are a sufficient number of neutrophils for analysis (*see* **Note 12**).
5. The protocol for labeling bacteria with Tag-it™ Violet was adapted from the manufacturer's protocol for labeling eukaryotic cells. Incubating *Neisseria gonorrhoeae* in PBS pH 7.4 with 5 mM $MgSO_4$ prevents bacterial autolysis. The labeling cannot be performed in RPMI + 10% FBS. We did not test any other buffers or solutions besides those described in this protocol. To date multiple strategies for labeling bacteria and other particles have been used and validated for imaging flow cytometry [10–12, 14, 16]. The use of Tag-it Violet™ is advantageous for the following reasons: (1) It is excited with a violet laser at 405 nm and emits at 455 nm. This fluorescence profile provides flexibility in using other channels for multiplex analysis with additional markers; (2) unlike the other violet spectrum dye, DAPI, Tag-it Violet™ covalently binds to bacterial proteins and does not diffuse out of the bacteria, producing minimal background in host cells. Here we use a multiplicity of infection (MOI) equal to 1 *Neisseria gonorrhoeae* bacterium per neutrophil. The optimal MOI or particle-to-neutrophil ratio should be determined experimentally (*see* **Note 17**).
6. Prechilling neutrophils and the centrifugation step after addition of bacteria helps to synchronize the infection. We do not change the media following centrifugation since the neutrophils are loosely attached under these conditions. If the

potential for loss of neutrophils is not a concern, investigators may choose to change the media after **step 4** to remove any bacteria that are not cell associated.

7. Adding paraformaldehyde to wells without removing the incubation medium is a particularly useful approach if lifting of neutrophils during the course of the experiment is a concern. Cell fixation prior to scraping also prevents dissociation of bound bacteria from the neutrophils during subsequent processing steps. The 16% paraformaldehyde solution should be aliquoted and stored at −20 °C to avoid freeze-thaw.
8. The shape of the 15-ml conical tubes and spinning in the swing bucket rotor allows for maximum recovery of the neutrophils during washing and staining steps.
9. Normal goat serum is used to block nonspecific interaction of goat IgG with the human neutrophils to minimize background staining with the goat anti-*Neisseria gonorrhoeae* antibody used next. The choice of blocking solution depends on the investigator's staining method.
10. Additional staining with antibodies for neutrophil surface markers can be performed at this step.
11. If staining with intracellular markers required neutrophils can be permeabilized and stained after Subheading 3.5, **step 8**.
12. The recommended concentration of cells for imaging flow cytometry analysis is 2×10^7 cells per ml in a minimum volume of 20 μl, or a minimum of 4×10^5 cells. When planning the experiment, we account for some loss of cells during sample collection and processing. Normally, two 25-mm coverslips per condition as described in **Note 4** yields a sufficient number of cells for analysis.
13. Data analysis manuals and tutorials can be found on the Amnis® Customer portal.
14. Gating on population based on the additional staining with neutrophil surface intracellular markers could be performed after this step (*see* **Notes 10** and **11**).
15. Neutrophils with high cytoplasmic DL650 staining are assumed to not be intact and are excluded from downstream analysis by this gating strategy.
16. Spot_Count features are based on the masks defining the individual spots. The software calculates Spot_Count features based on a manually assigned truth population, containing neutrophils with high and low spot count for each fluorophore. The accuracy of the resulting mask should be verified by viewing populations of the neutrophils with different numbers of spots. If required, redefining of the truth population or manual adjustment of the spot count mask can be performed.

17. In our experience, the most accurate identification of individual spots was with 1 to 3 spots per neutrophil. The MOI should be determined experimentally to ensure that most neutrophils will fall into this range. The sensitivity and specificity of the antibody used to detect bound bacteria must also be determined experimentally. In our experiments, 99% of bacteria are recognized by the antibody.
18. Using an antibody with high specificity for the bacteria and at the lowest possible concentration for detection ensures that the DL650-positive spots are colocalized with TIV-positive spots, for accurate analysis of phagocytosis. If masking is incorrect, the number of DL650-positive spot will be greater than the number of TIV positive spots per cell. Neutrophils in the 0, 1, 2, 3…spot count bins should be visually inspected, and the mask should be corrected either manually or through redefining the truth population. Representative neutrophils for each population are shown in Figs. 3 and 4b.
19. Here, TIV and DL650 spot counts are performed on all intact single focused neutrophils to determine the percent of neutrophils with internalized bacteria. Figure 4 shows a schematic of a gate that includes the population of neutrophils with up to 5 spots per cell. The gate could be extended to include all neutrophils, as shown in Fig. 2f. The percent of neutrophils where the spot count for DL650 exceeds the spot count for TIV will correspond to an analysis error. In our experience $\leq$5% of cells are in this category (*see* **Note 16** and [14]).
20. The analysis strategy should be validated in experimental conditions known to affect bacterial binding and/or internalization by neutrophils. Conditions may include infection of prefixed or dead neutrophils, treatment with inhibitors of phagocytosis, and infection at low temperature [14].

Acknowledgments

We thank Lacie Werner for technical assistance with the experiment presented in Fig. 2. This work has been supported by R01 AI097312 to A.K.C. and NIH SIG-1S10RR031633 for the ImageStream^{X} Mk II (T. Bender).

References

1. Elliott MR, Ravichandran KS (2010) Clearance of apoptotic cells: implications in health and disease. J Cell Biol 189:1059–1070
2. Fond AM, Ravichandran KS (2016) Clearance of dying cells by phagocytes: mechanisms and implications for disease pathogenesis. In: Gregory CD (ed) Apoptosis in cancer pathogenesis and anti-cancer therapy: new perspectives and opportunities. Springer International Publishing, Cham, pp 25–49. https://doi.org/10.1007/978-3-319-39406-0_2
3. Lim JJ, Grinstein S, Roth Z (2017) Diversity and versatility of phagocytosis: roles in innate immunity, tissue remodeling, and homeostasis. Front Cell Infect Microbiol 7:191–191
4. Sarantis H, Grinstein S (2012) Subversion of phagocytosis for pathogen survival. Cell Host Microbe 12:419–431
5. Hampton MB, Winterbourn CC (1999) Methods for quantifying phagocytosis and bacterial killing by human neutrophils. J Immunol Methods 232:15–22
6. Jersmann HP, Ross KA, Vivers S et al (2003) Phagocytosis of apoptotic cells by human macrophages: analysis by multiparameter flow cytometry. Cytometry A 51:7–15
7. Lehmann AK, Sornes S, Halstensen A (2000) Phagocytosis: measurement by flow cytometry. J Immunol Methods 243:229–242
8. Criss AK, Katz BZ, Seifert HS (2009) Resistance of *Neisseria gonorrhoeae* to non-oxidative killing by adherent human polymorphonuclear leucocytes. Cell Microbiol 11:1074–1087
9. Agerer F, Waeckerle S, Hauck CR (2004) Microscopic quantification of bacterial invasion by a novel antibody-independent staining method. J Microbiol Methods 59:23–32
10. Haridas V, Ranjbar S, Vorobjev IA et al (2017) Imaging flow cytometry analysis of intracellular pathogens. Methods 112:91–104
11. Johansson J, Karlsson A, Bylund J et al (2015) Phagocyte interactions with *mycobacterium tuberculosis*--simultaneous analysis of phagocytosis, phagosome maturation and intracellular replication by imaging flow cytometry. J Immunol Methods 427:73–84
12. Phanse Y, Ramer-Tait AE, Friend SL et al (2012) Analyzing cellular internalization of nanoparticles and bacteria by multi-spectral imaging flow cytometry. J Vis Exp:e3884
13. Ploppa A, George TC, Unertl KE et al (2011) ImageStream cytometry extends the analysis of phagocytosis and oxidative burst. Scand J Clin Lab Invest 71:362–369
14. Smirnov A, Solga MD, Lannigan J et al (2015) An improved method for differentiating cell-bound from internalized particles by imaging flow cytometry. J Immunol Methods 423:60–69
15. Kuhn J, Smirnov A, Criss AK et al (2019) Quantifying CEACAM targeted liposome delivery using imaging flow cytometry. Mol Pharm 16(6):2354–2363
16. Smirnov A, Solga MD, Lannigan J et al (2017) High-throughput particle uptake analysis by imaging flow cytometry. Curr Protoc Cytom 80:11.22.11–11.22.17
17. Boyum A (1968) Isolation of mononuclear cells and granulocytes from human blood. Isolation of monuclear cells by one centrifugation, and of granulocytes by combining centrifugation and sedimentation at 1 g. Scand J Clin Lab Invest Suppl 97:77–89
18. Stohl EA, Criss AK, Seifert HS (2005) The transcriptome response of *Neisseria gonorrhoeae* to hydrogen peroxide reveals genes with previously uncharacterized roles in oxidative damage protection. Mol Microbiol 58:520–532
19. Kellogg DS Jr, Peacock WL Jr, Deacon WE et al (1963) *Neisseria gonorrhoeae*. I. Virulence genetically linked to clonal variation. J Bacteriol 85:1274–1279
20. Criss AK, Seifert HS (2008) *Neisseria gonorrhoeae* suppresses the oxidative burst of human polymorphonuclear leukocytes. Cell Microbiol 10:2257–2270
21. Stevens JS, Criss AK (2018) Pathogenesis of *Neisseria gonorrhoeae* in the female reproductive tract: neutrophilic host response, sustained infection, and clinical sequelae. Curr Opin Hematol 25:13–21

Chapter 11

Visualization and Quantification of Phagocytosis by Neutrophils

Gaelen Guzman and Fikadu G. Tafesse

Abstract

Phagocytosis by phagocytes such as neutrophils is a crucial part of the host innate immune response against invading pathogens. Phagocytosis is a complex process that initiates with the binding of the particles on the cell surface of the phagocytes through the interaction of pattern recognition receptors with ligands on the surface of the pathogens. During this process, phagocytes undergo extensive membrane reorganization and cytoskeleton rearrangement at their cell surface. To gain better insight about the molecular mechanisms of this dynamic cellular process, visualization and quantification in a high-throughput manner is essential. Here, we describe a microscope-based method to visualize and quantify phagocytic uptake of pathogens (such as bacteria and fungi) and model particulates that are larger than 0.5 μm (such as Zymosan A and IgG-coated beads).

Key words Neutrophils, Phagocytes, Phagocytosis, High-content imaging, High-throughput, Phenotypic analysis

1 Introduction

Neutrophils represent a first line of defense when the body encounters a pathogen. These cells are capable of rapidly releasing chemotaxis to sites of bodily damage—they are some of the fastest cells in the mammalian body, and are commonly referred to as "the first responders" upon the initiation of an inflammatory signal [1, 2]. Once at an inflamed site, neutrophils are highly versatile in their antipathogenic response: they produce cytokines to induce a hostile environment for invading pathogens, they play roles in the activation and direction of the adaptive immune response, they modulate inflammation, and they directly act with potent microbicidal effect [1–5]. Understanding the mechanisms through which neutrophils respond to pathogenic invasion can give insight into how an infection is resolved—and understanding these mechanisms may allow for the implementation of therapeutic tools to enhance or direct the neutrophil response.

Mark T. Quinn and Frank R. DeLeo (eds.), *Neutrophil: Methods and Protocols*, Methods in Molecular Biology, vol. 2087, https://doi.org/10.1007/978-1-0716-0154-9_11,

Of particular importance, neutrophils wield phagocytosis to destroy would-be pathogens, as well as effete cells and debris [6]. Phagocytosis is a form of receptor-mediated cellular uptake that allows a cell to ingest particles greater than 0.5 μm in size [7]. This cellular process has been well studied, but there remain open questions regarding the mechanisms and signaling events that drive the dramatic changes in membrane reorganization, cell morphology, cytoskeleton remodeling, and membrane lipid composition that must ensue for the complete ingestion of a pathogen or particulate. A host of techniques allow for the assessment of phagocytic capacity, but many are limited to gross overviews of internalization (e.g., antibiotic protection assays) or lack the speed necessary for high-throughput phenotypic screens.

Here, we present a methodology for rapidly visualizing and quantifying the phagocytic capacity of neutrophils using high-content imaging microscopy (Fig. 1). As the HL-60 neutrophil-like cell line is a classical staple of neutrophil study [2, 8, 9], we focus on this line herein—however, these techniques are equally applicable to the study of primary neutrophils isolated from mouse or human peripheral blood. Similarly, any number of fluorophore-conjugated model particle or fluorescent reporter strain of pathogenic microbe may be used under this paradigm [10, 11]. Through the use of fluorescent phagocytic particles and a cell membrane stain, one may rapidly acquire visual data sufficient to calculate the rate of phagocytic uptake of many hundreds or thousands of cells across experimental conditions. Combining a well-plate format and automated, high-content image acquisition, this technique allows for high-throughput phenotypic analysis following genetic or chemical screens.

2 Materials

1. HL-60 peripheral blood promyeloblasts (ATCC; *see* **Note 1**).
2. Growth medium: RPMI 1640 supplemented with 10% fetal calf serum, nonessential amino acids, and optional antibiotics/antimycotics.
3. Phosphate-buffered saline (PBS).
4. Differentiation medium: Growth medium supplemented with 1.2% DMSO (v/v; *see* **Note 1**).
5. Imaging surface:
   (a) Option 1: Senso Microplate glass-bottom 96-well plate (Greiner Bio-One).
   (b) Option 2: Microscope coverslips(1.2 cm diameter) and standard 24-well plate.

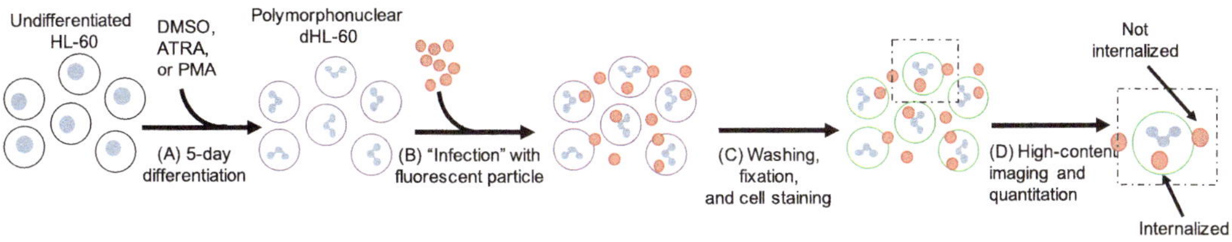

Fig. 1 An overview of the experimental workflow described in this methodology. (**a**) Induce polymorphonuclear differentiation of cultured HL-60 cells using 5-day treatment with one of the following: DMSO, ATRA, or PMA. Plate on polylysine-coated imaging surface. (**b**) Add fluorescent phagocytic particles to cells at a multiplicity of infection of approximately 1:10. (**c**) Wash unbound particles, fix using 4% paraformaldehyde, and stain using fluorescent cell boundary marker and nuclear stain. (**d**) Collect high-content images of "infected" cells using fluorescence microscopy. Uptake rate = (Number of InternalizedParticles)/(Number of NucleatedCells)

6. 20× poly-L-lysine coating solution: 50 mg poly-L-lysine in 50 mL diH_2O, filter through a 0.22 μm filter. Dilute in H_2O for a 1× working stock.
7. Model phagocytic particles (*see* **Note 2**):
   (a) Option 1: Zymosan A (*S. cerevisiae*) Bioparticles™ Alexa Fluor™ conjugate (Thermo Fisher Scientific; multiple fluorescence choices).
   (b) Option 2: Fluoro-Max™ Fluorescent Carboxylate-Modified Particles (Thermo Fisher Scientific).
   (c) Option 3: *Mycobacterium tuberculosis* H37Rv (ATCC 25618) expressing mCherry reporter (or another fluorescent reporter microbe).
8. 4% paraformaldehyde solution
9. Permeabilization/blocking buffer: 1% bovine serum albumin (w/v) and 0.1% Triton-X100 (v/v) in PBS.
10. Cell surface stain (*see* **Note 3**): Phalloidin Alexa Fluor™ (e.g., Thermo Fisher Scientific; multiple fluorescence options) in permeabilization/blocking buffer at approximately 1:50 dilution (titration may be necessary).
11. Nuclear stain (*see* **Note 3**): DAPI (4′,6-diamidino-2-phenylindole) in PBS at approximately 1μg/mL (titration may be necessary).

3 Methods

3.1 Differentiation of HL-60 Cells into Granulocytic Neutrophils

1. Grow HL-60 cells in growth media using standard tissue culture techniques. When cultured cells reach a density of approximately 1×10^6 cells/mL, replace growth media with an equal volume of differentiation media (*see* **Note 1**).
2. Incubate cells at 37 °C and 5% CO_2 for at least 5 days to obtain fully differentiated HL-60 (dHL-60) cells (*see* **Note 4**).

3.2 Immobilization of dHL-60 Cells to Glass-Bottom Plate or Coverslip

1. Coat desired imaging surface (96-well plate, coverslips, etc.) according to steps below (*see* **Note 5**) [12].
2. Add 1× polylysine solution to sterile multiwell plates or coverslips and incubate at 37 °C for 1 h.
3. Aspirate polylysine solution and leave to dry. Plates and coverslips may be kept at 4 °C.
4. From cultured dHL-60 cells, isolate an appropriate number of cells to your preferred imaging surfaces (*see* **Notes 6** and **7**).
5. Resuspend cells in the minimum volume to cover surface of the well (approximately 50 μL for 96-well and 250 μL for 24-well plate), and add cells to the appropriate wells.
6. Centrifuge cells in plate for 5 min at 250 × *g*, and allow the cells to recover in differentiation media at 37 °C and 5% CO_2 for at least 1 h (or overnight) to maximize immobilization.

3.3 Phagocytosis of Phagocytic Particles by dHL-60

1. Isolate an appropriate number of phagocytic particles for a multiplicity of infection (MOI) of approximately ten particles per cell.
2. Resuspend in growth medium or PBS at the minimum volume to cover the surface of the well (approximately 50 μL or 250 μL for 96- and 24-well respectively).
3. Particle solution may be sonicated for 5–10 min in a water bath or passed through a narrow-gauge syringe needle 10–20 times to reach a single-particle suspension.
4. Aspirate growth media from cells affixed to well-plate, wash gently with PBS, and add resuspended particle solution to the cells. Centrifuge plate for 1 min at 250 × *g* to sediment the particles to the bottom of the well.
5. Incubate cells and particles at 37 °C and 5% CO_2 for 10 min (*see* **Note 8**).
6. Gently wash cells twice with PBS to remove unbound particles.
7. Fix cells using ice-cold 4% paraformaldehyde solution for 30 min on ice. After fixation, remove formaldehyde solution and hydrate wells with PBS. Plate may be stored at 4 °C for up to a week after fixation, although it is recommended to proceed with staining as early as possible.

3.4 Fluorescent Cell Staining

1. If using phalloidin (*see* **Note 3**): aspirate PBS, and treat cells for 15 min with permeabilization/blocking buffer.
2. Remove permeabilization/blocking buffer, add cell surface stain, and incubate for approximately 1 h at room temperature under cover from light (*see* **Note 9**).
3. Remove cell surface stain solution and wash twice with PBS.

4. Stain nuclear DNA using nuclear staining solution for 10 min at room temperature under cover from light (*see* **Note 3**).
5. Remove nuclear staining solution and wash twice with PBS. Leave cells in PBS under cover from light until imaging is completed.
6. If using a 96-well plate, imaging may be performed with no further preparation. If using glass coverslips, follow the steps below to mount the slips to microscopy slides [13].
7. On clean microscopy slide, pipette a single droplet (~15 μL) of antifade mounting medium and carefully place coverslip cell-side down on top of this droplet, taking care to minimize air bubbles.
8. It is recommended to use a lacquer to glue down the coverslips and prevent drying. Standard clear nail polish is a sufficient affixation agent.
9. Allow lacquer to fully dry before imaging.

3.5 Image Acquisition

1. Regardless of acquisition setup (*see* **Note 10**), optimize parameters for excitation and exposure times for the respective fluorophores used in the experimental workflow.
2. We recommend image collection at 10× or 20× magnification to maximize the total number of cells captured per image. It is best to capture at least 500 cells across the majority of the well or coverslip per technical replicate per experimental condition.
3. For optimal quantification, images should be collected in a plane such that the nucleus, cell boundary, and any internalized beads are all clearly distinguishable across the relevant channels with minimal background and inter-channel fluorescence.

3.6 Image Analysis (Quantification)

1. To analyze images, identify the total number of cells captured per image and the number of fluorescent particles fully internalized (*see* **Note 11**).
2. If using a manual counting method, count the number of nuclei cells and internalized beads per image.
3. If using an automated or semiautomated counting method, define threshold parameters for each fluorescent channel, and apply the same parameters for every experimental condition within a biological replicate.
4. Calculate rate of phagocytic uptake by aggregating the total number of internalized particles and dividing by the total number of isolated cells for every technical replicate across every experimental condition. Technical replicates may be aggregated to report a single uptake rate for every biological replicate.

5. Normalization between biological replicates may be necessary to compensate for inter-experimental variability.
6. Uptake rates may be reported as box- or dot-plots and differences between experimental condition may be assessed using a two-tailed, unpaired Student's T-test.
7. Alternatively, one may depict uptake rate using a histogram or violin-plot to represent the number of beads internalized per cell in order to visualize the distribution of uptake events across experimental conditions.
8. Differences between experimental conditions may be assessed using a discrete nonparametric test, such as the Chi-squared test or the Kolmogorov–Smirnov Test.

4 Notes

1. HL-60 cells are derived from peripheral blood leukocytes isolated from a patient with acute promyelocytic leukemia, and can be differentiated into granulocytic neutrophil-like cells using 5-day treatments with DMSO, all trans-retinoic acid (ATRA), or phorbol 12-myristate 13-acetate (PMA) [2, 8, 14, 15]. It is advisable to titrate concentrations of activating agent to identify the optimal differentiation conditions in one's own laboratory. Spontaneous, nongranulocytic differentiation can occur in the absence of activating reagents, and it is recommended to limit passage number to avoid spurious differentiation [8].
2. One may select from a range of model phagocytic particles. This methodology requires only that the selected model particle is fluorophore-conjugated/labeled. Common models include: Zymosan A (which consists of a fungal glycan that acts as ligand to the pathogen recognition receptor Dectin-1), opsonized and nonopsonized silica beads (which act as ligands for FCγR and CR3, respectively), and live or heat-inactivated fluorescent reporter strains of bacteria such as *Salmonella enterica* Serovar Typhimurium or *Mycobacteria tuberculosis* (which are both decorated with ligands for a host of pathogen recognition receptors) [16, 17].
3. It is necessary to select cellular and nuclear stains with compatibility with the fluorescently labeled model phagocytic particle. We recommend phalloidin, a fungus-derived toxin that effectively stains filamentous actin. Phalloidin may be purchased as a fluorophore-conjugated cell boundary marker in multiple ranges of excitation/emission.
4. Differentiated HL-60 cells will remain in suspension, and previous reports suggest that dHL-60 cells will die through

apoptosis beginning on day 8 of treatment [14, 15]. If performing genetic or chemical screens, it is recommended that transfection/transduction or treatment begin during the 5-day differentiation period such that cells are prepared for phagocytosis analysis by day 5.

5. Poly-L-lysine coating may be used to affix dHL-60 cells to imaging glassware. Such coated products are available for purchase (*see* Subheading 2), or may be prepared in-house as described previously [12].
6. We recommend seeding approximately 2×10^4 cells when using a 96-well glass-bottom imaging plate, and approximately 5×10^4 cells when using 24 mm diameter coverslips in a standard 24-well plate.
7. It is highly recommended to perform this uptake assay with multiple technical replicates, with differentiation and uptake assays performed on independent days for each biological replicate. Experimental conditions should be assayed in parallel for every biological replicate, as significant variation may occur between replicates.
8. Exact uptake time may need to be optimized: 10 min may be too long if cells are highly activated, and one is characterizing a weak phenotype, or too short if the model phagocytic particle is taken up poorly. It is recommended to optimize this timing using a three- or four-point time course on wild-type cells.
9. As with many staining protocols, it may be necessary to modulate both the concentration of staining reagent and the time to maximize signal and minimize background.
10. Images for phagocytic analysis may be collected on a range of acquisition platforms—so long as the microscope has the excitation/detection capacities to capture the fluorophores used, it ought to be sufficient for this methodology. We recommend the KEYENCE BZ-X700 all-in-one fluorescence microscope (KEYENCE) due to its capacity for semiautomated image acquisition and multifield image stitching.
11. A number of image analysis software tools with compatible cell counting protocols are available, including CellProfiler (open access), FIJI (open access), Imaris (Oxford Instruments), and KEYENCE BZ-X Analyzer (KEYENCE). It is notable that a semiautomated cell counting tool is preferable to manual counting for enhanced throughput and reduced user bias.

References

1. Aulakh GK (2018) Neutrophils in the lung: the first responders. Cell Tissue Res 371:577–588
2. Hauert AB, Martinelli S, Marone C, Niggli V (2002) Differentiated HL-60 cells are a valid model system for the analysis of human neutrophil migration and chemotaxis. Int J Biochem Cell Biol 34:838–854
3. Cassatella MA (2017) Human mature neutrophils as atypical APC. Blood 129:1895–1896
4. Lee WL, Harrison RE, Grinstein S (2003) Phagocytosis by neutrophils. Microbes Infect 5:1299–1306
5. Koup RA, Liang F, Loré K et al (2017) Neutrophils acquire the capacity for antigen presentation to memory CD4 + T cells in vitro and ex vivo. Blood 129:1991–2001
6. Herant M, Heinrich V, Dembo M (2006) Mechanics of neutrophil phagocytosis: experiments and quantitative models. J Cell Sci 119:1903 LP–1901913
7. Stuart LM, Ezekowitz RAB (2005) Phagocytosis: elegant complexity. Immunity 22:539–550
8. Fleck RA, Romero-Steiner S, Nahm MH (2005) Use of HL-60 cell line to measure opsonic capacity of pneumococcal antibodies. Clin Vaccine Immunol 12:19–27
9. Collins SJ (1987) The HL-60 promyelocytic leukemia cell line: proliferation, differentiation, and cellular oncogene expression. Blood 70:1233–1244
10. Tafesse FG, Rashidfarrokhi A, Schmidt FI et al (2015) Disruption of Sphingolipid biosynthesis blocks phagocytosis of *Candida albicans*. PLoS Pathog 11:1–27
11. Rashidfarrokhi A, Richina V, Tafesse FG (2017) Visualizing the early stages of phagocytosis. J Vis Exp 120:54646
12. Polylysine-coated tissue culture surfaces. In: Protoc. https://www.protocolsonline.com/recipes/stock-solutions/polylysine-coated-tissue-culture-surfaces/. Accessed 20 May 2019
13. Protocol for the Preparation and Fluorescent ICC Staining of Cells on Coverslips. https://www.rndsystems.com/resources/protocols/protocol-preparation-and-fluorescent-icc-staining-cells-coverslips. Accessed 20 May 2019
14. Martin SJ, Bradley JG, Cotter TG (1990) HL-60 cells induced to differentiate towards neutrophils subsequently die via apoptosis. Clin Exp Immunol 79:448–453
15. Millius A, Weiner OD (2010) Manipulation of neutrophil-like HL-60 cells for the study of directed cell migration. Methods Mol Biol 591:147–158
16. Futosi K, Fodor S, Mócsai A (2013) Reprint of neutrophil cell surface receptors and their intracellular signal transduction pathways. Int Immunopharmacol 17:1185–1197
17. Agramonte-Hevia J (2002) Gram-negative bacteria and phagocytic cell interaction mediated by complement receptor 3. FEMS Immunol Med Microbiol 34:255–266

Chapter 12

Analysis of Neutrophil Bactericidal Activity

Nicholas J. Magon, Heather A. Parker, Louisa V. Ashby, Reuben J. Springer, and Mark B. Hampton

Abstract

This chapter describes three methods for measuring the bactericidal activity of neutrophils. All utilize colony counting techniques to quantify viable bacteria. A simple "one-step" protocol provides a composite measure of phagocytosis and killing, while a "two-step" protocol that separates extracellular and intracellular bacteria allows calculation of rate constants for both of these processes. We also present a method for selectively monitoring the long-term survival of bacteria within the phagosome. This may have application in identifying resistant strains and searching for compounds that sensitize pathogens to destruction.

Key words Neutrophil, Phagosome, Bacteria, Bactericidal activity, *Staphylococcus aureus*

1 Introduction

Neutrophils are the immune system's key defenders against bacterial infection, their primary function being to destroy invading pathogens. One of the ways they achieve this is by engulfing the pathogen into an intracellular compartment, the phagosome, then subjecting it to an array of both oxygen-dependent and oxygen-independent killing mechanisms. These involve the release of antimicrobial and proteolytic peptides and proteins from intracellular granules into the phagosomal space, along with the production of reactive oxygen species by an NADPH oxidase complex that assembles in the membrane. Release of the granule enzyme myeloperoxidase (MPO) into the phagosome results in production of the potent antimicrobial agent hypochlorous acid [1]. Despite such a harsh environment, some microbes are able to survive this assault [2]. It may be possible to develop compounds that sensitize

Electronic supplementary material: The online version of this chapter (https://doi.org/10.1007/978-1-0716-0154-9_12) contains supplementary material, which is available to authorized users.

Nicholas J. Magon and Heather A. Parker contributed equally to this work.

Mark T. Quinn and Frank R. DeLeo (eds.), *Neutrophil: Methods and Protocols*, Methods in Molecular Biology, vol. 2087, https://doi.org/10.1007/978-1-0716-0154-9_12,

pathogens to destruction within the phagosome. Accurate methods for quantifying neutrophil bactericidal activity are important for this effort.

Bactericidal activity is measured as the loss in viability of bacteria cocultured with neutrophils. The most common method for assessing viability is measuring the ability of bacteria to form colonies on nutrient agar. Alternative methods are available; fluorescent dyes such as acridine orange can distinguish viable from dead bacteria by intercalating into the less structurally organized DNA of the dead bacteria, while DiOC2 is a positively charged lipophilic green fluorescent dye that accumulates in cells or organelles maintaining a negative membrane potential [3, 4]. In this chapter, we focus on colony-forming methodology, using *Staphylococcus aureus* for illustrative purposes. The methods are applicable to other species, but it is important to note that each bacterial species requires its own method optimization and controls.

Three separate procedures are described. In the "one-step" protocol, neutrophils and bacteria are incubated together and the numbers of total viable bacteria remaining at different times are measured (Fig. 1). This simple procedure gives a composite measure of both phagocytosis and killing. The "two-step" protocol incorporates a differential centrifugation step that separates extracellular bacteria from those ingested by the neutrophils (Fig. 1). The proportions of viable extracellular and intracellular bacteria are measured at various times and a kinetic analysis is then undertaken, allowing separate pseudo first-order rate constants to be calculated for phagocytosis and killing. The one-step protocol requires less manual sample processing and can be used when it is not critical to distinguish which function (phagocytosis or killing) is affected. It is also useful for screening numerous samples or conditions. If differences are detected, samples can then be investigated in more depth using the two-step protocol. The two-step protocol is preferable for elucidating which aspect of the bactericidal process is defective. Finally, we have introduced a "survivor" protocol for *S. aureus* in which neutrophils are allowed to ingest bacteria for a short period, before extracellular bacteria are removed by differential centrifugation and exposure to lysostaphin that does not penetrate into the phagosome. Neutrophils and their phagosomal bacteria are subsequently monitored with time, enabling measurement of bacteria that can survive more effectively, and the screening of compounds that may impact on survival.

2 Materials

1. 5000 IU/mL sodium heparin solution.
2. Endotoxin-free sterile water.

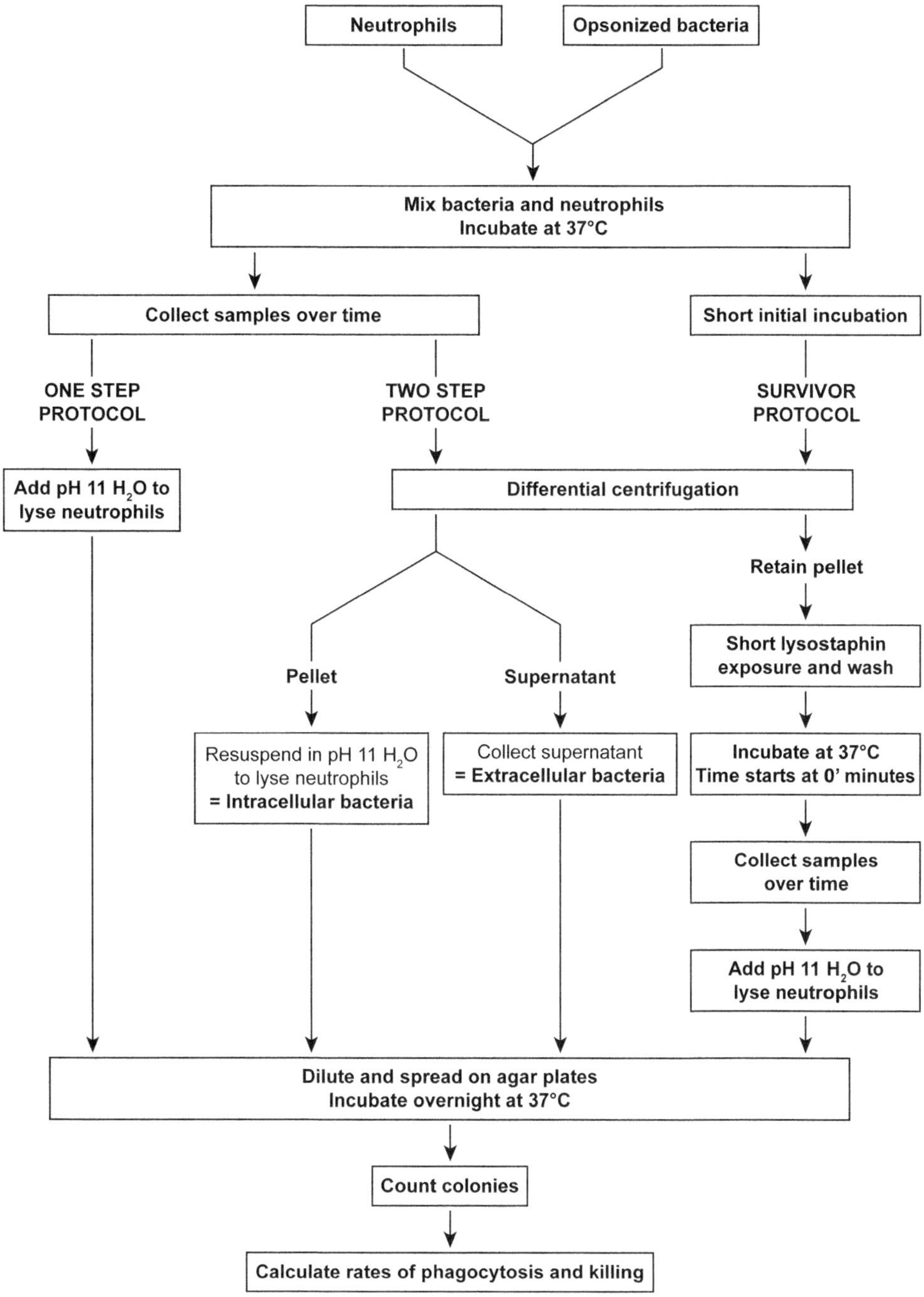

Fig. 1 Outline of the different methods for measuring neutrophil bactericidal activity

3. Phosphate buffered saline (PBS), endotoxin-free (Sigma).
4. Hank's balanced-salt solution (HBSS), endotoxin-free (Sigma).
5. 0.5 M NaCl solution, endotoxin free: dilute 5 M endotoxin-free NaCl solution (Sigma) in endotoxin-free water.

6. Sterile 0.9% NaCl solution: prepare in endotoxin-free water and sterilize or purchase commercially.
7. 5% dextran solution: Dissolve dextran (m_w 200,000–300,000; MP Biomedicals) in sterile 0.9% NaCl solution.
8. Ficoll-Paque™ PLUS (GE Healthcare).
9. RPMI 1640 + L-Glutamine (Gibco).
10. Tryptic soy broth: prepare according to the manufacturer's directions and autoclave before use.
11. Autologous human serum (*see* Subheading 3.1 **step** 7 for preparation).
12. pH 11 water: bring sterile water to pH 11.0 by adding NaOH.
13. DNase mix: combine equal quantities of 1 M Tris–HCl, pH 7.4, 100 mM $CaCl_2$, 50 mM $MgCl_2$, and 10,000 U/mL DNase (Roche).
14. 10 mM diphenyleneiodonium chloride (DPI): dissolve DPI in DMSO, store at −20 °C.
15. 0.5 mg/mL lysostaphin from *Staphylococcus staphylolyticus* (Sigma) ($\geq$500 U/mg): dissolve lysostaphin in PBS, store at −20 °C.
16. Columbia sheep blood agar plates (Fort Richards): these are suitable for the fastidious growth requirements of *S. aureus.*

3 Methods

These protocols have been optimized for use with *Staphylococcus aureus.* Other micro-organisms can be used, but sampling times, dilution factors and lysis conditions may need adapting. For the survivor protocol with bacteria other than Staphylococci, lysostaphin would need to be substituted with an alternative agent.

3.1 Neutrophil Preparation

1. Prepare neutrophils based on a modified method of Nauseef [5] (also *see* Chapter 3 of this volume).
2. Collect venous blood from healthy donors into heparin at a final concentration of 10 IU of heparin per mL of blood.
3. Dilute the blood in endotoxin-free PBS (*see* **Note 1**) followed by dextran sedimentation. For 25 mL of blood, add 15 mL PBS and 10 mL of 5% dextran solution, and incubate at room temperature for approximately 20 min. Transfer the leukocyte-rich erythrocyte-poor upper layer to fresh 15 mL tubes (maximum of 11 mL/tube) and then underlay with 3 mL Ficoll-Paque. Separate neutrophils from peripheral blood mononuclear cells (PBMCs) by centrifugation through the Ficoll (1000 × *g*, 20 min, without brake).

4. Remove plasma, PBMC, and Ficoll layers, suspend the erythrocyte/neutrophil pellet in a small volume of PBS, and transfer to a fresh 50 mL tube (*see* **Note 2**). Use PBS to make the total volume of this suspension 10 mL. Remove contaminating erythrocytes by the addition of 20 mL of water for 2 min with gentle mixing. Quickly restore tonicity by adding 10 mL of 0.5 M NaCl solution and mix gently. Pellet cells (450 × *g*, 5 min).
5. Suspend neutrophils at a final concentration of 1×10^7/mL in HBSS containing 10% autologous serum. Keep at room temperature until ready to use.
6. Neutrophil yield can vary considerably between donors, but most preparations would result in 50–200 × 10^6 cells per 50 mL of blood. Purity is checked by flow cytometry. Granulocytes (neutrophils and eosinophils) typically constitute 98% of the final white blood cell counts, with 2% contaminating PBMCs.
7. To prepare approximately 1 mL of autologous serum, reserve 4–5 mL of blood (without anticoagulant) from the donor whose neutrophils have been isolated. Allow the blood to clot at 37 °C for ~1 h, and loosen the clot from the sides of the tube by running a sterile pipette tip around the side of the tube. Centrifuge to separate serum from the clot (1200 × *g*, 2 min) and transfer the serum to a clean tube. Centrifuge at 12,000 × *g* for 30 s to remove contaminating erythrocytes.

3.2 Preparation of Bacteria

1. Inoculate 10 mL of sterile tryptic soy broth with a single colony of *S. aureus* grown on sheep blood agar. Culture overnight at 37 °C in a shaking incubator at 200 rpm.
2. Take a 1 mL sample of the overnight culture and centrifuge at 12,000 × *g* for 4 min to pellet the bacteria. Wash the pellet twice in PBS and resuspend in 1 mL of HBSS. Centrifuge at 100 × *g* for 5 min to remove any clumped bacteria (*see* **Note 3**).
3. Calculate the concentration of bacteria in the sample by measuring the optical density at 550 nm and relating to previously established standard curves of optical density vs. colony forming units (CFU).
4. Opsonize the bacteria by suspending 1×10^8 CFU/mL in HBSS containing 10% autologous serum in a 1.5 mL Eppendorf tube. Rotate end-over-end (6 rpm) for 20 min at 37 °C, then use immediately (*see* **Note 4**).

3.3 One-Step Assay for Neutrophil Bactericidal Activity

1. Prepare two tubes for each condition to be tested: an experimental and a control. For the experimental tube, add 500 μL (*see* **Note 5**) of freshly prepared neutrophils in HBSS containing 10% autologous serum (1×10^7 neutrophils/mL). Prepare

Table 1
Dilution guidelines for the one-step protocol

Dilutions for one-step protocol	Sample time			
	0 min	5 min	10 min	20 min
Approximate undiluted concentration/mL[a]	5×10^7	4×10^7	3×10^7	2×10^7
Dilution due to sample processing[b]	20×	20×	20×	20×
Dilution incurred by plating 20 μL[b]	50×	50×	50×	50×
Additional dilution required to get approximately 100 colonies per 20 μL[c]	500×	500×	300×	200×
Total dilution factor	5×10^5	5×10^5	3×10^5	2×10^5

[a]If the numbers of bacteria detected are too high or low to make accurate colony counts, the additional dilutions should be adjusted accordingly

[b]These are dilutions that will be automatically incurred through following the protocol as written

[c]These are the further dilutions required in the second to last step of the one-step protocol

an identical "control" tube, replacing the neutrophils with 500 μL of HBSS with 10% serum. The control allows direct measurement of the starting number of bacteria, their growth over the time course of the experiment, and the effects of opsonization and experimental drugs on bacterial growth.

2. Any experimental drugs or inhibitors are added at this stage. Incubate tubes at 37 °C for 10 min to prewarm.
3. Add 500 μL of freshly opsonized bacteria to the experimental tube(s) to start the reaction. The final ratio of bacteria to neutrophils is 10:1, with a final neutrophil concentration of 5×10^6/mL and a final serum concentration of 10% (*see* **Note 6**). Likewise, add 500 μL of bacteria to the control tubes.
4. Incubate the tubes at 37 °C with end-over-end rotation (6 rpm) (*see* **Note** 7).
5. For a time course, take 50 μL samples from the experimental tubes at each time point (e.g., 5, 10, and 20 min, *see* **Note 8**) and from the controls at the first and last time points and place on melting ice (*see* **Note 9**). Dilute into 950 μL of pH 11 water at room temperature (*see* **Note 10**).
6. Allow water lysis of the neutrophils to occur for 5 min, then vortex vigorously for approximately 5 s to disperse the bacteria.
7. Dilute in pH 11 H_2O to give a concentration of approximately 5000 CFU per mL (refer to Table 1). Spread 20 μL on half of an agar plate, giving approximately 100 CFU per half plate (*see* **Note 11**). Plate at least four half plates per sample (*see* **Note 12**). Incubate the plates overnight at 37 °C and count the number of colonies formed.

8. Convert colony counts to bacterial concentrations by multiplying by the appropriate dilution factor (refer to Table 1). Adjust bacterial numbers to account for growth over the time course of the assay by plotting the number of "control" bacteria at the start and end of the experiment. Calculate expected bacterial numbers at each time point from the equation of the line and then calculate the ratio of the extrapolated number of bacteria at a particular time to that at time zero. Divide the number of bacteria counted after incubation with neutrophils for a particular time by the ratio calculated for that time. This accounts for any growth of bacteria during the assay. These calculations are based on the assumption that bacteria grow at the same rate in the presence of neutrophils.

3.3.1 Data Analysis for the One-Step Assay

1. Plot values of viable bacteria against time or convert to percentage killing or percentage survival relative to the number of bacteria at the corresponding time point. Expression of data as a percentage normalizes any variation in the initial concentration of bacteria and enables experiments on different days to be compared.
2. Different experiments can also be compared by obtaining the slope of a semi-log plot or fitting an exponential curve to the data ($y = 100e^{-kx}$) to provide a single rate which is a composite measure of phagocytosis and killing. The half-life for a bacterium can be calculated from $t_{1/2} = \ln(2)/k$. Figure 2a shows an example of time course data fitted to an exponential curve. In Fig. 2b the percentage of viable bacteria is shown after 20 min incubation. Killing is inhibited by treatment with diphenyleneiodonium (DPI), an inhibitor of the NADPH oxidase, or the MPO inhibitor thioxanthine-1 (TX1) [6].

3.4 Two-Step Assay for Neutrophil Bactericidal Activity

1. The two-step assay protocol initially follows that of the one-step assay (**steps 1–4** of Subheading 3.3). At the point of sample collection, however, remove a 50 μL sample of neutrophils with bacteria into 950 μL of ice-cold PBS, halting neutrophil activity. Subsequent handling of the samples should also remain at low temperature, except for the water lysis which is done at room temperature.
2. Samples of bacteria alone are diluted in pH 11 water and plated, as described in **step 8** (see below).
3. Centrifuge samples with neutrophils at 100 × *g* for 5 min at 4 °C using a swing-out rotor (*see* **Note 13**).
4. Collect the supernatant, being careful not to disturb the neutrophil pellet (*see* **Note 14**). We advise leaving a small (~50 μL) meniscus over the pellet.

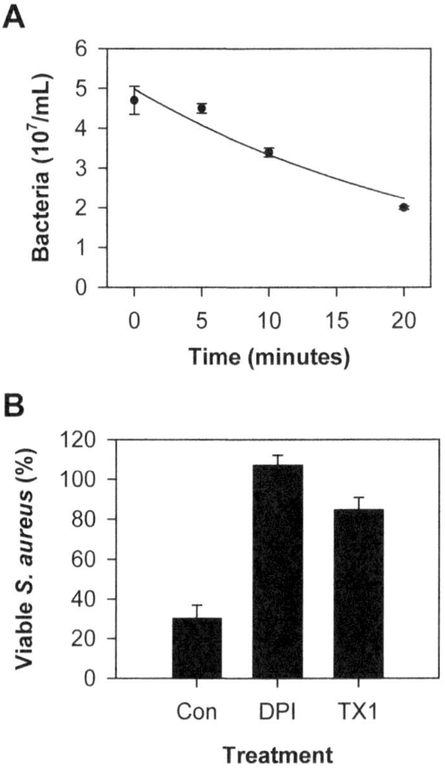

Fig. 2 Measurement of the bactericidal activity of neutrophils with the "one-step" protocol. (**a**) *S. aureus* were incubated with neutrophils at a ratio of 10:1 for 0–20 min. (**b**) *S. aureus* were incubated for 20 min with control neutrophils (Con) or neutrophils pretreated with 10 μM of either the NADPH oxidase inhibitor DPI or the myeloperoxidase inhibitor TX1. All samples were processed as described in the text using the one-step protocol. Data are means ± SE of 3–5 separate experiments using neutrophils from different donors. For (**a**), data are fitted to an exponential decay curve (2 parameter) using the curve fitting program in SigmaPlot

5. Wash the pellet twice more with 950 μL ice-cold PBS (100 × *g*, 5 min, 4 °C), pooling the supernatants. The pooled supernatants contain the bacteria that have not been phagocytosed (*extracellular* bacteria), while the phagocytosed bacteria are in the neutrophil pellet (*intracellular* bacteria).
6. Resuspend the pellets in 950 μL pH 11 water for 5 min at room temperature and then vortex briefly (*see* **Notes 10** and **15**).
7. If larger volumes of the initial sample have been taken (*see* **Note 13**), DNA release from the neutrophils upon water lysis may interfere with the assay. Bacteria may adhere to the strands of DNA resulting in plating of clumps of bacteria leading to an underestimation of the number of viable bacteria. To degrade the released neutrophil DNA, add 40 μL of DNase mix (*see*

Table 2
Dilution guidelines for the two-step protocol

Dilutions for two-step protocol	Sample							
	Control		Extracellular			Intracellular		
	0 min	20 min	5 min	10 min	20 min	5 min	10 min	20 min
Approximate undiluted concentration/mL[a]	5×10^7	5×10^7	2×10^7	1.5×10^7	5×10^6	1×10^7	1×10^7	5×10^6
Dilution due to sample processing[b]	20×	20×	60×	60×	60×	20×	20×	20×
Dilution incurred by plating 20 μL[b]	50×	50×	50×	50×	50×	50×	50×	50×
Additional dilution required to get approx 100 bacteria per 20 μL[c]	500×	500×	66.7×	50×	16.7×	100×	100×	50×
Total dilution factor	5×10^5	5×10^5	2×10^5	1.5×10^5	5×10^4	1×10^5	1×10^5	5×10^4

[a]If the numbers of bacteria detected are too high or low to make accurate colony counts, the additional dilutions should be adjusted accordingly
[b]These are dilutions that will be automatically incurred through following the protocol as written
[c]These are the further dilutions required in the second to last step of the two-step protocol

Note 16). Combine by inverting the tubes and then incubate at 37 °C for 10 min (*see* **Note 17**). Incubate the control bacteria under the same conditions.

8. Dilute each sample, including supernatants and controls, in ice-cold pH 11 water to give a bacterial concentration of approximately 5000 CFU per mL (refer to Table 2). Spread 20 μL on half an agar plate, giving approximately 100 CFU per half plate (*see* **Note 11**). Plate at least 4 half plates per sample (*see* **Note 12**). Incubate the plates overnight at 37 °C and count the number of colonies formed.
9. We have developed an Excel file that calculates the rate constants for phagocytosis and killing. This can be found in the online Supplementary material or by contacting the corresponding author. This file contains instructions for its use. Add the raw data and the dilution factors to the appropriate cells in the Excel spreadsheet.

3.4.1 Kinetic Basis for Data Analysis of the Two-Step Assay

1. Using data obtained with the two-step protocol, killing can be quantified using a kinetic analysis. As both phagocytosis and killing approximate first order processes [7], rate constants for these processes can be calculated.

2. The number of extracellular bacteria (A) decreases with time, while the number of viable intracellular bacteria (B) initially increases but then decreases as the bacteria are killed (C). These two events can be represented as occurring in series.

$$A \xrightarrow{k_p} B \xrightarrow{k_k} C$$

By obtaining values for A and B at different time points, rate constants for phagocytosis (k_p) and killing (k_k) can be calculated.

3. Phagocytosis and killing are represented by Eqs. (1 and 2), which can be integrated to give Eqs. (3 and 4), where A_0 = the initial number of bacteria added to the system and t = time.

$$\frac{\partial[A]}{\partial t} = k_p[A] \quad (1)$$

$$\frac{\partial[B]}{\partial t} = k_p[A] - k_k[B] \quad (2)$$

$$A = A_0 e^{-k_p t} \quad (3)$$

$$B = \frac{A_0 k_p}{k_k - k_p}\left(e^{-k_p t} - e^{-k_k t}\right) \quad (4)$$

4. The calculation of rate constants requires that both processes follow pseudo first-order kinetics (*see* **Note 18**). Solving for k_p from Eq. (3) involves obtaining the slope of a semi-log plot of A with time. A linear fit confirms that phagocytosis follows pseudo first-order kinetics. To calculate k_k, Eq. (4) can be rearranged and solved using the Lambert W function ($W(X)$). The solution to the Lambert W function can be determined using a mathematical software package or can be calculated as described in the Supplementary material. The k_k values for each time point are then averaged, giving an overall k_k value. k_p and k_k can also be converted to half-lives of phagocytosis and killing using the equation $t_{1/2} = \ln(2)/k$.

5. Figure 3 shows the graphical outputs obtained after several steps of the calculation are completed. Data for these analyses were obtained in a two-step assay using neutrophils from a healthy and an MPO-deficient donor. While the rate of phagocytosis is similar between donors, the rate of killing is reduced in MPO-deficient neutrophils.

3.5 Intraphagosomal Survival Assay

1. For measurement of long term intraphagosomal survival of bacteria, we have adapted the standard two-step assay to allow a short incubation time for phagocytosis and the use of lysostaphin to ensure the removal of all extracellular *S. aureus*. The decline in viability of the intracellular bacteria is then monitored.

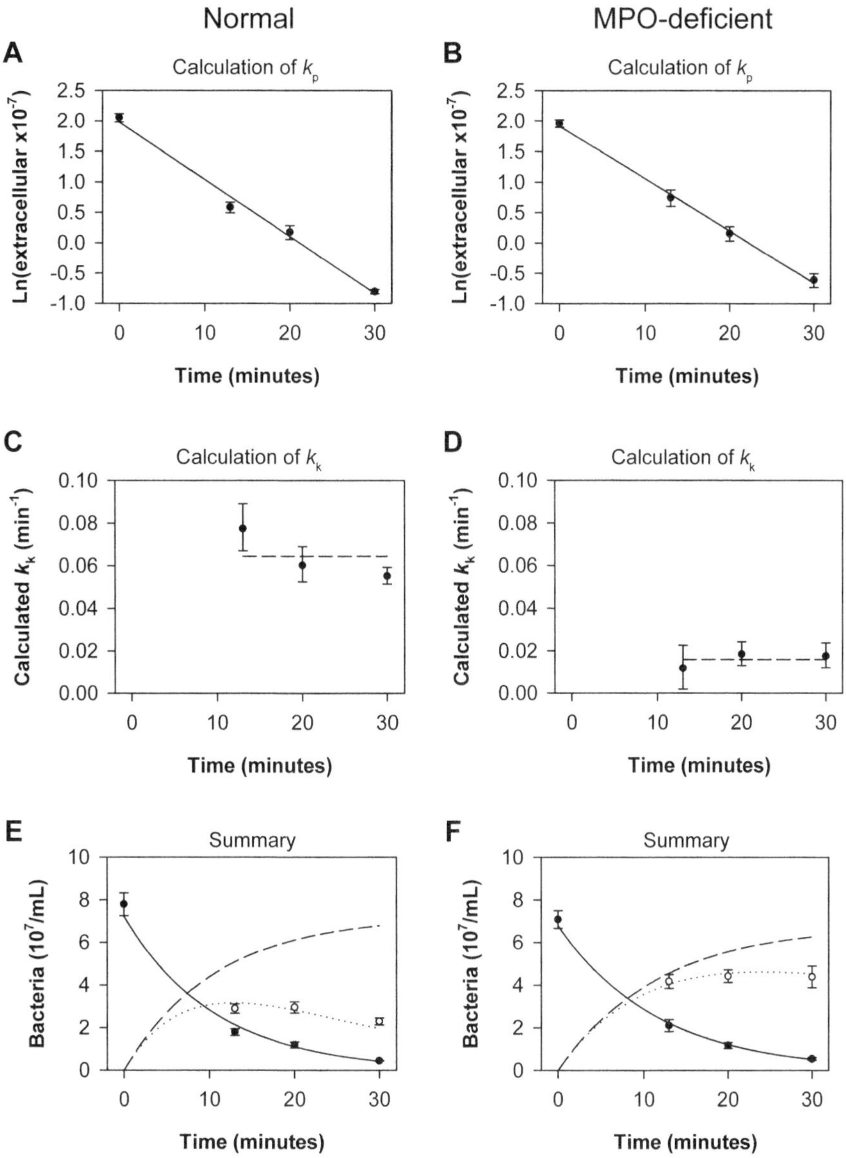

Fig. 3 Calculation of rate constants for phagocytosis (k_p) and killing (k_k) of *S. aureus* using two-step protocol results. Colony counts of control, extracellular and intracellular bacteria (means $\pm$ SD) were obtained using the two-step protocol with neutrophils from a healthy and an MPO-deficient donor. These have been converted back to bacterial concentrations using the appropriate dilution factors and adjusted at each time according to the growth of control bacteria. (**a** and **b**) k_p was determined from a semi-log plot of extracellular bacteria with time, where the slope of the regression line is equal to k_p. In this example $k_p = 0.094$ min^{-1}, $t_{1/2} = 7.4$ min and $k_p = 0.085$ min^{-1}, $t_{1/2} = 8.1$ min for normal and MPO-deficient, respectively. (**c** and **d**) Rate constants of killing (k_k) calculated for each sampling time. The dashed line is the mean k_k. In this example $k_k = 0.064$ min^{-1}, $t_{1/2} = 10.8$ min and $k_k = 0.016$ min^{-1}, $t_{1/2} = 43.5$ min for normal and MPO-deficient, respectively. The error bars in this graph reflect the error in the intracellular colony counts and the degree of phagocytosis. When limited phagocytosis has occurred, a small variation in the intracellular colony counts will have a much larger effect on the calculated k_k value than if a lot of bacteria have been phagocytosed. (**e** and **f**) Theoretical curves of extracellular (solid) and intracellular bacteria (dotted) were generated using the calculated rate constants and the experimentally obtained values for both extracellular (●) and intracellular (○) bacteria. The dashed line represents the expected number of viable intracellular bacteria if none were killed

2. As illustrated in Fig. 1, neutrophils and bacteria are incubated together as for the other protocols (**steps 1–4** of Subheading 3.3), but for only 10 min at a ratio of between 10:1 and 20:1 bacteria to neutrophils, to optimize the amount of viable intracellular bacteria. For this modified protocol, we use an incubation volume of 500 μL (with neutrophils at 1×10^7/mL) for the experiment tubes, and the entire volume is subjected to differential centrifugation ($100 \times g$, 5 min, at 4 °C using a swing-out rotor).
3. Collect the supernatant and wash the neutrophil pellet once with ice-cold PBS (as in Subheading 3.4, **step 4**), pool these supernatants to estimate the non-ingested population, then resuspend the neutrophil pellet in 1 mL PBS containing 50 μg/mL lysostaphin (*see* **Note 19**) and incubate at 37 °C for 5 min. We use prewarmed PBS and a heat-block for this incubation. Pellet the neutrophils again and wash once with 1 mL PBS, each time centrifuging at $100 \times g$ for 5 min at 4 °C (*see* **Note 20**).
4. Resuspend the neutrophils (with ingested bacteria) in 500 μL RPMI media containing 10% autologous serum and incubate at 37 °C, with end-over-end rotation (6 rpm). This is designated time point 0′, the beginning of a second incubation. RPMI is used as a more nutritious media for the neutrophils to maintain viability.
5. Remove 50 μL aliquots at selected times for neutrophil lysis in pH 11 water. As bacterial numbers are low in the extended incubation, good recovery of intracellular bacteria is ensured by adding saponin at 0.05% (w/v) to the pH 11 water and syringing the lysate five times through a 25G needle.
6. Quantification of phagosomal survivors is by plating, as described in Subheadings 3.3 and 3.4. The appropriate plating dilutions (made in pH 11 water) for the modified protocol of this section are 2×10^5 (for 0' min) and 2×10^4 (for 30', 90', and 180' min).
7. Figure 4 shows typical results of *S. aureus* phagosomal survival in the absence or presence of DPI (*see* **Note 21**). It demonstrates that although a significant number of the phagosomal *S. aureus* are killed during the initial phagocytosis period and subsequent step to remove extracellular bacteria, some viable phagosomal bacteria are still obtained. These bacteria decrease in number upon further incubation, but a small number survive. Phagosomal survivors increase when neutrophils are treated with DPI, indicating the importance of oxidative mechanisms in the killing of *S. aureus*.

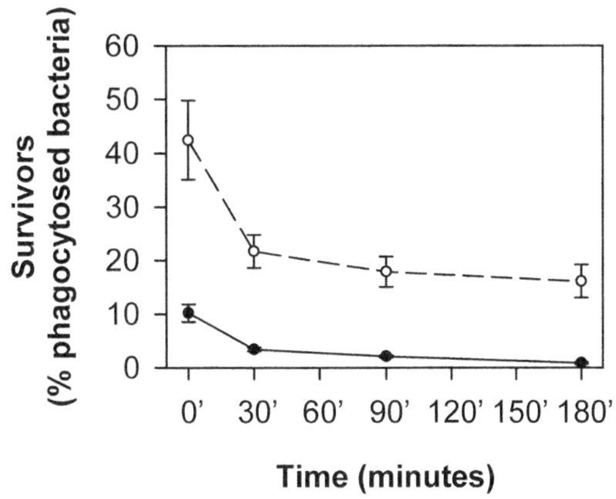

Fig. 4 Measurement of *S. aureus* survival in the neutrophil phagosome. *S. aureus* were opsonized in 10% autologous serum before being incubated for 10 min with freshly prepared human neutrophils, which had been pretreated with (○, dashed) or without (●, solid) 10 μM DPI. Following the removal of extracellular bacteria and lysostaphin treatment, the neutrophils and their intracellular bacteria were incubated for a further 3 h, starting at 0' min. Samples were removed at various time points; the neutrophils were lysed, and the numbers of viable bacteria determined. Data are means and SD of quadruplicate colony counts from a single representative experiment. In this experiment, the initial concentration of bacteria was 1.8×10^8/mL (18:1 ratio of bacteria to neutrophils), and 0.6×10^8/mL were measured as remaining in the extracellular medium. The number of viable phagosomal bacteria presented in the figure is expressed as a percentage of the 1.2×10^8/mL bacteria that are calculated to have been ingested by the neutrophils

4 Notes

1. Endotoxin-free reagents and solutions are used during neutrophil isolation to minimize the risk of priming these cells.
2. Cells are carefully transferred to a fresh tube for hypotonic lysis to reduce contamination by PBMCs that can stick to the sides of the tube during Ficoll centrifugation.
3. It is important to remove clumped bacteria that otherwise can pellet with the neutrophils during the differential centrifugation step in the two-step protocol.
4. Opsonins are gradually lost from *S. aureus* after approximately 20 min at room temperature or 37 °C. Therefore, it is important to use the opsonized bacteria immediately. For some bacteria it may be necessary to opsonize with heat-inactivated serum.
5. Larger or smaller volumes can be used, depending on how many time points are to be sampled. Adjust the volume of serum and bacteria accordingly.

6. If different bacteria to neutrophil ratios are used, dilutions for the colony counting assay must be adjusted accordingly.
7. Continuous, slow, end-over-end rotation of the tubes is important to prevent the neutrophils from sedimenting and clumping, and to ensure continual contact between neutrophils and noningested bacteria. The rotation speed must not be too vigorous or neutrophil function is disrupted.
8. The time points at which viable bacteria are measured should be selected to coincide with the period when most of the phagocytosis and killing occurs. For *S. aureus*, sampling up to 30 min will show the most change [7]. Samples taken at longer times generally show little further reduction in bacterial numbers as the rate of killing begins to decrease. The killing mechanisms at these later times may differ from those functioning during the initial period when the bulk of the bacteria are killed. If using other bacterial species or ratios it may be necessary to adjust the sampling times. Under conditions where phagocytosis is slow, rates of killing at early time points are variable (*see* "k_k variability at different time points" in the Supplementary material). Therefore, under these conditions it is better to start measurements later (e.g., 10–15 min).
9. Melting ice rather than fresh ice cools the samples rapidly to halt further phagocytosis and killing.
10. Dilution into pH 11 water results in osmotic lysis of the neutrophils without affecting the viability of *S. aureus*. The sensitivity of other bacteria to pH 11 water should be checked. When water at neutral pH is used, there is incomplete lysis of the neutrophils and ineffective dispersal of bacteria associated with the cell debris [8]. This results in an over-estimation of bactericidal activity due to multiple bacteria being counted as only a single colony. In this situation, partial defects may be overlooked, and complete defects may appear as only partial. To overcome this problem, pH 11 water is used [8, 9]. Decleva et al. [8] found this to be more effective than other lysis methods, including dilution into neutral pH water or into detergent-containing buffer. Experiments in our laboratory have confirmed that pH 11 water is more effective than saponin. In addition, use of pH 11 water for all subsequent dilution steps seems to avoid the clumping of *S. aureus* that is sometimes seen with PBS.
11. If counts are too variable from plating 20 μL/half plate we recommend plating 50-100 μL/plate. Adjust dilutions accordingly.
12. The greatest variation in the assay is incurred during plating. Therefore, it is better to plate more aliquots from a single replicate than to plate less aliquots of two or more replicates.

13. This is a low-speed centrifugation to sediment the neutrophils and their ingested bacteria but not extracellular bacteria. A barely discernible diffuse neutrophil pellet may be seen. If bigger pellets are desired, a larger volume from the experiment tubes can be taken and a larger volume of ice-cold PBS added to keep the dilution constant.
14. As the centrifugation speed is low, the pellet does not adhere strongly to the tube wall and care must be taken to avoid disturbing it as the supernatant is removed.
15. If the residual volume over the pellets is greater than 100 μL, the amount of pH 11 water added may need to be increased to maintain the pH at or near to 11. This needs to be accounted for when calculating the final dilutions.
16. The lysate solution is brought to an appropriate pH, and magnesium and calcium are added for optimized DNase activity. The final concentration of DNase is 100 U/mL.
17. We find that the DNA degrades better over this time if the samples are mixed by pipetting half-way through the incubation.
18. There may be some conditions when this pseudo first-order relationship does not hold. Results can then be presented as percentages of bacteria phagocytosed, and percentages of phagocytosed bacteria that have been killed at any time point. Impaired killing will be evident, but the degree of quantification is much less than with kinetic analysis, as it will be different at each time point.
19. Five minutes at 37 °C with 50 μg/mL lysostaphin is effective at killing any residual extracellular *S. aureus*. Intracellular bacteria are not affected as lysostaphin does not permeate neutrophils. If bacteria other than Staphylococci are being examined, a different agent will be required to kill extracellular bacteria.
20. Although handling of neutrophils is as gentle as possible, it is acknowledged that there may be some losses through the wash steps. It is paramount, however, that lysostaphin is thoroughly removed so there is no carry over into the extended time course and plating. Also, the lysostaphin treatment and washing is done as swiftly as possible to minimize bactericidal activity before starting the subsequent time course.
21. DPI is a flavocytochrome inhibitor that does not affect *S. aureus* viability at the concentration used.

References

1. Winterbourn CC, Kettle AJ, Hampton MB (2016) Reactive oxygen species and neutrophil function. Annu Rev Biochem 85:765–792
2. Gresham HD, Lowrance JH, Caver TE et al (2000) Survival of *Staphylococcus aureus* inside neutrophils contributes to infection. J Immunol 164:3713–3722
3. Shapiro HM (2008) Flow cytometry of bacterial membrane potential and permeability. Methods Mol Med 142:175–186
4. Parker HA, Magon NJ, Green JN et al (2014) Analysis of neutrophil bactericidal activity. Methods Mol Biol 1124:291–306
5. Nauseef WM (2014) Isolation of human neutrophils from venous blood. Methods Mol Biol 1124:13–18
6. Tiden AK, Sjogren T, Svensson M et al (2011) 2-thioxanthines are mechanism-based inactivators of myeloperoxidase that block oxidative stress during inflammation. J Biol Chem 286:37578–37589
7. Hampton MB, Vissers MC, Winterbourn CC (1994) A single assay for measuring the rates of phagocytosis and bacterial killing by neutrophils. J Leukoc Biol 55:147–152
8. Decleva E, Menegazzi R, Busetto S et al (2006) Common methodology is inadequate for studies on the microbicidal activity of neutrophils. J Leukoc Biol 79:87–94
9. Gargan RA, Brumfitt W, Hamilton-Miller JM (1989) Failure of water to lyse polymorphonuclear neutrophils completely. Role of pH and implications for assessment of bacterial killing. J Immunol Methods 124:289–291

Part IV

Biochemistry, Biology, and Signal Transduction of Neutrophils

Chapter 13

Assessment of Neutrophil Apoptosis

Nicole D. Barth, Marc Vendrell, David A. Dorward, Adriano G. Rossi, and Ian Dransfield

Abstract

The process of neutrophil apoptosis has an important role in the resolution of acute inflammation. Apoptotic cell death is characterized by a coordinated sequence of cellular alterations that serve to uncouple neutrophil effector functions whilst maintaining plasma membrane integrity. In this way the release on neutrophil intracellular contents, including proteases, glycosidases, and reactive oxygen species, is limited during apoptosis. In addition, plasma membrane alterations associated with neutrophil apoptosis provide molecular cues that enable recognition by phagocytic cells, including macrophages. The recognition and uptake of apoptotic neutrophils by macrophages dampens proinflammatory responses to pathogen- or damage-associated molecular patterns and triggers release of proresolution mediators, that further promote resolution of inflammation. The key cellular and molecular events that act to control neutrophil apoptosis and subsequent macrophage phagocytosis have been characterized by in vitro studies, unveiling potential therapeutic targets for the manipulation of these regulatory pathways. In this chapter, we outline some of the key assays that are used to assess neutrophil apoptosis in vitro, together with methods to assess activation of the apoptotic machinery and phagocytic clearance of apoptotic neutrophils.

Key words Neutrophil apoptosis, Flow cytometry, Caspases, Mitochondria, DNA fragmentation, Phosphatidylserine, Phagocytosis

1 Introduction

One of the hallmarks of the acute inflammatory response is the recruitment of neutrophils to sites of inflammation [1], a process that is critical for the control of the growth of invading microorganisms. Neutrophils are armed with a formidable arsenal of destructive enzymes that can destroy proteins, carbohydrates, and lipids to limit microbial growth [2]. In addition, once activated, neutrophils can release reactive oxygen and nitrogen species and form extracellular traps (NETs) that limit microbial spread and exert antimicrobial effects [3]. However, the controlled removal of inflammatory neutrophils represents an essential component of the process by which acute inflammation resolves [4]. The process of apoptosis functionally isolates the neutrophil from its

Mark T. Quinn and Frank R. DeLeo (eds.), *Neutrophil: Methods and Protocols*, Methods in Molecular Biology, vol. 2087, https://doi.org/10.1007/978-1-0716-0154-9_13,

proinflammatory surroundings and prevents the release of destructive intracellular components that would potentially exacerbate resolution of inflammation [5, 6]. Importantly, neutrophil apoptosis is associated with plasma membrane alterations that enable recognition and subsequent phagocytosis of apoptotic cells by macrophages and other phagocytic cells [7]. Dysregulation of either apoptosis or phagocytic clearance pathways would be predicted to contribute to prolongation of the inflammatory response and may be key in the establishment of chronic inflammation associated with many different inflammatory diseases.

Neutrophils that are isolated from peripheral blood are terminally differentiated cells. When cultured in vitro, neutrophils undergo apoptosis spontaneously in a manner that is dependent on the culture conditions [8]. A number of exogenous factors can act to delay (e.g., bacterial lipopolysaccharides (LPS) [9], hypoxia [10], pH, temperature [11], cell density and cytokines such as G-CSF) or accelerate (e.g., TNF or Fas) engagement [12] of the apoptotic machinery. For example, ligation of receptors that trigger apoptosis, including Fas (CD95) and TNF-related apoptosis inducing ligand (TRAIL) can result in assembly of a death-inducing signalling complex (DISC) [13]. Alternatively, cellular stress can alter the balance of pro- and antiapoptotic regulators within the cells. In neutrophils, the prosurvival B cell lymphoma 2 (Bcl-2) family member myeloid cell leukemia-1 (Mcl-1) is particularly important for engagement of the intrinsic apoptosis pathway, and acts to sequester proapoptotic Bcl-2 family members [14]. Intrinsic apoptosis pathways lead to increased mitochondrial membrane permeability and cytoplasmic translocation of cytochrome c. The interaction of cytochrome c with apoptotic protease activating factor-1 (APAF-1) leads to activation of the cysteine aspartic acid (caspase) protease cascade, following cleavage of procaspase 9 to its active form [15]. Caspase 9 is one of the apoptosis initiator caspases (caspase 8, 9, and 10). Both extrinsic and intrinsic apoptosis pathways converge at the point of activation of the executioner caspase, caspase 3, which then induces apoptosis [16]. For neutrophils, upregulation of antiapoptotic proteins, such as Mcl-1, markedly delays apoptosis, whereas Mcl-1 degradation is associated with accelerated apoptosis [8]. The balance of anti- and proapoptotic factors is crucial for the regulation of apoptosis. Thus, it is important to carefully consider the methods used for neutrophil isolation and culture when interpreting experimental data.

The merits of different methods of neutrophil isolation have been considered elsewhere [17, 18]. In our experience, consistent results of in vitro experiments relating to neutrophil apoptosis are critically dependent on the isolation of pure cell populations. In particular, the presence of small numbers of contaminating eosinophils or monocytes can dramatically affect the extent of constitutive neutrophil apoptosis during in vitro culture [19, 20]. In addition, minimizing preparation-induced neutrophil activation is also critical when investigating whether factors are able to act as regulators of apoptosis. Since neutrophil apoptosis is exquisitely sensitive to a wide variety of external factors as described above, including cell culture conditions (pH, oxygen tension, presence or absence of serum, cell density, cell culture environment, adherent or suspension culture), should also be carefully considered when designing experiments.

Similar considerations apply to the study of phagocytosis of apoptotic neutrophils. Many human studies use monocyte-derived macrophages as a convenient source of "macrophages" [21]. However, activation and differentiation of monocytes during in vitro culture can profoundly influence their transcriptional profile, including receptors involved in the recognition and internalization of apoptotic targets [22–24]. In addition, many of the receptors mediating apoptotic cell uptake act indirectly, through "bridging" ligands that bind to the aminophospholipid phosphatidylserine (PtdSer) that is exposed on the apoptotic cell membrane [25, 26]. The precise pathway used for internalization may therefore depend on the presence of exogenously added ligand. Furthermore, the repertoire of apoptotic cell recognition receptors that are utilized is likely to affect the subsequent macrophage responses in terms of suppression of proinflammatory cytokine production or the generation of anti-inflammatory mediators [27–29].

Monitoring apoptosis in neutrophils can be achieved by examining the hallmark features of apoptosis such as morphological changes, including chromatin condensation and DNA fragmentation associated with a pyknotic nuclear appearance [21]. In addition, although neutrophils are relatively resistant to apoptosis induced fragmentation into multiple apoptotic bodies, they do exhibit cellular shrinkage and cytoplasmic vacuolation, which can be detected by electron microscopy [30]. The plasma membrane of neutrophils is characterized by specific proteolytic loss of certain receptors, including CD16 [31, 32]. These membrane alterations are accompanied by one of the hallmark features of apoptosis, the loss of plasma membrane asymmetry and the presence of PtdSer in the outer leaflet of the plasma membrane [26]. The presence of PtdSer can be detected through binding of the Annexin family of molecules [33], via phagocytic "bridging" ligands [34] or with novel small cyclic peptides [35].

In this chapter, which is an updated version of our previous article [36], we describe diverse methods that can be used to investigate neutrophil apoptosis, including phagocytosis of apoptotic neutrophils by macrophages.

2 Materials

2.1 In Vitro Culture of Human Neutrophils

1. Iscove's Modified Dulbecco's Medium (IMDM).
2. Penicillin/Streptomycin 100×.
3. 5% autologous serum (*see* **Note 1**).
4. 96-well flat-bottom plates.

2.2 Cytocentrifuge Preparations for Light Microscopy

1. Cytocentrifuge chambers, filter cards, glass slides, and coverslips.
2. Methanol, Diff-Quik™ stains, and DPX mounting medium.

2.3 Preparation of Neutrophils for Electron Microscopy

1. 3% (0.1 M) glutaraldehyde: dilute 25% stock solution in 0.1 M sodium cacodylate buffer, pH 7.3.
2. 1% osmium tetroxide in 0.1 M sodium cacodylate.
3. 50%, 70%, 90%, and 100% normal grade acetones and analytical grade acetone.
4. Araldite resin.
5. Reichert OMU4 ultramicrotome.

2.4 Plasma Membrane Alterations

1. 96-well flat-bottom plate.
2. Alexa Fluor 647-conjugated Annexin V (ThermoFisher Scientific).
3. Annexin binding buffer: 20 mM HEPES pH 7.4, 140 mM NaCl, 2.5 mM $CaCl_2$ containing 0.1% (w/v) bovine serum albumin (BSA). Store at 4 °C.
4. Propidium iodide stock solution: 1 mg/mL propidium iodide in sterile H_2O.
5. Cytometry buffer: 20 mM HEPES pH 7.4, 140 mM NaCl, containing 0.1% BSA. Store at 4 °C.
6. APC-Cy7 anti-human CD16 (1:100 Biolegend).
7. Nonbinding murine isotype (IgG1) control (Biolegend).

2.5 Mitochondrial Permeability

1. MitoCapture™ Mitochondrial Apoptosis Detection Fluorometric Kit (Biovision, Milpitas, CA 95035 USA) contains: MitoCapture™ reagent (Store at −20 °C), and incubation buffer (Store at 4 °C.)

2.6 Western Blotting for Regulators of Apoptosis

1. Tris-buffered saline (10× TBS): 87.66 g of NaCl, 24.22 g of Tris base, 800 mL of distilled water (ddH_2O). Adjust the pH to 7.4 with HCl and add ddH_2O to 1 L. Store at room temperature.
2. 1× TBS: Dilute 10× TBS 1:10 with ddH_2O.

3. Protease inhibitor buffer (*see* **Note 2**): 0.3 mL of TBS, 0.5 mL of stock protease inhibitor cocktail for use with mammalian cell and tissue extracts (Sigma), 20 μL of 400 mM stock 4-(2-aminoethyl)benzenesulfonyl fluoride hydrochloride (AEBSF, stock in H_2O), 20 μL of 0.15 μM stock aprotinin (stock in H_2O), 20 μL of 20 mM stock leupeptin (stock in H_2O), 40 μL of 0.75 μM stock pepstatin A (stock in methanol), 20 μL of 1 M stock sodium vanadate (stock in H_2O, pH 10, boiled), 20 μL of 0.5 M stock benzamidine (stock in H_2O), 20 μL of 2 M stock levamisole (stock in H_2O), and 60 μL of 3.33 M stock β-glycerophosphate (stock in H_2O).
4. 10% Nonidet P-40 (NP-40) in TBS for cell lysis.
5. BCA protein assay.
6. Sample Buffer (for 10 mL of 4×): 4 mL of 50% glycerol, 4 mL of 20% SDS, 2.5 mL of 1 M Tris–HCl (pH 6.8), 20 μL of 1% (w/v in ethanol) bromophenol blue. Add 400 μL of β-mercaptoethanol in a fume hood.
7. Prestained molecular weight standards (e.g., Invitrogen).
8. 12% SDS-PAGE gel.
9. Running buffer (10×): 121 g of Tris base, 10 g of SDS, 238 g of HEPES, ddH_2O up to 1 L. Dilute in 1:10 ddH_2O prior to use.
10. 10× Transfer buffer: 30.3 g of Tris base, 144.12 g of glycine, ddH_2O up to 1 L.
11. 1× Transfer buffer: 100 mL of 10× transfer buffer, 200 mL of methanol, 700 mL of ddH_2O.
12. Polyvinylidene difluoride (PVDF) membrane (Millipore).
13. Blocking buffer: TBS, 0.1% Tween 20 (polysorbate 20), 5% dried milk powder.
14. Primary antibodies: Mcl-1 (1:500; Santa Cruz Biotechnology), GAPDH (1:10,000; Sigma), cleaved caspase-3 (1:1000, Cell Signaling), cleaved caspase-9 (1:1000; Cell Signaling).
15. Secondary antibody: corresponding horseradish peroxidase-conjugated antibody (1:2500, Dako).
16. ECL prime (GE Healthcare), light-sensitive film, X-ray developer.

2.7 Fluorimetric Caspase Kit

1. Homogeneous Caspase Assay Kit (Sigma-Aldrich) (*see* **Note 3**) contains: 1× incubation buffer, stock caspase substrate solution (500 μM DEVD-R110 in DMSO), positive control (lysate from apoptotic camptothecin-treated U937 cells), and R110 standard for calibration curve construction (1 mM in DMSO).

2.8 Caspase Profiling Plate

1. ApoAlert™ Caspase Profiling Plate (Clontech) contains: 96-well microplate with immobilized substrates for caspase 2 (VDVAD-AMC), caspase 3 (DEVD-AMC), caspase 8 (IETD-AMC), and caspase 9 (LEHD-AMC) in 24 wells each, lysis Buffer, 2× Reaction Buffer, 100× DTT solution, and inhibitors of caspases 2, 3, 8, and 9.

2.9 Gel Electrophoresis for DNA Fragmentation

1. Wizard® genomic DNA purification kit (Promega).
2. 0.5× TBE running buffer (5×): 54 g of Tris base, 27.5 g of boric acid, 20 mL of 0.5 M EDTA, up to 1 L of ddH_2O, pH 8.0.
3. SeaKem LE agarose (Cambrex) for DNA electrophoresis.
4. Ethidium bromide solution: 10 mg/mL in dH_2O.
5. GelRed™ Nucleic Acid Gel Stain (Jencons, Lutterworth, England).
6. 6× blue/orange loading dye (Promega).

2.10 Propidium Iodide Staining for Hyplodiploid Nuclei

1. 96-well flat-bottom plate.
2. PI solution: 250 μL of propidium iodide stock solution (10 mg/mL in ddH_2O), 2.2 mL of sodium citrate (2.2 g in 10 mL of ddH_2O), 50 μL Triton X-100, make up to 50 mL with ddH_2O. Store solution at 4 °C in the dark.

2.11 TUNEL Staining

1. 96-well flat-bottom flexible plate.
2. In Situ Cell Death Detection Kit, Fluorescein (Sigma-Aldrich) contains 10× enzyme solution (TdT) in storage buffer and 1× labelled nucleotide mixture in reaction buffer. This protocol also requires PBS (wash buffer), 3% H_2O_2 in methanol (blocking solution), 4% paraformaldehyde in PBS at pH 7.4 (fixation buffer; freshly prepared) and 0.1% Triton X-100 in 0.1% sodium citrate (permeabilization buffer; freshly prepared).

2.12 Plate-Based Phagocytosis Assay

1. 2.5% glutaraldehyde in PBS.
2. Dimethoxybenzidine (*o*-diansidine): 0.1 mg/mL of dimethoxybenzidine (*o*-diansidine) in PBS, made up fresh from 1 mg/mL frozen stock.
3. 30% H_2O_2 solution.

2.13 Flow Cytometry-Based Phagocytosis Assay

1. Cell Trace™ Far Red cell labeling kit (Thermofisher Scientific).
2. pHrodo™ Red succinimidyl ester (Invitrogen).
3. Trypsin–EDTA solution: 0.25% trypsin, 1 mM EDTA.

3 Methods

3.1 Analysis of Neutrophil Morphology

3.1.1 Analysis of Neutrophil Morphology by Light Microscopy

Apoptotic neutrophils can be identified by their characteristic condensation of nuclear material, prominent nucleoli and vacuolation of the cytoplasm (Fig. 1a, d). Morphological characterization of neutrophils by microscopy analysis allows rapid, cheap and accurate assessment of neutrophil apoptosis (*see* **Note 4**).

1. Suspend freshly isolated peripheral blood neutrophils (at least 97% purity – perform cytocentrifuge preparation, as described below) at 1×10^7 cells/mL in IMDM supplemented with 5% autologous serum and penicillin/streptomycin (1×) (*see* **Note 1**).

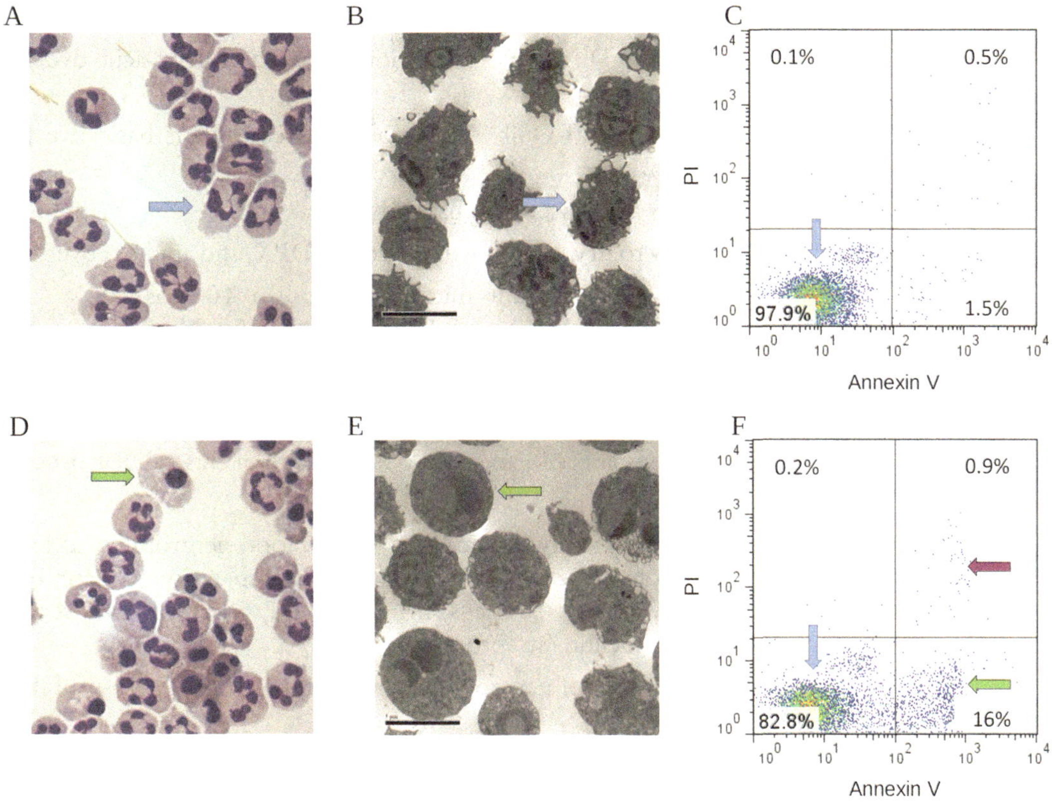

Fig. 1 Assessment of neutrophil apoptosis by microscopy and flow cytometry. Following 6 h in vitro culture isolated neutrophils are predominantly viable with characteristic multilobed nucleus and irregular cell membrane (blue arrow) as seen by light microscopy ((**a**) ×1000 magnification) and electron microscopy ((**b**) ×980 magnification). Viable cells are impermeable to propidium iodide (PI) and phosphatidylserine is not externalized, therefore no Annexin V (AnnV) binding occurs and an AnnV^{-ve}/PI^{-ve}, (blue arrow) is revealed by flow-cytometric analysis (**c**). In contrast drug treatment with apigenin (50 μM) induces neutrophil apoptosis with characteristic morphological changes of nuclear condensation and cell shrinkage (green arrow) visible by both light microscopy (**d**) and electron microscopy (**e**). Flow-cytometric analysis (**f**) reveals the presence of apoptotic cells as AnnV^{+ve}/PI^{-ve} (green arrow). Necrotic cells with loss of cell membrane integrity are permeable to PI and are detected as AnnV^{+ve}/PI^{+ve} (magenta arrow)

2. Add 75 μL of neutrophil suspension to wells of a 96-well flat-bottom plate.
3. To each well, add 15 μL of apoptosis-modifying agents (10× concentration) or buffer control and 60 μL of IMDM/5% serum. If two agents are used in combination, only 45 μL of IMDM is required.
4. Cover with a lid. Incubate at 37 °C in a 5% CO_2 incubator for the desired length of time.
5. Vigorously pipet the well to dislodge adherent cells and load a cytospin chamber with 100 μL of aged neutrophil suspension.
6. Cytocentrifuge at 30 × *g* for 3 min.
7. Air-dry for 5 min.
8. Fix in methanol for 2 min. Drain.
9. Stain in Diff-Quik™ solution 1 or equivalent acid dye for 1 min. Drain.
10. Stain in Diff-Quik™ solution 2 or equivalent basic dye for 1 min.
11. Drain and rinse with distilled water.
12. Allow to dry, mount with a drop of DPX, and apply a coverslip.
13. View using a light microscope with a 40× or 100× (oil) objective and count >300 cells per slide (*see* **Note 4**).

3.1.2 Analysis of Neutrophil Morphology by Electron Microscopy

Detailed structural analysis of the morphological changes that occur in apoptosis are observed best by electron microscopy and formed the basis of the initial observations of this cellular process [17] (Fig. 1b, d).

1. Suspend freshly isolated peripheral blood neutrophils (at least 97% purity—perform cytocentrifuge preparation as described below) at 1×10^7 cells/mL in IMDM/5% autologous serum and penicillin/streptomycin.
2. Add 75 μL of neutrophil suspension to wells of a 96-well flat-bottom plate.
3. To each well add 15 μL of apoptosis-modifying agents (10× concentration) or buffer control and 60 μL IMDM with 5% serum. If two agents are used in combination, only 45 μL of IMDM is required.
4. Cover with a lid and incubate at 37 °C in a 5% CO_2 incubator for the desired length of time.
5. Vigorously pipet the well to dislodge adherent cells and pool multiple 5 wells together into 500 μL Eppendorf and centrifuge at 300 × *g* for 5 min.
6. Resuspend in 3% glutaraldehyde in 0.1 M sodium cacodylate buffer, pH 7.3 for 2 h.

7. Centrifuge at 300 × *g* for 5 min and resuspend in 0.1 M cacodylate. Incubate for 10 min and repeat three times.
8. Post-fix in 1% osmium tetroxide in 0.1 M sodium cacodylate for 45 min.
9. Centrifuge at 300 × *g* for 5 min and resuspend in 0.1 M cacodylate. Incubate for 10 min and repeat three times.
10. Dehydrate in 50%, 70%, 90% and 100% normal grade acetones for 10 min each, then for a further two 10-min changes in analytical grade acetone.
11. Embed in Araldite resin.
12. Cut 1 μm sections on an ultramicrotome, stain with toluidine blue, and view in a light microscope to select suitable areas for further study.
13. Cut 60 nm ultrathin sections from those areas identified and stain in uranyl acetate and lead citrate.
14. Viewed in a transmission electron microscope.

3.2 Analysis of Cell Membrane Changes Associated with Neutrophil Apoptosis

3.2.1 Annexin V/Propidium Iodide Staining

Apoptosis is associated with the loss of phospholipid asymmetry and the externalization of phosphatidylserine on the outer surface of the plasma membrane. Annexin V (AnnV) binds specifically to phosphatidylserine in the presence of Ca^{2+}, and its fluorescent derivatives can be used to identify apoptotic cells. A range of conjugated fluorophores are available to suit different experimental designs. We have used AnnV-AF647 to provide a good signal in multiparameter flow-cytometric analysis and compatibility with use of propidium iodide (PI), a nuclear dye, which is excluded from cells with an intact plasma membrane. Apoptotic cells in culture ultimately lose membrane integrity and become necrotic. The combination of PI with AnnV is thus able to discriminate viable, apoptotic and necrotic cells by a simple and rapid flow-cytometric method (Fig. 1c, f).

1. Suspend freshly isolated peripheral blood neutrophils (at least 97% purity—perform cytocentrifuge preparation as described below) at 1×10^7 cells/mL in IMDM supplemented with 5% autologous serum and penicillin/streptomycin.
2. Add 75 μL of neutrophil suspension to wells of a 96-well flat-bottom plate. To each well add 15 μL of apoptosis-modifying agents (10× concentration) or buffer control and 60 μL of IMDM/5% serum. If two agents are used in combination, only 45 μL of IMDM is required.
3. Cover with a lid, and incubate at 37 °C in a 5% CO_2 incubator for the desired length of time.

4. Vigorously pipet the well to dislodge adherent cells and transfer 75 μL of cells into a flow tube containing 250 μL of AnnV diluted in AnnV buffer (*see* **Note 5**).
5. Incubate for 10 min on ice.
6. Immediately prior to running each sample on a flow cytometer add PI (1 μL of 1 mg/mL stock solution).
7. Analyze by flow cytometry using FL-4/FL-2 channel analysis following appropriate compensation. Live cells are AnnV negative and PI negative; apoptotic cells are AnnV positive and PI negative; necrotic cells are both AnnV and PI positive (Fig. 1c, f).

3.2.2 Loss of CD16 Expression

Neutrophil apoptosis is associated with a marked downregulation of FcγRIII (CD16) [18]. A simple flow-cytometric method using an anti-CD16 monoclonal antibody will reliably discriminate apoptotic from nonapoptotic neutrophils in a mixed cell population. However, it should be noted that eosinophils express low levels of CD16 and that neutrophil expression of CD16 can be regulated during cellular activation, so it is important to verify that CD16 expression correlates with apoptosis using additional methods (*see* **Note 6**).

1. Suspend freshly isolated peripheral blood neutrophils (at least 97% purity) at 2×10^6 cells/mL in IMDM/ 5% autologous serum.
2. Add 90 μL of neutrophil suspension to wells of a 96-well flat-bottom. To each well add 10 μL of apoptosis-modifying agents (10× concentration) or buffer control, cover with a lid, and incubate at 37 °C in a 5% CO_2 incubator for the desired length of time.
3. Transfer 2×10^5 neutrophils (40 μL of a 5×10^6/mL suspension) to a 96-well U-bottom plate and centrifuge at 200 × *g* for 2 min at 4 °C. Discard the supernatants.
4. Briefly vortex the plate to disrupt the pelleted cells.
5. Incubate neutrophils with APC/Cy7 anti-CD16 (clone 3G8) or nonbinding control IgG1 in 50 μL of cytometry buffer on ice for 30 min.
6. Add 75 μL of cytometry buffer. Centrifuge at 200 × *g* for 2 min at 4 °C. Discard the supernatants.
7. Resuspend cells in 125 μL of cytometry buffer. Centrifuge at 200 × *g* for 2 min at 4 °C. Discard the supernatants and vortex the plate for 5 s to disrupt the pelleted cells.
8. Analyze cells using a flow cytometer. Apoptotic neutrophils have low CD16 expression fluorescence; nonapoptotic neutrophils have high CD16 fluorescence.

3.3 Measurement of Mitochondrial Membrane Potential Using MitoCapture™

Mitochondrial membrane potential ($\Delta\Psi M$) is lost during apoptosis with the formation of pores which allow the passage of proteins from the mitochondria into the cytosol, leading to activation of caspase 9 and subsequently caspase 3. Changes in neutrophil mitochondrial membrane potential may be measured using MitoCapture™, a cationic dye which exhibits membrane potential-dependent accumulation in mitochondria. In viable cells with intact mitochondrial membrane potential, MitoCapture™ is able to enter mitochondria and polymerize, where it fluoresces in the red (FL-2) channel indicated by a fluorescence emission shift from green (525 nm) to red (590 nm). However, when $\Delta\Psi M$ is lost during apoptosis, MitoCapture™ is unable to accumulate in the mitochondria and remains in the monomeric form in the cytosol and fluoresces in the green (FL-1) channel. Apoptosis-associated mitochondrial depolarization is quantified flow-cytometrically by an increase in FL-1 fluorescence (Fig. 2a) or alternatively as a plate-based fluorometric assay with a decrease in the red/green fluorescence intensity ratio (*see* **Note 7**).

1. This protocol assumes the use of a MitoCapture™ mitochondria permeability detection kit (Biovision).
2. For each experimental sample dilute 0.5 μL of MitoCapture™ reagent in 500 μL of prewarmed (37 °C) MitoCapture™ Incubation buffer in a 1.5-mL Eppendorf tube.
3. Suspend freshly isolated peripheral blood neutrophils (at least 97% purity) at 1×10^7 cells/mL in IMDM/5% autologous serum.
4. Add 75 μL of neutrophil suspension to wells of a 96-well flat-bottom plate.
5. To each well add 15 μL of apoptosis-modifying agents (10× concentration) or buffer control and 60 μL of IMDM/5% serum. If two agents are used in combination, only 45 μL of IMDM is required.
6. Cover with a lid, and incubate at 37 °C in a 5% CO_2 incubator for the desired length of time.
7. Add 150 μL of cell suspension to 500 μL of diluted MitoCapture™ reagent.
8. Incubate on shaking heat block at 300 rpm, 37 °C, 15 min.
9. Centrifuge at 300 × *g* for 5 min.
10. Resuspend in 300 μL of MitoCapture™ Incubation buffer.
11. Analyze on flow cytometer with an increase in fluorescence in FL-1 channel (Excitation/Emission = 488/530 ± 30 nm) indicating loss of $\Delta\Psi M$.

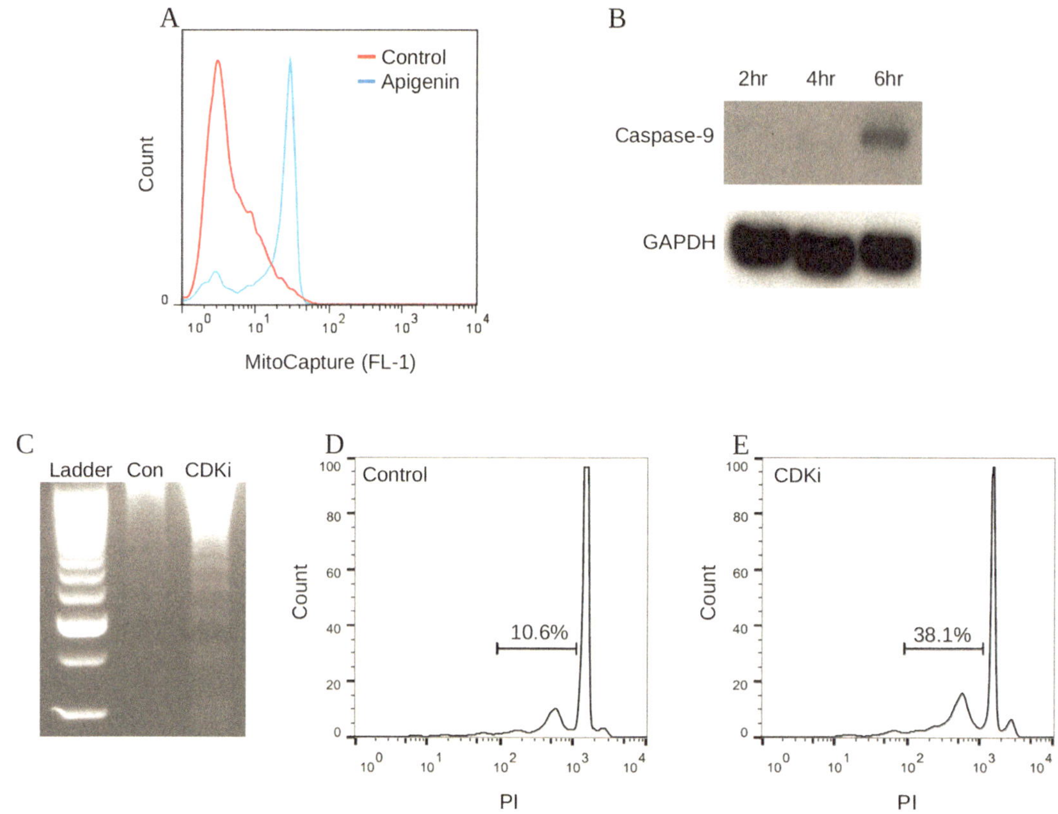

Fig. 2 Analysis of intracellular apoptotic events. Loss of mitochondrial membrane potential is demonstrated by increased fluorescence in MitoCapture™-labelled neutrophils after 2 h incubation with apigenin (50 μM), a known inducer of apoptosis, relative to vehicle control. ((**a**) red line, control, blue line, apigenin). Constitutive apoptosis is associated with a time-dependent accumulation of intracellular cleaved caspase 9 within neutrophils, GAPDH loading control (**b**). Increased apoptosis is also demonstrated by DNA fragmentation in cyclin-dependent kinase inhibitor (CDKi) treated cells visible on gel electrophoresis with "laddering" ((**c**) 4 h). Hypodiploid DNA is also visible when analyzed by flow cytometry ((**d**) control; (**e**) CDKi 20 h)

3.4 Analysis of Caspase Activation

Caspase signalling is central to both the initiation and execution of the apoptotic pathway with detection by Western blot of the active forms of the protein, or disappearance of the inactive precursor, in response to apoptotic stimuli. An example of Western blotting for active caspase 9 is shown in Fig. 2b (*see* **Note 8**).

3.4.1 Western Blotting for Caspases and Apoptotic Proteins

1. These instructions assume the use of an XCell SureLock™ Mini-Cell Electrophoresis System (Invitrogen).
2. Suspend freshly isolated peripheral blood neutrophils (at least 97% purity) at 5×10^6 cells/mL in IMDM/5% autologous serum. Dispense 1 mL of neutrophil suspension into 2-mL round-bottomed polypropylene tubes and incubate ± apoptosis-modifying agents at 37 °C for the desired length of time.

3. Centrifuge the neutrophil suspensions at 13,000 × *g* for 1 min. Discard the supernatants.
4. Resuspend the cell pellets in protease inhibitor buffer (90 μL per 5 × 10^6 cells) and incubate on ice for 10 min.
5. Add 10% NP-40 in TBS (10 μL per 5 × 10^6 cells), immediately vortex for 5 s, and incubate for a further 10 min on ice.
6. Centrifuge at 13,000 × *g* for 20 min at 4 °C, and retain the detergent-soluble supernatant containing the cytosolic and membrane fractions. Samples can be stored at this point (−20 °C) prior to further analysis.
7. Calculate protein concentration of each sample using BCA protein assay per manufacturer's instructions.
8. Transfer volume equivalent to 30 μg of protein into fresh Eppendorf tubes and make up to total volume of 30 μL with PBS (without cations) and 10 μL of 4× sample buffer.
9. Heat at 95 °C for 5 min.
10. Load samples (30 μL per lane) onto a 12% polyacrylamide gel alongside molecular weight standards and run at 110 V until the dye front reaches the bottom of the gel.
11. Transfer proteins from the gel to the PVDF membrane at 80 V for 1 h at 4 °C.
12. Wash the membrane in TBS/0.1% Tween 20 for 5 min.
13. Block the membrane in 10 mL of 5% dried milk powder in TBS/0.1% Tween 20 at room temperature on a rocking platform for 1 h.
14. Wash the membrane three times in TBS/0.1% Tween®20 for 5 min.
15. Incubate with the primary antibody at the indicated concentrations in TBS/0.1% Tween 20 containing 5% dried milk powder (5 mL volume). Incubate in a 50 mL conical polypropylene tube overnight at 4 °C on rollers.
16. Wash membrane three times in TBS/0.1% Tween® 20 each for 5 min at room temperature on a rocking platform.
17. Incubate with the secondary antibody (HRP-conjugated polyclonal goat anti-rabbit Ig (Dako) diluted 1:2500 in blocking buffer (5 mL). Incubate in a 50 mL conical polypropylene tube for 2 h at room temperature on a rocking platform.
18. Wash membrane three times in TBS/0.1% Tween® 20 each for 5 min at room temperature on a rocking platform.
19. Develop using enhanced chemiluminescence (ECL Prime; GE Healthcare) according to the manufacturer's instructions.
20. Strip and reprobe blot with β-actin or GAPDH as a loading control.

3.4.2 Fluorometric Homogeneous Caspase Assay

Apoptosis usually occurs via caspase-dependent processes, and an increase in caspase activity gives a general indication of the occurrence of apoptosis. Commercial kits are available in which total caspase activity can be measured (homogeneous caspase assay). However, these kits do not dissect out precisely which caspases are active. They utilize cleavage of a fluorogenic nonspecific caspase substrate (e.g., VAD-fmk, DEVD) to generate a fluorescent product (e.g., FITC, rhodamine 110) so that fluorescence correlates with the extent of total caspase activity.

1. These instructions assume the use of a Homogeneous Caspases Assay Kit (Sigma-Aldrich). (*see* **Note 3**).
2. Suspend freshly isolated peripheral blood neutrophils (at least 97% purity) at 1×10^6 cells/mL in IMDM/10% autologous serum.
3. Plate neutrophils (1×10^5 per well, *see* **Note 9**) and incubate at 37 °C ± apoptosis-modifying agents for the required time in a black 96-well microplate (total volume 100 μL).
4. Dilute stock caspase substrate 1:10 in incubation buffer. Add 100 μL of freshly prepared caspase substrate to each well, plus duplicate wells containing medium alone (negative control) and duplicate wells containing positive control lysate.
5. Cover the plate and incubate at 37 °C for at least 1 h.
6. Measure fluorescence with a plate reader (excitation: 470–500 nm, emission: 500–560 nm).

3.4.3 Caspase Profiling Assay

Fluorometric assays for specific caspases work in a similar way to the homogeneous caspase assays but exploit a certain degree of specificity between the substrates of individual caspases or groups of caspases. Fluorogenic substrates specific for certain caspases (e.g., 2, 3, 8, and 9) are immobilized in a 96-well plate. When cell lysates are added to the wells and incubated with the substrates, the amount of fluorescence generated correlates with the activation of that particular caspase. They can therefore be used to study particular pathways of apoptosis by looking for activity of a caspase protease that is specific for a particular pathway or cell type (e.g., caspase 8 in death receptor-mediated death).

1. These instructions assume the use of the ApoAlert™ Caspase Profiling Plate (Clontech) (*see* **Notes 10** and **11**).
2. Suspend freshly isolated peripheral blood neutrophils (at least 97% purity) at 2×10^6 cells/mL in IMDM/10% autologous serum. Dispense 1 mL of neutrophil suspension (2×10^6 cells) into 2-mL round-bottomed polypropylene tubes and incubate ± apoptosis-modifying agents at 37 °C for the desired length of time.

3. Centrifuge at 220 × *g* for 5 min at 4 °C. Discard the supernatants.
4. Resuspend the cells in 400 μL of ice-cold lysis buffer and incubate for 10 min on ice.
5. Meanwhile, add 10 μL DTT per 1 mL of 2× reaction buffer, then add 50 μL of this mixture to each well of the 96-well caspase profiling plate. Cover the plate with film and incubated for 5 min at 37 °C.
6. Vortex the neutrophil lysates, then add 50 μL from each lysate to duplicate wells of each caspase substrate.
7. Cover the plate with film and incubate for 2 h at 37 °C.
8. Measure fluorescence using a plate reader (excitation: 380 nm, emission 460 nm).

3.5 Analysis of Nuclear Changes Associated with Neutrophil Apoptosis

3.5.1 Gel Electrophoresis of DNA

A characteristic event of apoptosis is the endonuclease-mediated cleavage of DNA at regular intervals along its length, thus generating single-nucleosome fragments of around 180 base pairs, or oligonucleosomal fragments at multiples thereof, following earlier large-scale (50–200 kbp) degradation. Such ordered fragmentation produces discrete sized lengths of DNA with a distinct "laddering" pattern on DNA gel electrophoresis, in contrast to necrotic cell death in which DNA is cleaved randomly, thus producing a smear on a DNA gel (Fig. 2c).

1. Suspend freshly isolated peripheral blood neutrophils (at least 97% purity) at 5 × 10^6 cells/mL in IMDM/10% autologous serum. Dispense 1 mL of neutrophil suspension into 2-mL round-bottomed polypropylene tubes and incubate ± apoptosis-modifying agents at 37 °C for the desired length of time.
2. Extract the genomic DNA using a Wizard® Genomic DNA Purification Kit (Promega).
3. DNA (23 μL DNA plus 7 μL of loading dye) is electrophoretically resolved on a 2% agarose gel containing either ethidium bromide (2.5 μg/mL) or GelRed (5 μL in 50 mL) in TBE buffer at 10 V/cm. Visualize under ultraviolet illumination. DNA from apoptotic cells exhibits a characteristic ladder pattern.

3.5.2 Hypodiploid DNA Content

DNA fragmentation also leads to an apparent reduction in nuclear DNA content of Triton X-100-permeabilized cells, so that staining with a DNA-intercalating dye such as PI allows detection of a "hypodiploid" cell population. This technique works particularly well with neutrophils, because they are terminally differentiated cells which do not undergo proliferation, and consequently generate only two peaks when DNA content is measured: diploid (viable) cells and hypodiploid (apoptotic) cells (Fig. 2d, e).

1. Suspend freshly isolated peripheral blood neutrophils (at least 97% purity) at 1×10^7 cells/mL in IMDM/5% autologous serum and penicillin/streptomycin.
2. Add 75 μL of neutrophil suspension to wells of a 96-well flat-bottom plate. To each well add 15 μL of apoptosis-modifying agents (10× concentration) or buffer control and 60 μL of IMDM/5% serum. If two agents are used in combination, only 45 μL of IMDM is required.
3. Cover with a lid, and incubate at 37 °C in a 5% CO_2 incubator for the desired length of time.
4. Vigorously pipet the well to dislodge adherent cells and transfer 50 μL into flow tubes containing 250 μL of PI solution.
5. Incubate at 4 °C for 15 min in the dark.
6. Analyze by flow cytometry (FL2 channel) to determine the percentage of cells with hypodiploid DNA content.

3.5.3 TUNEL Staining for DNA Breaks

The presence of DNA strand breaks can be assessed by enzymatic methods, since DNA breaks create acceptor sites for enzymes such as terminal deoxyribonucleotidyltransferase (TdT). Addition of TdT together with fluorescein-12-2′deoxyuridine-5′-triphosphate is used to reveal DNA fragmentation in the TUNEL technique.

1. These instructions assume the use of the In Situ Cell Death Detection Kit, Fluorescein (Sigma-Aldrich).
2. Suspend freshly isolated peripheral blood neutrophils (at least 97% purity) at 2×10^7 cells/mL in IMDM/10% autologous serum and penicillin/streptomycin.
3. Add 90 μL of neutrophil suspension to wells of a 96-well flat-bottom plate. To each well add 10 μL of apoptosis-modifying agents (10× concentration) or buffer control.
4. Cover with a lid, and incubate at 37 °C in a 5% CO_2 incubator for the desired length of time.
5. Transfer 100 μL of neutrophil suspension to a 96-well U-bottom flexible plate and centrifuge at 200 × *g* for 2 min at 4 °C. Discard the supernatants.
6. Wash the cells three times by adding 100 μL of PBS per well, centrifuging the plate at 200 × *g* for 3 min at 4 °C, discarding the supernatants, and vortexing the plate for 5 s.
7. Add 100 μL of fixation solution to each well.
8. Incubate on a shaker for 60 min at room temperature.
9. Added 200 μL of PBS to each well then centrifuge the plate at 200 × *g* for 10 min at 4 °C and discard the supernatants.
10. Resuspend the cells in permeabilization solution for 2 min on ice.

11. Add 50 μL of nucleotide mixture to the two negative control wells. The TUNEL reaction mixture is then made up by mixing the enzyme solution (50 μL) with the remaining 450 μL of nucleotide mixture.
12. The two positive control wells are treated for 10 min at room temperature with DNase I to introduce DNA strand breaks.
13. Wash twice in PBS (200 μL per well) then resuspend in TUNEL reaction mixture (50 μL per well).
14. Cover the plate and incubate at 37 °C for 60 min in the dark.
15. Wash twice in PBS (200 μL per well) then transfer to flow cytometry tubes for analysis of fluorescence levels (FL-1).

3.6 Analysis of Macrophage Phagocytosis of Apoptotic Neutrophils

3.6.1 Plate-Based Assay for Phagocytosis of Apoptotic Neutrophils

Macrophage phagocytosis of apoptotic neutrophils may be assessed using minor modifications of a serum-free phagocytosis assay first described by Newman et al. in 1982. This method uses adherent human monocyte-derived macrophages (MDMφ) which are most efficient at ingesting apoptotic cells, but it has also been used successfully with murine peritoneal and bone marrow-derived monocytes. Depending on the phagocytosis pathway being examined, one could add specific opsonins such as C1q, MFG-E8, or other ligands as required. NB: It is important to wash neutrophils in HBSS containing 2 mM EDTA prior to assay to ensure they are free of serum opsonins (e.g., Protein S).

1. This method assumes the use of adherent MDMφ in 48 well TC-treated microplates.
2. Suspend 10^8 freshly isolated peripheral blood neutrophils (at least 97% purity) in 20 mL of IMDM/5% autologous serum. Dispense the neutrophil suspension into a 75 cm^2 cell culture flask and stand the flask on its end in an incubator for 20 h at 37 °C in a 5% CO_2 atmosphere.
3. Harvest the neutrophil suspension into a 50 mL conical polypropylene tube and wash once in warm IMDM containing 2 mM EDTA (25 mL) by centrifuging at 220 × g for 5 min and discarding the supernatant. After washing, gently resuspend the neutrophil pellet in 1 mL of warm (37 °C) IMDM using a plastic pipette to avoid clumping of the cells. Count the cells using a hemocytometer and finally resuspend the aged neutrophils at 6 × 10^6 cells/mL in warm (37 °C) IMDM (no serum).
4. Rinse the MDMφ monolayer with warm (37 °C) IMDM containing 2 mM EDTA to remove nonadherent cells.
5. Overlay the MDMφ monolayer with 0.5 mL of the suspension of pHrodo™-labelled neutrophils (3 × 10^6 cells; an MDMφ–apoptotic target ratio of ~1:6) in IMDM (serum-free), and incubate for 40 min at 37 °C in a 5% CO_2 atmosphere. For

examination of specific apoptotic cell clearance pathways, it may be necessary to add apoptotic cell opsonins (e.g., Protein S or Milk Fat Globule Epidermal Growth Factor 8 (MFG-E8)) as appropriate (*see* **Note 12**).

6. Wash each well with 0.5 mL of ice-cold IMDM to minimize loss of MDMφ via detachment. Use an inverted microscope to check that noningested neutrophils have been largely removed.
7. Repeat the wash step as necessary, verifying by microscopy to ensure that the MDMφ monolayer has not been disrupted. Usually, 3–4 washes are sufficient.
8. Fix in 2.5% glutaraldehyde for 30 min and rinse with PBS.
9. Stain for myeloperoxidase (MPO) with 0.1 mg/mL dimethoxybenzidine and 0.03% (v/v) H_2O_2 in PBS for 60 min at room temperature.
10. Count the percentage of MDMφ (MPO-negative) that have phagocytosed one or more apoptotic neutrophils (MPO-positive) by examination with an inverted microscope of at least five fields (minimum 400 cells), and record as the mean percent phagocytosis of duplicate or triplicate wells.

3.6.2 Flow Cytometry-Based Phagocytosis Assay

This modification of the plate-based phagocytosis assay utilizes a fluorescent label (CellTrace™ Far Red) to identify MDMφ and avoids loss of cells as a result of vigorous washing of the MDMφ monolayer following interaction with aged neutrophils. In addition, it eliminates potential observer bias when counting MDMφ. It is less laborious than the plate-based counting method and offers a method that is free of observer bias and therefore particularly suitable for repeated measurements and screening assays. Use of a pH-sensitive succinimidyl ester (pHrodo™) that is weakly fluorescent at a neutral pH but fluoresces brightly in acidic conditions to identify apoptotic neutrophils present within the acidic phagosomal compartment of MDMφ.

1. This method assumes the use of adherent MDMφ in 48-well TC-treated microplates.
2. Suspend freshly isolated peripheral blood neutrophils (at least 97% purity) at 2×10^7 cells/mL in IMDM/5% autologous serum and penicillin/streptomycin in a 15-mL conical polypropylene tube.
3. Centrifuge at $220 \times g$ for 5 min and discard supernatant. Resuspend in PBS and centrifuge again at $220 \times g$ for 5 min.
4. Resuspend cells at 4×10^6 cells/mL in IMDM/5% autologous serum and penicillin/streptomycin. Dispense the neutrophil suspension into a 75 cm^2 cell culture flask and stand the flask on its end in an incubator for 20 h at 37 °C in a 5% CO_2 atmosphere.

5. Harvest the neutrophil suspension into a 50-mL conical polypropylene tube and wash in warm IMDM containing 2.5 mM EDTA by centrifuging at 220 × *g* for 5 min and discarding the supernatant. After washing, gently resuspend the neutrophil pellet in 1 mL of warm IMDM using a plastic pipette to avoid clumping of the cells. Count the cells using a hemocytometer and resuspend the aged neutrophils at 6×10^6 cells/mL in warm (37 °C) HBSS.
6. Incubate cells with 1 μM pHrodo™ for 30 min at room temperature on a rotary mixer.
7. Centrifuge at 220 × *g* for 5 min and discard supernatant. Resuspend in PBS and centrifuge again at 220 × *g* for 5 min.
8. Resuspend cells at 6×10^6/mL in warm (37 °C) IMDM (serum free).
9. Prior to the assessment of MDMφ phagocytosis of apoptotic cells, label MDMφ with CellTrace™ Far Red (1 μM in IMDM) for 30 min at 37C and then wash in prewarmed (37 °C) IMDM.
10. Rinse the MDMφ monolayer with warm (37 °C) HBSS without divalent cations containing 2.5 mM EDTA to remove nonadherent cells and potential apoptotic cell opsonins.
11. Overlay the MDMφ monolayer with 500 μL of the suspension of pHrodo™-labeled neutrophils (3×10^6 cells; a MDMφ–apoptotic target ratio of ~1:6) in IMDM (serum-free), and incubate for 40 min at 37 °C in a 5% CO_2 atmosphere.
12. Gently aspirate the neutrophil suspension from the wells and wash each well once with 0.5 mL of IMDM.
13. Detach the MDMφ by incubation with 250 μL 0.25% trypsin–EDTA solution for 5 min at 37 °C followed by 10 min at 4 °C (*see* **Note 13**).
14. Harvest the detached MDMφ by vigorous pipetting and place on ice.
15. Analyze the entire samples (unfixed) immediately using a flow cytometer, aiming to collect at least 5000 events in the MDMφ gate.
16. The MDMφ population can be distinguished from apoptotic cells on the basis of their distinct fluorescence emission at 660 nm (*see* Fig. 3). The percentage of the $pHrodo^{+ve}$ events within the MDMφ gate represents the proportion of MDMφ that have internalized apoptotic cells.

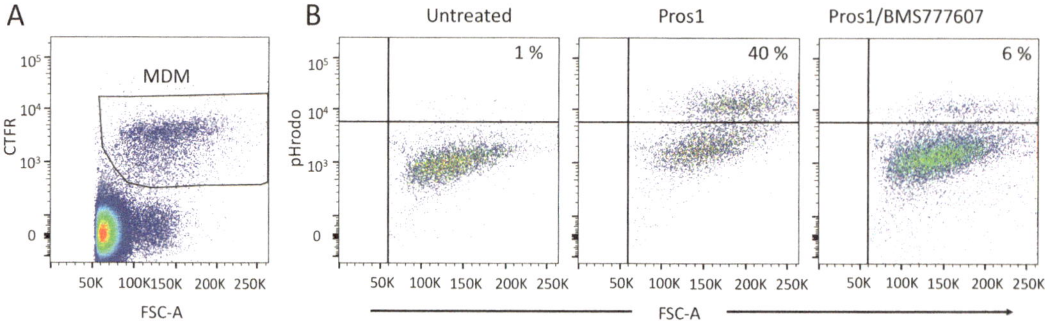

Fig. 3 Analysis of macrophage phagocytosis of apoptotic cells by flow cytometry to define molecular mechanism. Dexamethasone treated monocyte-derived macrophages (250 nM for 5 days) were labelled with CellTrace™ Far Red and then incubated with pHrodo™-labelled apoptotic neutrophils in the absence or presence of 250 nM Protein S for 30 min prior to detachment with trypsin–EDTA and analysis by flow cytometry. (**a**) Monocyte-derived macrophages were distinguished from apoptotic neutrophils on the basis of CellTrace™ Far Red (CTFR) fluorescence. (**b**) Macrophages that had internalized apoptotic cells in the absence (Untreated) or presence of Protein S (Pros1) was calculated on the basis of increased fluorescence at 585 nm. (**c**) Pretreatment of monocyte-derived macrophages with the tyrosine kinase inhibitor BMS777607 was used to block phagocytosis that was dependent on the Mer/Protein S pathway (Pros1/BMS777607)—in this example phagocytosis was reduced from 40% to 6%, equivalent to 85% reduction in phagocytosis in the presence of BMS777607

4 Notes

1. We culture neutrophils in IMDM containing 10% serum. We routinely use autologous platelet rich plasma-derived serum, prepared from platelet-rich plasma (harvested after centrifugation of citrate-anticoagulated blood) by recalcification with 20 mM $CaCl_2$ for 1 h at 37 °C in glass tubes. However, bovine serum may also be used. Alternatively, neutrophils may be cultured in the absence of serum with a small amount of added protein (e.g., 0.5% (w/v) serum albumin), but apoptosis will proceed more rapidly. Cell saver pipette tips should be used when pipetting neutrophils to minimize cell damage.
2. Neutrophil granules contain high concentrations of proteases. Extreme care must be taken to keep all samples on ice during the preparation of neutrophil lysates in order to prevent the protein of interest being degraded before blotting. Higher protease inhibitor concentrations are required for neutrophils than for other cell types. The protocol describes high protease inhibitor concentrations that we have used successfully in our laboratory; however, some experimentation with regard to manipulation of concentrations may be required. Recent evidence demonstrates that preincubation of neutrophils with 1 mM phenylmethanesulfonyl fluoride (PMSF) for up to 1 h prior to lysis may be required to prevent neutrophil protease-

mediated cleavage of intracellular proteins following addition of lysis buffer.

3. The protocol described is based on the homogeneous caspase kit available from Sigma-Aldrich, and represents the simplest method for fluorometric detection of total caspase activity. However, other commercial kits are available, e.g. Apo-ONE Homogenous Caspase 3/7 Assay (Promega) which may be based on either the same or slightly different protocol.
4. Some investigators add extra serum to the cells in the cytospin chamber to avoid artifacts caused by cell breakage during centrifugation. We have observed that adding serum reduces the effectiveness of deposition of neutrophils that have progressed to a late stage of apoptosis ("late apoptotic" or postapoptotic cells), so potentially underestimating the true rate of neutrophil apoptosis. Late apoptotic neutrophils appear as cell ghosts with little or no evidence of nuclear staining, having undergone "nuclear evanescence." Furthermore, late apoptotic cell ghosts in a cell population may be overlooked because the lack of nuclear material means that they stain very faintly with Romanowsky stains such as Diff-Quik™. In addition, late apoptotic cells may be inadvertently gated out as "debris" when analyzed by flow cytometry.
5. Dilutions prior to flow cytometry should be performed using AnnV binding buffer, because in the absence of Ca^{2+} AnnV will rapidly dissociate from PtdSer on the apoptotic cell surface.
6. Eosinophils present in the granulocyte population also have low levels of CD16 expression, so eosinophil counts (by analysis of cell morphology in cytocentrifuge preparations or CD16 expression) should be performed at baseline.
7. Another way of analyzing mitochondrial changes associated with apoptosis is Western blotting of the Bcl-2 family proteins that are important in the control of the intrinsic (mitochondrial) apoptotic pathway. The expression profile of Bcl-2 family proteins in peripheral blood neutrophils has been analyzed and shows expression of Bak, Bad, Bcl-w, and Bfl-1 in these cells but relatively little Bcl-2, Bcl-xL, Bik, and Bax [19]. Therefore, the more abundant proteins should be considered first for analysis for roles in neutrophil apoptosis, although some proteins, such as Bax, may be transcriptionally upregulated in response to certain apoptotic stimuli under the control of p53, and may therefore still be important in neutrophil apoptosis. For a detailed Western blotting protocol, please refer to Subheading 3.4.1 as the method is identical to that for caspases.
8. The 12% gel recommended in this protocol is based on blotting for cleaved caspase 3 (14–21 kDa). However, gel percentages may be altered according to the size of the protein of interest.

In general, low percentage gels are used to blot for high MW proteins, and high percentage gels for low MW proteins. Therefore, the composition of the gel may have to be altered depending on the size of the protein of interest. Some proteins that are important in apoptosis are of low size similarly to cleaved caspase 3 (e.g., Bax (21 kDa) and truncated Bid (15 kDa)) and therefore a similar percentage gel will probably be appropriate, whereas other proteins such as procaspase 8 (55 kDa) and procaspase 3 (32 kDa) are larger and may require a lower percentage gel for optimal resolution.

9. The protocol for this kit recommends a cell number of 4×10^4 cells per well in a volume of 100 μL. However, in assays involving neutrophils, cell numbers are often boosted in order to amplify the signals obtained from the assay, particularly as neutrophils are so abundant. On the other hand, cell density may affect the rate of spontaneous apoptosis of these cells during culture, with a high cell density promoting survival, especially at densities of 8×10^6 cells/mL and above [37]. Therefore, some experimentation may be required, to manipulate the cell density, volume and time of culture in order to gain the best results with this assay.
10. Colorimetric assays for single-caspase activity are also available. These are similar to the fluorometric assays, and follow a similar protocol, but instead utilize cleavage of a chromophore (e.g., *p*-nitroanilide) from caspase substrates as a measure of caspase activation. Development of color can be monitored using a spectrophotometer or microplate reader (405 nm), and activity quantified by comparison with a calibration curve constructed using known standards.
11. This protocol applies to any cell type, as caspase activation is a general event of apoptosis and is not unique to neutrophils. However, differences may exist between the expression profiles of the various caspases in different cell types, and this must be borne in mind when selecting an assay to use in neutrophils. Caspases 1, 3, 4, and 7–10 are expressed in neutrophils [15]. In contrast, it has been reported that caspase 2 is absent from peripheral blood neutrophils, although it is expressed in HL-60 cells.
12. Some molecular pathways of phagocytosis are opsonin-independent (e.g., those involving direct PtdSer recognition molecules such as BAI-1) and others are dependent on the presence of specific apoptotic cell opsonins (e.g., Integrin- or Mer-dependent phagocytosis). Our described method aims to specifically remove apoptotic cell opsonins that are present in serum and give a reliable estimate of basal levels of phagocytosis. Depending on the phagocytosis pathway being examined, it will be necessary to add specific apoptotic cell opsonins such as Protein S, Gas-6, C1q, MFG-E8, or others as required.

13. Treatment with trypsin–EDTA may lead to clumping of cells leading to blockage of the flow cytometer's sample intake nozzle. Clumping may be minimized by adding 50 μL of bovine serum to each well following incubation with trypsin–EDTA.

Acknowledgments

The authors would like to acknowledge funding from the Medical Research Council UK (MR/KO13386/1: AGR), Engineering and Physical Sciences Research Council and MRC Centre for Doctoral Training in Optical Imaging (OPTIMA) (EP/L016559; NDB), European Research Council (ERC Consolidator Grant 771443, MV). The facilities and staff of the Queen's Medical Research Flow Cytometry Facility are gratefully acknowledged. Figures 1 and 2 are reprinted by permission from Springer, Methods in Molecular Biology; Assessment of neutrophil apoptosis. Dorward DA, Rossi AG, Dransfield I, Lucas CD (2014) 1124:159–180. https://doi.org/10.1007/978-1-62703-845-4_10

References

1. Nourshargh S, Alon R (2014) Leukocyte migration into inflamed tissues. Immunity 41:694–707
2. Borregaard N, Sørensen OE, Theilgaard-Mönch K (2007) Neutrophil granules: a library of innate immunity proteins. Trends Immunol 28:340–345
3. Nauseef WM, Borregaard N (2014) Neutrophils at work. Nat Immunol 15:602–611
4. Robb CT, Regan KH, Dorward DA et al (2016) Key mechanisms governing resolution of lung inflammation. Semin Immunopathol 38:425–448
5. Whyte MK, Meagher LC, MacDermot J et al (1993) Impairment of function in aging neutrophils is associated with apoptosis. J Immunol 150:5124–5134
6. Dransfield I, Stocks SC, Haslett C (1995) Regulation of cell adhesion molecule expression and function associated with neutrophil apoptosis. Blood 85:3264–3273
7. Hart SP, Ross JA, Ross K et al (2000) Molecular characterization of the surface of apoptotic neutrophils: implications for functional downregulation and recognition by phagocytes. Cell Death Differ 7:493–503
8. Duffin R, Leitch AE, Fox S et al (2010) Targeting granulocyte apoptosis: mechanisms, models, and therapies. Immunol Rev 236:28–40
9. Colotta F, Re F, Polentarutti N et al (1992) Modulation of granulocyte survival and programmed cell death by cytokines and bacterial products. Blood 80:2012–2020
10. Hannah S, Mecklenburgh K, Rahman I et al (1995) Hypoxia prolongs neutrophil survival in vitro. FEBS Lett 372:233–237
11. Pryde JG, Walker A, Rossi AG et al (2000) Temperature-dependent arrest of neutrophil apoptosis. Failure of Bax insertion into mitochondria at 15 °C prevents the release of cytochrome c. J Biol Chem 275:33574–33584
12. Murray J, Barbara JA, Dunkley SA et al (1997) Regulation of neutrophil apoptosis by tumor necrosis factor-alpha: requirement for TNFR55 and TNFR75 for induction of apoptosis in vitro. Blood 90:2772–2783
13. Scheel-Toellner D, Wang K, Craddock R et al (2004) Reactive oxygen species limit neutrophil life span by activating death receptor signaling. Blood 104:2557–2564
14. Moulding DA, Quayle JA, Hart CA et al (1998) Mcl-1 expression in human neutrophils: regulation by cytokines and correlation with cell survival. Blood 92:2495–2502
15. Geering B, Simon H-U (2011) Peculiarities of cell death mechanisms in neutrophils. Cell Death Differ 18:1457–1469

16. Maianski NA, Maianski AN, Kuijpers TW et al (2004) Apoptosis of neutrophils. Acta Haematol 111:56–66
17. Haslett C, Guthrie LA, Kopaniak MM et al (1985) Modulation of multiple neutrophil functions by preparative methods or trace concentrations of bacterial lipopolysaccharide. Am J Pathol 119:101–110
18. Dooley DC, Simpson JF, Meryman HT (1982) Isolation of large numbers of fully viable human neutrophils: a preparative technique using Percoll density gradient centrifugation. Exp Hematol 10:591–599
19. Dorward DA, Lucas CD, Alessandri AL et al (2013) Technical advance: autofluorescence-based sorting: rapid and nonperturbing isolation of ultrapure neutrophils to determine cytokine production. J Leukoc Biol 94:193–202
20. Sabroe I, Prince LR, Dower SK et al (2004) What can we learn from highly purified neutrophils? Biochem Soc Trans 32:468–469
21. Savill JS, Wyllie AH, Henson JE et al (1989) Macrophage phagocytosis of aging neutrophils in inflammation. Programmed cell death in the neutrophil leads to its recognition by macrophages. J Clin Invest 83:865–875
22. Sporn SA, Eierman DF, Johnson CE et al (1990) Monocyte adherence results in selective induction of novel genes sharing homology with mediators of inflammation and tissue repair. J Immunol 144:4434–4441
23. Liu Y, Cousin JM, Hughes J et al (1999) Glucocorticoids promote nonphlogistic phagocytosis of apoptotic leukocytes. J Immunol 162:3639–3646
24. Michlewska S, Dransfield I, Megson IL et al (2009) Macrophage phagocytosis of apoptotic neutrophils is critically regulated by the opposing actions of pro-inflammatory and anti-inflammatory agents: key role for TNF-alpha. FASEB J 23:844–854
25. Lemke G (2019) How macrophages deal with death. Nat Rev Immunol. In press
26. Nagata S, Suzuki J, Segawa K et al (2016) Exposure of phosphatidylserine on the cell surface. Cell Death Differ 23:952–961
27. Fadok VA, Bratton DL, Konowal A et al (1998) Macrophages that have ingested apoptotic cells in vitro inhibit proinflammatory cytokine production through autocrine/paracrine mechanisms involving TGF-beta, PGE2, and PAF. J Clin Invest 101:890–898
28. Girkontaite I, Urbonaviciute V, Maseda D et al (2007) Apoptotic cells selectively suppress the Th1 cytokine interferon gamma in stimulated human peripheral blood mononuclear cells and shift the Th1/Th2 balance towards Th2. Autoimmunity 40:327–330
29. Poon IKH, Lucas CD, Rossi AG et al (2014) Apoptotic cell clearance: basic biology and therapeutic potential. Nat Rev Immunol 14:166–180
30. Hébert MJ, Takano T, Holthöfer H et al (1996) Sequential morphologic events during apoptosis of human neutrophils. Modulation by lipoxygenase-derived eicosanoids. J Immunol 157:3105–3115
31. Dransfield I, Buckle AM, Savill JS et al (1994) Neutrophil apoptosis is associated with a reduction in CD16 (Fc gamma RIII) expression. J Immunol 153:1254–1263
32. Homburg CH, de Haas M, von dem Borne AE et al (1995) Human neutrophils lose their surface Fc gamma RIII and acquire Annexin V binding sites during apoptosis in vitro. Blood 85:532–540
33. Reutelingsperger CPM, Dumont E, Thimister PW et al (2002) Visualization of cell death in vivo with the annexin A5 imaging protocol. J Immunol Methods 265:123–132
34. Dransfield I, Zagórska A, Lew ED et al (2015) Mer receptor tyrosine kinase mediates both tethering and phagocytosis of apoptotic cells. Cell Death Dis 6:e1646
35. Subiros-Funosas R, Mendive-Tapia L, Sot J et al (2017) A Trp-BODIPY cyclic peptide for fluorescence labelling of apoptotic bodies. Chem Commun (Camb) 53:945–948
36. Dorward DA, Rossi AG, Dransfield I et al (2014) Assessment of neutrophil apoptosis. Methods Mol Biol 1124:159–180
37. Hannah S, Nadra I, Dransfield I et al (1998) Constitutive neutrophil apoptosis in culture is modulated by cell density independently of β_2-integrin-mediated adhesion. FEBS Lett 421:141–146

Chapter 14

Optical Methods for the Measurement and Manipulation of Cytosolic Calcium Signals in Neutrophils

Maurice B. Hallett, Rhiannon E. Roberts, and Sharon Dewitt

Abstract

The measurement and manipulation of cytosolic free Ca^{2+} of neutrophils is crucial for investigating the mechanisms within living neutrophils which generate Ca^{2+} signals and the cellular responses triggered by them. Optical methods for this are the most applicable for neutrophils, and are discussed here, especially the use of fluorescent indicators of Ca^{2+} and photoactivation of reagents involved in Ca^{2+} signaling. Both of these synthetic agents can be loaded into neutrophils as lipid-soluble esters or can be microinjected into the cell. In this chapter, we outline some of the techniques that have been used to monitor, visualize, and manipulate Ca^{2+} in neutrophils.

Key words Calcium signaling, Cytosolic calcium flux, Fluorescent calcium indicator dye, Photoactivation, Calcium imaging

1 Introduction

The measurement and manipulation of cytosolic free Ca^{2+} permits the investigation of the mechanisms of generation of the Ca^{2+} signal and cellular responses to these Ca^{2+} signals within living neutrophils. The optical methods most applicable to neutrophils, which will be discussed here, are (1) the use of fluorescent indicators of Ca^{2+} and (2) photoactivation of reagents involved in Ca^{2+} signaling. Both of these synthetic agents can be loaded into neutrophils as lipid-soluble esters or can be microinjected into the cell. In this chapter, we will outline some of the techniques that have been used to monitor, visualize, and manipulate Ca^{2+} in neutrophils. See additional reference for further details on some of these methods [1] or a discussion of other methods not discussed here, such as the expression of luminescent and fluorescent indicators of Ca^{2+} [2].

Mark T. Quinn and Frank R. DeLeo (eds.), *Neutrophil: Methods and Protocols*, Methods in Molecular Biology, vol. 2087, https://doi.org/10.1007/978-1-0716-0154-9_14,

1.1 Principle of Action

The fluorescent Ca^{2+} chelator probes, developed by the Nobel laureate RY Tsein, have opened up the study of cytosolic free Ca^{2+} in small cells, including neutrophils [3, 4]. The ester-method of loading the probe into the cytosol of neutrophils works successfully. These indicators can be synthesized or purchased with the ester groups, which both "mask" the Ca^{2+} binding part of the molecule and makes them lipid soluble. Thus, the ester-derivative readily crosses the plasma membrane and enters the neutrophil cytosol. Here, esterases cleave the ester bond to generate the acid form of the probe, which becomes entrapped within the cell as it is hydrophilic and thus unable to easily cross the plasma (or other) membrane. It is the acid form of the probe that is also the Ca^{2+} sensing form. There are two potential problems with this approach, (1) the accumulation of indicator into organelles, and (2) partial (rather than full) hydrolysis of the probe to generate products with increased hydrophilicity, but that are insensitive to Ca^{2+} [5]. However, these problems are easily avoided in the protocols given and the measurement methodologies would detect such problems. It is, however, important to limit the amount of probe loaded in this way because (1) as the indicator is a Ca^{2+} chelator, it will buffer cytosolic free Ca^{2+} changes and "blunt" the very response that is being measured and (2) there is a toxicity associated with the products of de-esterification of the probe (namely formaldehyde and H^+ ions), which can cause a reduction in ATP levels in neutrophils [6] and may have other toxic effects on the cell, including stimulation of cell aggregation. Some of these problems can be overcome by microinjection of nonesterified forms of the probes or high molecular weight forms that do not accumulate in organelles [1, 7], thereby circumventing the problems of location and toxicity associated with ester-loading. SLAM-injection (simple-lipid-assisted microinjection) and electroinjection (*see* Chapter 9) works well with neutrophils and are the methods of choice [8].

2 Materials

1. Purified neutrophils (e.g., *see* Chapters 3 and 4 in this volume).
2. 1–4 mM ionomycin stock solution: Prepare 1–4 mM ionomycin stock solution in DMSO so that it can be diluted 1:1000 into the final assay mixtures. When added to cell suspensions, the final concentration should be 1–4 μM so that the final DMSO concentration is acceptable (i.e., 0.1%) (*see* Subheading 3.2).
3. 1.5 mM digitonin stock solution: digitonin is readily soluble in aqueous media, and a stock solution of 1.5 mM can be made in the desired experimental medium. Note that commercially

Table 1
Fluorescent Ca^{2+} probes useful in neutrophils

	K_d (Ca^{2+} nM)	Excitation (nm)	Emission (nm)	Mode
Calcium Crimson	205	588	611	S
Calcium Green-1	189	506	534	S
Calcium Green-2	574	506	531	S
Calcium Green-5N	3300	506	531	S
Calcium Orange	328	554	575	S
FFP-18	400	340 and 380	505	R
Fluo3	864	506	526	S
Fluo3FF	62,000	506	526	S
Fluo4	345	494	516	S
Fura Red	133	420 and 480	640	R
Fura2	224	340 and 380	505	R
FuraFF	38,000	340 and 380	504	R
Indo-1	250	340	410 and 485	R
Mag-Fura-2	25,000	340 and 380	504	R
Mag-Fura-5	28,000	340 and 380	504	R
MOMO	nd	~495	~515	S
Quin2	114	340	490	S
Rhod-2	1000	553	576	S

All these probes are available as cell-permeant esters. The mode of use is given as single wavelength (*S*) or the ratio of two wavelengths (*R*)

available sources of digitonin are rarely pure, and the fractional purity should be used when calculating the correct concentration.

4. Dry dimethyl sulfoxide (DMSO) in sealed vials (e.g., Sigma).
5. 25% pluronic (w/v) in DMSO.
6. 20 mM EGTA solution in deionized H_2O.
7. Fluorescent probes: the properties of fluorescent Ca^{2+} probes suitable for use in neutrophils are summarized in Table 1. These reagents and those for manipulation of cytosolic free Ca^{2+} are available from Invitrogen/Molecular Probes, Sigma, Calbiochem, Teflabs, or their agents.

3 Methods

3.1 Selecting the Appropriate Fluorescent Ca^{2+} Probe

3.1.1 Ca^{2+} Affinity

There are now a myriad of commercially available Ca^{2+} probes, many of which are available as ester derivatives suitable for cytosolic loading into neutrophils (*see* Table 1). When selecting which probe to use, it is important to consider its Ca^{2+} affinity. As in other cell types, neutrophils have resting cytosolic free Ca^{2+} concentrations of ~100 nM. On stimulation, this rises transiently to ~1 μM. As the fluorescent signal depends upon the binding of Ca^{2+} to the probe, the Ca^{2+} dissociation constant, K_d, will define the range over which the probe can be usefully employed. A probe with a K_d of 300 nM will be only ~25% saturated in the resting cell, and thus, 75% of its dynamic range will be available to monitor a rise in cytosolic free Ca^{2+}. It will, however, be difficult to measure cytosolic free Ca^{2+} concentrations above 1–3 μM, as the probe will be more than 90% saturated with Ca^{2+}. In contrast, a probe with a K_d of 1 μM would be more useful for higher Ca^{2+} changes. The K_d for Ca^{2+} will also determine its Ca^{2+} buffering effect within the cytosol. Fortunately, neutrophils have high endogenous Ca^{2+} buffering, with estimates ranging from 1:1000 to 1:3000 [6, 9]. An intracellular fura2 concentration of 25–50 μM is thus estimated to increase the Ca^{2+} buffering by only about 10%. With probes of higher K_d such as fluo3 or fluo4 the buffering effect would be even less.

3.1.2 Single- or Dual-Wavelength Ca^{2+} Probes

There are two types of fluorescent Ca^{2+} probes: (1) those that change their signal at two wavelengths, either on excitation or emission (ratiometric dyes) and (2) those that change their signal at only one wavelength (*see* Table 1). The single wavelength, non-ratiometric indicators, such as fluo3 or fluo4, can also be excited at visible wavelengths (such as 488 nm, a wavelength produced by laser light). Techniques involving confocal microscopy are powerful for locating the Ca^{2+} change within neutrophils and also can give information on fast time scales (1–10 ms). However, as the indicators produce only a single intensity change on binding Ca^{2+}, caution must be exercised in interpreting an increase in intensity as being solely due to an increase in cytosolic free Ca^{2+} concentration. For example, if a brightly loaded organelle moves into the confocal imaging plane, this would give an increased fluorescent signal unrelated to cytosolic free Ca^{2+} concentration. For this reason, it is recommended that Ca^{2+} measurements are initially performed using ratiometric indicators where there is certainty that the ratio change would be caused by a Ca^{2+} signal. The excitation spectrum of perhaps the most widely used indicator, fura2, shifts on binding Ca^{2+}, so that there is significant fluorescence from both the Ca^{2+}-bound and the Ca^{2+}-free forms of the probe. This permits monitoring of both the Ca^{2+}-free and the Ca^{2+}-bound forms of the indicator. By monitoring two wavelengths, one on either side of the isoemissive point (usually 340 nm and 380 nm), a reassuring fluorescent "Ca^{2+} signature" is observed, with the signal at one wavelength rising while the signal at the other wavelength declining.

3.2 "Ester Loading" Procedure for Neutrophils

It is crucial (*see* **Note 1**) that the loading medium contains Ca^{2+} (e.g., 1–2 mM) because as the Ca^{2+} chelator is generated within the cytosol it will bind Ca^{2+} (the extent depending on its K_d). Without additional extracellular Ca^{2+} to replace this, Ca^{2+} will be displaced from Ca^{2+} stores within the cell (or worse, not replaced at all). Another important point in the method is the 1000-fold dilution of the ester stock solution. This gives an acceptably low concentration the solvent DMSO. DMSO can have toxic effect on neutrophils, and it is therefore important that it is kept to a minimum. At 0.1% (i.e., 1/1000), the effects of DMSO are minimal.

1. Dissolve ester in dry DMSO (*see* **Note 1**) to give a stock (1–5 mg/mL). Store at −20 °C.
2. 2 (optional). Mix 2.5 μL of 25% pluronic with 5 μL of ester stock. This step may assist transfer into the neutrophils, pluronic being a "dispersing agent," which assists in keeping the ester in solution.
3. Add 1 μL of ester to 1 mL of neutrophil suspension (*see* **Note 2**) of 1–50 × 10^6 cell/mL in Ca^{2+} containing medium.
4. Incubate neutrophils for 20–60 min (room temperature or 37 °C). Note that for FFP18-AM, several hours are required to generate sufficient FFP on the inner surface of the plasma membrane to be useful.
5. Resuspend neutrophils in fresh medium.
6. If desired, neutrophils can be placed on ice to reduce leakage of Ca^{2+} probe.
7. It is recommended that before accepting that loading has been successful, the following simple checks are made.
   (a) Check that fluorescence at 360 nm excitation (505 nm emission) is significantly higher than in nonloaded cells. Record excitation or emission spectrum (quartz cuvette or optics) of loaded neutrophils to ensure that conversion of ester to its acid is complete (e.g., for fura2, the ester and the acid have clearly different spectra, see, for example, Molecular Probes Handbook).
   (b) Treat neutrophils with either 150 μM digitonin or Ca^{2+} ionophore (nonfluorescent ionomycin or Br A23187). Check that the spectral change is consistent with Ca^{2+} saturation for the probe (e.g., for fura2 the excitation spectrum peaks at 340 nm). Add 20 mM EGTA solution and check that the spectral change is consistent with zero

saturation for the probe (e.g., for fura2, the excitation peak is shifted toward 380 nm with the two spectra crossing at ~360 nm).

(c) Observe the fluorescence microscopically (e.g., Zeiss filters LP420/G365 or similar) to check loading has no obvious nonuniformity (i.e., in granules or phagosomes).

3.3 Measurement of Cytosolic Free Ca^{2+}

For a single wavelength indicator, the cytosolic free Ca^{2+} concentration can be calculated from: $Ca^{2+} = K_d(F - F_{min})/(F_{max} - F)$, where F is the fluorescent signal through the experiment, and F_{min} and F_{max} are the minimum and maximum obtainable signal from the indicator in the presence and absence of saturating amounts of Ca^{2+}. F_{min} and F_{max} are obtained at the end of the experiment by permeabilizing the cell (with ionophores) in the presence of Ca^{2+} and then chelating the Ca^{2+} with EGTA. The single wavelength approach may be used confocally or flow cytometrically but is not advised for conventional fluorometry or imaging. For these latter procedures, ratio dyes are preferable. The ratio of the two signals will be independent of the concentration (or amount) of probe, and the free Ca^{2+} required to give this signal ratio can be calculated from: $Ca^{2+} = K_d \times \beta(R - R_{min})/(R_{max} - R)$, where $\beta = Sf_2/Sb_2$, where S_{xy} is the emission signal at wavelength y ($y = 1$ for 340 nm and 2 for 380 nm) from the fully Ca^{2+} saturated indicator ($x = b$), totally Ca^{2+} free indicator ($x = f$) or variable Ca^{2+} saturation in the cell during the experiment ($x = v$), and $R = Sv_1/Sv_2$, $R_{max} = Sb_1/Sb_2$ and $R_{min} = Sf_1/Sf_2$. The characteristic "Ca^{2+} signature" can be confirmed to originate solely from a change in Ca^{2+} by noting that the sum, $ASv_1 + Sv_2$, (where $A = (Sf_2 - Sb_2)/(Sb_1 - Sf_1)$ will be constant at all Ca^{2+} concentrations [6]. The R_{max} and R_{min} values are essential for the calculation and are usually taken at the end of the experiment by the addition of ionomycin or digitonin (to allow Ca^{2+} saturation) and then EGTA (to remove Ca^{2+} from the probe) (*see* **Note 2**).

3.4 Measurement Protocols

3.4.1 Fluorometry

1. Add cell suspension to quartz or UV transmissible cuvette. Keep stirred to ensure that cells remain in the illumination beam. Set temperature (37 °C).
2. Set fluorometer to illuminate and record emission at the appropriate wavelengths. Record single- or dual-wavelength data, as appropriate (*see* Table 1).
3. Once data points are constant (i.e., after warm-up time), add stimulus while recording (*see* Fig. 1).
4. At the end of the experiment, add 1–4 μM ionomycin or 150 μM digitonin to saturate the probe. The stock 1.5 mM digitonin solution is diluted 1:10 into the cell suspension to provide a final concentration of 150 μM. This will cause the

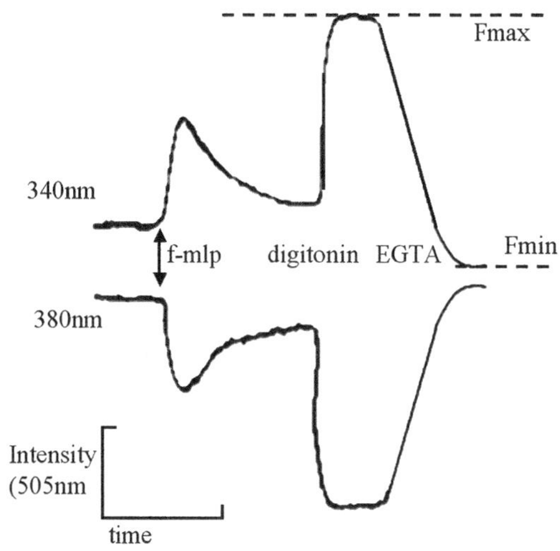

Fig. 1 A typical experiment in which cytosolic free Ca^{2+} concentration is monitored by Fura2. The emission at 505 nm is recorded for excitation at 340 nm and 380 nm, as indicated. A stimulus is added which produces the "Ca^{2+} signature." The addition of digitonin and then EGTA is for calibration. For a signal wavelength probe, the values of F_{max} and F_{min} are used. For a dual-wavelength probes (as here) R_{max} and R_{min} are calculated from the two signals (Reproduced from Hillson, et al. [14] with permission from Humana Press)

solution to become cloudy due to the presence of light-scattering micelles, as the critical micelle concentration is about 0.5 mM.

5. Once saturation has occurred, add 20 mM EGTA (*see* Fig. 1).
6. Note parameters listed below and calculate the cytosolic free Ca^{2+} concentration throughout the time course using the appropriate equation given earlier and in Fig. 1.

3.4.2 Microfluorometry

By optically coupling a wavelength changer to the input port and a photomultiplier tube to the output port of a fluorescent microscope, the procedure given above can be used with individual cells. In order to reduce background signal from areas of the field not occupied by the cells (or by other cells), a pinhole in the focal plane can be useful. This could be either fixed, so that the cell is moved to the pinhole by the microscope stage, or moveable. However, as neutrophils, by their nature, tend to move as part of their response, it is preferable to replace the photomultiplier tube with an intensified CCD camera. A "virtual mask" can be set up by binning the data from the region of the image that includes the cell of interest (*see* **Note 4**). This approach has several advantages: (1) the cell is visualized, so artifacts caused by cell movement are immediately apparent and (2) as the masks are electronic, several "masks" can be defined that read out the ratio values of more than one individual cell in the field.

Digitonin should not be used in microfluorometry to determine F_{max}, as the Ca^{2+} indicator will be lost from the neutrophils (*see* **Note 3**). When this happens, the indicator will no longer be within the defined area, and the information gathered will not be meaningful. It is, therefore, recommended that ionomycin is used to provide the maximum and minimum signals under these conditions.

3.4.3 Flow Cytometry

Measurement of cytosolic free Ca^{2+} concentration within individual cells as a cell population is possible using flow cytometry. Time course of Ca^{2+} changes can be achieved by addition of a stimulus to the cell population as it passes through the machine. Caution must be exercised in the interpretation of these data as, although individual cells are being interrogated, no single cell is followed through the time course, and the observed changes are merely population average changes. These often do not reflect changes at the single-cell level. For example, unless stimulus-induced Ca^{2+} spikes are synchronized within the population, these will not be observed by flow cytometry. With a UV laser cytometer, Indo-1, which can be used ratiometrically (*see* Table 1), is the probe of choice. However, single wavelength indicators, such as fluo3 and calcium green, have been successfully used with conventional 488 nm lasers.

3.5 Imaging Ca^{2+} in Individual Neutrophils

3.5.1 Ratiometric Imaging

The fluorescent intensity at any point within the 2D microscopic image of the cell will be proportional to the amount of probe in that "line of sight." Thus, it will be brighter at the center of the cell, where it is thicker, than at the edge, where the cell may become progressively thinner. Thus, it is helpful to use a ratiometric probe and take a ratio of two images to provide a "Ca^{2+} map" of the cytosolic free Ca^{2+} (*see* **Note 5**). In order to achieve this, changing excitation wavelength must be synchronized with the acquisition of images, often by a spinning filter wheel, optical chopper, or rapid-changing monochromator. These are commercially available from many sources including PTI and Cairn Instruments. Imaging is best achieved by coupling the wavelength changer to the "fluorescent input" of the microscope and either a digital camera or a video camera to the "output." Digitization of the signal to provide an array of pixels each with a value that corresponds to the intensity of the fluorescent image in that region of the field is then recorded. After background subtraction, the ratio is calculated, and a look-up table (LUT) is used to provide color on the image corresponding to the cytosolic free Ca^{2+} concentration (Ca^{2+} map). With an intensified video camera, the speed of acquisition is probably 25–30 frames/s, so that one ratio image would take ~80 ms. The time taken to change wavelengths can be minimized by using fast filter wheels, optical choppers, or rapid changing monochromators. However, the image quality is often poor at high speed, and it is usually necessary to average a number of frames, thus increasing the signal but decreasing the time resolution, or bin pixels to increase the signal but decreasing spatial resolution (*see* **Note 4**). The need to do either is reduced by using image intensification, but in practice, useful ratio images can rarely be acquired faster than about 1 s^{-1}. The speed barrier is overcome by rapid confocal imaging (*see* Subheading 3.6). Other problems associated with Ca^{2+} imaging include photobleaching of the probe (*see* **Note 4**) and light-induced activation of the neutrophil under view (*see* Subheading 3.8 and **Note 6**). Fuller details of Ca^{2+} imaging artifacts and their solution may be found in [2].

3.5.2 Confocal Imaging

The single wavelength probes (e.g., fluo4 and calcium green) can only confidently be used with confocal imaging, because unlike conventional microscopy, confocal microscopy images only an optical section through the cell of a defined thickness. Therefore, the problems associated with cell thickness artifacts are eliminated. This optical sectioning enables Ca^{2+} to be monitored within the nucleus, through the cell perpendicular (*xz*-plane) to the normal viewing plane, or at any predefined locus [10]. However, it is important to be aware that the cytoplasm of living cell is neither homogeneous nor static [11, 12]. This can give differences in the intensity of the fluorescence signal observed which are unrelated to Ca^{2+} concentration [11]. This problem is particularly apparent in granular cells, such as neutrophils, which undergo chemotaxis. The leading edge and pseudopodia (e.g., during phagocytosis) can often be devoid of granules and gives significantly higher fluorescent signals [11–13]. One solution to this problem is to double-label the neutrophils with both a Ca^{2+}-sensitive and Ca^{2+}-insensitive probe. The ratio of the two images will be independent of the spatial optical artifact and so give a spatial pure distribution of Ca^{2+} signaling despite changes in the cell morphology.

3.6 Rapid Ca^{2+} Imaging in Neutrophils

Many of the Ca^{2+} signals in neutrophils have time-scales that can be measured over 10–100 s. However, it is becoming more evident that these Ca^{2+} events may be composed of faster events occurring on the msec timescale [14]. Confocal laser scanning of a single line repeatedly (*xt* scanning) through individual fluo3-loaded neutrophils can be used to accumulate data at a rate of at least 80 lines/s, giving a time resolution of greater than 12.5 ms with events in the cell distinguishable to about 0.1–0.2 μm lateral resolution [15]. Conventional confocal laser scanning in both *x* and *y* directions is necessarily slower, but resonant scanning in the *x* direction can generate useful images at 17.5 ms/frame and a rotating Nipkow disk (a series of pinpoints in the rotating disk) scans multiple laser beams across the field at up to 360 frames/s (3 ms/frame).

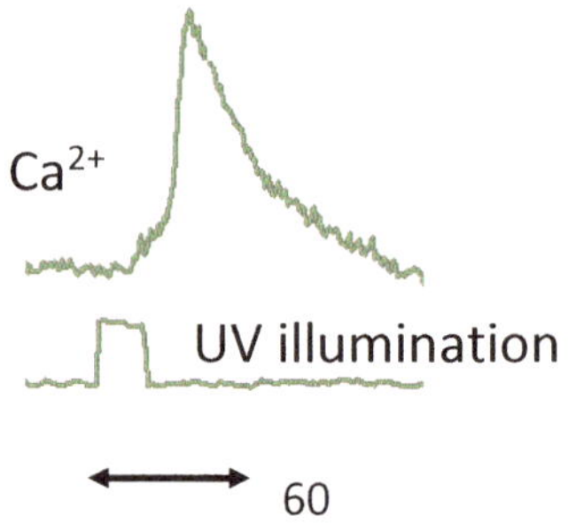

Fig. 2 Effect of uncaging IP_3 on neutrophil cytosolic free Ca^{2+}. The upper trace shows the change in cytosolic free Ca^{2+}, monitored by fluo4 intensity and the lower trace the pulse of UV illumination. There is a lag of several seconds before the Ca^{2+} begins to rise slowly. This is not observed when the cells are stimulated by fMLF or similar against. When the cytosolic Ca^{2+} reaches a critical point, a robust Ca^{2+} signal is fired. The latter signal is sufficient to trigger a cell spreading response

3.7 Near Membrane Ca^{2+} in Neutrophils

An analogue of fura2, FFP-18, with a long hydrophobic tail, accumulates in the membranes rather than the cytosol, and has been used to monitor near plasma membrane Ca^{2+} in neutrophils [16, 17]. More recently, an equivalent "near membrane" probe excitable by visible light, MOMO (Table 1) has become available. However, this probe is more hydrophilic and partitions into the nuclear membrane of neutrophils (Fig. 2) and is excluded from the nuclear lobes. It is useful for locating Ca^{2+} signals that originate near the nuclear boundary. In contrast, FFP-18—AM is hydrophobic and can be loaded into neutrophils from its AM-ester [16, 17]. The esterified probe binds quickly to the neutrophil membrane and there follows a slow process (presumably limited by the rate of "flip-flop diffusion of the probe across the cell membrane) where the AM ester is cleaved on the cytosolic facing leaflet of the plasma membrane [17]. Although the process is slow and requires hours to load the dye, it may be a useful approach for visualizing near membrane Ca^{2+} events. Furthermore, it can be used nonratiometrically with excitation from a UV diode laser emitting at 410 nm. With confocal imaging, the signal increase due to near membrane Ca^{2+} events can easily be seen. However, extreme caution must be exercised, as the dye equilibrates across a number of membranes in the neutrophil, especially nuclear envelope. Without ratiometric measurement, it is not possible to distinguish high florescence that results from high dye content from that resulting from high Ca^{2+}. Although measurements can be made from neutrophil populations, the intensity of probe at the cell edge is very low and, in our hands, it has not been possible to image near membrane Ca^{2+} with these lipophilic probes. However, an alternative approach has been developed based on the use of low affinity Ca^{2+} genetically encoded Ca^{2+}

indicators or GECIs [18]. Unlike small molecule Ca^{2+} indicators, GECIs can be engineered to locate themselves to specific locations within the cell, such organelles, based on specific targeting sequences that locate GECIs to intracellular organelles [19]. We have adopted a different " targeting" strategy to locate the probe at the cell periphery by coupling the Ca^{2+} indicator to ezrin, a protein which, in myeloid cells, locates almost exclusively at the cell periphery. This construct gives a sufficiently bright signal to image Ca^{2+} at the cell edge [20, 21]. Of course, primary human neutrophils are not easily transfected, but this approach could be used myeloid cell lines, and human or mouse myeloid stem cells [11, 22] to generate mature neutrophils expressing the construct. This promises to be a productive approach to visualizing periphery Ca^{2+} in neutrophils and myeloid cells.

3.8 Simultaneous Fluorescence and Phase Contrast Imaging

It is often necessary to visualize the phase contrast image to undertake microinjection, directed phagocytosis (*see* Chapter 9 of this volume) or to provide conformation of the morphological activity of the neutrophil whilst recording the fluorescence signal from a Ca^{2+} probe. With scanning confocal microscopy, near synchronous images can often be obtained with rapid laser scanning by acquiring data alternately line-by-line from a sensor of transmitted light and a sensor of the emitted fluorescent light. In widefield (nonconfocal) ratio imaging, absolute synchronous phase contrast and fluorescent images can be achieved by a simple optical trick. A far-red filter in position at the transmission illuminator provides a "red phase contrast image" without exciting fluorescence. The emission from the fluorescent Ca^{2+} indicator, usually at 500 nm, and the red transmitted light are allowed to exit the microscope together and are separated into two images by an additional dichroic mirror (Fig. 3). A camera can be attached to each output or both outputs spatially separated and captured by a single camera.

3.9 Manipulating Cytosolic Ca^{2+} in Neutrophils by Photolysis

Cytosolic free Ca^{2+} within neutrophils can be manipulated on demand by photorelease of "caged Ca^{2+}" (e.g., nitr-5), "caged Ca^{2+} chelator" (e.g., diazo-2), or caged IP_3 at defined times. These three "caged" compounds can be loaded into cells from their available acetoxymethyl-esters. Other (nonesterified) caged compounds can also be introduced into cells by microinjection techniques (*see* Chapter 9). The "caged" compound is inert until photolysis. In the case of "caged Ca^{2+}", the affinity of the chelator for Ca^{2+} changes dramatically on photolysis at 360 nm. In this way, the cytosolic free Ca^{2+} concentration can be elevated on photolysis [23]. Caged-IP_3 AM can be loaded into neutrophils in the same way as described for fluorescent probes (Subheading 3.4). The IP_3 concentration within the cytosol is then stepped up on photolysis (Fig. 2). The efficiency of the uncaging system can be monitored using caged fluorescein as an easily quantifiable output. However,

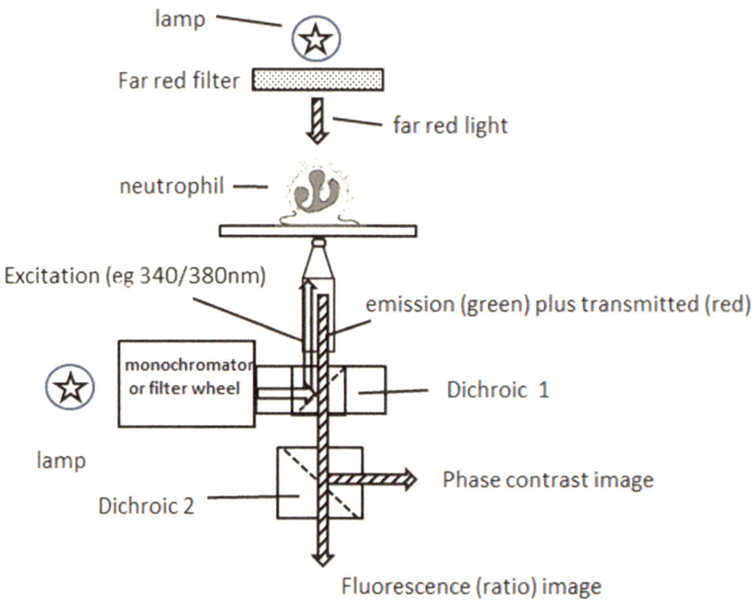

Fig. 3 A typical arrangement of filters and dichroics to give simultaneous fluorescence and phase contrast imaging. In this diagram, ratio excitation is achieved by either a spinning filter wheel or a rapid changer monochromator. This optical arrangement is ideal for an inverted microscope (usual used for microinjection, etc.) where Dichroic 1 is part of the standard fluorescence microscope filter block, with the excitation and some "barrier" filters removed, Dichroic 2 intercepts the light before being captured and spits the beam into the far-red transmitted light (phase contrast) image and the fluorescence image

the optical situation within the neutrophil cytosol is more complex. Another approach is to use nitr5 (caged Ca^{2+}) and to monitor the output (elevation of cytosolic Ca^{2+}. Our system causes uncaging of Ca^{2+} (nitr5) to elevate cytosolic free Ca^{2+} in neutrophils at a rate of about 10 nM/s, equivalent to 1 μM/s total Ca^{2+} released (taking into account a cytosolic Ca^{2+} buffering capacity of about 100:1). Since the ratio of uncaging efficacy (uncaging index = $\varepsilon\phi$) of the photosensitive bonds in nitr5 to that in caged IP_3 is 0.24 [24], its rate of IP_3 uncaging can be estimated at 4 μM/s. We estimate that during the delay between the onset of uncaging and the Ca^{2+} signal, the IP_3 concentration in the neutrophils thus rises to about 20 μM. Also, since the cytoplasm is optically nonuniform [11–13], the location of uncaging may be important. It is possible to compare the efficiencies of UV delivery to the cytosol using the photoconversion of dihydroethidium to the ethidium, which fluoresces on binding to the neutrophil nucleus and acts a surrogate monitor of cytosolic UV exposure [25]. Of course, care must be taken to perform appropriate uncaging controls in neutrophils, such as performing sham photolysis (i.e., no caged compound) or photolysis of presumed biologically inert compounds (e.g., fluorescein).

4 Notes

1. De-esterification that is catalyzed enzymatically in the cell, will also occur spontaneously (at a slow rate). If this occurs before presentation to the cells, the probe either will not be able to enter the cells or enter the cells in the partially de-esterified, but in the non-Ca^{2+} sensitive form. Hydrolysis in the stored stock solution is slowed by reduced temperature and the exclusion of water. As DMSO is hydroscopic and will readily absorb water from the air, standard laboratory-grade DMSO is not recommended. Dry DMSO can be purchased in sealed containers in smaller volumes (Sigma). Another precaution that can reduce the water content of the solution is to be sure that when taking the container from the freezer (at –20 °C) to let it warm up to room temperature before opening it. If this is not done, water from the air may condense on the inner surface of the container, contaminate the stock solution and increase ester hydrolysis.
2. Extracellular Ca^{2+} must be present during the time that the Ca^{2+} chelating probe is loaded into the cells so that it can replace Ca^{2+}, which will be bound to the probe. Otherwise, Ca^{2+} will be removed from intracellular sites. However, provided these precautions are taken, Ca^{2+} chelating probes have little obvious (adverse) effect on Ca^{2+} signaling. On the other hand, omission of extracellular Ca^{2+} during loading has been used as a deliberate strategy for "depleting cell Ca^{2+}" in order to establish an intracellular role for this ion in a particular cell activity.
3. Digitonin is often preferred when measuring cytosolic free Ca^{2+} concentration in a cell population in suspension, as it ensures that all fura2 gains access to the high Ca^{2+} concentration in the extracellular medium. If ionomycin is used (e.g., during Ca^{2+} imaging), it is important that sufficient ionomycin is added to produce a truly maximal fluorescence signal, since it is possible to elevate cytosolic free Ca^{2+} with ionophores to concentrations that are <μM. Under these latter conditions, the ionomycin signal will not correspond with R_{max}, and cytosolic free Ca^{2+} concentration cannot be correctly calculated. The problem can be reduced by also increasing the extracellular Ca^{2+} concentration during the addition of ionomycin.
4. The information in a single pixel is limited by the noise associated with the detection system of the imager. A technique often used, called binning, is to collect the data from a pixel plus its neighbors, and to treat the combined signal as though arising from a larger single pixel. Binning of information on neighboring pixels will increase the signal but decrease the spatial resolution. A statistical approach can be used to

determine whether small areas within the cell truly have raised Ca^{2+}. Areas of interest in the image (with n pixels) can be chosen for comparison of their cytosolic free Ca^{2+} concentration. The areas will have statistically significant cytosolic free Ca^{2+} concentrations when

$$n > 2[\sigma(Z_{2\alpha} + Z_{2\beta})/\delta]^2 \text{ or } \delta > [\sqrt{(2/n)}][\sigma(Z_{2\alpha} + Z_{2\beta})],$$

where n is the number of pixels in each of the two areas of the image, σ is the standard deviation of the distribution of Ca^{2+} values in the pixel arrays, Z_x is the standard normal deviate exceeded with probability x, β is the significance level for the test, $(1 - \beta)$ is the power of the test, and δ is the difference in cytosolic free Ca^{2+} concentrations [26]. The ability to detect small and localized changes in Ca^{2+} depends on both the magnitude of the Ca^{2+} change and the area it occupies. Detection of both very small and very localized Ca^{2+} changes is thus difficult and ultimately limited by the image noise (i.e., the variance or standard deviation of individual pixel values).

5. The major problems associated with fluorescent imaging are photobleaching and image noise. Photobleaching arises where excessive excitation results in the destruction of the fluorescent molecules and hence a reduction in the emission intensity (bleaching). Each fluorescent molecule emits about 10^4 to 10^5 photons/molecule before photolysis. With ratiometric methods, photobleaching is less of a problem as the ratio will remain constant during the bleaching provided that the pairs of images are taken close together in time (when no significant bleaching has occurred). With nonratiometric confocal Ca^{2+} imaging, bleaching during the time of the experiment can be avoided by attenuating the laser light and increasing the detector (photomultiplier) sensitivity so that the minimum usable emission intensity is employed.
6. Ratiometric dyes require excitation near the UV region, which stimulate fluorescence from endogenous molecules such as NADPH within neutrophils, and consequently the signal–noise ratio of the Ca^{2+} probe is reduced.

References

1. Hallett MB, Hodges R, Cadman M et al (1999) Techniques for measuring and manipulating free Ca^{2+} in the cytosol and organelles of neutrophils. J Immunol Methods 232:77–88
2. Tepikin AV (2000) Calcium signalling: a practical approach, 2nd edn. Oxford Univ. Press, Oxford, U.K, p 230
3. Pozzan T, Lew DP, Wollheim CB et al (1983) Is cytosolic ionized calcium regulating neutrophil activation? Science 221:1413–1415
4. Hallett MB, Davies EV, Pettit EJ (1996) Fluorescent methods for measuring and imaging the cytosolic free Ca^{2+} in neutrophils. Methods 9:591–606
5. Scanlon M, Williams DA, Fay FS (1987) A Ca^{2+}-insensitive form of fura2 associated with polymorphonuclear leukocytes-assessment and accurate Ca^{2+} measurement. J Biol Chem 262:6308–6312
6. Al-Mohanna FA, Hallett MB (1988) The use of fura 2 to determine the relationship between

intracellular free Ca^{2+} and oxidase activation in rat neutrophils. Cell Calcium 8:17–26
7. Laffafian I, Hallett MB (1998) Lipid-assisted microinjection: introducing material into the cytosol and membranes of small cells. Biophys J 75:2558–2563
8. Laffafian I, Hallett MB (2000) Gentle micro-injection for myeloid cells using SLAM. Blood 95:3270–3271
9. Von Tscharner V, Deranleau DA, Baggiolini M (1986) Calcium fluxes and calcium buffering in human neutrophils. J Biol Chem 261:10163–10168
10. Pettit EJ, Hallett MB (1996) Localized and global cytosolic Ca^{2+} changes in neutrophils during engagement of CD11b/CD18 integrin visualized using confocal laser scanning reconstruction. J Cell Sci 109:1689–1694
11. Dewitt S, Darley R, Hallett MB (2009) Translocation or just location? Pseudopodia affect fluorescent signals. J Cell Biol 184:197–203
12. Dewitt S, Hallett MB (2011) Optical complexities of living cytoplasm—implications for live cell imaging and photo-micromanipulation techniques. J Microsc 241:221–224
13. Hallett M, Dewitt S (2011) A trick of the light: the optical properties of living cytoplasm which can mislead. Integr Biol 3:180–184
14. Hillson EJ, Hallett MB (2007) Localised and rapid Ca^{2+} micro-events in human neutrophils: conventional Ca^{2+} puffs and global waves without peripheral-restriction or wave cycling. Cell Calcium 41:525–536
15. Pettit EJ, Hallett MB (1995) Early Ca^{2+} signalling events in neutrophils detected by rapid confocal laser scanning. Biochem J 310:445–448
16. Davies EV, Hallett MB (1996) Near membrane Ca^{2+} changes resulting from store release in neutrophils: detection by FFP-18. Cell Calcium 19:355–362
17. Davies EV, Hallett MB (1998) High micromolar Ca^{2+} beneath the plasma membrane in stimulated neutrophils. Biochem Biophys Res Commun 248:679–683
18. Sun XR, Badura A, Pacheco DA et al (2013) Fast GCaMPs for improved tracking of neuronal activity. Nat Commun 4:2170
19. Suzuki J, Kanemaru K, Ishii K et al (2014) Imaging intraorganellar Ca^{2+} at subcellular resolution using CEPIA. Nat Commun 5:4153
20. Roberts RE, Vervliet T, Bultynck G et al (2017) Dynamics of ezrin location at the plasma membrane: relevance to neutrophil spreading. Eur J Clin Investig 47:148
21. Roberts RE (2017) The μ-calpain-ezrin axis: A potential target for therapy in inflammatory disease. PhD Thesis: Cardiff University. http://orca.cf.ac.uk/id/eprint/108477
22. McDonald JU, Cortini A, Rosas M et al (2011) In vivo functional analysis and genetic modification of in vitro-derived mouse neutrophils. FASEB J 25:1972–1982
23. Pettit EJ, Hallett MB (1998) Release of "caged" cytosolic Ca^{2+} triggers rapid spreading of human neutrophils adherent via integrin engagement. J Cell Sci 111:2209–2215
24. Ellis-Davies GCR (2007) Caged compounds: photorelease technology for control of cellular chemistry and physiology. Nat Methods 4:619–628
25. Brasen JC, Dewitt S, Hallett MB (2010) A reporter of UV intensity delivered to the cytosol during photolytic uncaging. Biophys J 98: L25–L27
26. Armitage P, Berry G (1987) Statistical methods in medical research, 2nd edn. Blackwell Scientific, Boston, MA, pp 181–182

Chapter 15

Labeling Acidic Compartments of Neutrophils with Cresyl Violet

Philip P. Ostrowski, Ziv Roth, and Sergio Grinstein

Abstract

We introduce the acidotropic marker cresyl violet to stain acidic granules in live neutrophils. Cresyl violet is less phototoxic, more photostable, and more cost-effective than other commercially available acidotropic markers. Additionally, it does not photoconvert to fluorescent species of a different color, a limitation of other commonly used acidotropic markers. Staining can be readily detected by fluorescence microscopy or by flow cytometry, and can be used as a readout of degranulation in activated neutrophils.

Key words Neutrophil, Lysosome, Granule, Degranulation, Cresyl violet, Acidotropic, Microscopy, Flow cytometry

1 Introduction

Neutrophils are an essential component of the innate immune system, serving as the first active responders to danger signals [1]. To combat foreign threats, they deploy a specialized assortment of intracellular organelles known as secretory granules. These membrane-bound compartments store an assortment of proteins and peptides that have antimicrobial and lytic activities, including the ability to generate reactive oxygen species [2]. When stimulated, neutrophils can engulf foreign particles into vacuoles (phagosomes) that fuse with secretory granules. The contents of the granules can also be released extracellularly by exocytosis, a process known as degranulation that enables the neutrophils to extend their microbicidal activity to the surrounding microenvironment.

In addition to containing microbicidal agents, neutrophil granules are generally acidic. The acidic pH is required for the import, processing, and optimal activity of some of the granules' contents [3]. The acidic nature of the granules provides a unique opportunity to visualize them and to assess their mobilization in intact cells by fluorescence determinations. Specifically, they can be stained with acidotropic dyes conventionally used for lysosomal

Mark T. Quinn and Frank R. DeLeo (eds.), *Neutrophil: Methods and Protocols*, Methods in Molecular Biology, vol. 2087, https://doi.org/10.1007/978-1-0716-0154-9_15,

visualization. Acidotropic dyes are fluorescent weak bases that are membrane-permeable and, as such, can cross the plasma and granule membranes. Importantly, the weak bases become protonated, and thus cationic, in the acidic lumen of the granules, becoming much less permeant and hence trapped within the compartment. Their ensuing net accumulation results in a marked increase in intragranular fluorescence. Acidotropic markers include acridine orange and LysoTracker Red®, which have been used previously to stain granules in neutrophils [4, 5]. These dyes, however, are limited by significant weaknesses. Acridine orange is phototoxic: when excited by light it produces reactive oxygen species that can damage and perforate the acidic compartment [6]. LysoTracker Red® suffers from poor photostability, bleaching quickly during imaging [7]. Furthermore, LysoTracker Red® also undergoes a photoconversion reaction upon excitation, generating a green fluorescent species that can confound multicolor labeling experiments [8].

We have recently demonstrated that cresyl violet, a weak base commonly used in histology, can be used as an effective marker of acidic compartments in a multitude of cell types from a variety of species, from yeast to human [7]. Not only is cresyl violet considerably less expensive than other lysosomotropic markers, it is not phototoxic, experiences little bleaching during imaging, and does not photoconvert. Here we describe the application of cresyl violet to stain acidic organelles in live neutrophils (*see* **Note 1**) for detection by fluorescence microscopy imaging (Fig. 1) and by flow cytometry (Fig. 2) and document the ability of the dye to measure neutrophil degranulation following treatment with stimuli.

2 Materials

2.1 Neutrophil Isolation

1. Freshly collected blood from a healthy donor (*see* **Note 2**).
2. Endotoxin-free density gradient solution for polymorphonuclear (PMN) cells (e.g., Polymorphprep™).
3. Centrifuge that can be used for 15 and 50 mL conical and round-bottom centrifuge tubes.
4. Sterile Hank's Buffered Salt Solution (HBSS) with calcium and magnesium, pH 7.2.

2.2 Neutrophil Staining with Cresyl Violet

1. 5 mL round-bottom polystyrene tubes (FACS tubes) or 15 mL conical centrifuge tubes.
2. Centrifuge that can be used for 5 mL round-bottom centrifuge tubes.
3. Sterile Hank's Buffered Salt Solution (HBSS) with calcium and magnesium, pH 7.2.

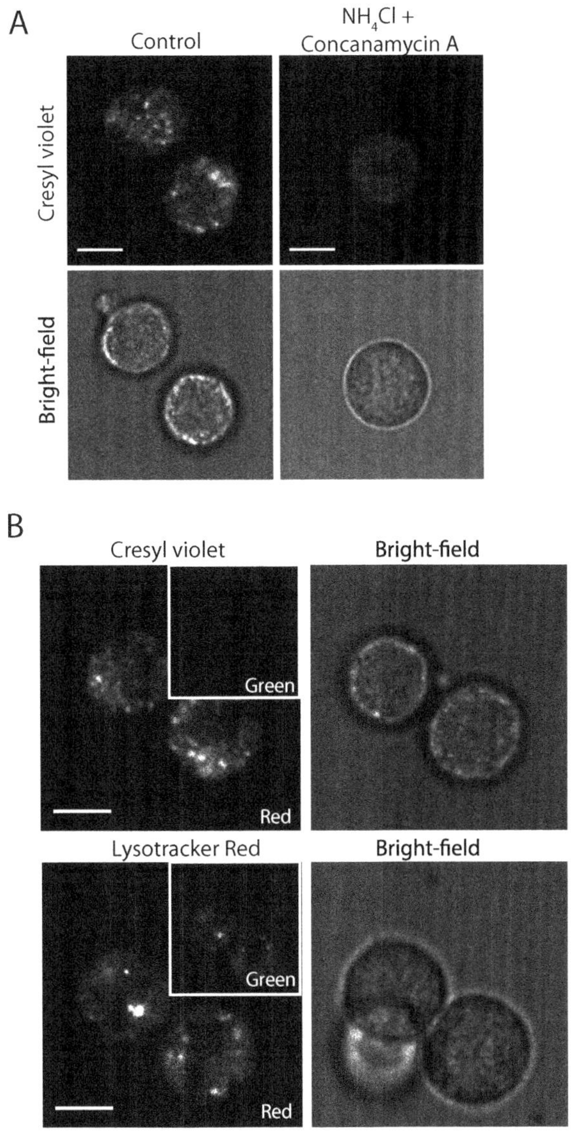

Fig. 1 Neutrophil staining with cresyl violet and LysoTracker Red®: fluorescence microscopy. (**a**) Neutrophils stained with cresyl violet were imaged by laser scanning confocal microscopy in red channel (top) and bright-field (bottom). Otherwise untreated cells are shown on left, while cells previously alkalinized using ammonium chloride (NH_4Cl) and concanamycin A before addition of cresyl violet are shown on right. Scale bars = 5 μm. (**b**) Neutrophils stained with cresyl violet (top) or with LysoTracker Red® (bottom) according to manufacturer's protocol. Fluorescence was imaged in red channel (left—main panel) and green channel (left—inset) to demonstrate photoconversion of LysoTracker Red®, leading to appearance of spurious fluorescence in green channel. Corresponding bright-field images are shown on right. Scale bars = 5 μm

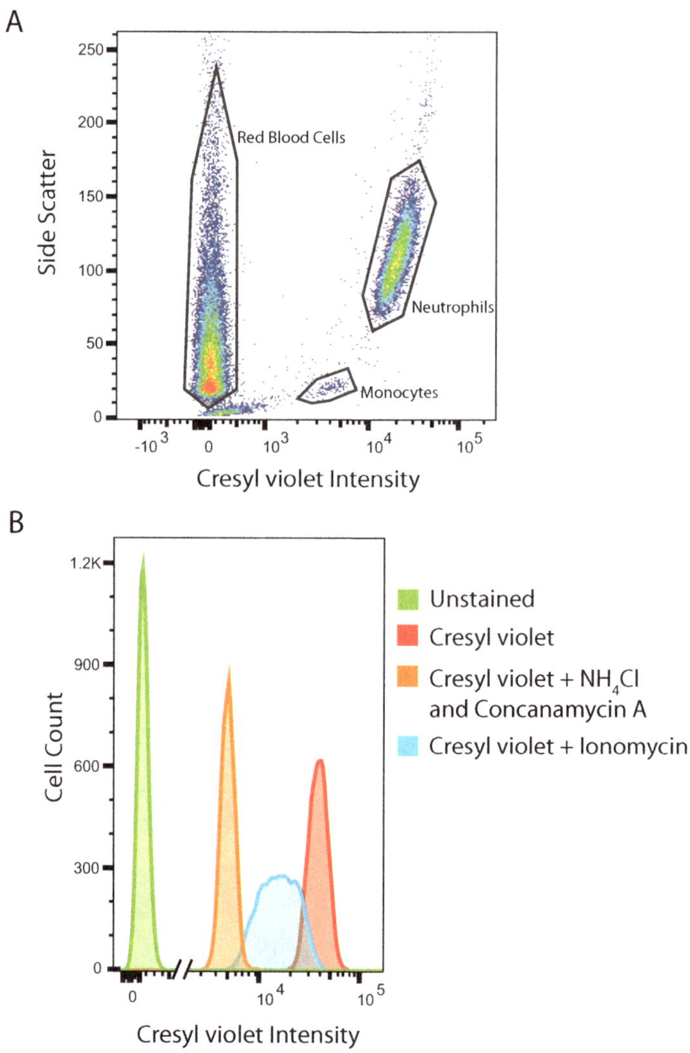

Fig. 2 Cresyl violet staining of neutrophils detected by flow cytometry. (**a**) Sample flow cytometry data of cresyl violet loaded cells according to protocol described. The cresyl violet fluorescence intensity was plotted against the side scatter of the cells. Side scattering enabled the identification of red blood cell, monocyte, and neutrophil populations. (**b**) Cresyl violet intensity of unstained neutrophils (green), and of cresyl violet-stained control neutrophils that were otherwise untreated (red), treated with ionomycin to trigger degranulation (blue) or alkalinized using ammonium chloride (NH_4Cl) and concanamycin A prior and during cresyl violet staining (orange)

4. Cresyl violet stock solution: 1 mM cresyl violet in autoclaved double-distilled H_2O.
5. Degranulation solution: HBSS with calcium and magnesium and 1 μM ionomycin.

6. Granule alkalinization solution: HBSS, 500 nM concanamycin A, and 10 mM ammonium chloride.

2.3 Neutrophil Microscopy

1. Poly-L-lysine.
2. Glass coverslips for live microscopy.
3. Phosphate-buffered saline (PBS), pH 7.4.

3 Methods

3.1 Neutrophil Isolation

1. Isolate 5 mL of blood from a healthy donor (*see* **Note 3**).
2. Add 5 mL of Polymorphprep™ to 15 mL conical tube (*see* **Note 4**).
3. Gently layer 5 mL of blood on top of Polymorphprep™ in conical tube using Pasteur pipette (*see* **Note 5**).
4. Centrifuge sample for 30–35 min at 500 × *g* at room temperature using gentle acceleration and without imposed (brake) deceleration (*see* **Note 6**).
5. Identify three layers formed following centrifugation. The top layer consists of mononuclear cells, the intermediate layer contains the polymorphonuclear cells, while the red blood cells sediment to the bottom.
6. Gently transfer the polymorphonuclear cell layer using a Pasteur pipette to a 50 mL Falcon tube.
7. Dilute the cells to 50 mL using HBSS at room temperature.
8. Centrifuge cells for 10 min at 400 × *g*.
9. Aspirate supernatant and gently resuspend in 1–5 mL HBSS (*see* **Note 7**).
10. Count cells using Coulter Counter or hemocytometer, if desired.

3.2 Neutrophil Staining

1. Transfer desired number of neutrophils to 5 mL FACS tube.
2. Add cresyl violet to a final concentration of 0.5 μM (1:2000 from cresyl violet stock solution) (*see* **Note 8**).
3. Gently mix cell suspension.
4. Immediately centrifuge cells for 3 min at 300 × *g* (*see* **Note 9**).
5. Aspirate and resuspend in HBSS.
6. Repeat washing procedure twice more.

3.3 Neutrophil Microscopy

1. Coat sterile glass coverslips with 100 μg/mL poly-L-lysine for 30–60 min at 37 °C.
2. Wash three times with PBS.
3. Transfer coated coverslip to microscope.

4. Treat neutrophils with cresyl violet and any other desired markers/conditions (*see* **Note 10**).
5. Allow stained neutrophils to settle onto poly-L-lysine-coated glass coverslip (up to 0.5×10^6 cells per 10 mm coverslip).
6. Image cresyl violet in Texas red or red channel settings using red excitation laser (*see* **Note 11**) (Fig. 1).

3.4 Flow Cytometric Measurement of Cresyl Violet Fluorescence

1. Resuspend neutrophils labelled with cresyl violet (and any other desired markers) in a 5 mL FACS tube (*see* **Note 10**).
2. Perform flow cytometry according to device protocol, gating out contaminating red blood cells based on their differential forward and side scattering or using other markers (*see* **Note 12**).
3. Image cresyl violet with Texas red or red channel settings (*see* **Note 11**) (Fig. 2).

4 Notes

1. As for other acidotropic dyes, retention of cresyl violet is dependent on the maintenance of its acidic pH. Neutralization of the pH will result in loss of the dye. As a result, this class of dyes is generally not compatible with fixation protocols. Even those acidotropic dyes containing a reactive amino moiety fix rather poorly.
2. Collect blood in the presence of heparin to prevent clotting and use shortly after isolation to minimize cell death.
3. Maintaining a sterile, lipopolysaccharide-free environment is important to minimize neutrophil activation and degranulation.
4. This protocol can be scaled to larger volumes. For the first centrifugation use equal amounts of Polymorphprep™ and blood in an appropriately sized tube.
5. Be careful not to mix layers, as this may decrease the purity of the resultant cells.
6. Use of rapid acceleration or deceleration can create excess sheer forces that will disrupt the layering of the sample.
7. Neutrophils need to be treated gently to protect them from degranulation and should not be vortexed, pipetted, or otherwise mixed harshly.
8. Cresyl violet is lipophilic and can stain membranes or other less acidic compartments nonspecifically at higher concentrations.

9. Cresyl violet will continue to stain cells during centrifugation and, if standardized staining is desired, should be aspirated immediately once the centrifugation is completed.
10. It is recommended to ensure that cresyl violet is staining primarily acidic compartments. To this end, acidic compartments should be alkalinized using 10 mM ammonium chloride and 500 nM of the vacuolar H^+-ATPase inhibitor concanamycin A for 10 min before and during cresyl violet staining and washing. Alternatively, neutrophil granules can be released by degranulation by treating the cells with 1 μM ionomycin for 10 min following cresyl violet staining. These treatments should largely eliminate staining.
11. Cresyl violet has maximum excitation at 585 nm and maximum emission at 627 nm.
12. Lysing red blood cells with osmotic shock or ammonium chloride can perturb granules/acidic compartments. Red cell contaminants can be gated by forward/side scatter and/or by staining with a specific antibody. Human red cells do not contain acidic organelles and will therefore not stain with cresyl violet.

Acknowledgments

P.P.O is supported by a Studentship from the Canadian Institutes of Health Research (CIHR). Work in the authors' laboratory is supported by grant FDN-143202 from CIHR to S.G.

References

1. Ley K, Hoffman HM, Kubes P et al (2018) Neutrophils: new insights and open questions. Sci Immunol 3:4579
2. Yin C, Heit B (2018) Armed for destruction: formation, function and trafficking of neutrophil granules. Cell Tissue Res 371:455–471
3. Nanda A, Brumell JH, Nordstrom T et al (1996) Activation of proton pumping in human neutrophils occurs by exocytosis of vesicles bearing vacuolar-type H+-ATPases. J Biol Chem 271:15963–15970
4. Bassoe CF, Li N, Ragheb K et al (2003) Investigations of phagosomes, mitochondria, and acidic granules in human neutrophils using fluorescent probes. Cytometry B Clin Cytom 51:21–29
5. Abrams WR, Diamond LW, Kane AB (1983) A flow cytometric assay of neutrophil degranulation. J Histochem Cytochem 31:737–744
6. Pierzynska-Mach A, Janowski PA, Dobrucki JW (2014) Evaluation of acridine orange, LysoTracker Red, and quinacrine as fluorescent probes for long-term tracking of acidic vesicles. Cytometry A 85:729–737
7. Ostrowski PP, Fairn GD, Grinstein S et al (2016) Cresyl violet: a superior fluorescent lysosomal marker. Traffic 17:1313–1321
8. Freundt EC, Czapiga M, Lenardo MJ (2007) Photoconversion of Lysotracker Red to a green fluorescent molecule. Cell Res 17:956–958

Chapter 16

Neutrophil Degranulation of Azurophil and Specific Granules

Samia Bedouhène, Pham My-Chan Dang, Margarita Hurtado-Nedelec, and Jamel El-Benna

Abstract

Neutrophils play a pivotal role in innate immunity and in the inflammatory reactions. Upon activation, neutrophils release several toxic molecules directed against microbial pathogens into the phagosome. These molecules include reactive oxygen species (ROS), myeloperoxidase, glucosidases, proteases, and antibacterial peptides. In resting cells these proteins and the enzyme responsible for ROS production (NOX2) are stored inside or at the membranes of different granules called azurophil or primary, specific or secondary, gelatinase or tertiary, and the secretory vesicles. Each granule has a specific density, content, and markers. Myeloperoxidase (MPO) is the azurophil granule marker, and the neutrophil-gelatinase-associated lipocalin (NGAL) is the specific granule marker. After cell activation by different stimuli, granule contents are released into the phagosome or in the extracellular space through a process called degranulation. Also during this process, membrane granules fuse with the phagosome and plasma membrane allowing expression of new markers at the cell surface. The degranulation can be assessed by measuring either the release of different proteins by neutrophils or the expression of granule markers at the plasma membrane. In this chapter, we describe the techniques used to measure degranulation of azurophil and specific neutrophil granules using different approaches such as measurement of MPO enzymatic activity and detection of MPO and NGAL proteins by SDS-PAGE and Western blot.

Key words Neutrophils, Degranulation, Azurophil granules, Specific granules, Myeloperoxidase, NGAL

1 Introduction

Polymorphonuclear neutrophils (PMN) constitute more than 60% of circulating leukocytes and they play a key role in host defense against pathogens and inflammation [1–4]. Neutrophils are the first cells to migrate out of the circulation to the infection site [4–6]. Once at the infectious site, neutrophils recognize the pathogen via different receptors and ligands followed by engulfment of the pathogen in a vacuole called the phagosome [7, 8]. Pathogens are then killed by release of highly toxic agents such as reactive oxygen

Mark T. Quinn and Frank R. DeLeo (eds.), *Neutrophil: Methods and Protocols*, Methods in Molecular Biology, vol. 2087, https://doi.org/10.1007/978-1-0716-0154-9_16,

species (ROS), myeloperoxidase, glucosidases, proteases, and antibacterial peptides into the phagosome [9–12].

In resting cells these toxic agents and their sources are stored in different granules that have different composition and density [13, 14]. Azurophil or primary granules contain myeloperoxidase (MPO), elastase, proteinase-3, cathepsins, glucuronidase, lysozyme, defensins, and so on. Specific or secondary granules contain lactoferrin, lipocalin/NGAL, collagenase, gelatinase, histaminase, lysozyme, membrane receptors (for fMLF, TNF, integrins, fibronectin, etc.), cytochrome b558 (or NOX2), and so on. Gelatinase or tertiary granules essentially contain gelatinase, acetyltransferase, lysozyme, membrane receptors (for fMLF, TNF, integrins, fibronectin, etc.), cytochrome b558 (or NOX2), etc., and the highly mobilizable secretory vesicles contain plasma proteins, alkaline phosphatase, membrane receptors (for fMLF, integrins, etc.), cytochrome b558 (NOX2), and so on [13, 14]. The release of these granules upon cell activation is called degranulation and is an important neutrophil function for host defense against pathogens and inflammation [4, 8].

In this chapter, we describe the techniques used to measure neutrophil degranulation. The first historical technique detects released MPO by measuring its enzymatic activity in cell supernatants [15, 16]. As some agents may interfere with MPO activity, it is essential to use a second technique to ensure that proteins are released from azurophil and specific granules such SDS-PAGE and Western-Blots [17, 18].

2 Materials

1. SDS-PAGE and Western blotting equipment and reagents.
2. Rabbit polyclonal anti-MPO and anti-lipocalin/NGAL antibodies (e.g., Abcam).
3. Horseradish peroxidase (HRP)-conjugated goat anti-rabbit and HRP-conjugated goat anti-mouse antibodies.
4. Luminol reagents.

2.1 Buffers and Solutions

1. Sterile phosphate-buffered saline (PBS): commercially available.
2. Sterile Hanks balanced-salt solution (HBSS) containing Ca^{2+}, Mg^{2+}, and D-glucose: commercially available.
3. 0.2%, 0.9%, and 1.6% NaCl solutions: dissolve the appropriate amounts of NaCl (w/v) in sterile H_2O.
4. 2% Dextran solution: Dissolve 2 g of Dextran T500 in 100 ml of 0.9% NaCl solution and filter-sterilize. Store the solution at 4 °C for up to 4 weeks (*see* **Note 1**).
5. Ficoll solution: commercially available.

6. 0.005% H_2O_2 solution: Dilute stock 30% H_2O_2 solution in PBS (*see* **Note 2**).
7. 10 mM *N*-formyl-methionyl-leucyl-phenylalanine (fMLF) solution: dissolve fMLF in sterile DMSO, aliquot, and stored at −20 °C. Prepare working solutions in PBS prior to use and keep on ice.
8. 5 mg/ml cytochalasin B solution: dissolve cytochalasin B in sterile DMSO, aliquot, and store at −20 °C.
9. 1.67 mg/ml ortho-dianisidine dihydrochloride solution: dissolve ortho-dianisidine dihydrochloride powder in PBS (*see* **Note 3**).
10. 2× Laemmli sample buffer: 125 mM Tris–HCl (pH 6.8), 6% SDS, 8% β-mercaptoethanol, 20% glycerol, 5 mM EDTA, 5 mM EGTA, 10 μg/ml leupeptin, 10 μg/ml pepstatin, 10 μg/ml aprotinin, 10 mM NaF, 5 mM $NaVO_3$, and 2 mM *p*-nitrophenyl phosphate. Store buffer at −20 °C for up to 12 months (*see* **Note 4**).
11. Transfer buffer: 50 mM Tris base, 95 mM glycine, 0.08% SDS, and 20% methanol.
12. TBS–Tween (TBST) buffer: 25 mM Tris–HCl pH 7.5, 150 mM NaCl, 0.05% Tween 20.
13. TBST + 1% and 5% milk solutions: dissolve nonfat dried milk in TBST buffer.

3 Methods

3.1 Isolation of Human Neutrophils

Collect blood from healthy adult volunteers using citrate dextrose as the anticoagulant. Isolated neutrophils by a classical technique [19, 20] using Dextran sedimentation and Ficoll density gradient centrifugation as follows:

1. Mix 25 ml of whole blood and 25 ml of the 2% Dextran solution (1% final) in 50 ml tubes. Gently mix by inverting the tubes several times and allow the cells to sediment at 4 °C for 20–40 min (*see* **Note 5**).
2. Gently collect the upper layer containing the leukocytes into centrifuge tubes and discard the pellet containing red cells.
3. Centrifuge the collected upper layer at 400 × *g* for 8 min at 22 °C (*see* **Note 6**).
4. After centrifugation, the pellets contain leukocytes and some contaminating erythrocytes while the supernatants contain plasma, dextran and platelets. Discard the supernatants by gently inverting the tubes, resuspend each pellet in 5 ml of PBS and pool several pellets from the same donor.

5. Prepare a gradient by gently layering the cells on top of the Ficoll solution (1–3 volumes of cell suspension/1 volume Ficoll) avoiding mixing of the two layers, and centrifuge the tubes at 400 × *g* for 30 min at 22 °C without the centrifuge brake (*see* **Note 7**).
6. After centrifugation, discard the upper layers and mononuclear cells, which are at the interface of the buffer and Ficoll using suction.
7. Gently disperse the pellet containing neutrophils and contaminating erythrocytes (*see* **Note 8**).
8. Remove contaminating erythrocytes by hypotonic lysis: add 15 ml of ice-cold sterile 0.2% NaCl for 40 s, mix gently and restore isotonicity by adding 15 ml of 1.6% NaCl (*see* **Note 9**).
9. Centrifuge the tubes at 400 × *g* for 8 min at 4 °C. Aspirate the red supernatant with gentle suction, and resuspend the neutrophil pellet in 5–10 ml of PBS.
10. Dilute the cells in trypan blue (1/100) and count them using a hemocytometer.

3.2 Neutrophil Degranulation

1. Resuspend neutrophils at 5 × 10^6 in 500 μl of HBSS in sterile 1.5 ml Eppendorf tubes and preincubate at 37 °C for 5 min.
2. Add 5 μg/ml of the cytochalasin B solution, and incubate for 5 min.
3. Stimulate by addition of 1 μM fMLF, and incubate for 2 min at 37 °C.
4. Stop the reaction by centrifugation for 30 s at 12,000 × *g*.
5. Collect supernatants in new tubes.
6. Centrifuge the supernatants again at 15,000 × *g* for 10 min at 4 °C and collect the supernatants in new tubes (*see* **Note 10**). The supernatants are used to measure MPO activity and for SDS-PAGE. Keep the neutrophil pellets, which are used as controls.
7. For SDS-PAGE of the supernatants, add 100 μl of 2× Laemmli sample buffer to 100 μl of the supernatants and boil at 100 °C for 3 min.
8. Resuspend the cell pellets from **step 6** in 500 μl of cold PBS + 0.5% HTAB, sonicate 2 × 10 s, and centrifuge at 15,000 × *g* for 10 min at 4 °C.
9. For SDS-PAGE of the cell pellet lysates, mix 100 μl of the lysates from **step 8** with 100 μl of 2× Laemmli sample buffer and boil at 100 °C for 3 min.

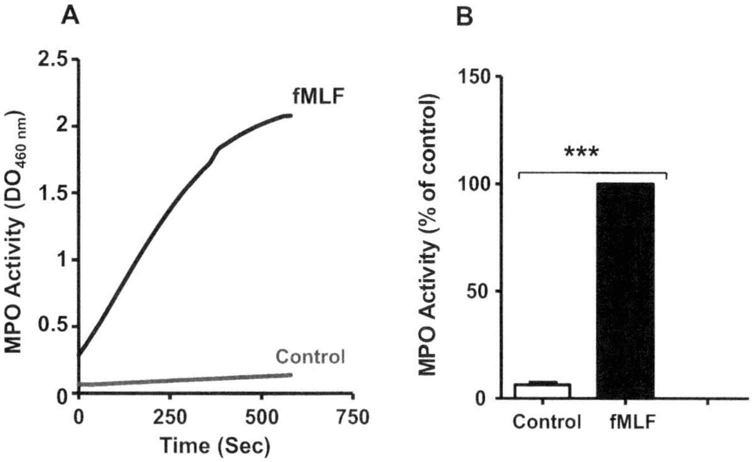

Fig. 1 Release of myeloperoxidase (MPO) by human neutrophils as measured by its activity. Neutrophils (5×10^6/0.5 ml Hanks buffer) were pretreated with 5 μg/ml cytochalasin B for 5 min, before stimulation with 1 μM fMLF for 2 min. After centrifugation, the supernatants were recovered to measure MPO activity using the spectrophotometric method. (**a**) The graph shows a representative experiment of kinetics of MPO activity of resting and fMLF-stimulated cells. (**b**) Data from three different experiments are expressed as % of control (the activity of fMLF-stimulated cells represents 100%). Data are mean ± SEM of three separate experiments ($*P < 0.0001$)

3.3 Measurement of Myeloperoxidase (MPO) Activity

MPO activity is measured using a modified spectroscopic method, as described previously [14–17].

1. In a spectrophotometer cuvette, mix 50 μl of the centrifuged supernatant (from **step 6**) or 50 μl of the cell pellet lysates (control from **step 8**) with 350 μl of PBS (*see* **Note 11**).
2. Add 50 μl of the 1.67 mg/ml ortho-dianisidine dihydrochloride solution and mix.
3. Add 50 μl of the 0.005% H_2O_2 solution.
4. Mix the cuvette and monitor the change in absorbance at 460 nm at 22 °C for 10 min (Fig. 1a).
5. The results are expressed as percent of control (Fig. 1b).

3.4 Detecting Myeloperoxidase and NGAL by SDS-PAGE and Western Blotting

1. Denature proteins in supernatants and neutrophil pellets as indicated above (**steps** 7 and **10**, respectively). The samples are then subjected to 10% SDS-PAGE using standard techniques [19, 20]. The separated proteins are electrotransferred to nitrocellulose using the transfer buffer [19, 20]. After transfer:
2. Incubate the membrane in TBST + 5% milk for 30 min at room temperature on a shaker to block any nonspecific protein binding sites.

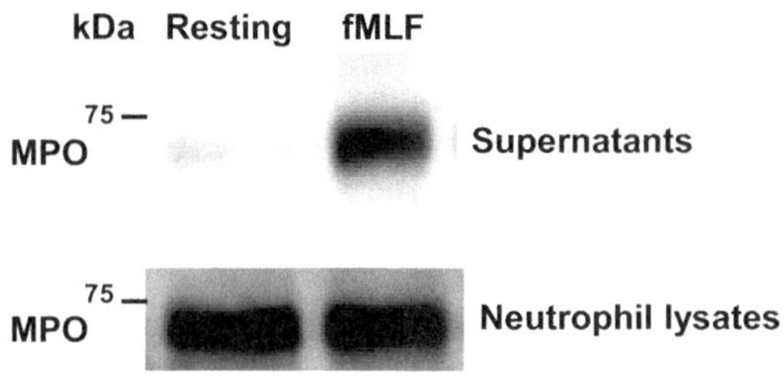

Fig. 2 Release of myeloperoxidase (MPO) by human neutrophils as detected by SDS-PAGE and Western blot. Neutrophils (5 × 10^6/0.5 ml Hanks buffer) were pretreated with 5 μg/ml cytochalasin B for 5 min, followed by stimulation with 1 μM fMLF for 2 min. After centrifugation, the supernatants and cell pellets were recovered and denatured. Samples were subjected to SDS-PAGE followed by immunoblot analysis with anti-MPO. Representative of three different experiments

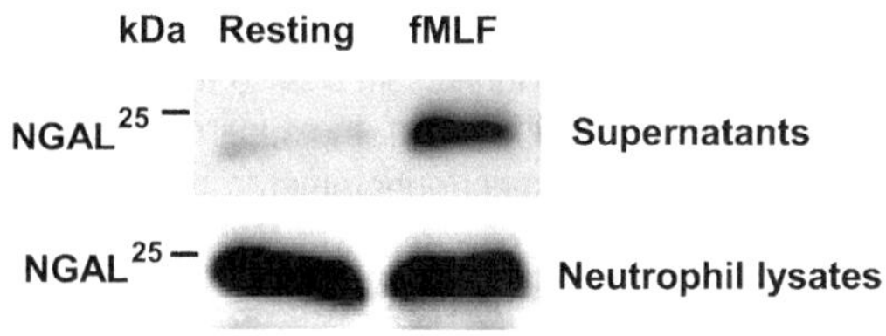

Fig. 3 Release of NGAL by human neutrophils as detected by SDS-PAGE and Western Blot. Neutrophils (5 × 10^6/0.5 ml Hank's buffer) were pretreated with 5 μg/ml cytochalasin B for 5 min, before stimulation with 1 μM fMLF for 2 min. After centrifugation, the supernatants and cell pellets were recovered and denatured. Samples were subjected to SDS-PAGE followed by immunoblot analysis with anti-NGAL. Representative of three different experiments

3. Add the anti-MPO antibody (1:10,000 dilution) or the anti-NGAL antibody (1:10,000 dilution) in TBST + 1% milk buffer and incubate for 1 h at room temperature (*see* **Notes 12** and **13**).
4. Wash three times with TBST for 10 min each wash.
5. Add HRP-conjugated goat anti-rabbit or anti-mouse antibodies (1:20,000 dilution) in TBST + 1% milk and incubate for 1 h at room temperature.
6. Wash three times with TBST for 10 min each wash.
7. Develop with the luminol reagent (HRP) or other ECL kit (Figs. 2 and 3).

4 Notes

1. All solutions used for cell preparation and cell manipulation, such as Dextran and 0.9% NaCl are prepared with sterile pyrogen-free water.

2. Prepare H_2O_2 solution fresh just before use.
3. Prepare ortho-dianisidine dihydrochloride solution fresh just before use.
4. Solutions for protein analysis must be prepared in very pure water (resistivity:18.2 MΩ-cm).
5. We found that incubation at 4 °C yields less primed neutrophils.
6. This centrifugation step is used to concentrate the cells and thus to use less Ficoll. If only a small volume of blood is used, the supernatant can be layered directly onto the Ficoll.
7. When layering the cells on Ficoll, take care to avoid mixing the two media.
8. Never lyse the erythrocytes without resuspending the pellet first, as the cells will tend to aggregate. The tube can be gently tapped on the bench to dislodge the pellet prior to lysis.
9. Lysis in 0.2% NaCl yields less primed neutrophils than lysis in H_2O or ammonium chloride.
10. Supernatants can be immediately used or stored at −80 °C for several days.
11. We used classical PBS pH 7.4, but other buffers with acidic pH can be used.
12. Since the molecular weight of MPO and NGAL are very different, the same nitrocellulose membrane can be cut around the 40 kDa marker, and the upper part can be used to detect MPO and the lower gel can be used to detect NGAL.
13. Lactoferrin is also a specific marker of specific granules and can be used instead of NGAL.

Acknowledgments

This work was supported by grants from the Institut National de la Santé et de la Recherche Médicale (INSERM) and the Centre National de la Recherche Scientifique (CNRS), Université Paris Diderot, Labex Inflamex, and Association Vaincre la Mucoviscidose (VLM).

References

1. Summers C, Rankin SM, Condliffe AM et al (2010) Neutrophil kinetics in health and disease. Trends Immunol 31:318–324
2. Mantovani A, Cassatella MA, Costantini C et al (2011) Neutrophils in the activation and regulation of innate and adaptive immunity. Nat Rev Immunol 11:519–531
3. Nauseef WM, Borregaard N (2014) Neutrophils at work. Nat Immunol 15:602–611

4. Malech HL, Deleo FR, Quinn MT (2014) The role of neutrophils in the immune system: an overview. Methods Mol Biol 1124:3–10
5. Soehnlein O, Lindbom L, Weber C (2009) Mechanisms underlying neutrophil-mediated monocyte recruitment. Blood 114:4613–4623
6. Mócsai A (2013) Diverse novel functions of neutrophils in immunity, inflammation, and beyond. J Exp Med 210:1283–1299
7. Witko-Sarsat V, Rieu P, Descamps-Latscha B et al (2000) Neutrophils: molecules, functions and pathophysiological aspects. Lab Investig 80:617–653
8. Borregaard N (2010) Neutrophils, from marrow to microbes. Immunity 33:657–670
9. Hampton MB, Kettle AJ, Winterbourn CC (1998) Inside the neutrophil phagosome: oxidants, myeloperoxidase, and bacterial killing. Blood 12:3007–3017
10. Nauseef WM (2007) How human neutrophils kill and degrade microbes: an integrated view. Immunol Rev 219:88–102
11. El-Benna J, Dang PM, Gougerot-Pocidalo MA et al (2005) Phagocyte NADPH oxidase: a multicomponent enzyme essential for host defenses. Arch Immunol Ther Exp 3:199–206
12. El-Benna J, Dang PMC, Hurtado-Nedelec M et al (2016) Priming of the neutrophil respiratory burst : role in host defense and inflammation. Immunol Rev 273:180–193
13. Faurschou M, Borregaard N (2003) Neutrophil granules and secretory vesicles in inflammation. Microbes Infect 5:1317–1327
14. Cowland JB, Borregaard N (2016) Granulopoiesis and granules of human neutrophils. Immunol Rev 273:11–28
15. Bradley PP, Christensen RD, Rothstein G (1982) Cellular and extracellular in pyogenic inflammation. Blood 60:618–622
16. Bradley PP, Priebat DA, Christensen RD et al (1982) Measurement of cutaneous inflammation: estimation of neutrophil content with an enzyme marker. J Invest Dermatol 78:206–209
17. Boudiaf K, Hurtado-Nedelec M, Belambri SA et al (2016) Thymoquinone strongly inhibits fMLF-induced neutrophil functions and exhibits anti-inflammatory properties in vivo. Biochem Pharmacol 104:62–73
18. Bedouhene S, Moulti-Mati F, Dang PM et al (2017) Oleuropein and hydroxytyrosol inhibit the N-formyl-methionyl-leucyl-phenylalanine-induced neutrophil degranulation and chemotaxis via AKT, p38, and ERK1/2 MAP-kinase inhibition. Inflammopharmacology 25:673–680
19. El-Benna J, Dang PM (2007) Analysis of protein phosphorylation in human neutrophils. Methods Mol Biol 412:85–96
20. Belambri SA, Dang PM, El-Benna J (2014) Evaluation of p47phox phosphorylation in human neutrophils using phospho-specific antibodies. Methods Mol Biol 1124:427–433

Chapter 17

Influence of Oxygen on Function and Cholesterol Composition of Murine Bone Marrow-Derived Neutrophils

Katja Branitzki-Heinemann, Graham Brogden, and Maren von Köckritz-Blickwede

Abstract

During inflammation and infection, invading pathogens as well as infiltrating neutrophils locally consume oxygen and reduce the present oxygen level. Since oxygen is an elementary component of the microenvironment required for cell activity and alters multiple cellular functions, it is important to study neutrophil functionality and phenotype at characteristic pathophysiological oxygen levels that reflect the hypoxic phenotype during infection and inflammation. Here, we describe methods to study murine neutrophils under hypoxic compared to normoxic conditions, including analysis of cholesterol content as a key lipid involved in biological functions.

Key words Hypoxia, Normoxia, Reactive oxygen species, Cholesterol, Negative selection, Neutrophil isolation

1 Introduction

In vitro research on innate immune cells such as neutrophils is usually performed at atmospheric oxygen levels under standard culture conditions: ~21% O_2 (159 mmHg; called normoxia), 5% CO_2, 37 °C. Indeed, physiologically relevant oxygen concentrations are substantially lower: physiological oxygen levels depend on the tissue location and range from around 13.2% O_2 (100 mmHg) in arterial blood to 7% O_2 (53 mmHg) in intestinal mucosa and submucosa [1]. During pathophysiological processes that arise as a consequence of infection and inflammation, local consumption of O_2 by invading pathogens and transmigrating immune cells leads to levels of around 1% O_2 (7 mmHg; called hypoxia) or even lower, subsequently affecting cellular functions [2–5].

To better understand the function of neutrophils and to identify reliable in vitro data for the development of new therapeutic or prophylactic intervention strategies, we strongly recommend the consideration of (patho)physiologically relevant O_2 conditions in

Mark T. Quinn and Frank R. DeLeo (eds.), *Neutrophil: Methods and Protocols*, Methods in Molecular Biology, vol. 2087, https://doi.org/10.1007/978-1-0716-0154-9_17,

the experimental setup. Hypoxia glove boxes that allow well-controlled adjustment of sustained O_2 values inside an enclosed workstation support the goal of closely mimicking physiological conditions in vitro. Dissolved O_2 in a neutrophil suspension culture can be measured noninvasively and without consumption using specific measurement systems (e.g., PreSens Precision Sensing GmbH) and reveal that neutrophil cultures maintain a constant O_2 level, reflecting the atmospheric environment when cultured under normoxic conditions. In contrast, hypoxic incubation decreases the dissolved O_2 level in the culture to lower than 13.3 mmHg (1.8% O_2), indicating that manually applied experimental settings closely reflect pathophysiological O_2 conditions that may occur in infected tissue. Interestingly, in vitro conditions for most epithelial cell cultures grown in a monolayer (e.g., intestinal epithelial Caco-2-cells [6]) show distinctly reduced O_2 levels compared to suspension cultures of neutrophils or mast cells [7, 8]. Thus, cell-type-specific O_2 adjustments are necessary to reflect pathophysiological O_2 conditions in vitro.

Monitoring and adapting O_2 levels during in vitro studies is of special importance when investigating neutrophil antimicrobial function as the first line of host defense of the innate immune system against various pathogens including bacteria, fungi and protozoa. Besides degranulation and intracellular killing of microbes, neutrophils are able to entrap and kill pathogens by the release of extracellular structures, so called neutrophil extracellular traps (NETs) [9]. It is still not entirely clear when and why a neutrophil decides to from a NET. A wide range of proinflammatory stimuli, including interferon-α, interleukin-8 (IL-8), the pharmacological agent phorbol 12-myristate 13-acetate (PMA), as well as numerous microbes, induce the formation of NETs [reviewed in 10]. NET formation and NET-based antimicrobial activity partially depend on the formation of reactive oxygen species (ROS) through the membrane-bound NADPH oxidase enzyme complex; thus, blocking the NADPH oxidase inhibits ROS generation and subsequent NET release [11, 12]. Moreover, NET formation can be correlated with lipid alterations. Decreased levels of cholesterol mediated by methyl-β-cyclodextrin in primary blood-derived neutrophils leads to increased spontaneous NET formation [13], and pharmacological treatment of neutrophils with statins that block cholesterol synthesis also induce formation of NETs [14]. Especially cholesterol, among other lipids, has been shown to play an important role in infection and inflammation. For example, cholesterol depletion can reduce the risk of invasion from several pathogens [15]. Using an HPLC-based approach [16], we recently investigated the cholesterol level of neutrophils under hypoxia compared to normoxia [8]. Interestingly, we showed, that human neutrophils have a significantly higher cholesterol level after incubation under hypoxia compared to normoxia, which correlates with a reduced spontaneous NET-formation under hypoxia [8]. In this chapter, we describe an adjusted method to analyze cholesterol levels in murine bone-marrow-derived neutrophils incubated under hypoxic versus normoxic O_2 levels.

2 Materials

2.1 Isolation of Murine Bone Marrow-Derived Neutrophils

1. 10 ml syringe and 25-gauge needle.
2. 100 μm mesh nylon strainer.
3. 70% ethanol.
4. 1× PBS.
5. Preparation media: RPMI +10% fetal calf serum (FCS) + 1× Pen/Strep (for 50 ml: 44.5 ml of RPMI +5 ml of FCS + 500 μl of 100× Pen/Strep), prepare 30–40 ml for each animal.
6. Isolation media: RPMI +10% FCS + 2 mM EDTA (for 50 ml: 43.8 ml of RPMI +5 ml of FCS + 1.2 ml of 82.5 mM EDTA), prepare 30–40 ml for each animal.
7. 0.2% NaCl solution for erythrocyte lysis.
8. 1.6% NaCl solution.
9. Supplemented PBS: 1× PBS + 2% FCS + 1 mM EDTA (for 50 ml: 48.4 ml of 1× PBS + 1 ml of FCS + 0.6 ml of 82.5 mM EDTA).
10. EasySep™ Mouse Neutrophil Enrichment Kit from STEMCELL Technologies.
11. 0.4% trypan blue solution.
12. Primary antibodies for purity determination: FITC-conjugated CD11b (rat anti-mouse CD11b, Clone M1/70), PE-conjugated Ly6G (rat anti-mouse Ly-6G, Clone 1A8), FITC-conjugated Ly6G/C (rat anti-mouse Ly-6G and Ly-6C, Clone RB6-8C5).
13. Isotype control antibodies: FITC-conjugated rat IgG2b, PE-conjugated rat IgG2a.

2.2 Stimulation of Murine Neutrophils and Measurement of Specific O_2 Conditions

1. Hypoxia chamber: To maintain physiological relevant O_2 conditions, a consistent atmosphere is required for an accurate experimental performance. A hypoxia glove box (e.g., Coy Laboratory Products) is recommended to induce the hypoxic environment. O_2 is dispelled by flushing the chamber with N_2 and is constantly monitored by an O_2-sensor on the inside of the chamber. A real-time feedback loop system continuously controls and adjusts not only O_2 but also CO_2, temperature, and humidity.
2. O_2 sensor spots (3 mm, PSt-7, PreSens Precision Sensing GmbH) glued on the bottom of well plates (special glue based on silicone, PreSens).

3. Multi-well plates (e.g., 24- or 48-well).
4. Cell culture medium: RPMI without phenol red.
5. Oxy-1-ST measurement system (e.g., PreSens Precision Sensing GmbH).

2.3 Lipid Isolation and Cholesterol Analysis of Murine Neutrophils

1. HPLC grade methanol, chloroform, and H_2O.
2. 15 ml glass tubes with PTFE seal.
3. Sample rotator.
4. Vacuum centrifuge and a vacuum pump connected to a tube with a glass Pasteur pipette.
5. Amber glass 1.5 ml vials with PTFE/White Silicone lids.
6. Chromolith HighResolution RP-18 endcapped 100-4.6 mm column (e.g., Merck).
7. High Performance Liquid Chromatograph (e.g., Hitachi; parameters are dependent on individual HPLC), including UV detector, column oven, and auto sampler.

3 Methods

3.1 Isolation of Murine Bone Marrow-Derived Neutrophils

1. Keep femur and tibia in preparation medium and clean them from muscles and tendons, briefly dip in 70% ethanol, and wash three times in 1× PBS.
2. Mouse neutrophils are isolated by flushing bone marrow cells into isolation medium using a syringe equipped with a 25-gauge needle. Disperse remaining clumps by gently passing the cell suspension through a 100 μm mesh nylon strainer.
3. After centrifugation at 360 × *g* for 7 min at 4 °C discard the supernatant.
4. To remove erythrocytes from the cell suspension, resuspend the pellet in 10 ml of 0.2% NaCl. The lysis is stopped after 20 s by adding the same volume of 1.6% NaCl.
5. After another centrifugation step, wash the cell pellet with isolation medium and finally, resuspend in ice cold supplemented PBS.
6. For neutrophil enrichment, follow the instructions of the Easy-Sep™ Mouse Neutrophil Enrichment Kit from STEMCELL technologies (*see* **Note 1**).
7. After enrichment, examine cell viability by trypan blue staining, and determine the number of cells using a hemocytometer.
8. Purity of the isolated neutrophil suspension is assessed by flow cytometry, identifying cells which are positive for CD11b, Ly6G, and Ly6G/C. Therefore, incubate 2×10^5 cells with either 500 ng FITC-conjugated CD11b and 200 ng

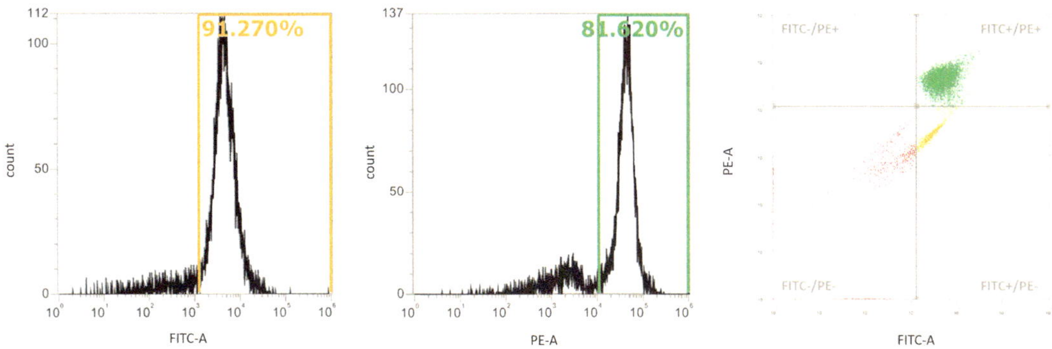

Fig. 1 Representative FACS analysis of FITC-CD11b (FITC-A) and PE-Ly6G (PE-A)-double stained neutrophils after purification with the EasySep™ Mouse Neutrophil Enrichment Kit. Percentage of positive stained cells from the total cell amount is highlighted

PE-conjugated Ly6G, or with 200 ng FITC-conjugated Ly6G/C, and with the respective isotype controls in supplemented PBS for 45 min in the dark at RT. Afterwards, centrifuge samples and keep cells in the dark on ice in supplemented PBS until FACS analysis (*see* **Note 2**) (Fig. 1).

3.2 Stimulation of Murine Neutrophils and Measurement of Specific O_2 Conditions

The Oxy-1- measurement system by PreSens Precision Sensing GmbH reveals the O_2 concentration in the neutrophil suspension culture. Using optical sensors that are fixed on glass coverslips placed in the wells or directly on the bottom of the wells of the experimental setup, the dissolved O_2 level in the cell culture medium is measured based on O_2-dependent quenching of phosphorescent probes without consumption. The sensors are covered with a distinct fluorescent indicator dye that gets excited by light from an LED to emit fluorescence. In the presence of O_2 molecules, the emitted fluorescence signal is decreased or quenched. The resulting signal shift is detected and used for calculating the amount of O_2.

1. Prepare and glue 3 mm sensor spots on the bottom of well-plates, ensuring the black side will face the sample during the experiments. Let the glue cure for 24 h before seeding cells.
2. Adjust the cell suspension to 1×10^7 cells/ml in 24 h pre-equilibrated cell culture medium (*see* **Note 3**). If needed, pool the cell harvest from several animals.
3. Prepare individual wells for measuring O_2: (a) medium only and (b) cell suspension (with and without stimulation), which are equipped with sensor spots on glass coverslips.
4. Seed 6.7×10^5 cells/ml (e.g., 2×10^5 cells in a single well of a 48-well plate or 1×10^6 cells in a 1.5 ml tube) and incubate in a hypoxic chamber under hypoxia (1% O_2) or in the normal cell culture incubator under normoxia (21% O_2).

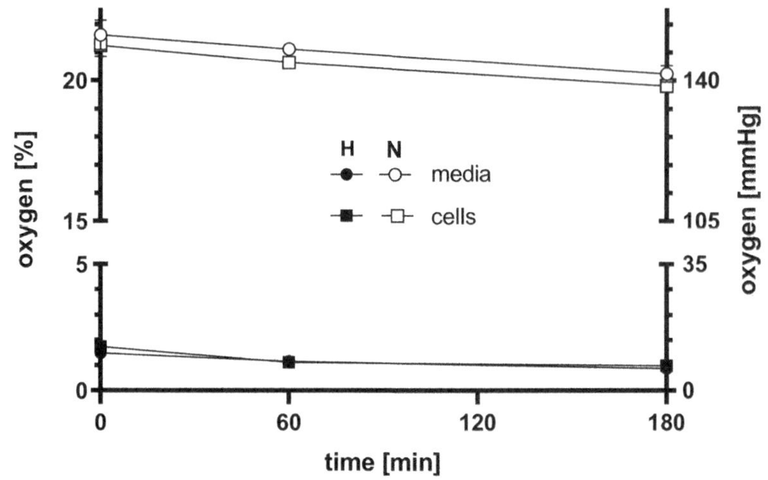

Fig. 2 Oxygen levels in murine neutrophil suspension cultures. O_2 levels were measured in media-only control and in medium containing murine neutrophils incubated under normoxia (N, 21% O_2) or hypoxia (H, 1% O_2). Dissolved O_2 levels were monitored at time point zero and after 1 and 3 h of incubation. Plotted values represent the mean ± SEM of 6 values and are displayed as % O_2 on the left *y*-axis and mmHG O_2 on the right *y*-axis

5. Stimulate selected samples pharmacologically with 25 nM PMA in a final concentration for 3 h (*see* **Note 4**).
6. At time point zero, after 1 h of incubation, and at the final point after 3 h, measure O_2 in each individual sample in duplicate with the Oxy-1 ST device:
   (a) Put the well with the sensor spot directly above the LED-detector.
   (b) If necessary, adjust air pressure and temperature manually.
   (c) Start the measurement and collect at least two values.
   (d) Perform O_2 measurement with the other samples, respectively.
   (e) Sampling for further control readouts (e.g., oxidative burst or pH) (*see* **Notes 5** and **6**, Fig. 2).

3.3 Lipid Isolation and Cholesterol Analysis of Murine Neutrophils

3.3.1 External Standard

Prepare cholesterol dilutions dissolved in chloroform–methanol (1:1) in the following concentrations: 1 mg/ml, 500 ng/ml, 200 ng/ml, 100 ng/ml 50 ng/ml, 20 ng/ml, 10 ng/ml, 5 ng/ml, and 2 ng/ml. Prepare at least five independent dilutions and measure using the protocol described below.

3.3.2 Sample Preparation

1. Centrifuge the cell suspension for 10 min at 400 × *g* at 4 °C, remove the medium, wash once with cold 1× PBS, and resuspend in 350 μl cold H_2O (HPLC grade).
2. Transfer the samples to 15 ml screw cap glass tubes with PTFE seals on ice.

3. Add 2 ml of methanol and then incubate the samples at room temperature for 1 min. Add 1 ml of chloroform and shake the samples for 1 min.
4. Rotate the samples at ~10 rpm at room temperature for 30 min.
5. Centrifuge the samples at 7 °C at 1952 × *g* for 10 min to pellet the protein fraction.
6. Transfer the supernatant to a new 15 ml glass tube with a PTFE lid, leaving the protein-containing pellet behind. This protein-containing pellet can be kept for future analysis.
7. Add 1 ml of chloroform, incubate for 1 min, then add 1 ml of H_2O and invert the samples several times to mix.
8. Centrifuge the samples at 7 °C at 1952 × *g* for 10 min, then carefully remove and discard the upper phase using a glass Pasteur pipette connected to a vacuum pump, down to, but not including the cloudy layer.
9. If a compact clear cloudy layer is not visible, an optional further purification step can be performed by repeating **step 8** again.
10. Preheat the vacuum dryer to 60 °C, place the samples in the rotor, and dry for approximately 45 min, or until completely dry. The samples can then be frozen and stored at −20 °C until required.
11. Finally, resuspend the samples in 250 μl chloroform–methanol (1:1) in amber colored 1.5 ml glass vials with a PTFE/White silicone lid.

3.3.3 Cholesterol Analysis

Lipid analysis is performed firstly by using a form of reverse phase liquid chromatography, in this case HPLC, followed by a method of detection, here UV was used. A mobile phase carries the samples through a column, which acts as the stationary phase, separating the lipid mixture based on polarity and retention time.

1. Configure the HPLC machine using the following parameters: Chromolith® HighResolution RP-18 endcapped 100-4.6 mm column coupled to a 5–4.6 mm guard cartridge (both Merck) heated to 32 °C. Methanol is used as the mobile phase at a flow rate of 1 ml/min at 22 bar, and a UV detector measuring at 203 nm to quantify the amount of cholesterol in each sample. The peak corresponding to cholesterol has a retention time of approximately 4.9 min. Therefore, a run time of 6 min is required (Fig. 3). Prior to analysis, the HPLC should be cleaned and prepared according to the manufacturer's guidelines.

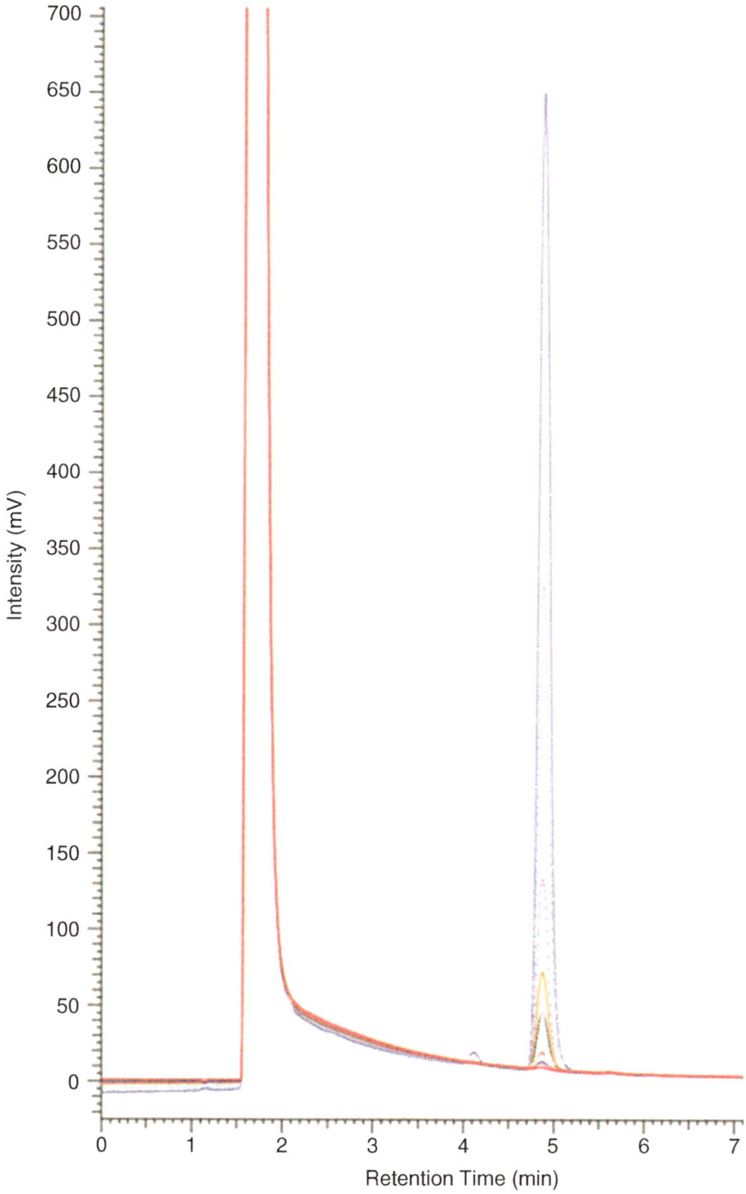

Fig. 3 Representative chromatogram showing serial cholesterol dilutions analyzed by HPLC. Cholesterol concentrations ranging from 1000 μg/ml to 2 μg were dissolved in chloroform/methanol (1:1) and measured giving a cholesterol-specific peak at 4.9 min. The peak with a retention time of 1.9 min is produced from chloroform emitting a single peak at 203 nm

2. Results are initially given as area under the curve. These values can be quantified against the equation obtained from the linear calibration curve. The value obtained can be expressed as the total cholesterol amount per 10^6 cells.

4 Notes

1. To avoid prestimulation or labeling of isolated neutrophils, negative enrichment is recommended by which non-neutrophils are targeted with a combination of biotinylated monoclonal antibodies and magnetic particles which allows separation within a magnetic field. Final neutrophil purity is suggested to reach values up to 89%, as determined by flow cytometry based on the detection of specific surface markers (e.g., CD11b, Ly6G/C, and Ly6G). Myeloid derived cells from the bone marrow express CD11b and Ly6G/C (Gr-1). Although an exclusive marker for mouse bone marrow neutrophils has not been identified, Ly6G is known to be mainly found in the granulocytic population, whereas Ly6C is highly expressed in the monocytic fraction [17].
2. FACS analysis was performed with an Attune N×T Flow Cytometer from ThermoFisher Scientific with blue/red laser configuration using the Invitrogen Attune NxT Software. The blue lasers (BL) 1 and 2 were chosen with an intensity of 310 V (BL-1) or 390 V (BL-2), and a flow through rate of 100 μl/min. Singlet cells were gated using forward scatter signal (190 V), giving an amount of more than 95%, and further analyzed for specific antibody signal.
3. To ensure that the cells face hypoxia from the beginning, cell culture medium is pre-equilibrated under hypoxia over night to reach low O_2 concentrations. Measuring O_2 conditions in the medium directly after the incubation is initiated to confirm adaption of the medium to 1% O_2.
4. Phorbol 12-myristate 13-acetate (PMA) is a well-know and globally used pharmacological NET-inducer. In human neutrophil cultures, PMA stimulates NET-formation within 4 h in almost 100% of the cells. The stimulatory effect of PMA in murine neutrophils is less; however, NET-formation is also induced. It should be taken into consideration, that concentration and/or stimulation time may need to be adjusted.
5. Oxidative burst can be analyzed as control sampling of cellular behavior under hypoxia. ROS are produced by neutrophils and play a key role in antimicrobial activity. ROS production is pharmacologically stimulated by PMA via activation of the membranous NADPH-oxidase. Ex vivo/in vitro, cellular ROS can be measured by using 2′,7′-dichlorofluorescein (DCF) diacetate (DCFDA; prepare 10 mM stock in DMSO). DCFDA is deacetylated by cellular esterases to a nonfluorescent compound but is oxidized by ROS into the highly fluorescent DCF, which can be detected by fluorescence spectroscopy (e.g., with flow cytometry). For ROS analysis, 10 min prior to

the finished incubation time, add DCFDA at a final concentration of 10 μM to the respective samples, and after 20 min of further incubation analyze by flow cytometry. Cells without DCF serve as background control; PMA-stimulated cells are used as positive controls. Under hypoxia, cells show a reduced oxidative burst in response to PMA.

6. In addition to O_2 as an important physiological marker, osmolarity or pH are further features of in vitro experiments which at one point need to be addressed when studying dynamics in cell culture systems. The dual lifetime referenced method from PreSens Precision Sensing GmbH uses optical pH sensors, which enable external monitoring of the pH in neutrophil suspension cultures. Similar to the O_2 measurements, sensor spots are covered with pH sensitive fluorescent dyes. Luminescence lifetime changes are detected and compared to an internal reference indicator, facilitating pH determination. pH sensor spots are introduced into the cell culture system like those for O_2 measurements. However, a different device, such as the pH-1 mini, is used for pH analysis.

Acknowledgments

This work was supported by R2N Project funded by the Federal State of Lower Saxony.

References

1. Carreau A, El Hafny-Rahbi B, Matejuk A et al (2011) Why is the partial oxygen pressure of human tissues a crucial parameter? Small molecules and hypoxia. J Cell Mol Med 15:1239–1253
2. Lone AG, Atci E, Renslow R et al (2015) *Staphylococcus aureus* induces hypoxia and cellular damage in porcine dermal explants. Infect Immun 83:2531–2541
3. Melican K, Boekel J, Månsson LE et al (2008) Bacterial infection-mediated mucosal signalling induces local renal ischemia as a defense against sepsis. Cell Microbiol 10:1987–1998
4. Campbell EL, Bruyninckx WJ, Kelly CJ et al (2014) Transmigrating neutrophils shape the mucosal microenvironment through localized oxygen depletion to influence resolution of inflammation. Immunity 40:66–77
5. Schaffer K, Taylor CT (2015) The impact of hypoxia on bacterial infection. FEBS J 282:2260–2266
6. Zeitouni NE, Fandrey J, Naim HY et al (2015) Measuring oxygen levels in Caco-2 cultures. Hypoxia 9:53–66
7. Möllerherm H, Branitzki-Heinemann K, Brogden G et al (2017) Hypoxia modulates the response of mast cells to *Staphylococcus aureus* infection. Front Immunol 11:541
8. Branitzki-Heinemann K, Möllerherm H, Völlger L et al (2016) Formation of neutrophil extracellular traps under low oxygen level. Front Immunol 7:518
9. Brinkmann V, Reichard U, Goosmann C et al (2004) Neutrophil extracellular traps kill bacteria. Science 303:1532–1535
10. von Köckritz-Blickwede M, Nizet V (2009) Innate immunity turned inside-out: antimicrobial defense by phagocyte extracellular traps. J Mol Med 87:775–783
11. Fuchs TA, Abed U, Goosmann C et al (2007) Novel cell death program leads to neutrophil extracellular traps. J Cell Biol 176:231–241
12. Stoiber W, Obermayer A, Steinbacher P et al (2015) The role of reactive oxygen species

(ROS) in the formation of extracellular traps (ETs) in humans. Biomol Ther 5:702–723

13. Neumann A, Brogden G, Jerjomiceva N et al (2014) Lipid alterations in human blood-derived neutrophils lead to formation of neutrophil extracellular traps. Eur J Cell Biol 93:347–354
14. Chow OA, von Köckritz-Blickwede M, Bright AT et al (2010) Statins enhance formation of phagocyte extracellular traps. Cell Host Microbe 8:445–454
15. Gilk SD, Cockrell DC, Luterbach C et al (2013) Bacterial colonization of host cells in the absence of cholesterol. PLoS Pathog 9: e1003107
16. Brogden G, Neumann A, Husein DM et al (2017) Methods to study lipid alterations in neutrophils and the subsequent formation of neutrophil extracellular traps. J Vis Exp (121):54667
17. Zhao Y, Wu T, Shao S et al (2016) Phenotype, development, and biological function of myeloid-derived suppressor cells. Onco Immunology 5(2)

Chapter 18

In Vitro Assay for Sensitive Determination of Human Blood PMN Responses

Noah Fine, William Khoury, and Michael Glogauer

Abstract

Polymorphonuclear neutrophils (PMNs) are the most common leukocytes in the circulation and exhibit a wide range of distinct cellular phenomena as part of their microbicidal killing activities, including degranulation, phagocytosis, reactive oxygen species (ROS) production, adhesion, chemotaxis and production of PMN Extracellular Traps (NETs). As a simple in vitro test of PMN functional responses in human blood we have developed a multicolor flow cytometry-based assay of PMN cluster of differentiation (CD) surface marker expression. Short incubations of whole human blood can be performed in the presence of a wide range of agonists or inhibitors, followed by sensitive detection of changes in CD marker expression. This protocol has the advantage that small amounts of human blood are necessary and there are no PMN isolation steps, which can alter PMN activation status.

Key words PMNs, Flow cytometry, CD marker

1 Introduction

PMNs are the most abundant leukocyte subset in human blood and act as critical first responders during acute inflammation [1]. They express a wide range of surface receptors to various target ligands, including: pathogen associated molecular patterns (PAMPs), cytokines, chemokines, and adhesion receptors. As an important aspect of their ability to quickly and effectively respond to a wide range of insults, PMNs display exquisite sensitivity to extracellular stimuli including PAMPs, cytokines, chemokines, mechanosensory stimuli, and other components of the innate and adaptive immune system, such as target bound antibody, complement proteins, macrophages, and platelets. During priming and activation PMNs undergo degranulation [2], one consequence of which is the rapid surface upregulation of a number of Cluster of Differentiation (CD) markers, which in turn alter downstream PMN functionality. Due to the sensitivity of PMNs to in vitro manipulation [3–5], accurate flow cytometric determination of CD marker expression

Mark T. Quinn and Frank R. DeLeo (eds.), *Neutrophil: Methods and Protocols*, Methods in Molecular Biology, vol. 2087, https://doi.org/10.1007/978-1-0716-0154-9_18,

on these cells has been problematic. Here we describe a method to determine steady state PMN CD marker expression levels in human blood samples with high fidelity, and to accurately measure upregulation of degranulation products in response to short term in vitro stimulation of PMNs in whole blood, by multicolor flow cytometry. Fidelity of the CD marker signal is achieved by fixing fresh blood samples with formaldehyde immediately following a short stimulation at 37 °C. This assay can be used to assess the effects of a wide range of pro- and anti-inflammatory factors on PMN responses [2, 6].

2 Materials

1. N-Formyl-Met-Leu-Phe (*f*MLF). Prepare 10 mM stock in DMSO, and store aliquots at −20 °C.
2. Phosphate-buffered saline (PBS): 137 mM NaCl, 2.7 mM KCl, 10 mM Na_2HPO_4, 1.8 mM KH_2PO_4, pH 7.4.
3. Human recombinant tumor necrosis factor alpha (TNF-α). Prepare 100 ng/100 μl stock in phosphate-buffered saline (PBS) and store aliquots at −20 °C.
4. Red blood cell lysis buffer such as BD Pharm lyse solution.
5. Hank's balanced-salt solution (HBSS): 138 mM NaCl, 5.3 mM KCl, 0.34 mM Na_2HPO_4, 0.44 mM KH_2PO_4, 4.2 mM $NaHCO_3$, 5.6 mM glucose, pH 7.4.
6. FACS buffer: HBSS with 2 mM EDTA, 1% BSA, filtered.
7. Vacutainer tubes containing sodium citrate.
8. 16% paraformaldehyde (PFA) ampule.
9. 1 mg/ml rat serum.
10. 1 mg/ml mouse IgG.
11. Mouse anti-human CD16-PE (Clone: 3G8).
12. Mouse anti-human CD66a-APC (Clone: CD66a-B1.1).
13. Mouse anti-human CD11b-APC-Cy7 (Clone: ICRF44).
14. Mouse anti-human CD18-BV421 (Clone: 6.7).
15. Flow cytometry compensation beads (e.g., OneComp eBeads from eBiosciences).
16. Sonicator.
17. Coulter counter or hemocytometer.
18. Vacuum aspirator.

3 Methods

All samples should be maintained on ice unless otherwise noted, and all centrifugations are at 4 °C.

3.1 Preparation of Working Solutions of fMLF and TNF-α

1. Dilute stock *f*MLF 10× using sterile PBS, followed by serial dilutions of 1:10. Dilute stock TNF-α 1:500. Diluted working solutions will be diluted a further 1:10 into blood samples (*see* **Note 1**).

3.2 Acquisition of Human Blood Samples

1. Blood should be drawn by a trained phlebotomist (*see* **Note 2**).
2. Blood samples are drawn from the median cubital vein into a Vacutainer containing 0.1 volumes of sodium citrate anticoagulant.
3. Maintain the blood on ice until needed and mix by gentle inversion just prior to aliquoting (*see* **Note 3**).

3.3 Blood PMN Stimulation

1. Aliquot 100 μl of whole blood (which should contain ~0.5×10^6 leukocytes) into polystyrene flow cytometry tubes.
2. To stimulate, add *f*MLF (10 μl), TNF-α (10 μl), or vehicle (10 μl) to the blood and immediately vortex the tubes gently to mix (Fig. 1) (*see* **Note 4**).
3. Cover the tubes with paraffin wax and incubate them at 37 °C for 30 min with periodic agitation (*see* **Note 5**).
4. Prepare one unstimulated tube (vehicle only) and keep it on ice, as a control for metabolic activation of the PMNs.
5. Prepare extra tubes for unstained and fluorescence minus one (FMO) controls if desired.
6. At endpoint, add 1/tenth volume of concentrated PFA (~12.2 μl, 1.6% final concentration) to each tube to fix the PMNs (*see* **Note 6**).
7. Gently vortex each tube immediately after adding PFA and incubate them on ice for 15 min.
8. Dilute the fixative with 1 ml of PBS to wash the cells.
9. Centrifuge the tubes for 5 min at 1000 × *g* and 4 °C.
10. Carefully aspirate the supernatant.
11. Lyse the red blood cells (RBCs) by resuspending each pellet in 1 ml of red blood cell lysis buffer.
12. Incubate the tubes on ice for 5 min.
13. Centrifuge the tubes for 5 min at 1000 × *g* and 4 °C.
14. Carefully aspirate the supernatant.
15. Repeat the RBC lysis steps (**steps 9–12**) until pellets are white (*see* **Note 7**). After the second lysis step, reduce the volume of lysis buffer used to 0.5 ml.
16. Resuspend the pellets in 1 ml of FACS buffer.
17. Use a Coulter counter or hemocytometer to count the cells that appear in the size range of 7–12 μm (*see* **Note 8**).

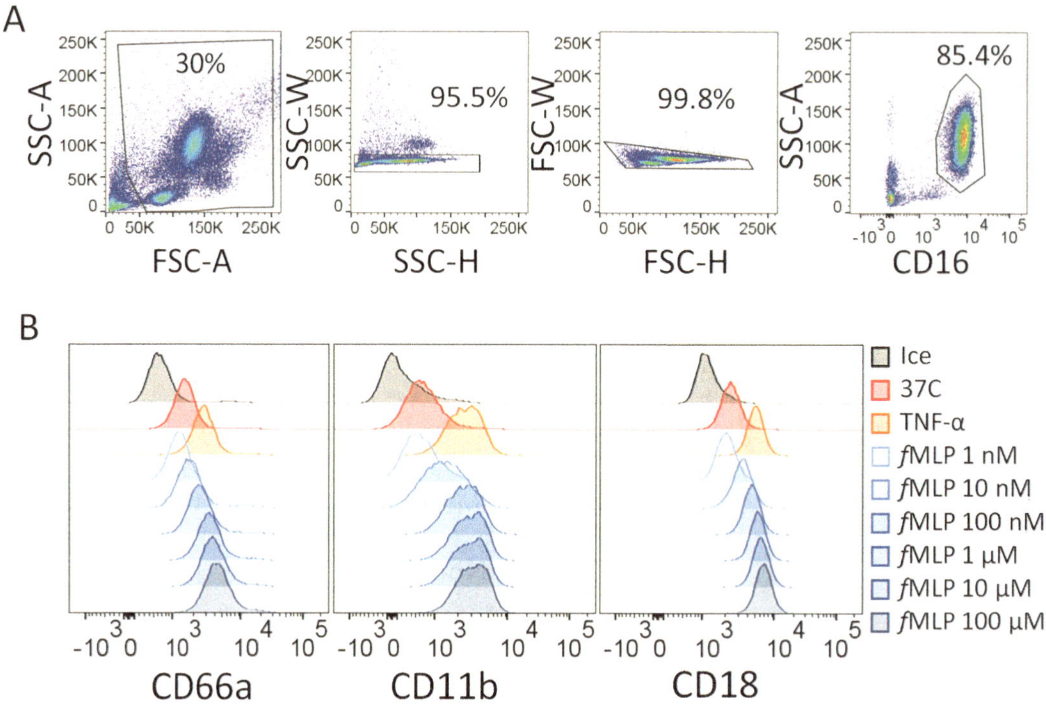

Fig. 1 Human blood PMN stimulation. (**a**) Whole blood leukocytes were analyzed by multicolor flow cytometry, and PMNs were gated according to the strategy shown here. FSC-A × SSC-A was used to gate out debris and RBCs. Doublets were excluded using side scatter height (SSC-H) × side scatter width (SSC-W) and forward scatter height (FSC-H) × forward scatter width (FSC-W). PMNs were gated in whole blood based on high SSC-A and high expression of CD16. The percentage of the parent population is shown for each gated population. (**b**) Blood of one healthy human volunteer was maintained on ice for 30 min or stimulated in vitro at 37 °C in the presence or absence of TNF-α and increasing concentrations of *f*MLF, for 30 min. PMN gating was performed as in (**a**). Representative histograms of CD66a, CD11b and CD18 expression are shown. At least 2×10^4 gated neutrophil events were acquired. Flow cytometric analysis was performed using FlowJo X. Increased PMN surface CD marker expression correlated with the presence and increasing concentration of the in vitro stimuli

18. Centrifuge the tubes for 5 min at 1000 × *g* at 4 °C.
19. Carefully aspirate the supernatant.

3.4 Fluorescent Staining

1. Resuspend the cells in each tube in a small volume of FACS buffer. Use an appropriate volume of FACS buffer so that the total volume after addition of antibodies will be 50 μl.
2. Add 1 μl rat serum and 2 μl mouse IgG to each tube to block Fc-receptors, vortex gently, and incubate on ice for 20 min.
3. Prepare the master mix of fluorescently conjugated antibodies (*see* **Notes 9** and **10**).
4. Add the antibody master mix, vortex gently and incubate for 30 min on ice protected from light (*see* **Note 11**).
5. Wash each pellet with 1 ml of FACS buffer.

6. Vortex briefly and pellet the cells at 1000 g for 5 min at 4 °C.
7. Decant the supernatant.
8. Repeat the wash steps (**steps 5–7**) two additional times.
9. Resuspend the pellets in 250 μl of FACS buffer and vortex briefly.
10. Prepare single stained compensation beads controls for each fluorescent channel according to manufacturers' instructions.
11. Resuspend labeled compensation beads in 400 μl of FACS buffer and vortex briefly.
12. Cover the tubes with paraffin wax and aluminum foil to shield the tubes from the light and store in the fridge at 4 °C.

3.5 Sample Acquisition

1. Perform instrument setup according to the protocols of your flow cytometry facility.
2. Vortex the first sample and load onto the cytometer.
3. Adjust the forward scatter (FSC) and side scatter (SSC) settings so that the population of interest (PMNs) is displayed in the center of the scatterplot.
4. Ensure fluorescent signals are on scale (*see* **Note 12**).
5. Perform automated compensation using single stained bead controls.
6. Run each sample and acquire at least 2×10^4 gated events.
7. Export FCS files.
8. Data can be analyzed using FlowJo or other commercially available software.

4 Notes

1. Final concentrations after dilution in blood were 100 pg/110 μl for TNF-α and a range of 10 nM to 100 μM for *f*MLF.
2. Research Ethics Board approval is required and signed informed consent should be obtained from all participants in human studies.
3. Fresh blood can be maintained on ice prior to stimulation at 37 °C.
4. The total volume of stimuli added should be kept to a minimum. If possible, add ≤10% (10 μl) of the total blood volume.
5. Short incubations (≤30 min) are optimal, since PMNs become activated by incubation at 37 °C alone. Although longer incubations are possible, metabolic activation and significant cell death can become problematic.

6. There are typically very low (<5%) amounts of PMNs in fresh human blood samples, which tend to appear in the lower left quadrant of FSC-A × SSC-A plots and are therefore easily excluded based on these parameters. Although not necessary for most applications, a fixable live/dead marker such as eFluor506 (Thermo Fisher Scientific) can be added directly to fresh blood samples prior to fixation. Human blood is incubated with 1 μl of eFluor506 for 15 min on ice, washed once with $PBS^{-/-}$, fixed and processed as above. Heat killed (5 min at 64 °C) cells, labeled with eFluor506, and unstained cells will be necessary as compensation controls.
7. Lysis of RBCs usually takes 2–3 lysis steps for whole blood. It is not necessary to completely eliminate the RBCs, since residual RBCs can be gated out by flow cytometry.
8. One hundred μl of human blood typically contains ~0.5×10^6 leukocytes, which is sufficient to acquire a minimum of 20,000 gated PMNs.
9. Although CD16+ monocytes are present, CD16 × SSC-A plots are sufficient to distinguish PMNs in whole leukocyte preparations (Fig. 1a). A wide range of PMN CD markers of activation can be included for multicolor flow cytometry applications [2, 7, 8] and for gating on alternate leukocyte populations, including monocytes [9] and lymphocytes.
10. Antibody concentrations should be titrated appropriately. Prepare enough mastermix for 1.1 times the number of tubes to be labeled. For compensation, a control tube with eFluor506 positive cells and an unstained control should be prepared. Depending on the percentage of cells that are eFluor506 positive in the single stained control and the event rate during sample acquisition, unstained sample can be added appropriately to prepare a compensation control where half of the PMNs are eFluor506 positive. Commercially available beads, such as OneComp eBeads™ (ThermoFisher Scientific) can be used for compensation of fluorescent antibody channels. If desired fluorescence minus one (FMO) controls can be prepared.
11. Fluorescent tagged antibodies should be protected from the light whenever possible. When using tandem dyes, it is advisable to test single stained samples periodically to avoid tandem degradation.
12. Refer to best practices when designing multicolor flow cytometry panels.

References

1. Borregaard N (2010) Neutrophils, from marrow to microbes. Immunity 33:657–670
2. Fine N, Barzilay O, Sun C et al (2019) Primed PMNs in healthy mouse and human circulation are first responders during acute inflammation. Blood Adv 3:1622–1637
3. Fearon DT, Collins LA (1983) Increased expression of C3b receptors on polymorphonuclear leukocytes induced by chemotactic factors and by purification procedures. J Immunol 130:370–375
4. Macey MG, McCarthy DA, Vordermeier S et al (1995) Effects of cell purification methods on CD11b and L-selectin expression as well as the adherence and activation of leucocytes. J Immunol Methods 181:211–219
5. Hamblin A, Taylor M, Bernhagen J et al (1992) A method of preparing blood leucocytes for flow cytometry which prevents upregulation of leucocyte integrins. J Immunol Methods 146:219–228
6. Oveisi M, Shifman H, Fine N et al (2019) Novel assay to characterize neutrophil responses to oral biofilms. Infect Immun 87(2).
7. Fine N, Hassanpour S, Borenstein A et al (2016) Distinct oral neutrophil subsets define health and periodontal disease states. J Dent Res 95:931–938
8. Lakschevitz FS, Hassanpour S, Rubin A et al (2016) Identification of neutrophil surface marker changes in health and inflammation using high-throughput screening flow cytometry. Exp Cell Res 342:200–209
9. Fine N, Sheikh Z, Al-Jaf F et al (2019) Differential response of human blood leukocytes to brushite, monetite, and calcium polyphosphate biomaterials. J Biomed Mater Res B Appl Biomater.

Chapter 19

Fast and Accurate Quantitative Analysis of Cytokine Gene Expression in Human Neutrophils by Reverse Transcription Real-Time PCR

Nicola Tamassia, Marco A. Cassatella, and Flavia Bazzoni

Abstract

Polymorphonuclear neutrophils, traditionally viewed as short-lived effector cells, are nowadays regarded as important components of effector and regulatory circuits in the innate and adaptive immune systems. Most of the physiological functions of neutrophils as crucial players in the host immune response, able not only to act in the early phases of acute inflammation but also to condition the progression of the inflammatory reaction and the subsequent initiation of the specific immune response, relies on their capacity to produce and release a number of proinflammatory and immunoregulatory cytokines. This fact has reevaluated the importance, the role, and the physiological and pathological significance of neutrophils in the pathogenesis of inflammatory, infectious, autoimmune, and neoplastic diseases and has identified neutrophils as an important potential target for selective pharmacological intervention to both promote and restrain inflammation. In this context, understanding the mechanisms of modulation of neutrophil-derived cytokines and chemokines represents a critical step toward a better understanding of how neutrophils may influence pathophysiological processes in vivo. Herein, we describe and discuss an updated version of the methods that we have developed to rapidly and precisely characterize the pattern of cytokine expression in in vitro-activated human neutrophils. The validation of the reverse transcription quantitative real-time PCR assay as a suitable strategy for an accurate, sensitive, reliable, and bona fide analysis of cytokine gene expression in human neutrophils overcomes several problems strictly specific to neutrophils and offers an important tool, in the neutrophil research area, to test many experimental conditions for gene expression analysis.

Key words Neutrophils, RT-qPCR, Cytokines, SYBR green, Gene expression

1 Introduction

Neutrophils constitute a first line of defense against pathogens by virtue of their ability to release a set of preformed cytotoxic enzymes and generate reactive oxygen-derived species. Neutrophils can also produce, upon appropriate stimulation in vitro and in vivo, a variety of proteins, including cytokines, chemotactic molecules, and other mediators that are involved in various effector functions [1]. This latter function is currently the subject of a new wave of

Mark T. Quinn and Frank R. DeLeo (eds.), *Neutrophil: Methods and Protocols*, Methods in Molecular Biology, vol. 2087,
https://doi.org/10.1007/978-1-0716-0154-9_19,

enthusiastic research, in that it shows that neutrophils may also act as key regulators of the in vivo cross talk between immune, endothelial, stromal, and parenchymal cells. For a fairly exhaustive description of experimental conditions leading to the production of individual cytokines by neutrophils, as well as the molecular regulation and potential biological relevance of this process, the reader is referred to recent reviews [2].

There is an increasing interest in the identification and molecular characterization of protein/cytokine expression and production by neutrophils. In most studies aimed at characterizing neutrophil-derived proteins, granulocytes are isolated from fresh blood or buffy coats, suspended in culture medium containing up to 10% serum, and then cultured at 37 °C in a 5% CO_2 humidified atmosphere. Other investigators suspend neutrophils in serum-free medium or in buffered solutions containing various concentrations of serum or bovine serum albumin. Although this is appropriate for short-term incubations, it is not recommended for experiments requiring several hours in culture, such as the determination of neutrophil gene expression or protein release. In our studies, highly purified neutrophil populations are routinely suspended at no more than 10^7 cells/mL in medium containing 10% low-endotoxin fetal bovine serum. Under these conditions, granulocytes can be cultured in tissue culture plasticware for up to 44 h if appropriate neutrophil survival factor(s) are added to the cultures.

Cytokine synthesis and/or release by polymorphonuclear neutrophils (PMN) can be measured in cell-free supernatants (or in cell pellets) by using different methods, such as enzyme-linked immunosorbent assay (ELISA), radioimmunoassay (RIA), immunoprecipitation after metabolic labeling, bioassays, and also by intracellular staining followed by flow cytometry and immunohistochemistry. However, the reliability of the latter two methodologies is influenced by technical issues, particularly in human neutrophils [2]. Because the induction of cytokine production and release by neutrophils is usually preceded by an increased accumulation of the related mRNA transcripts, a number of molecular biology techniques, such as Northern blotting, ribonuclease protection assays (RPA), reverse-transcription (RT)-PCR, in situ hybridization, microarray analysis and RNA-Seq can be also performed (e.g., *see* Chapter 21).

Northern blots and RPA have the advantage of making differences in cytokine mRNA expression quantitatively measurable, but the very low RNA content of neutrophils (average 1 μg/10^7 cells) represents a limiting factor in the analysis of gene expression modulation when many experimental conditions are tested. The development of reverse transcription quantitative Real-Time PCR (RT-qPCR) has revolutionized the measurement of gene expression, because it requires less RNA template than other methods of gene expression analysis and allows a reliable quantification of

RT-PCR products. RT-qPCR assays are more sensitive than RPA [3], dot blot [4] and can even detect a single copy of a specific transcript [5]. This feature represents a notable advantage, in that the expression of several (up to 10) cytokine/chemokine genes can be analyzed in RT-qPCR starting from as little as 0.1 μg of total RNA (Fig. 1a). For its accuracy, RT-qPCR is also the method of choice for the validation of microarray [6] or RNA-Seq [7] results (Fig. 1b, c). On the other hand, because of its extremely high sensitivity, RT-qPCR can easily amplify cytokine mRNA from a few (<0.5%) contaminating monocytes, lymphocytes, or eosinophils. This represents a major concern, because peripheral blood mononuclear cells (PBMCs) possess 10–20 times more RNA per cell than neutrophils, so that a contamination of only 1% PBMCs can translate into an up to 20% of contamination at the RNA level. Recently a novel neutrophil purification procedure based on negative immunoselection with magnetic beads (EasySep neutrophil enrichment kit, StemCell Technologies), that allow to achieve very high level of purity (>99.5%) have been developed [8]. The use of this procedure is therefore crucial to generate uncontroversial evidence that cytokine mRNA expression can be directly attributed to neutrophils, excluding the potential contribution of PBMCs or even eosinophils. Based on reports from our group [9–11] as well as from other laboratories [12, 13], the absence of detectable IFNβ, IL-10 or IL-6 mRNA in neutrophils stimulated with up to 100 ng/mL ultrapure lipopolysaccharide (LPS) represents a good control of neutrophil purity (Fig. 2). In addition, correct methodological analysis and detailed investigation of the pattern of cytokine gene expression by neutrophils in vitro represent the required and important steps.

In this chapter, we describe validated methods currently used in our laboratory for the optimal application of the RT-qPCR strategy in the analysis of human neutrophil cytokine gene expression. There are several types of detection chemistry available for the quantitative real-time PCR (qPCR) assay (DNA binding dyes, hybridization probes, hydrolysis probes, molecular beacons, scorpions, etc.). Among the two most commonly used (SYBR Green and TaqMan probes), we describe here the SYBR Green-based assay. SYBR Green is a fluorescent DNA intercalator dye that binds to double-stranded PCR products that accumulate during cycling. With well-designed primers, SYBR Green produces results with the sensitivity and range of quantification comparable to those obtained with TaqMan, making this strategy very flexible and less expensive as compared to other chemistries. Finally, and most importantly, the data obtained by this RT-qPCR strategy have been validated by comparable results obtained with other methods for gene expression analysis (Fig. 1).

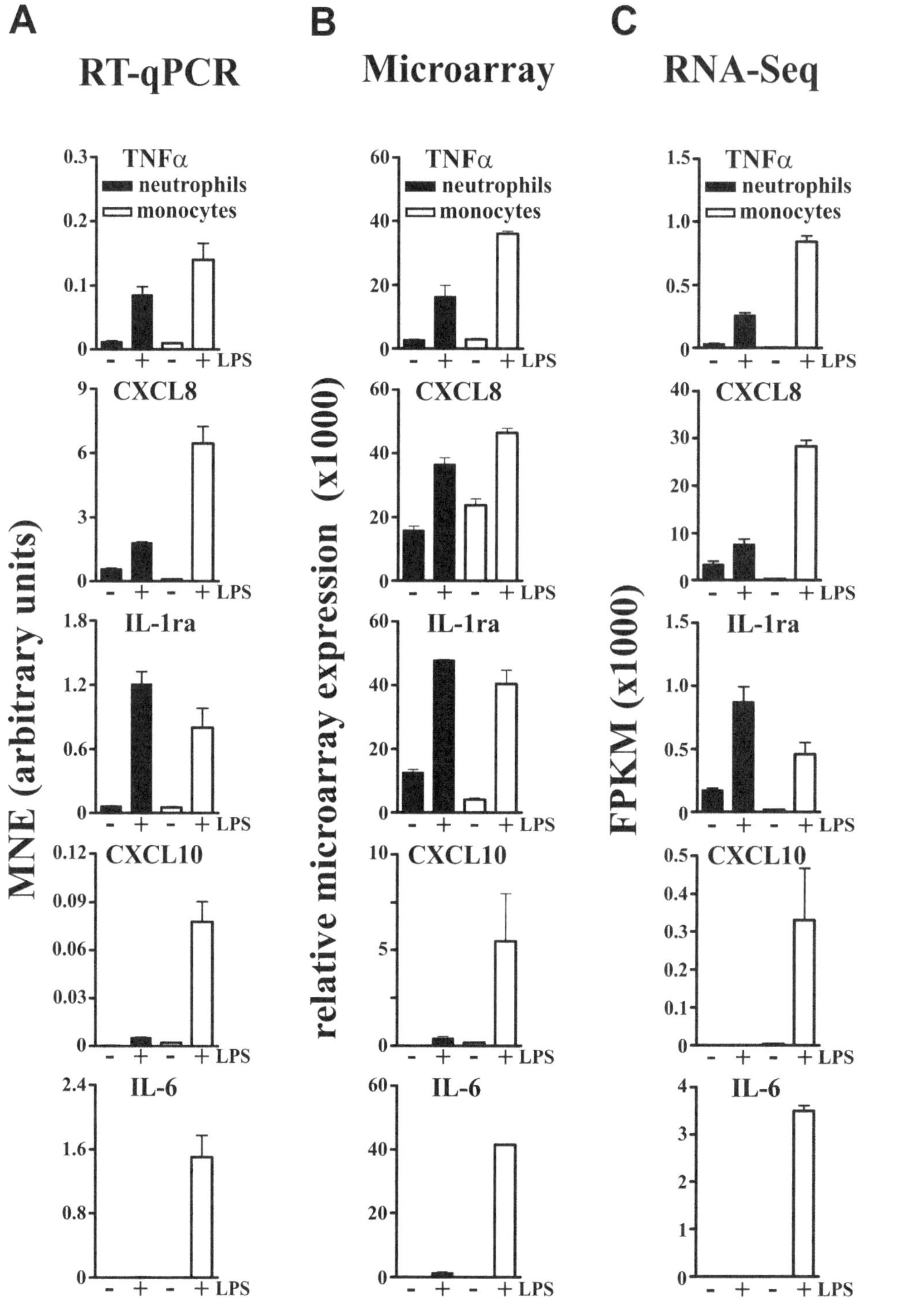

Fig. 1 Quantitative analysis of cytokine gene expression in activated neutrophils and monocytes performed in RT-qPCR (**a**), microarray (**b**), and RNA-Seq (**c**). Neutrophils and autologous monocytes were cultured with 100 ng/mL ultrapure LPS. After 4 h, total RNA was extracted and analyzed for TNF-α, CXCL8, IL-1ra, CXCL10, and IL-6 mRNA accumulation. (**a**) 1 μg RNA/condition was processed by RT-qPCR (described in this chapter).

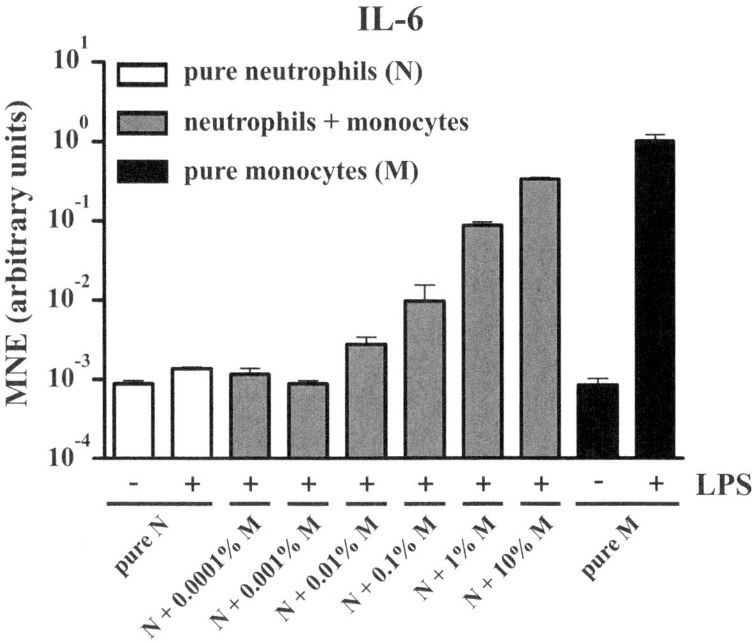

Fig. 2 The presence of few contaminating monocytes in LPS treated neutrophils can lead to a considerable IL-6 mRNA induction. Neutrophils (N), purified by negative immunomagnetic selection, monocytes (M), purified by CD14 positive immunomagnetic selection, or neutrophils to which were added increasing percentages of monocytes (ranging from 0.0001% to 10%) were treated or not with 100 ng/mL ultrapure LPS for 2 h. Total RNA was then extracted, and levels of IL-6 transcript were analyzed by RT-qPCR, as described in this chapter. Gene expression is depicted as MNE units, calculated after GAPDH normalization (*see* **Note 15**). (Reproduced from Tamassia et al. [21] with permission from Humana Press)

2 Materials

2.1 Miscellaneous

1. RPMI-1640 medium supplemented with 10% low-endotoxin fetal bovine serum (FBS; endotoxin content must be lower than 5–10 pg/mL, certified by the company or verified by a Limulus amebocyte lysate [LAL] assay).

Fig. 1 (continued) Expression levels of cytokine mRNAs is reported as mean normalized expression (MNE), calculated after GAPDH normalization (*see* **Note 15**). (**b**) 10 μg of total RNA/sample were processed and analyzed by NimbleGen oligonucleotide microarray (Roche) according to the manufacturer's protocol. Data are reported as mean relative microarray expression ($n = 3$) after quantile normalization. (**c**) 1 μg of total RNA/sample was processed for library preparation using the TruSeq RNA sample preparation kit (Illumina) according to the manufacturer's protocol. Data are reported as Fragments Per Kilobase of transcript per Million mapped reads (FPKM) ($n = 2$) after normalization using DESeq2 [20]. Graphs demonstrate that these three techniques produce similar quantitative results

2. Ultrapure LPS: extracted by successive enzymatic hydrolysis steps and purified by the phenol-TEA-DOC extraction (commercially available from different sources).
3. All plasticware utilized to work with RNA (pipet tips, polypropylene tubes, syringes, 20-G needles) must be RNase-free, autoclaved, and kept in a separate dedicated area. All disposable plastics utilized for RT and qPCR, including pipet tips or tubes, must be certified for the absence of DNA contamination, RNase, DNase, or other PCR inhibitors, and kept in a dedicated area.
4. Sterile, pyrogen- and RNase-free H_2O approved for clinical use. Prepare 1 mL aliquots.
5. 70% (v/v) ethanol: dilute ethanol in RNase-free H_2O and use it for RNA manipulation only.

2.2 Total RNA Purification

1. RNeasy mini kit (Qiagen): contains RNeasy mini spin columns, RLT Buffer (lysis buffer) (supplement with 143 mM β-mercaptoethanol immediately before use), RW1 Buffer (first washing buffer), and RPE Buffer (second washing buffer) (dilute with 4 vol of 96–100% ethanol immediately before use) (*see* **Note 1**).
2. RNase-Free DNase Set (Qiagen): contains RNase-free water; DNase I, which must be dissolved in 550 μL of RNase-free water, aliquoted, and stored at −20 °C (*see* **Note 2**); RDD Buffer for on-column DNA digestion (stored at 4 °C); and DNase I incubation mix, which must be prepared by adding 10 μL of a DNase I stock solution to 70 μL of RDD buffer (*see* **Note 1**).

2.3 RNA Quantification

1. RiboGreen RNA Quantification Kit (Molecular Probes): contains RiboGreen RNA quantification reagent (store at −20 °C protected from the light); 20× TE (200 mM Tris–HCl, 20 mM ethylenediamine tetraacetic acid [EDTA], pH 7.5), which must be stored at 4 °C and diluted to 1× with RNase-free water before use; ribosomal RNA Standard (2 μg/mL), which must be stored at 4 °C or −20 °C for short- or long- term periods, respectively (*see* **Note 1**).
2. Black, 96-well microplates optimized for fluorescence readings (Packard Instruments).
3. Fluorescence microplate reader equipped with approx. 480-nm excitation and approx. 520-nm emission filters (e.g., VICTOR^{3}™ Multilabel Counter, Perkin Elmer).

2.4 RNA Reverse Transcription

1. Reverse Transcriptase: Superscript™ III Reverse Transcriptase (200 U/μL), supplied with 5× First-Strand Buffer, and 0.1 M dithiothreitol (DTT) (Life Technologies). Store all reagents at −20 °C.

2. Recombinant Ribonuclease Inhibitor: RNaseOUT™ Recombinant enzyme (acidic protein) from Life Technologies. Store at −20 °C.
3. Keep SuperScript III and RNaseOUT in a refrigerated bench cooler (−20 °C) when performing reactions.
4. 10 mM dNTP mix: aliquot and store at −20 °C.
5. Random primers: dilute at 100 ng/μL with RNase-free H_2O. Aliquot and store at −20 °C.

2.5 Quantitative Real-Time PCR

1. Thermocycler for qPCR: The protocol described here is optimized for the ViiA™ 7 real-time PCR system, including the ViiA™ 7 software (Life Technologies) (*see* **Note 3**).
2. SYBR Green qPCR master mix (2× concentrated solution; e.g., Fast SYBR Green Master Mix [Life Technologies]). Store at −20 °C protected from light. Once thawed, store master mix at 4 °C. This solution contains all necessary reagents to perform a qPCR (i.e., appropriate buffer, dNTPs, SYBR Green I dye, ROX™ dye passive reference, and AmpliTaq® Fast DNA Polymerase), except primers (*see* **Note 4**).
3. Forward and reverse primers: dissolve in RNase-free H_2O at a final concentration of 100 μM. Dilute primers at the working concentration of 20 μM. Aliquot and store at −20 °C.
4. 96- or 384-well microplates for fast qPCR equipped with optical adhesive film suitable for a thermocycler. All plastics must be stored in a clean place to avoid any possible contamination by external DNA.

3 Methods

3.1 Total RNA Purification

The extraction of total RNA from neutrophils can be performed by different methods. However, depending on the method chosen, the quantity and/or the quality of the RNA extracted from neutrophils may greatly differ. It is essential for the RNA used in RT-qPCR not to be degraded, to contain as little contaminating salts as possible, and to be free of genomic DNA. For RT-qPCR, small quantities of total RNA are required (from 1 ng to 1 μg), unlike other methodologies, including the Northern Blotting or RPA, which need at least 5–10 μg of RNA for each condition. Thus, the low yield of RNA usually purified from neutrophils does not represent a limiting factor for this technique.

RNA purification methods that rely on selective RNA binding properties of silica-gel-based membranes are, in our opinion, the best because they are fast, reliable, allow rapid and simultaneous processing of many samples at once, and do not need a phenol/chloroform extraction step. If one desires to obtain 1 μg of total

RNA per sample (which is the optimum RNA amount that can be used for RT-qPCR studies), the optimal starting cell number is 10^7 neutrophils/condition, given that 10^6 neutrophils contain approx. 0.1 μg of total RNA. 10^7 cells correspond to the maximum number of cells per sample that can be processed by an RNeasy mini kit (*see* **Note 5**). We process and purify our samples exactly as described in the protocol "purification of total RNA from animal cells using spin technology" reported in the RNeasy mini kit instruction manual. The protocol is detailed, comprehensive, and easy to follow, so that no detailed explanation needs to be given herein. The volumes used to resuspend cells in the initial steps of the procedures are specified as follows.

1. Resuspend purified neutrophils at 5 × 10^6 cells/mL in RPMI supplemented with 10% low-endotoxin FBS (*see* **Note 6**).
2. Plate 2 × 10^6 cells in a 24-well tissue culture plate (0.4 mL/well), stimulate as appropriate, and incubate at 37 °C in a 5% CO_2 atmosphere for the desired time.
3. Collect neutrophils into 1.5 mL tubes, and pellet at 300 g for 5 min.
4. Carefully remove culture supernatants and loosen cell pellets thoroughly by flicking the tube several times.
5. Resuspend pellets in 350 μL of RLT buffer, and vortex (*see* **Note 7**).
6. Homogenize lysates by passing them through a 20-G needle fitted to a syringe at least five times or until the solution becomes less dense. At this stage, lysates can be further processed or stored at −70 °C for several months without any change in RNA integrity (*see* **Note 8**).
7. We do not perform the optional **step 6** of the RNeasy mini kit protocol because we perform the on-column DNase digestion (*see* **Note 9**). Instead, the optional **step 9** of the RNeasy mini kit protocol is performed. Otherwise, there are no further changes to the procedures described in the manual supplied with the kit.
8. Elute RNA from the column with 30 μL of RNase-free water. The expected RNA yield is ~ 0.1 μg per 10^6 neutrophils.
9. RNA can be quickly concentrated up to 12 μL using vacuum concentrator system. From this step onward, RNA must be kept on ice or stored at −20° to −70 °C.

3.2 RNA Quantification

RNA quantification is a crucial step for optimal RT-qPCR results, and we recommend performing the RT reaction with equivalent amounts of RNA from each sample. This limits differences in RT efficiency among the samples and favors subsequent normalization of RT-qPCR results. If extracting RNA from less than 3 × 10^6

human neutrophils we do not recommend absorbance-based quantification because is not precise for detecting low target concentrations even with new generation spectrophotometer like NanoDrop (Thermo scientific). The use of more sensitive methods, such as staining of nucleic acid with fluorescent dyes, overcomes this problem. We routinely use the RiboGreen RNA quantification reagent that consists of an ultrasensitive fluorescent nucleic acid stain. This reagent detects as little as 1 ng/mL of RNA in solution. The following method is adapted from the instructions included in the RiboGreen RNA quantification kit (*see* **Note 1**).

1. Prepare a standard curve by diluting the standard 2 μg/mL RNA solution (provided with the kit) with TE. Prepare 1000, 500, 50, and 20 ng/mL RNA standards.
2. Pipet 100 μL of the prepared RNA standard dilutions in duplicate into wells of 96-well black microtiter plates.
3. Dilute RNA samples directly into the microtiter plates by adding 1 μL of RNA sample to 99 μL of TE. Make one or more replicates.
4. Dilute (200-fold) the RiboGreen reagent with TE and add 100 μL of the diluted RiboGreen reagent to each well.
5. Gently shake the plate, and incubate it in the dark for 5 min.
6. Measure the fluorescence in each well with a fluorescence microtiter plate reader (excitation ~480 nm, emission ~520 nm) (*see* **Note 10**).
7. Subtract the blank signal from the fluorescence value of each sample and generate a standard curve of fluorescence vs. RNA concentration.
8. Determine the RNA concentration of the various samples by linear regression using the standard curve as reference.

3.3 RNA Reverse Transcription

RT is carried out with SuperScript III Reverse Transcriptase following the protocol included in the manual for this enzyme (*see* **Notes 1** and **11**)

1. For each sample, prepare the following RNA/primer mix directly in a 0.2-mL tube (keep samples on ice) (*see* **Note 12**): From 1 μg to 0.1 μg of total RNA, 1 μL of random primers from 100 ng/μL stock, and 1 μL of dNTP mix from 10 m*M* stock. Make up to 14 μL final volume with RNase-free H_2O.
2. Prepare identical samples for controls that will not receive reverse transcriptase (−RT) (*see* **Note 13**).
3. Centrifuge the tubes briefly (quick-spin).
4. For optimal RNA/random primer denaturation, incubate samples at 65 °C for 5 min in a preset thermocycler.

5. Prepare RT reaction master mix. For each reaction add: 4 μL of 5× First-Strand Buffer, 1 μL of 0.1 M DTT solution, 0.5 μL RNaseOUT from 40 U/μL stock, and 0.5 μL of SuperScript III from 200 U/μL stock). Prepare a volume of master mix greater than what is necessary. For the −RT controls, prepare the RT reaction master mix in the same manner, but omit SuperScript III.
6. At the end of the denaturation step, pause the thermocycler, put tubes on ice for at least 1 min and add 6 μL of RT reaction master mix to each tube. Mix by pipetting and, if bubbles form, perform a quick-spin with the tubes.
7. For RNA/random primer annealing incubate tubes at 25 °C for 5 min and then at 42 °C for 1 h for RT. Subsequently heat-inactivate the reactions at 95 °C for 5 min.
8. At this stage, samples are single-strand cDNA and can be stored at −20 °C or immediately used as template for amplification in PCR.

3.4 Quantitative Real-Time PCR

The following qPCR procedure is adjusted for use of Fast SYBR Green Master Mix as qPCR master mix in combination with a ViiA™ 7 thermocycler (*see* **Notes 3** and **4**). The use of qPCR instruments and reagents enabled for fast PCR protocol allows a significant overall reduction of the reaction time from the typical 90–120 min to less than 40 min. For normalization strategy, in addition to the target genes, we always quantify also endogenous reference genes. Thermocycling conditions are constant for all assays.

1. Edit the following program in the real-time thermocyler:
   (a) Step 1: 95 °C for 20 s (initial denaturation) (*see* **Note 4**).
   (b) Step 2: 95 °C for 1 s (denaturation); 60 °C for 20 s (annealing/extension); fluorescence read. Repeat for 45 cycles. An extension step is not required, because all of the PCR products are 50–250 bp.
   (c) Step 3: Melting Curve, set the final sample heating from 60 °C to 90 °C and have the fluorescence reading recorded every 0.5 °C (*see* **Note 14**).
2. Prepare qPCR master mixes based upon the number of genes to be analyzed. Each sample must be tested in triplicate, and negative controls must be included. For every experiment the analysis of the expression of at least two reference genes must be always performed for accurate normalization (*see* **Note 15**).
3. For a 10 μL-reaction, use the following components for each well or tube (*see* **Note 16**): 5 μL of 2× Fast SYBR Green Master Mix, 0.1 μL of forward primer from 20 μM stock, 0.1 μL of reverse primer from 20 μM stock, 1.8 μL of autoclaved H_2O (7 μL final volume).

4. Dilute all cDNA samples to be analyzed to a concentration of 1 ng/μL (*see* **Notes 17** and **18**).
5. Aliquot sequentially 7 μL of qPCR master mix and 3 μL of sample or H_2O (for the "no template" control) or −RT (for the negative control) into each well or tube. This step must be carried out very rapidly, as SYBR Green is light-sensitive.
6. Carefully close the wells, with optical adhesive film, using the appropriate sealer, paying great attention not to touch, or scratch the upper part of the film.
7. Briefly centrifuge the plate or the tubes, and then start the reaction in the real-time thermocycler.
8. Optional: remove the plate or the tubes from the thermocycler after the qPCR is finished. Separate 5 μL of each sample by electrophoresis on a 2% agarose gel to check for quality of PCR products. If the reaction proceeded correctly, a single amplified product of the expected length should be visible in the gel.
9. At the end of the reaction, check the melting curve analysis to verify whether abnormal amplification plots or any bimodal dissociation curves are present (*see* **Note 14** and Fig. 3a).
10. At the end of the run, the ViiA 7 system software returns raw data (fluorescence value (ΔR_n: normalized reporter value) of each sample for each real-time PCR cycle) that must be analyzed.

3.5 Analysis of Real-Time PCR Data

ViiA7 software calculates the RT-qPCR results using the ΔΔCt method [14]. This method does not correct RT-qPCR data for amplification efficiency of the reaction that is an important consideration when performing relative quantification. To overcome this problem, we calculate by LinRegPCR software [15], the average amplification efficiency for each gene analyzed (including reference genes) from each single sample. Subsequently we calculate the RT-qPCR results using the Q-Gene software application [16] that will give back the expression level of the gene of interest relative to an endogenous reference gene mRNA. A detailed description is referred in each software manual or original article [16, 15].

1. Set the baseline on the analysis settings of ViiA 7 system software. The initial cycles of PCR where there is little fluorescence signal are termed "baseline" and should be eliminated (Fig. 3b). The baseline should be wide enough to eliminate background, but should not overlap with the area in which the amplification signal begins to rise above background (*see* **Note 19**).
2. Set the threshold on the analysis settings of ViiA 7 system software. The threshold is an arbitrary fluorescence value that

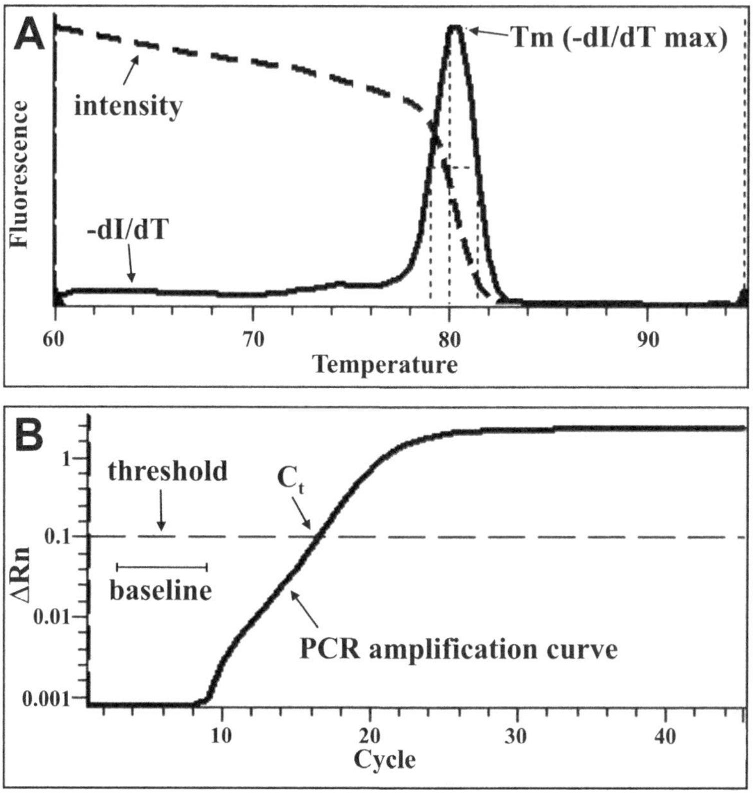

Fig. 3 Melting and PCR amplification curves as shown by the ViiA 7 system software. (**a**) The melting curve is plotted as fluorescence intensity (dot line) or as first derivative (−dI/dT) (solid line) of the PCR product as a function of temperature. The melting temperature (T_m) of the PCR product is defined as the temperature at which −dI/dT reaches the maximum value. (**b**) PCR amplification curve. The fluorescence intensity (ΔR_n) is plotted as a function of cycle number. The baseline, threshold, and threshold cycle are indicated. (Reproduced from Tamassia et al. [21] with permission from Humana Press)

must intercept the PCR amplification curve in the exponential phase of the reaction (*see* **Note 20** and Fig. 3b).

3. From the ViiA 7 system software export Amplification data and Results data of the qPCR run.
4. Calculate the efficiency of qPCR reaction for each gene assayed using amplification data. There are several methods of calculating amplification efficiency on raw data collected during PCR. We routinely calculate the amplification efficiency with LinRegPCR software [15]. The program uses linear regression analysis to find the best-fit straight line through the PCR data set. From this line, PCR efficiency of each individual sample is calculated. The amplification efficiency value that we take for each gene is the average of the single efficiencies calculated with LinRegPCR (*see* **Note 21**).

5. From the results data export threshold cycles (C_t) values to the Q-Gene Excel datasheet. The cycle at which the fluorescence signal from the reaction crosses the threshold line is defined as the C_t. Once the threshold line is set, the real-time software automatically generates a C_t value for each sample. The C_t is a reliable indicator of the initial copy number of each amplified cDNA.
6. Perform the final quantification with Q-Gene software (*see* **Note 22**). Insert the following values in the Q-Gene Excel datasheet: (1) the C_t s generated for each sample, both for the target gene and for the reference gene by the real-time thermocycler software; (2) the calculated amplification efficiency for the gene of interest and for the reference gene; and (3) the calculation procedure 2, corresponding to Eq. (3) described by Muller and coauthors [16].
7. The Q-Gene software calculates the Mean Normalized Expression (MNE) with standard errors for each sample, corrected for amplification efficiencies. Examples of the results that can be obtained are shown in (Figs. 1 and 2).

4 Notes

1. See accompanying instruction manual for further details.
2. Once thawed, reconstituted DNase I is stable for up to 6 weeks at 4 °C. Do not refreeze the aliquots after thawing. Never vortex DNase I because this enzyme is ultrasensitive to physical denaturation.
3. If qPCR thermocyclers from other suppliers are going to be used, appropriate modifications to our protocol should be preliminarily determined (for instance, the reaction volumes, the PCR reaction settings, usage of reference dyes in the master mix, and so forth).
4. Many qPCR master mix solutions are commercially available. If using qPCR master mix purchased from a different company, we suggest carefully reviewing the accompanying product insert. Changes in the reaction preparation and/or in the PCR reaction (for instance, the initial denaturation step or the annealing/elongation step) may vary, as the length of the denaturation step depends on the polymerase included in the master mix.
5. Among the commercially available kits, we have experimentally verified that the "RNeasy mini kit" works well, because it facilitates purification of high-quality RNA (not degraded, salts- and genomic DNA-free), whose integrity and purity is easily assessed by denaturing agarose gel electrophoresis or, in

case of very low amount of material, by Agilent 2100 bioanalyzer (*see also* Chapter 21). In addition, this kit permits use of an on-column DNase digestion step, thereby avoiding potential DNase carryover.

6. We strongly recommend working with highly pure neutrophil preparations obtained by negative immunoselection with magnetic beads (EasySep neutrophil enrichment kit, StemCell Technologies) (*see* Chapter 4 of this volume). Preparation purity can be rapidly checked by flow cytometry for the high CD66b surface expression of neutrophil. Using this procedure contamination by eosinophils, that with conventional density gradient sedimentation method were copurified (varying from 1% to 8% depending on the donor), is excluded. The same also for mononuclear leukocytes (monocytes and lymphocytes) which presence, as low as 1%, may greatly influence the results attributed to neutrophils. Most importantly, the presence of as little as 0.01% contaminating monocytes, likely generating false-positive results, in the neutrophils culture can be revealed by the analysis of cytokines such as IL-6 (Fig. 2), IFNβ and/or IL-10 (not shown) expression.
7. Some neutrophil agonists, for instance LPS, other Toll like receptors (TLR) ligands or tumor necrosis factor (TNF)-α, cause strong cell adherence to the culture vessel, especially within the first 2 h of incubation. In this case, we suggest removing culture medium and adding RLT buffer directly into the well.
8. The quality of the cDNA template depends on the integrity of the RNA. All possible precautions must be taken in order to maintain an RNase-free environment when handling/manipulating RNA. Hands and dust particles may carry common sources of RNase contamination, such as bacteria and molds. Always wear gloves and change them frequently. Because RNases are less active at cold temperature, we suggest keeping samples chilled on ice, especially following RNA extraction.
9. DNA digestion is a necessary step in all cases in which contaminating genomic DNA can interfere with subsequent applications. The RNeasy kit offers the optional step of "on column" DNase treatment, which we always perform. If a different kit for RNA purification is used, genomic DNA should be removed by a DNase digestion step following RNA isolation. To avoid the risk of a possible DNase carry-over, which could be detrimental for the subsequent cDNA synthesis step, we recommend performing phenol-chloroform treatment of all RNA samples after DNase treatment.

10. The excitation maximum for RiboGreen reagent bound to RNA is approx. 500 nm and the emission maximum is approx. 525 nm.
11. Because even a small amount of contaminants can greatly impact results of the assay, use of molecular biology-grade water, RNase/DNase/nucleic acid-free tubes, aerosol-barrier pipette tips, and dedicated pipettors of all types (i.e., pipettors used only for RNA or PCR applications, which are kept out of areas used for plasmid or genomic DNA work) is strongly recommended for all steps. The use of a small PCR hood may be very helpful to avoid PCR contamination. If a hood is not available, perform PCR in a dedicated space, possibly at distance from the thermocycler or spaces in which the electrophoresis equipment are located.
12. We suggest processing all samples from each experiment at the same time, starting from the same amount of RNA for each sample. This reduces differences in the quality and quantity of cDNAs obtained and increases the reliability of the final RT-qPCR results. If RNA recovery is low following the RNA purification step, RT can be performed with <1 μg of RNA without major problems. However, we suggest that the RT reaction be performed with ≥0.1 μg of RNA.
13. Because most of the primers we use in RT-qPCR span an exon junction, the −RT control is not needed for every sample. We usually perform −RT controls only for new primer pairs, and occasionally on random samples.
14. Specificity of the qPCR reaction can be confirmed by melting curve analysis, where the presence of different PCR products is reflected in the number of first-derivative melting peaks. The melting curve analysis is set as final step of the qPCR program by gradually heating the qPCR product (from 60 °C to 95 °C) and recording the fluorescence value every 0.5 °C. As the temperature increases from 60 °C to 95 °C, the final double-strand PCR product is denatured, intercalating SYBR Green is released, and the fluorescence intensity decreases (*see* Fig. 3a). The fluorescence intensity vs. temperature plotted as first derivative (−dI/dT) generates the melting curve. The presence of more than one peak indicates that more than one PCR product has been formed and that the PCR reaction is not specific. The negative control, containing PCR primers but no template, could give rise to a peak in the melting curve that usually has a low T_m and is formed in the last PCR cycles. This peak is usually representative of primer dimer formation.
15. Normalization to an endogenous reference gene or genes is the preferred method to more accurately quantify the expression levels of a given mRNA. To select a stable reference gene, it is

very important to avoid acquisition of biologically irrelevant data. Unfortunately, is very hard to identify a single reference gene stable in all the possible experimental conditions that can be tested, therefore the best approach that we have found is to always measure 2–3 reference genes and use the more stable one for normalization. The genes that we commonly test are GAPDH (Glyceraldehyde 3-phosphate dehydrogenase reference), RPL32 (Ribosomal protein L32), PPIB (peptidylprolyl isomerase B), and B2M (β2-microglobulin).

16. Primer designing is a crucial step for performance of qPCR. Primers should be designed according to standard PCR guidelines. A good primer (1) is 19–24 bases in length; (2) preferably has a base composition of 50–60% in G + C; (3) preferably has a primer ending at the 3′-end that is a G or C, or CG or GC—this prevents "breathing" of ends and increases efficiency of priming; (4) has no runs of three or more C's or G's at the 3′-end of primers, which may promote mispriming at G or C-rich sequences (because of stability of annealing) and should be avoided; (5) has 3′-ends that are not complementary (i.e., base pair), as otherwise primer dimers will be synthesized preferentially to any other product; and (6) has no primer self-complementarity (ability to form 2° structures such as hairpins). To optimize efficiency, the product of qPCR must be approx. 80–250 bp. Melting temperature (T_m) of the primers should be 60–64 °C; it is important that all the primers function at the same annealing temperature, thereby allowing analysis of different targets in the same plate. When designing primers for RT-qPCR, we suggest identifying at least one primer that overlaps an exon–exon boundary in order to prevent the amplification of genomic DNA. Alternatively, the two primers should anneal to two different exons on each side of an intron. A public database for the selection of primer sequences suitable for RT-qPCR, RTPrimerDB, is available at the following website: http://www.rtprimerdb.org/index.php [17]. Our primers are designed with the PerlPrimer software [18], which is free at http://perlprimer.sourceforge.net. The specificity of the identified primers can be computationally tested by Primer-BLAST [19]. This software tool permits to identify all the possible PCR products that can be formed, on the whole transcriptome, using different primer pairs. If potential erroneous amplicons are detected using the Primer-BLAST default parameters, different primers pairs should be designed.

17. It is possible to use 1 pg to 100 ng of cDNA in a qPCR reaction, and the amount available will likely depend on the quantity of total RNA isolated from neutrophils. We experimentally determined that 3 ng of cDNA is an optimal amount

and permits quantification of low- and high-expressed transcripts.

18. Note that there is a risk for pipetting errors associated with manipulating very small vol (<2 μL) that could cause differential input of template.

19. For most of the cytokine genes, baseline can be set at cycle range 3–15. For low-expressed genes this value should be extended to 3–20.

20. In our conditions, the threshold is usually set at a fluorescence value (ΔR_n) of 0.1 for all the genes. Because plate variation introduced during the PCR run are minimal in the current generation of real-time PCR cyclers, the ViiA7 software enable the analysis of different genes on different plates, feature especially useful when many genes are analyzed using the same samples.

21. Amplification efficiency of the reaction is an important consideration when performing relative quantification. An efficiency value of 2 means that the PCR product doubles during every cycle within the exponential phase of the reaction. However, many PCR reactions do not have ideal amplification efficiencies, and calculation without an appropriate correction factor may overestimate starting concentration. We use as amplification efficiency for each gene analyzed (including reference genes) the average of amplification efficiencies calculated by LinRegPCR [15] for each single sample. Data input and output are through an Excel spreadsheet. The LinRegPCR software is freely available at http://LinRegPCR.nl.

22. Q-Gene is Excel-based software available free for download at http://www.gene-quantification.de/download.html#qgene [16].

Acknowledgments

This work was supported by grants from Associazione Italiana per la Ricerca sul Cancro, (IG-20339); Ministero dell'Istruzione, dell'Università e della Ricerca and (2015YYKPNN), and Fondazione CARIPLO (2015-0584) to M.A.C.

References

1. Ley K, Hoffman HM, Kubes P et al (2018) Neutrophils: new insights and open questions. Sci Immunol 3:4579
2. Tamassia N, Bianchetto-Aguilera F, Arruda-Silva F et al (2018) Cytokine production by human neutrophils: revisiting the "dark side of the moon". Eur J Clin Investig 48(Suppl 2):e12952
3. Wang T, Brown MJ (1999) mRNA quantification by real time TaqMan polymerase chain reaction: validation and comparison with

RNase protection. Anal Biochem 269:198–201

4. Malinen E, Kassinen A, Rinttila T et al (2003) Comparison of real-time PCR with SYBR green I or 5′-nuclease assays and dot-blot hybridization with rDNA-targeted oligonucleotide probes in quantification of selected faecal bacteria. Microbiology 149:269–277
5. Palmer S, Wiegand AP, Maldarelli F et al (2003) New real-time reverse transcriptase-initiated PCR assay with single-copy sensitivity for human immunodeficiency virus type 1 RNA in plasma. J Clin Microbiol 41:4531–4536
6. Arikawa E, Sun Y, Wang J et al (2008) Cross-platform comparison of SYBR green real-time PCR with TaqMan PCR, microarrays and other gene expression measurement technologies evaluated in the MicroArray quality control (MAQC) study. BMC Genomics 9:328
7. Nagalakshmi U, Wang Z, Waern K et al (2008) The transcriptional landscape of the yeast genome defined by RNA sequencing. Science 320:1344–1349
8. Pelletier M, Maggi L, Micheletti A et al (2010) Evidence for a cross-talk between human neutrophils and Th17 cells. Blood 115:335–343
9. Bazzoni F, Cassatella MA, Laudanna C et al (1991) Phagocytosis of opsonized yeast induces tumor necrosis factor-alpha mRNA accumulation and protein release by human polymorphonuclear leukocytes. J Leukoc Biol 50:223–228
10. Tamassia N, Le Moigne V, Calzetti F et al (2007) The MyD88-independent pathway is not mobilized in human neutrophils stimulated via TLR4. J Immunol 178:7344–7356
11. Davey MS, Tamassia N, Rossato M et al (2011) Failure to detect production of IL-10 by activated human neutrophils. Nat Immunol 12:1017–1018; author reply 1018-1020
12. Wang P, Wu P, Anthes JC et al (1994) Interleukin-10 inhibits interleukin-8-production in human neutrophils. Blood 83:2678–2683
13. Reglier H, Arce-Vicioso M, Fay M et al (1998) Lack of IL-10 and IL-13 production by human polymorphonuclear neutrophils. Cytokine 10:192–198
14. Livak KJ, Schmittgen TD (2001) Analysis of relative gene expression data using real-time quantitative PCR and the 2(−Delta Delta C (T)) method. Methods 25:402–408
15. Ruijter JM, Ramakers C, Hoogaars WM et al (2009) Amplification efficiency: linking baseline and bias in the analysis of quantitative PCR data. Nucleic Acids Res 37:e45
16. Muller PY, Janovjak H, Miserez AR et al (2002) Processing of gene expression data generated by quantitative real-time RT-PCR. Biotechniques 32:1372–1374. 1376, 1378–1379
17. Lefever S, Vandesompele J, Speleman F et al (2009) RTPrimerDB: the portal for real-time PCR primers and probes. Nucleic Acids Res 37: D942–D945
18. Marshall OJ (2004) PerlPrimer: cross-platform, graphical primer design for standard, bisulphite and real-time PCR. Bioinformatics 20:2471–2472
19. Ye J, Coulouris G, Zaretskaya I et al (2012) Primer-BLAST: a tool to design target-specific primers for polymerase chain reaction. BMC Bioinformatics 13:134
20. Love MI, Huber W, Anders S (2014) Moderated estimation of fold change and dispersion for RNA-seq data with DESeq2. Genome Biol 15:550
21. Tamassia N, Cassatella MA, Bazzoni F (2014) Fast and accurate quantitative analysis of cytokine gene expression in human neutrophils. Methods Mol Biol 1124:451–467

Chapter 20

Detection of Intact Transcription Factors in Human Neutrophils

Patrick P. McDonald and Richard D. Ye

Abstract

The crucial contribution of neutrophils to innate immunity extends well beyond their traditional role as professional phagocytes. Indeed, it is now well established that neutrophils generate a plethora of inflammatory cytokines and chemokines that are profoundly involved in the onset and evolution of the inflammatory reaction. Several recent studies have shown that neutrophils can represent an important source of inflammatory cytokines in pathophysiological settings. The inflammatory and immunomodulatory cytokines produced by neutrophils are generally encoded by immediate-early response genes, which in turn depend on the activation of transcription factors such as those belonging to the nuclear factor κB (NF-κB) and signal transducers and activators of transcription (STAT) families. We have shown in the past that the expression of such factors is induced in neutrophils stimulated by physiological agonists. However, the detection of *intact* (i.e., undegraded) transcription factors in neutrophils requires special precautions and a specially designed protocol, due to the huge amounts of endogenous proteases present in these cells. This protocol is the focus of this chapter.

Key words Transcription factors, NF-κB, STAT, Nuclear extracts, Electrophoresis mobility shift assay, Granulocytes

1 Introduction

Neutrophils are the most abundant circulating leukocytes and play a crucial role in innate immunity. They are usually the first cells to infiltrate inflamed tissues, where they exert a series of antimicrobial responses culminating in the elimination or inactivation of the infectious agent. In addition to their traditional role as professional phagocytes, neutrophils also express a plethora of genes in response to proinflammatory cytokines and bacterial products such as LPS and N-formyl peptides [1–5]. Examples include inflammatory cytokines such as TNFα and IL-1β, chemokines such as IL-8/CXCL8, Mip-1α/CCL3, Mip-1β/CCL4, MIG/CXCL9, IP-10/CXCL10, I-TAC/CXCL11, and receptors such as MHCII and the high-affinity IgG receptor FcγRI/CD64. In addition, neutrophil

Mark T. Quinn and Frank R. DeLeo (eds.), *Neutrophil: Methods and Protocols*, Methods in Molecular Biology, vol. 2087,
https://doi.org/10.1007/978-1-0716-0154-9_20,

transcriptional activities have been closely associated with cell apoptosis [6] and are regulated by environmental factors [7, 8]. Neutrophils also play a role in the secretion of immunomodulatory cytokines [9, 10].

Many of the aforementioned protein factors are encoded by immediate-early response genes, which are under the control of the NF-κB and/or STAT transcription factors. We have previously shown that the same stimulatory conditions allowing for the expression of the aforementioned mediators and receptors also lead to the activation of the NF-κB or STAT transcription factors in neutrophils [11–16].

Both NF-κB and STAT factors are normally sequestered in the cytoplasm and are mobilized to the nucleus upon cell activation, where they can bind to their cognate sequences on the promoter region of target genes, and thereby initiate or enhance gene transcription (for recent reviews, *see* [17–20]). The activation of NF-κB is a highly regulated process, which usually involves the engagement of cell surface receptors, thereby generating various intracellular signals that eventually converge on the κB kinase (IKK) complex. This results in the phosphorylation and activation of IKK subunits, which can in turn phosphorylate the inhibitor protein IκB-α, thereby targeting it for ubiquitin-dependent degradation. The proteolysis of IκB-α, which is coupled to NF-κB dimers in the cytoplasm of resting cells, unmasks nuclear localization sequences on NF-κB subunits that allow the NF-κB complexes to translocate to the nucleus. By comparison, the STAT proteins are latent cytoplasmic proteins that are rapidly recruited to SH2 domains within the cytoplasmic tail of cytokine receptors upon cell stimulation, where they become tyrosine-phosphorylated by Janus kinases (JAK). Tyrosine phosphorylated STAT proteins can then dimerize into various DNA-binding configurations, forming homo- or heterodimers, and translocate to the nucleus.

The detection of nuclear transcription factors involves cell disruption followed by the preparation of nuclear extracts, which are then incubated with an excess of radiolabeled oligonucleotide probe that contains the cognate DNA binding sequence for a given factor. The resulting mixtures are then migrated on native gels to separate transcription factor complexes bound to labeled probe from unbound probe. Conventional protocols typically achieve cell lysis by solubilizing the cytoplasmic membrane with nonionic detergents [21], or sometimes by repeated freeze–thaw cycles or homogenization. In neutrophils, however, these approaches are problematic because of the huge quantities of proteolytic enzymes that are present in neutrophil cytoplasmic granules. Indeed, repeated freeze–thaw cycles break open the protease-rich neutrophil granules [22, 23]. We have shown that detergent lysis similarly results in the solubilization of granule-bound proteases, resulting in partially degraded NF-κB or STAT complexes and individual constituent proteins, even in the presence of an elaborate cocktail of protease inhibitors [13]. Accordingly, transcription factor complexes that are detected following neutrophil disruption by conventional procedures almost invariably migrate faster than authentic complexes and no longer react with certain antibodies (extensively reviewed in [24]). In contrast, nitrogen cavitation represents a gentle, yet efficient way to disrupt human neutrophils while preserving the integrity of intracellular organelles such as granules and nuclei [22]. We have adapted this technique to the preparation of neutrophil nuclear extracts, and this approach consistently yields *intact* (i.e., undegraded) NF-κB and STAT complexes [11, 13, 15]. Our cavitation-based protocol is the focus of the present article.

2 Materials

2.1 Neutrophils

Peripheral blood neutrophils are obtained from healthy volunteers, whose informed consent is ensured in accordance with the relevant institutional review board. Neutrophils must be prepared under endotoxin-free conditions to avoid inadvertent activation during isolation. The isolation procedure we use [25] is one of the many variations of the original Boyum protocol [26]. Details on neutrophil isolation can be found in Chapter 3 of this Volume. Regardless of the exact variation, isolated neutrophils should contain less than 1% contaminating mononuclear cells and display at least 98% viability.

2.2 Buffers

1. Hanks' balanced-salt solution (HBSS).
2. Krebs-Ringer phosphate buffer containing dextrose (KRPD).
3. Relaxation buffer: 10 mM PIPES, pH 7.30, 30 mM NaCl, 3.5 mM $MgCl_2$, 0.5 mM EGTA, 0.5 mM EDTA, and 1 mM DTT. Supplement relaxation buffer with anti-protease cocktail (1 mM DFP, 1 mM AEBSF, 1 mM PMSF, and 10 μg/mL each of aprotinin, leupeptin, and pepstatin A, final concentrations) and phosphatase inhibitors (10 mM NaF, 1 mM Na_3VO_4, 10 mM $Na_4P_2O_7$). This buffer is designed for the generation of nuclear extracts from cavitated neutrophils (*see* **Note 1**).
4. Nuclear extraction buffer: Relaxation buffer, 10% (v/v) glycerol, anti-protease cocktail. Phosphatase inhibitors may also be included, with the exception of Na_3VO_4, which interferes with protein quantitation by the Bradford method [27].
5. Binding buffer: 20 mM Tris base, pH 7.50, 50 mM KCl, 1 mM EDTA, 1 mM DTT, 300 μg/mL acetylated BSA, 50 μg/mL poly(dI-dC), 0.1% NP-40 (v/v), and 5% glycerol (v/v). This is the buffer in which transcription factors present in nuclear extracts are allowed to interact with labeled DNA probes (*see* **Note 2**).

2.3 Nitrogen Cavitation Vessel

A cell disruption bomb of any model can be used. We favor Model 4635 (Parr Instrument Company, Moline, IL, USA), custom-fitted with four discharge valves, as this makes it possible to process many samples simultaneously.

2.4 Electrophoretic Mobility Shift Assay (EMSA)

1. For NF-κB binding, use any double-stranded oligonucleotide probe containing a consensus NF-κB sequence. In this chapter, we use an oligonucleotide containing tandemly repeated NF-κB sites (capitalized) identical to those of the HIV promoter (5′-gatca GGGACTTTCC gctg GGGACTTTCC-3′).
2. For the binding of STAT proteins, use various commercially available double-stranded oligonucleotide probes, depending on the nature of the STAT complex. In this chapter, we use an oligonucleotide probe corresponding to the gamma response region (GRR) of the CD64 promoter (5′-CTT TTC TGG GAA ATA CAT CTC AAA TCC TTG AAA CAT GCT-3′), which can bind STAT1-, STAT3- and STAT5-containing complexes.
3. [γ-^{32}P]ATP with specific activity of at least 3000 Ci/mmol for probe labeling.
4. Sephadex G-50 spin columns for the separation of labeled oligonucleotide probe from unincorporated [γ-^{32}P]ATP, used according to the manufacturer's instructions.

2.5 Native Polyacrylamide Gel Electrophoresis

1. TBE buffer: 89 mM Tris, 89 mM boric acid, 0.2 mM EDTA, pH 8.30. Use the same batch of TBE buffer for gel preparation and subsequent electrophoresis. A 5× TBE stock can be made and stored at RT (*see also* **Note 3**).
2. Nondenaturing polyacrylamide gel preparation: Using a 25:1 mixture of acrylamide–bisacrylamide ensures that large protein complexes migrate through the gel without undue hindrance. Although 5% gels are generally used, varying the total acrylamide content of the gels can yield better resolution. Use a freshly prepared ammonium persulfate solution to ensure optimal polymerization. Always let the gels completely polymerize (at least 1.5 h) to ensure the best possible resolution. Finally, it is advisable to use a gel size of 16 × 14 cm ($w \times h$), as it allows for better sample separation than mini-gels. A gel thickness of 1–1.5 mm works well.

3 Methods

Neutrophil stimulation with several physiological agonists results in a greatly enhanced detection of nuclear NF-κB-binding complexes in EMSA, and in the onset of nuclear STAT-containing

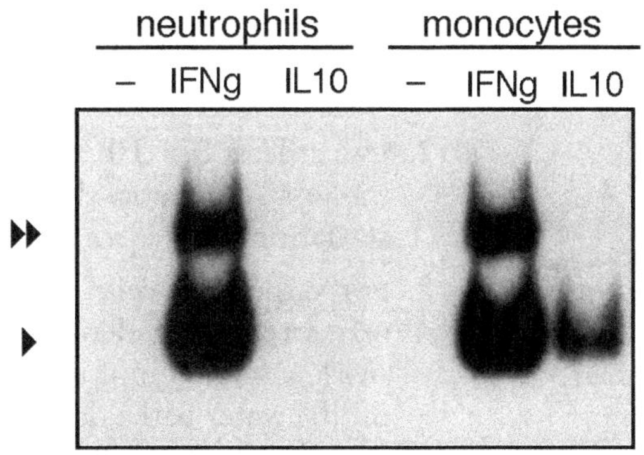

Fig. 1 Induction of STAT1 multimers in neutrophils and monocytes. Human neutrophils and autologous monocytes were stimulated with 100 U/mL IFNγ, 1000 U/mL IL-10, or diluent control, for 15 min at 37 °C. Cells were cavitated, and nuclear extracts were prepared and analyzed in EMSA using a GRR probe. The single arrowhead represents STAT1 dimers, and the double arrowheads represent STAT1 tetramers, as previously reported [15]. (Reproduced by permission of Humana Press©2007 [28])

DNA-binding activities (reviewed in [24]). However, the detection of intact (i.e., unproteolyzed) transcription factors in neutrophils requires an alternative protocol that we developed, which is based on nitrogen cavitation of the cells. This protocol consistently yields NF-κB and STAT complexes that are indistinguishable from those isolated from other cell types, such as human monocytes or Jurkat cells [13] (*see* also Fig. 1 for STAT multimers and Figs. 3 and 4 for NF-κB dimers). In the case of STAT complexes, conventional detergent-based methods additionally produce artifacts such as the detection of a constitutive DNA binding in nuclear extracts from resting neutrophils [13, 29], which are not observed when the neutrophils are cavitated instead [13] (*see* also Fig. 1). The above considerations make it clear that nitrogen cavitation offers unmatched advantages for the preparation of neutrophil nuclear extracts. Because few granules are ever broken, protease inhibitors can effectively neutralize the small quantities of proteases which may be released during preparation (as opposed to the huge amounts released using conventional approaches).

Over the years, several refinements have been made to the procedure that we originally described [11, 13], which we have incorporated in the following protocol. Because crucial aspects of the protocol are the disruption of neutrophils and preparation of nuclear extracts, we will mainly focus on these aspects. The subsequent EMSA analysis is essentially the same for neutrophil nuclear extracts as for extracts originating from other cell types.

3.1 Preparation of Neutrophil Nuclei

1. Stimulate neutrophils either in suspension or plated in tissue culture-treated plasticware and then stimulated (both approaches will yield a comparable outcome).
2. Use at least 3×10^7 cells per experimental condition to offset the loss of material due to the significant void volume of the cavitation device (*see* **Note 4**).
3. For suspension cells, a maximum concentration of 10^7 cells/mL is recommended. Keep the cells in suspension by gentle swirling at 2–3 min intervals (or with constant swirling in an orbital water bath) during the stimulation period.
4. For plated cells, neutrophils can be cultured in large tissue culture-treated petri dishes in an incubator. Cell density is less important, but care must be taken so that neutrophils do not stack on each other when they sediment to the bottom of the dish.
5. For suspension or plated cells, add an equal volume of ice-cold buffer to stop the incubation (the same buffer used for cell incubations, such as HBSS, KRPD, or culture medium containing 2 mM DFP and phosphatase inhibitors (10 mM NaF, 1 mM Na_3VO_4, 10 mM $Na_4P_2O_7$)).
6. For plated neutrophils, collect cells by gentle resuspension using a pipette fitted with a 1 mL tip that has been cut at the end to minimize turbulence.
7. Centrifuge collected cells (300 × *g*, 5 min, 4 °C) and resuspend in ice-cold Relaxation buffer supplemented with protease and phosphatase inhibitors in 100 × 16 mm round-bottom polypropylene centrifuge tubes. The final cell concentration should be 2×10^7/mL or less (Fig. 2, middle panel) to ensure total cell lysis.
8. Place samples into a nitrogen bomb containing crushed ice. This keeps the samples cold, and also helps position the four tubes into the four-outlet bomb (*see* **Note 4**).
9. Place a small magnetic stir bar in each tube, and place the nitrogen bomb itself a magnetic stirrer to prevent cells from aggregating during the cavitation process (*see* **Note 5**).
10. Pressurize samples with N_2 (350 psi) for 10 min at 4 °C (*see* **Note 6** and Fig. 2).
11. Collect each sample by inserting the bomb's outlet tube halfway into a 15 mL conical polypropylene centrifuge tube and carefully opening the discharge valve.
12. Collect cavitate dropwise into the tube, as opposed to releasing the sample all at once. This step requires some practice, because it must be performed correctly and swiftly to avoid leaving the last sample for too long in the bomb (*see* **Note 7**).

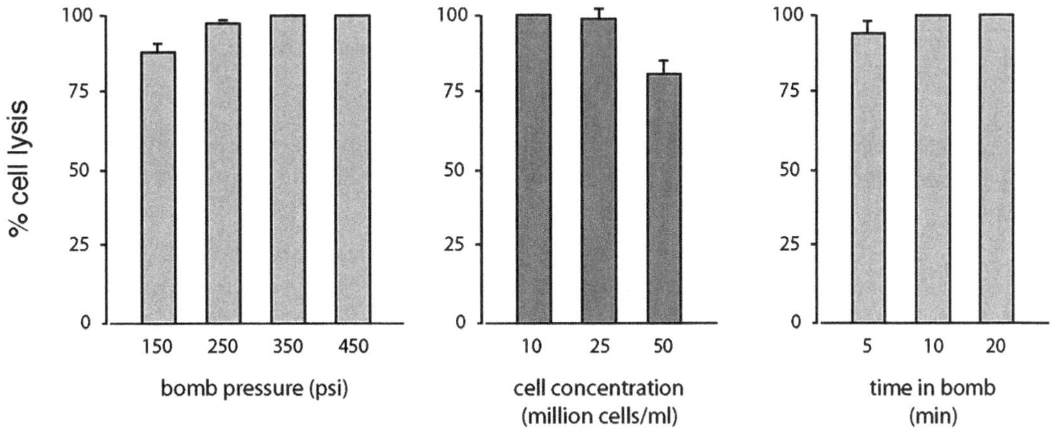

Fig. 2 Nitrogen cavitation set-ups for human neutrophils. Left panel, neutrophils (10^7/mL) were cavitated at the indicated pressures for 10 min. Middle panel, neutrophils were cavitated (350 psi, 10 min) at the indicated cell concentrations. Right panel, neutrophils (10^7/mL) were cavitated at 350 psi for the indicated times. Mean ± s.e.m. of at least three independent experiments. (Reproduced by permission of Humana Press©2007 [28])

13. Place tubes on ice immediately after each cavitate is collected.
14. Centrifuge the cavitate (1500 × *g*, 4 °C, 10 min) to pellet the nuclei.
15. Collect the resulting supernatants and recentrifuge under the same conditions to pellet the remaining nuclei.
16. Combine the nuclear pellets are combined, wash once with 500 μL Relaxation buffer, and used immediately for preparation of nuclear extracts (*see* **Note 8**).

3.2 Preparation of Nuclear Extracts

1. Gently resuspend nuclear pellets in Nuclear Extraction buffer using a 100 μL pipette tip that has been cut at the end to avoid breaking the fragile nuclei during pipetting. We found that using a volume (in μL) that is 1.5 times the number of million cell-equivalents in the nuclear pellet works best.
2. Add NaCl from a concentrated NaCl solution (made in Relaxation buffer) to yield a final concentration of 400 mM (*see* **Notes 9–11**).
3. Immediately mix samples by flicking the Eppendorf tubes a few times.
4. Perform nuclear extraction on wet ice for 20 min, with occasional flicking of the tubes (*see* **Note 12**).
5. Pellet samples in a microfuge (top speed, 10 min, 4 °C).
6. Set aside a small volume of the resulting supernatants (the nuclear extracts) for protein content determination, and immediately snap-freeze the extracts in liquid nitrogen.

7. Store frozen extracts at −80 °C (*see* **Note 13**). Nuclear extracts prepared in this manner should contain 1–1.5 μg/μL of protein.

3.3 Radiolabeling of Oligonucleotide Probe

1. End label annealed oligonucleotide probes with either T4 polynucleotide kinase or the Klenow fragment according to the supplier's instructions.
2. Proper labeling generates probes with at least 2 × 10^6 cpm/pmol probe (corrected for the reference date of the radiolabel).
3. A labeled probe has a useful life of about 4 weeks, although using freshly labeled probe obviously reduces the autoradiography exposure times.

3.4 Binding Reaction

The basic protocol for nuclear extract binding to a labeled probe is described first. This is followed by variations in which competition binding or antibody binding is also involved.

1. For each sample, mix 2–5 μg of the nuclear extract proteins with 10 μL of Binding buffer by gentle flicking of the Eppendorf tubes (*see* **Note 14**).
2. Adjust the final volume to 15 μL.
3. Incubate the samples at room temperature for 10 min.
4. Add approximately 20 fmol of ^{32}P-labeled oligonucleotide probe (30,000–50,000 cpm) in a volume of 2 μL, and incubate the reaction mixture for an additional 10 min at room temperature.
5. Immediately load the samples on native polyacrylamide gels.

3.5 Confirming of Specificity

To confirm the specificity of the interaction between the DNA-binding complexes and the DNA sequence of interest, the following competition experiments can be performed.

1. Allow the nuclear extracts to interact with unlabeled competitor probes for 10 min at room temperature in Binding buffer.
2. Add labeled probe, and incubate for another 10 min at room temperature.
3. For best results, add increasing amounts of unlabeled oligonucleotide probe to the binding mix (typically, 10×, 25×, and 50× excess cold probe) to compete with the radiolabeled probe for binding to the transcription factor. Any specific binding should be displaced using 10× or 25× cold probe (as shown for neutrophil NF-κB DNA-binding activities in Fig. 3).
4. As a negative control, perform the same competition experiments using an oligonucleotide with a mutated binding site for the transcription factor of interest, which should only displace nonspecific complexes (as shown in Fig. 3).

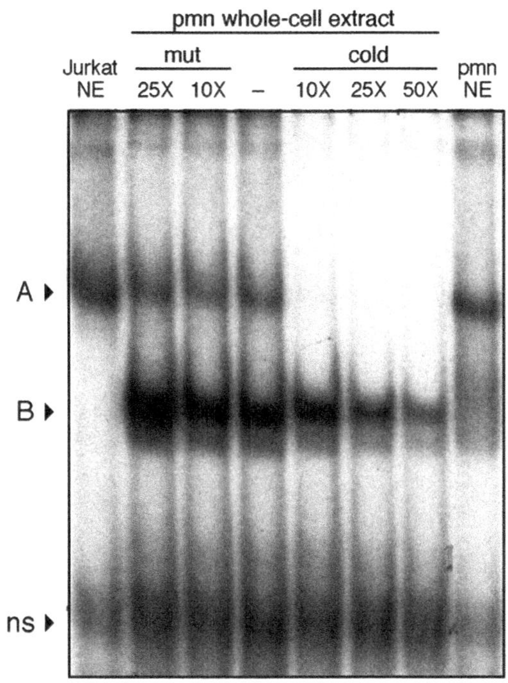

Fig. 3 Human neutrophils were stimulated for 10 min with 100 U/mL TNFα, and whole-cell extracts were prepared by supplementing raw cavitates with glycerol and NaCl (10% v/v and 400 mM final concentrations, respectively) and incubating on ice for 20 min, prior to centrifugation. Extracts were then incubated in Binding buffer in the absence ("–") or presence of increasing concentrations of either unlabeled NF-κB probe ("cold") or of a mutated NF-κB probe ("mut") in which the consensus sequence was changed to 5′-AATACTTTCC (the mutated nucleotides are underlined), prior to addition of labeled NF-κB probe and subsequent EMSA analysis. For comparison, a nuclear extract from the same neutrophil preparation was loaded in the last lane ("pmn NE"), and a nuclear extract from TNF-activated Jurkat cells ("Jurkat NE") was loaded in the first lane as a positive control. "A" denotes the specific, inducible NF-κB complex; "B" denotes a constitutive complex showing some specificity which is present in whole-cell extracts (but mostly absent from the corresponding nuclear extracts); "ns" denotes a constitutive nonspecific complex. (Reproduced by permission of Humana Press©2007 [28])

3.6 Identification of DNA-Binding Complex Constituents

To directly identify the constituents of a specific DNA-binding complex, antibodies raised against potential candidates can be included in the binding mix.

1. Allow the nuclear extracts to interact with one or more antibodies (about 2 μg each) for 20 min at room temperature in Binding buffer.
2. Add labeled probe, and incubate for another 10 min at room temperature.

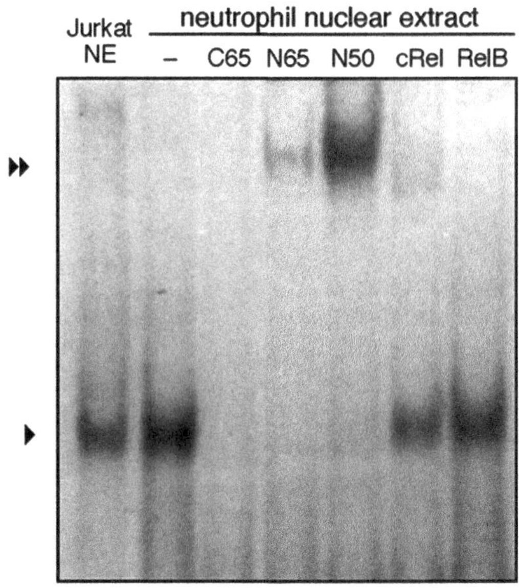

Fig. 4 Human neutrophils were stimulated for 10 min with 100 ng/mL LPS, and nuclear extracts were incubated in Binding buffer in the absence ("–") or presence of antibodies directed at a C-terminal sequence within RelA that abuts the Rel homology domain ("C65"), the N-terminus of RelA ("N65"), the N terminus of p50 ("N50"), the C-terminus of c-Rel ("cRel"), or the N-terminus of RelB ("RelB"), prior to addition of labeled NF-κB probe and subsequent EMSA analysis. A nuclear extract from TNF-activated Jurkat cells ("Jurkat NE") was loaded in the first lane as a positive control. The single arrowhead denotes the specific, inducible NF-κB complex; double arrowheads indicate the supershifted complexes. (Reproduced by permission of Humana Press©2007 [28])

3. As a negative control, use isotype-matched control antibodies at the same concentration (*see* **Note 15**).
4. Binding of antibody to its target protein, will increase the size of the DNA-protein complex, resulting in further retardation of its gel mobility or "supershift" (as shown in Figs. 4 and 5 for neutrophil NF-κB and STAT DNA-binding activities, respectively).
5. Sometimes, the antibody used binds at, or near to, a transcription factor's DNA-binding domain. In this particular case, the labeled DNA probe can no longer interact with the resulting DNA-binding complex, and there will consequently be no observable supershift, but rather a disappearance of the band (as shown in Fig. 4, lane 3).

3.7 Sample Analysis by Polyacrylamide Gel Electrophoresis and Autoradiography

1. Prerun the native gels made in TBE or in an alternative buffer (*see* **Note 3**) for 90 min at 10 V/cm. This ensures a completely isocratic buffer system for optimal migration.

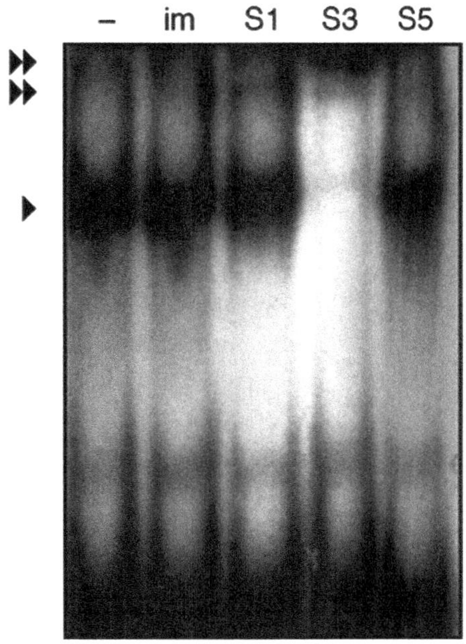

Fig. 5 Human neutrophils were stimulated for 10 min with 1000 U/mL G-CSF, and nuclear extracts were incubated in Binding buffer in the absence ("–") or presence of antibodies directed at STAT1 ("S1"), STAT3 ("S3"), STAT5 ("S5"), or an isotype-matched control Ab ("im"), prior to the addition of labeled GRR probe and subsequent EMSA analysis. The single arrowhead denotes the specific, inducible GRR complex; double arrowheads indicate the supershifted complexes. (Reproduced by permission of Humana Press©2007 [28])

2. Load the samples immediately after the binding reaction. Also load 10 μL xylene cyanol (0.1% w/v in Binding buffer) in the first and last lanes of the gel as a tracking dye.
3. Run the gels at 12–14 V/cm for an additional 2–3 h. The total run time depends on how far the tracking dye has migrated. Empirically determine what distance corresponds to the optimal migration of the transcription factor complexes under study).
4. Transfer the gels to Whatman DE-81 paper, and heat dry under vacuum in a gel dryer for 60 min, or until the gels are completely dry.
5. Expose the dried gels to autoradiographic film at −80 °C with intensifying screens. Alternatively, gels can be exposed to PhosphorScreens (GE Healthcare) at RT prior to being read on a STORM 860 PhosphorImager (Molecular Devices).

4 Notes

1. Our Relaxation buffer formulation represents a modification of the original Relaxation buffer described by Borregaard [22], which among other things contained too much salt, resulting in the extraction of nuclear proteins during cell disruption. Our formulation overcomes this limitation. We no longer include the protease inhibitors, phosphoramidon and bestatin, because they do not alter the degradation of NF-κB or STAT complexes by purified neutrophil granule proteases.
2. This formulation gives very consistent results with either NF-κB or STAT transcription factors. For other factors, various parameters may need to be changed, such as the amounts of DTT, acetylated BSA, and poly(dI-dC); the presence of $MgCl_2$; and so forth.
3. TGE buffer (50 mM Tris–HCl, pH 7.5, 380 mM glycine, 2 mM EDTA, pH 8.0) may be used instead of TBE. Again, use the same buffer for gel preparation and electrophoresis. Because of the high ionic strength of this buffer, the concentration of carrier DNA may be reduced.
4. The device we use for nitrogen cavitation is a Model 4635 Cell Disruption Bomb fitted with 4 outlets (*see* Subheading 2.3). The length of each dip tube used for sample collection is about 12 cm, with a diameter of 1 mm. This translates into a void volume of about 0.5 mL, which makes it difficult to work with small volumes without incurring a significant loss of cavitated sample. A smaller cell disruption device (Model 4639) is also available from the same company, featuring nearly no void volume, but it does not allow custom fitting of additional discharge valves for the simultaneous fractionation of multiple samples.
5. Small magnetic stir bars ("fleas") can be bought from Sigma-Aldrich (cat # Z11, 884-2).
6. As shown in Fig. 2, neutrophils are efficiently lysed over a wide pressure range. However, it must be pointed out that although 450 psi achieves total cell lysis, the resulting nuclear extracts usually contain fewer DNA-binding transcription factor complexes. Thus, cavitating between 250 and 350 psi is recommended. Figure 2 also shows that a cavitation time of 10 min is sufficient to achieve total cell lysis; 5 min also yields good results, albeit with significant variability.
7. One can practice using distilled water instead of a cell suspension. The entire step should not last more than 10 min for the collection of four samples. It is also sometimes necessary to inject more nitrogen into the bomb, if the pressure goes down to 250 psi after collecting 2–3 samples.

8. No further purification of the nuclei is required, as the current protocol does not yield any unbroken cells that would otherwise cosediment with the nuclei. Nuclei thus prepared are free from cytoplasmic contamination [11, 30].
9. The final NaCl concentration must not exceed 500–600 mM, as nuclear lysis is likely to occur.
10. We found that adding NaCl after the nuclear pellets have been resuspended (instead of including it in the Nuclear Extraction buffer, as with conventional protocols) improves the yield of extracted transcription factors significantly.
11. We observed that our modified Relaxation buffer, unlike the original formulation, results in the association of about 10–15% of the granules with the nuclei (as determined by presence of myeloperoxidase and lactoferrin). These granules are still intact, as evidenced by the consistent lack of transcription factor degradation achieved using our protocol. By comparison, resuspension of nuclear pellets in the presence of high NaCl in the Nuclear Extraction buffer results in some degradation of the NF-κB and STAT proteins. This possibly reflects the impossibility of adding enough protease inhibitors to neutralize the large quantities of proteases released in a very small volume as the nuclear-bound granules break, presumably during resuspension of the nuclei.
12. Nuclei should not be allowed to extract for longer times, as significant nuclear lysis already occurs at 30 min.
13. Simply placing the samples at −80 °C results in a noticeable loss of activity, and this loss increases after each subsequent freeze–thaw cycle. By contrast, snap-freezing almost completely prevents this loss.
14. The volume of nuclear extract in the binding mix must not exceed 5 μL or the NaCl concentration will be too high, resulting in unpalatable smearing on the gels. If the samples are too dilute, they can be reconcentrated and desalted using spin columns with a MW cutoff of about 100 kDa, such as the Microcon-100 (Millipore).
15. The signal can sometimes be a little stronger with the control antibody. This has been observed in many studies and is considered normal.

Acknowledgments

This work was supported by grants to PPMcD from the Canadian Institutes for Health Research and the Arthritis Foundation of Canada, and to RDY from the National Institutes of Health. PPMcD is a scholar of Fonds de la recherche en santé du Québec (FRSQ).

References

1. Cassatella MA (1999) Neutrophil-derived proteins: selling cytokines by the pound. Adv Immunol 73:369–509
2. Scapini P, Lapinet-Vera JA, Gasperini S et al (2000) The neutrophil as a cellular source of chemokines. Immunol Rev 177:195–203
3. Ellis TN, Beaman BL (2004) Interferon-γ activation of polymorphonuclear neutrophil function. Immunology 112:2–12
4. Galligan C, Yoshimura T (2003) Phenotypic and functional changes of cytokine-activated neutrophils. Chem Immunol Allergy 83:24–44
5. Cheng SS, Kunkel SL (2003) The evolving role of the neutrophil in chemokine networks. Chem Immunol Allergy 83:81–94
6. Leitch AE, Lucas CD, Marwick JA et al (2012) Cyclin-dependent kinases 7 and 9 specifically regulate neutrophil transcription and their inhibition drives apoptosis to promote resolution of inflammation. Cell Death Differ 19:1950–1961
7. Wither JE, Prokopec SD, Noamani B et al (2018) Identification of a neutrophil-related gene expression signature that is enriched in adult systemic lupus erythematosus patients with active nephritis: clinical/pathologic associations and etiologic mechanisms. PLoS One 13:e0196117
8. Agraz-Cibrian JM, Giraldo DM, Urcuqui-Inchima S (2019) 1,25-Dihydroxyvitamin D3 induces formation of neutrophil extracellular trap-like structures and modulates the transcription of genes whose products are neutrophil extracellular trap-associated proteins: a pilot study. Steroids 141:14–22
9. Zindl CL, Lai JF, Lee YK et al (2013) IL-22-producing neutrophils contribute to antimicrobial defense and restitution of colonic epithelial integrity during colitis. Proc Natl Acad Sci U S A 110:12768–12773
10. Zhang G, Liu J, Wu L et al (2018) Elevated expression of serum amyloid a 3 protects colon epithelium against acute injury through TLR2-dependent induction of neutrophil IL-22 expression in a mouse model of colitis. Front Immunol 9:1503
11. McDonald PP, Bald A, Cassatella MA (1997) Activation of the NF-κB pathway by inflammatory stimuli in human neutrophils. Blood 89:3421–3433
12. Bovolenta C, Gasperini S, Cassatella MA (1996) Granulocyte colony-stimulating factor induces the binding of STAT1 and STAT3 to the IFNγ response region within the promoter of the FcγRI/CD64 gene in human neutrophils. FEBS Lett 386:239–242
13. McDonald PP, Bovolenta C, Cassatella MA (1998) Activation of distinct transcription factors in neutrophils by bacterial LPS, interferon-γ, and GM-CSF and the necessity to overcome the action of endogenous proteases. Biochemistry 37:13165–13173
14. McDonald PP, Cassatella MA (1997) Activation of transcription factor NF-κB by phagocytic stimuli in human neutrophils. FEBS Lett 412:583–586
15. Bovolenta C, Gasperini S, McDonald PP et al (1998) High affinity receptor for IgG (FcγRI/CD64) gene and STAT protein binding to the IFN-γ response region (GRR) are regulated differentially in human neutrophils and monocytes by IL-10. J Immunol 160:911–919
16. McDonald PP, Russo MP, Ferrini S et al (1998) Interleukin-15 (IL-15) induces NF-κB activation and IL-8 production in human neutrophils. Blood 92:4828–4835
17. Hayden MS, Ghosh S (2012) NF-κB, the first quarter-century: remarkable progress and outstanding questions. Genes Dev 26:203–234
18. Gilmore TD, Wolenski FS (2012) NF-κB: where did it come from and why? Immunol Rev 246:14–35
19. Stark GR, Darnell JE Jr (2012) The JAK-STAT pathway at twenty. Immunity 36:503–514
20. Kiu H, Nicholson SE (2012) Biology and significance of the JAK/STAT signalling pathways. Growth Factors 30:88–106
21. Dignam JD, Lebovitz RM, Roeder RG (1983) Accurate transcription initiation by RNA polymerase II in a soluble extract from isolated mammalian nuclei. Nucleic Acids Res 11:1475–1489
22. Borregaard N, Heiple JM, Simons ER et al (1983) Subcellular localization of the b-cytochrome component of the human neutrophil microbicidal oxidase: translocation during activation. J Cell Biol 97:52–61
23. Nachman R, Hirsch JG, Baggiolini M (1972) Studies on isolated membranes of azurophil and specific granules from rabbit polymorphonuclear leukocytes. J Cell Biol 54:133–140
24. McDonald PP (2004) Transcriptional regulation in neutrophils: teaching old cells new tricks. Adv Immunol 82:1–48
25. Cloutier A, Ear T, Borissevitch O et al (2003) Inflammatory cytokine expression is independent of the c-Jun N-terminal kinase/AP-1 signaling cascade in human neutrophils. J Immunol 171:3751–3761
26. Boyum A (1968) Isolation of mononuclear cells and granulocytes from human blood. Isolation of monuclear cells by one centrifugation,

and of granulocytes by combining centrifugation and sedimentation at 1 g. Scand J Clin Lab Invest Suppl 97:77–89

27. Bradford MM (1976) A rapid and sensitive method for the quantitation of microgram quantities of protein utilizing the principle of protein-dye binding. Anal Biochem 72:248–254
28. McDonald PP, Ye RD (2007) Detection of intact transcription factors in human neutrophils. Methods Mol Biol 412:473–486
29. Epling-Burnette PK, Zhong B, Bai F et al (2001) Cooperative regulation of Mcl-1 by Janus kinase/stat and phosphatidylinositol 3-kinase contribute to granulocyte-macrophage colony-stimulating factor-delayed apoptosis in human neutrophils. J Immunol 166:7486–7495
30. Ear T, Cloutier A, McDonald PP (2005) Constitutive nuclear expression of the I κB kinase complex and its activation in human neutrophils. J Immunol 175:1834–1842

Chapter 21

Genome-Scale Transcript Analyses of Human Neutrophils

Scott D. Kobayashi, Adeline R. Porter, Sarah L. Anzick, Dan E. Sturdevant, and Frank R. DeLeo

Abstract

Transcriptome analyses of unicellular and multicellular organisms have changed fundamental understanding of biological and pathological processes across multiple scientific disciplines. Over the past 15 years, studies of polymorphonuclear leukocyte (PMN or neutrophil) gene expression on a global scale have provided new insight into the molecular processes that promote resolution of infections in humans. Herein we present methods to analyze gene expression in human neutrophils using Affymetrix oligonucleotide microarrays and next-generation sequencing. Notably, the procedures utilize commercially available reagents and materials and thus represent a standardized approach for evaluating PMN transcript levels.

Key words Neutrophil, Microarray, Gene expression, Affymetrix, Next-generation sequencing, Transcript, Phagocyte, Polymorphonuclear leukocytes

1 Introduction

Mature polymorphonuclear leukocytes (PMNs, granulocytes, or neutrophils) are fully equipped to destroy invading microorganisms, and inhibitors of transcription and translation fail to have an immediate effect on phagocytosis and microbicidal capacity [1–4]. As such, it was widely presumed that new gene transcription played no significant role in neutrophil function. However, extended incubation with actinomycin D ($\geq$1 h), an inhibitor of RNA synthesis, significantly reduces granulocyte phagocytosis, superoxide production, and bactericidal activity [1–4], suggesting that new gene transcription is important for maintaining PMN function.

Methods for measuring gene expression in human granulocytes have existed for at least 50 years. In 1966, Cline first reported that phagocytosis increases RNA synthesis in human granulocytes [4, 5], and subsequent studies by Jack and Fearon showed that peripheral blood neutrophils constitutively synthesize a limited repertoire of mRNA transcripts and protein [6]. These early studies

Mark T. Quinn and Frank R. DeLeo (eds.), *Neutrophil: Methods and Protocols*, Methods in Molecular Biology, vol. 2087, https://doi.org/10.1007/978-1-0716-0154-9_21,

used radiolabel-based techniques, such as incorporation of H^3-uridine or C^{14}-orotic acid into RNA, to estimate overall increases in transcript levels [5, 6]. Northern blotting has also been a widely used method for evaluating neutrophil mRNA levels [7, 8]. In as much as these methods are suited to evaluate a limited number of transcripts or provide an indication of general RNA synthesis, they are not appropriate for genome-wide analyses of transcript levels. Studies by Itoh et al. and Subrahmanyam et al. investigated patterns of gene expression in neutrophils by sequencing of 3′-directed cDNA libraries or cDNA display, respectively [9, 10]. Such studies were the first to use global approaches to investigate neutrophil gene expression. The work by Itoh et al. and Subrahmanyam et al. provided a segue into current PMN transcriptome analyses, which almost exclusively use microarray-based approaches [11–25]. Importantly, the microarray-based work has significantly enhanced our understanding of neutrophil biology and function. For example, studies of the PMN transcriptome during phagocytosis provided new insight into the role of PMNs in the resolution of inflammation and infection [26].

In this chapter, we first provide a method for analyzing human PMN gene expression with Affymetrix oligonucleotide microarrays, arguably the most universal platform for microarray-based research in human tissues. We also provide a method for using next-generation sequencing (NGS) to assess changes in neutrophil gene transcription. With the exception of human neutrophils, all of the materials are commercially available. Therefore, the method described in the following pages produces results that can be compared readily to those generated by other laboratories using the same methodology.

2 Materials

2.1 Isolation of Neutrophils and Analysis of Purity

1. Ficoll-Paque^{PLUS} (GE Healthcare).
2. Injection- or irrigation-grade water.
3. Dulbecco's phosphate-buffered saline, pH 7.2 (DPBS).
4. Injection- or irrigation-grade 0.9% NaCl solution (w/v).
5. 3% Dextran solution (w/v): mix 150 mL 20% dextran solution (500,000 MW) into 1 L (final volume) of injection- or irrigation-grade 0.9% NaCl solution and add 1.35 g RNase-free NaCl. Filter-sterilize using a 0.2-μm bottle-top filter and test for endotoxin.
6. 1.7% NaCl solution (w/v): dissolve 17 g of RNase-free NaCl into 1 L (final volume) of injection- or irrigation-grade water. Filter-sterilize using a 0.2-μm bottle-top filter, and test for endotoxin.

2.2 Purification of RNA from Neutrophils

1. RPMI 1640 medium.
2. Buffered RPMI 1640 medium: 5 mL of sterile, endotoxin-tested 1 *M* HEPES to a new 500 mL bottle of RPMI 1640 medium.
3. 12-well flat-bottom nonpyrogenic polystyrene cell culture plates (e.g., Corning).
4. Aerosol-resistant micropipette tips, RNase-free and nonpyrogenic.
5. 1.7 mL microcentrifuge tubes, RNase-free, nonstick.
6. RNeasy mini kit (Qiagen).
7. 14.5 M 2-mercaptoethanol (2-ME).
8. QiaShredder columns (Qiagen).
9. Diethylpyrocarbonate (DEPC)-treated or nuclease-free H_2O.
10. Baseline Zero DNase kit (Lucigen).

2.3 Determination of RNA Quantity

1. Quant-iT RiboGreen RNA Assay (Molecular Probes).
2. 96-well black microtiter plates.

2.4 Analysis of RNA Quality

1. RNA 6000 Nano LabChip (Agilent Technologies).
2. RNA 6000 Ladder (Invitrogen).

2.5 Generation of Labeled cRNA

1. Affymetrix GeneChip one-cycle target labeling and control reagents kit (Affymetrix) (*see* **Note 1**).
2. 1 M Tris–acetate solution: 121.14 g Trizma base in nuclease-free H_2O, adjust pH to 8.1 with glacial acetic acid.
3. 5× cRNA fragmentation buffer: mix 4.0 mL 1 M Tris–acetate solution, 0.64 g MgOAc (MilliporeSigma), and 0.98 g KOAc (MilliporeSigma). Make up to a final volume of 20 mL with nuclease-free H_2O. Filter-sterilize and store at room temperature.

2.6 Hybridization of cRNA to Affymetrix GeneChip

1. Control oligo B2, 3 n*M* (Affymetrix).
2. Amber microcentrifuge tubes.
3. Tough-Spots label dots (Diversified Biotech).
4. 5 *M* NaCl solution, RNase- and DNase-free.
5. 20× SSPE: 3 *M* NaCl, 0.2 *M* NaH_2PO_4, 0.02 *M* ethylenediamine tetraacetic acid (EDTA).
6. ImmunoPure Streptavidin (Pierce).
7. Herring sperm DNA (Promega).
8. EDTA solution: 0.5 M EDTA, disodium salt in nuclease-free H_2O or purchase premade.
9. Goat immunoglobulin (Ig)G, reagent grade.

10. Biotinylated goat anti-streptavidin antibody.
11. 12× MES stock solution: 64.61 g of MES hydrate, 193.3 g of MES sodium salt, 800 mL of nuclease-free water. Adjust volume to 1 L; the solution pH should fall between 6.5 and 6.7. Filter-sterilize and store shielded from light at 2–8 °C.
12. Wash buffer A (nonstringent): 6× SSPE (from 20× stock) and 0.01% Tween-20 (v/v) (from 10% stock; Surfact-Amps 20, Pierce Biotechnology). Filter-sterilize and store at room temperature.
13. Wash buffer B (stringent): 100 m*M* MES, 0.1 *M* NaCl, 0.01% Tween-20. This buffer is prepared from 12× MES, 5 *M* NaCl solution, and 10% Tween-20 stocks, passed through a 0.2 μm filter, and stored at room temperature.
14. 2× hybridization buffer: 200 m*M* MES, 2 *M* NaCl, 40 m*M* EDTA, 0.02% Tween-20. Use 12× MES, 5 *M* NaCl, 0.5 *M* EDTA solution, and 10% Tween-20 stocks to prepare this solution. The solution is stored between 2–8 °C and shielded from light.
15. 2× stain buffer: 200 m*M* MES, 2 *M* NaCl, 0.1% Tween-20. Solution is prepared from 12× MES, 5 *M* NaCl solution, and 10% Tween-20 stocks, passed through a 0.2 μm filter, and stored shielded from light between 2–8 °C.
16. Streptavidin–phycoerythrin solution (SAPE): prepare fresh by mixing 1× MES (from 2× stock), 2 mg/mL bovine serum albumin (BSA: from 50 mg/mL stock), and 10 μg/mL streptavidin–phycoerythrin (from 1 mg/mL stock prepared in DPBS; Molecular Probes).
17. Antibody solution: prepare fresh by mixing 1× MES (from 2× stock), 2 mg/mL BSA (from 50 mg/mL stock), 0.1 mg/mL goat IgG (from 10 mg/mL stock prepared in 150 m*M* NaCl), and 5 μg/mL biotinylated anti-streptavidin antibody (from 0.5 mg/mL stock).
18. Hybridization cocktail: Hybridization reaction (single-probe array; 200 μL/array) contains 1× hybridization buffer (from 2× stock), 50 pM B2 Control Oligo (from 3 nM stock), 0.1 mg/mL herring sperm DNA (from 10 mg/mL stock), 0.5 mg/mL BSA (from 50 mg/mL stock), 7.8% dimethylsulfoxide (DMSO), 15 μg fragmented and labeled cRNA, 20× spike-in.

2.7 GeneChip Processing, Scanning, and Conversion of Image Files

1. Affymetrix GeneChip Scanner 3000, enabled for high-resolution scanning.
2. Affymetrix GeneChip Command Console (AGCC) including AGCC Portal, AGCC Viewer, Data Exchange Console, AGCC Scan Control, AGCC Fluidic Control, and Expression Console.

2.8 Preparation of Library for Next-Generation Sequencing

1. TruSeq® Stranded mRNA Sample Preparation kit (Illumina).
2. AMPure XP purification beads (Beckman Coulter).
3. SuperScript II Reverse Transcriptase (Invitrogen). SuperScript II mix: 1 μL SuperScript II to 9 μL First Stand Synthesis Act D (per sample). Alternatively, 50 μL SuperScript II can be added to the First Stand Synthesis Act D mix and stored at −20 °C if no more than six freeze-thaw cycles are anticipated.
4. 80% Ethanol, freshly prepared.
5. DNA 1000 kit (Agilent Technologies).
6. KAPA Library Quantification Kit, Illumina Complete Universal (KAPA Biosystems).
7. NextSeq 500/550 Mid Output Kit v2.5 (Illumina).

3 Methods

3.1 Isolation of Neutrophils

1. Human PMN can be isolated using Hypaque-Ficoll gradient separation described by Nauseef [27] or Siemsen et al. [28] (also *see* Chapters 3 and 4 of this volume). All studies with human neutrophils were performed in accordance with a protocol (01-I-N055) approved by the Institutional Review Board for human subjects, National Institute of Allergy and Infectious Diseases. Human volunteers gave written informed consent prior to participation. The factors most critical for transcriptome analyses are high purity of PMN preparations, i.e., ≤1% contaminating mononuclear cells (typically lymphocytes and monocytes), and that cells are not primed or activated. Techniques to assess cell purity and viability are described in Chapters 2 and 3.

3.2 Purification of RNA from Neutrophils

1. PMN assays (e.g., phagocytosis) are routinely performed in 12-well tissue culture plates containing 10^7 PMNs/well. The PMNs are lysed directly in the wells by adding 600 μL RLT (Qiagen) + 143 mM 2-ME per 10^7 PMNs (*see* **Note 2**). Lysate should be pipetted to ensure complete lysis of PMN and to provide maximum protection to RNA. When performing time course experiments, lysate can be stored at −20 or −80 °C to facilitate sample processing.
2. Homogenize lysate by passage over QIAshredder column and centrifuge (30,000 × *g*) in a microcentrifuge for 2 min.
3. Add 600 μL of 70% ethanol and mix by pipetting. Load one-half of lysate onto a Qiagen RNeasy column and centrifuge for 15 s (30,000 × *g*). Decant flow-through and repeat with second half of lysate.

4. Place column in a new collection tube and add 500 μL of buffer RPE (Qiagen). Spin 15 s in microcentrifuge (30,000 × *g*), decant flow-through, and repeat.
5. Place the column in a new collection tube and centrifuge for 2 min (30,000 × *g*) to remove any residual wash. Place the column in a new RNase-free elution tube.
6. Elute RNA by addition of 50 μL of DEPC-treated H_2O (70 °C), allow to sit 1 min, and then centrifuge (30,000 × *g*) for 1 min. Repeat a second time for maximum elution of RNA and pool the two eluates (100 μL total). If the purified RNA is to be used for NGS, proceed to **step 7** to remove contaminating DNA. If the RNA is used for Affymetrix microarray analysis, proceed directly to **step 8**.
7. To the 100 μL of RNA eluate, add 10 μL of 10× Baseline-ZERO DNase reaction and 5 μL of Baseline-ZERO DNase for each sample. Incubate at 37 °C for 30 min. To terminate the DNase reaction, add 10 μL of 10× Baseline-ZERO DNase Stop Solution and incubate at 65 °C for 10 min.
8. Second purification of RNA is critical for complete cDNA synthesis (*see* **Note 3**). To the 100 μL of RNA eluate, add 350 μL of RLT + 2-ME (143 mM final). Add 250 μL of 100% ethanol and mix by pipetting. Apply mixture to a new RNeasy column and spin (30,000 × *g*) for 15 s.
9. Decant flow-through and add 500 μL of RPE. Spin 15 s (30,000 × *g*), decant flow-through, and repeat.
10. Place column in a new tube and spin 1 min at the maximum rpm to remove residual wash. Place the column in an elution tube, add 50 μL of DEPC-treated H_2O (70 °C), allow to sit for 1 min, and centrifuge (30,000 × *g*) for 1 min. Repeat a second time for maximum elution of RNA. Store RNA at −80 °C. The expected yield of total RNA prepared by this method is approximately 2.0–5.0 μg.

3.3 Determination of RNA Quantity

1. RNA quantity can be measured either by absorbance at 260 nm or by fluorescence. The following protocol assumes the use of a microplate spectrofluorometer (e.g., Molecular Devices Gemini XPS) and the Molecular Probes Quant-iT RiboGreen RNA Assay. The excitation maximum for RiboGreen bound to RNA is 500 nm and the emission maximum is 525 nm (*see* **Note 4**).
2. A 2× working stock of RiboGreen is prepared from a 1:200 dilution of the manufacturer's supplied stock and is used at a concentration of 1× per assay.
3. Each microtiter plate should contain two blank wells (1× RiboGreen) and a serial dilution of control rRNA (supplied). Each assay should contain 1 μL of PMN total RNA, 1× RiboGreen,

and 1× TE buffer. The expected final concentration of total RNA as prepared above (from 10^7 PMNs) is approx. 20 μg/mL (0.02 μg/mL in the assay). Thus, the final concentration of the rRNA for the standard curve in the RiboGreen assay should cover a range from 0 to 1 μg/mL.

4. The assay is incubated for 2 to 5 min at RT and protected from light.
5. Following incubation, the plate is scanned in the microplate reader and the concentration of PMN total RNA is calculated by extrapolating from the standard rRNA curve (*see* **Note 5**).

3.4 Analysis of RNA Quality

1. The isolation of high-quality RNA is essential for both microarray and NGS analysis. RNA integrity of low-yield samples can be assessed on microfluidics-based automated electrophoresis systems (Fig. 1). The following protocol assumes the use of an Agilent 2100 Bioanalyzer (Agilent Technologies) or similar device and the RNA 6000 Nano LabChip (Agilent).
2. The gel–dye mix is prepared by centrifuging 400 μL RNA gel matrix through the supplied spin filter at 1500 × *g* for 10 min. The filtered gel matrix is stored at 4 °C and must be used within 4 weeks. To make the working stock of gel matrix–dye reagent, 2 μL of RNA dye concentrate is added to 130 μL of the filtered gel mix and thoroughly vortexed. The remaining gel–dye mix is stored protected from light at 4 °C and should be used within 1 week.
3. Place a new chip on the priming station and load 9 μL prepared gel–dye mix into the well labeled 'G' (third row, black circle). Close the priming station and depress plunger until it engages with the clip. After 30 s, release the clip and check the back of the chip for air bubbles.
4. Add 9 μL prepared gel–dye mix to the remaining two 'G' wells (rows 1 and 2, no circle). Load 5 μL of the supplied marker buffer to the ladder well and all 12 sample wells. Add 1 μL heat-denatured (95 °C, 5 min) RNA 6000 ladder to the ladder well, and load 1 μL of purified PMN RNA to each sample well.
5. Place the loaded chip into the adapter in the supplied vortexer, and mix for 1 min. The chip must be run within 5 min.
6. The chip is read using the integrated eukaryotic total RNA algorithm, and RNA integrity is assessed by the appearance of two well-defined peaks denoting the 18S and 28S rRNA subunits (approx. 2:1 ratio) (Fig. 1).

3.5 Generation of Labeled cRNA for Microarrays

1. The volume of PMN total RNA is reduced to 8 μL in a centrifugal vacuum concentrator, or alternatively desiccated and suspended in 8 μL of RNase-free water. cDNA synthesis, cleanup, and cRNA labeling are performed with the GeneChip one-cycle target labeling and control reagents kit (*see* **Note 6**).

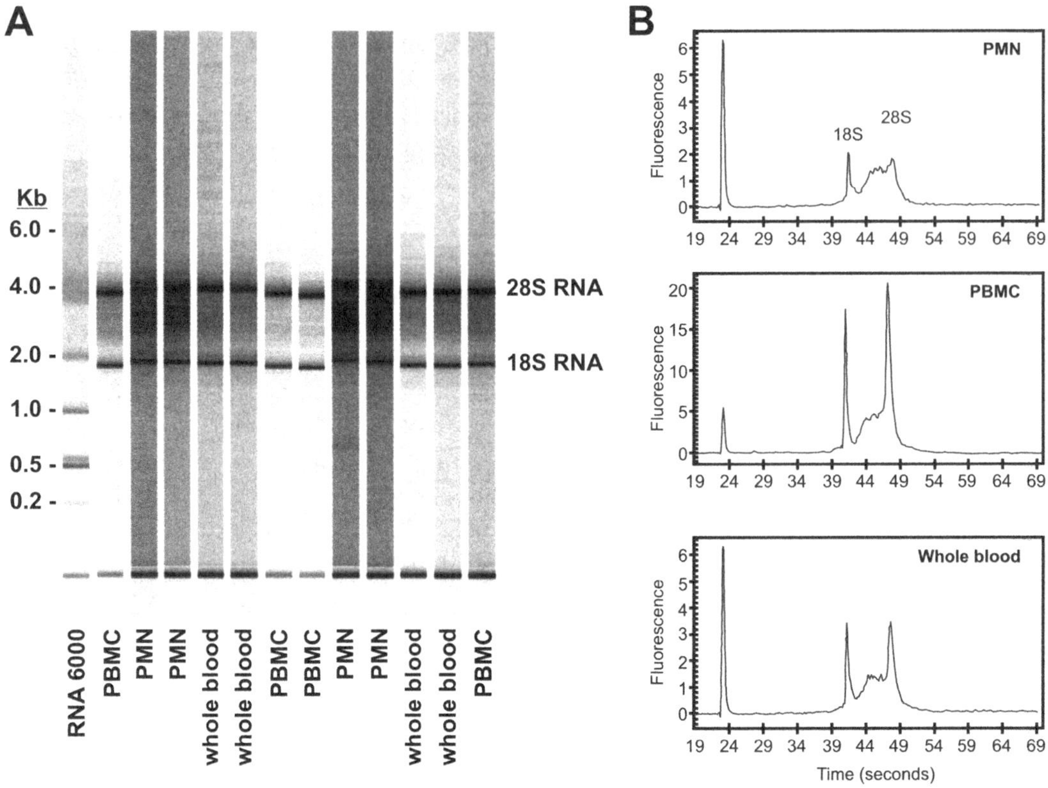

Fig. 1 Analysis of RNA quality using a microfluidics-based automated electrophoresis system. (**a**) 1 μL of total RNA (typically 20–30 ng) purified from whole blood, peripheral blood mononuclear cells (PBMC), or polymorphonuclear neutrophils (PMN) was evaluated with an Agilent 2100 Bioanalyzer. RNA 6000 indicates an RNA 6000 Ladder (Ambion) comprised of 6 transcripts of varied size as shown. (**b**) Histograms of representative RNA samples. (Reproduced from ref. 29 by permission of Humana Press © 2007)

2. Preparation of poly-A RNA spike-in controls. The RNA controls consist of four *Bacillus subtilis* genes (*lys*, *phe*, *thr*, and *dap*) that are absent from eukaryotic samples and are used to monitor the labeling efficiency of target RNA. The final spike-in control concentration is based on the amount of total PMN RNA (~1–2 μg). Three serial dilutions are prepared from the Affymetrix poly-A control stock. For the first dilution, add 2 μL poly-A control stock to 38 μL of poly-A control dilution buffer (1:20). The second and third dilutions are prepared at 1:50 into the dilution buffer. Two microliters of the final dilution are used in each cDNA synthesis reaction.
3. First-strand cDNA synthesis, primer hybridization. Add 2 μL T7-Oligo (dT) primer (50 μ*M*), 8 μL total PMN RNA and 2 μL diluted poly-A controls to a 0.5-mL RNase-free-tube. Incubate the reaction for 10 min at 70 °C and cool to 4 °C.
4. First-strand cDNA synthesis. Add 2 μL of 0.1 M DTT, 4 μL 5× first strand cDNA synthesis buffer, and 1 μL 10 mM dNTP and incubate for 2 min at 42 °C. Add 1 μL SuperScript III RT and incubate at 42 °C for 60 min.

5. Second-strand cDNA synthesis. To each tube from the first-strand cDNA reaction, add 130 μL of master mix (91 μL DEPC-treated H_2O, 30 μL second-strand buffer, 3 μL of 10 mM dNTPs, 1 μL of 10 U/μL DNA ligase, 4 μL of 10 U/μL DNA Polymerase I, and 1 μL of 2 U/μL RNase H). Incubate the reactions at 16 °C for 2 h. Add 2 μL T4 DNA polymerase to each reaction and incubate 16 °C for 15 min. Stop the reaction by addition of 10 μL of 0.5 *M* EDTA, pH 8.0.
6. Purification of cDNA, sample cleanup module. Add 600 μL of cDNA binding buffer to the double-stranded cDNA reaction mix and vortex for 3 s. Apply 500 μL of the sample to the cDNA cleanup spin column (save the remaining sample), and centrifuge for 1 min at 8000 × *g*. Discard the flow through. Reload the spin column with the remaining sample and centrifuge for 1 min at 8000 × *g*. Discard the flow-through and collection tube. Transfer the spin column to a new 2-mL collection tube. Add 750 μL of the cDNA wash buffer to the spin column. Centrifuge for 1 min at 8000 × *g*. Discard the flow through. Cut the cap off of the spin column (label the side of the column) and centrifuge for 5 min at >25,000 × *g*. Discard the flow–through and the collection tube. Transfer the spin column to a new collection tube and add 14 μL of cDNA elution buffer directly onto the spin column membrane. Incubate for 1 min at room temperature and centrifuge for 1 min at >25,000 × *g*. The expected volume of cDNA following cleanup is 12 μL.
7. Synthesis of labeled cRNA, GeneChip in vitro transcription (IVT) labeling. The entire 12 μL of cDNA (from 10^7 PMNs) is used in the IVT labeling reaction. Prepare a master mix containing 4 μL 10× IVT labeling buffer, 12 μL IVT labeling NTP mix, 8 μL RNAse-free water, and 4 μL IVT labeling enzyme mix per reaction. Add 28 μL of the IVT master mix to 12 μL cDNA in 0.5-mL PCR tubes. Incubate the reaction at 37 °C for 16 h. Store labeled cRNA at −70 °C or proceed to cleanup.
8. Cleanup of labeled cRNA. The sample cleanup module is used for this step. Add 60 μL of RNase-free water to the IVT reaction and mix by vortexing for 3 s. Add 350 μL IVT cRNA binding buffer to the sample and vortex for 3 min. Add 250 μL 100% ethanol and mix well by pipetting. Apply the sample to the IVT cRNA cleanup spin column and centrifuge for 15 s at 8000 × *g*. Discard the flow through and collection tube. Transfer the spin column to a new collection tube and pipe 500 μL IVT cRNA wash buffer onto the column. Centrifuge for 15 s at 8000 × *g* and discard flow-through. Pipet 500 μL of 80% (v/v) ethanol onto the spin column and

centrifuge for 15 s at 8000 × *g* and discard flow–though. Cut the cap off of the spin column (label the side of the column) and centrifuge for 5 min at >25,000 × *g*. Transfer the spin column to a new collection tube and pipe 11 μL of RNase-free water directly onto the spin column membrane. Centrifuge 1 min at >25,000 × *g* to elute. Pipet 10 μL of RNase-free water directly onto the spin column membrane and centrifuge for 1 min at >25,000 × *g*. Store cRNA at −70 °C.

9. Determination of cRNA quantity. The quantity of cRNA is determined similar to that of total RNA as described under Subheading 3.3. by use of RiboGreen. However, the expected yield of cRNA generated from the IVT reaction is ten-fold the starting amount of total RNA. Therefore, the cRNA sample should be diluted ten-fold in order to fall within the linear range of the rRNA standard curve. *Optional*: Labeled cRNA samples can be analyzed with Agilent 2100 Bioanalyzer as described under Subheading 3.4. A typical signature for PMN cRNA is a broad smear of rather than distinct bands (Fig. 2).
10. Fragmentation of labeled cRNA. Fragmentation of target cRNA prior to hybridization is necessary for optimal assay sensitivity. cRNA is desiccated in a centrifugal vacuum concentrator and suspended to 1 μg/μL in RNase-free water. Add 8 μL of 5× fragmentation buffer and 17 μL of RNase-free water to 15 μL of cRNA. The cRNA is fragmented by incubation at 94 °C for 35 min followed by cooling on ice.

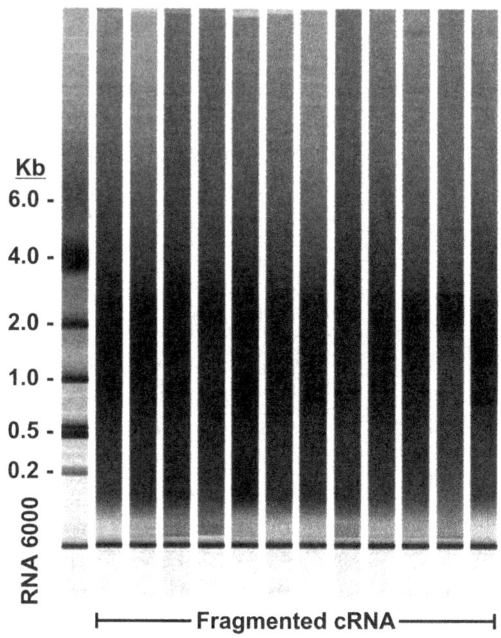

Fig. 2 Analysis of labeled polymorphonuclear neutrophil cRNA using an Agilent 2100 Bioanalyzer. (Reproduced from ref. 29 by permission of Humana Press © 2007)

3.6 Hybridization of cRNA to Affymetrix GeneChip

1. The following protocol is designed for the analysis of human PMN RNA transcripts on Affymetrix 49 format (standard) arrays such as the U133 Plus 2.0.
2. Remove chips from 4 °C storage and acclimate to room temperature (≥60 min) prior to hybridization. Remove reagents from cold storage and allow to thaw at room temperature. Set heat blocks to 65 and 99 °C. Set the hybridization oven to 45 °C.
3. Place the 20× Eukaryotic hybridization controls tube into the 65 °C heat block for 5 min.
4. Mix the hybridization cocktail for each chip, by using 15 μg of cRNA mix with 5 μL of control oligonucleotide B2 at 3 n*M*, 15 μL of 20× Eukaryotic hybridization controls that have been heated to 65 °C for 5 min, 3 μL of 10 mg/mL herring sperm, 3 μL of BSA (50 mg/mL), 150 μL of 2× hybridization buffer, 30 μL of DMSO, and 54 μL RNase-free water for a final volume of 300 μL.
5. Incubate the hybridization cocktail at 99 °C for 5 min.
6. Place the chips on the bench so that the back is facing up, insert a 200-μL pipette tip into one of the septa and fill the chip with 200 μL of 1× hybridization buffer through the remaining septum.
7. Incubate the filled chips in the 45 °C hybridization oven for 10 min at 60 rpm.
8. The hybridization cocktail is incubated at 45 °C for 5 min, and centrifuged for 5 min at maximum rpm in a microfuge.
9. Remove the chips from the hybridization oven and place a 200-μL pipette tip in one of the septa. Then using the other septum, remove the 1× buffer and add 200 μL of the appropriate hybridization cocktail.
10. Place the chips back into the 45 °C hybridization oven for 16 h at 60 rpm.

3.7 GeneChip Processing, Scanning, and Conversion of Image Files

1. The following protocol requires the use an Affymetrix GeneChip Scanner 3000, enabled for high-resolution scanning and the Affymetrix GeneChip Command Console (AGCC) software package.
2. It is important to process the Affymetrix GeneChip directly following hybridization. The preparation of staining and washing reagents and the priming of the Affymetrix workstation must occur prior to completion of the hybridization step.
3. Turn on the fluidics station and verify that tubing is appropriately connected to wash bottles A and B, water and waste.

4. Turn on the scanning workstation and start the AGCC Fluidics Control.
5. Select and prime the workstation(s) intended for use.
6. For each chip, using an amber tube, mix 600 µL of 2× stain buffer, 48 µL 50 mg/mL BSA, 12 µL streptavidin-phycoerythrin (SAPE 1 mg/mL), and 540 µL of distilled water. This is the SAPE solution mix.
7. Remove 600 µL of the SAPE solution and place into another amber tube.
8. For each chip, using a clear tube, mix 300 µL of 2× stain buffer, 24 µL of 50 mg/mL BSA, 6 µL of 10 mg/mL goat IgG stock (10 mg/mL), 3.6 µL of 0.5 mg/mL biotinylated antibody, and 266.4 µL of distilled water. This is the antibody solution mix.
9. Remove the chips from the hybridization oven and place a 200 µL pipette tip into one of the septa. Remove the hybridization cocktail and place into the appropriate sample tube. These samples can be run on other chips at a later date. Store at −70 °C.
10. Fill the chip with 250 µL of wash buffer A without air bubbles. If processing more chips than allowed in the fluidics station, the remaining filled chips can be stored up to 4 h at 4 °C.
11. In the AGCC Portal, select samples and register all samples to be run using the preferred method of single, quick, or batch.
12. Once all samples are registered, they can be viewed in AGCC Portal, Data, folder or project view.
13. In the AGCC Fluidics Control, select the appropriate protocol for each module, station, and select run.
14. Place the appropriate chip into the fluidics modules and place the SAPE solution mix tubes into sample holder positions 1 and 3 for each module.
15. Place the antibody solution mix tube into sample holder position 2 for each module.
16. The protocol should take approximately 90 min to finish. Once the fluidics station displays 'remove cartridge' take the chip out and inspect for air bubbles. If any are present place back into fluidics module and let it refill. If there are still bubbles then the chip must be filled manually with wash buffer A.
17. Place a tough spot over each septum of the chips to be scanned.
18. Gently wipe the glass of the chip with a Kimwipes in one direction. Do not change direction or the paraffin at the edges may coat the glass, making scanning difficult.

19. Scanning. The scanning protocol is written using the autoloader. If an autoloader is not present, the protocol must be altered to accommodate single chip scanning.
20. Press the start button on the scanner.
21. Open the AGCC Scan Control and verify that the connection between the scanner and workstation is complete.
22. Allow scanner to warm up for 10 min.
23. Place chips into the autoloader starting with position one.
24. In the AGCC Scan control select the "start scan" icon and while the chip is being scanned, the AGCC Viewer can be started to visualize previously scanned chips.
25. A successful scan will create the ∗.DAT, ∗.jpg, ∗.audit, and ∗. CEL files with verification of grid alignment of the ∗.DAT and correct ∗.CEL file creation is possible by using the AGCC viewer. Third party software can be used to analyze these files and Expression Console can be used to create the ∗.CHP files.

3.8 Purification of Poly-A mRNA for NGS Library Construction

1. Dilute 1.0 μg total RNA (kit specifications are for 0.1–4.0 μg) into a final volume of 50 μL nuclease-free ultrapure water.
2. Vortex the RNA purification beads to completely resuspend the oligo-dT beads and add 50 μL RNA purification beads to the total RNA. Gently pipette up and down to mix. To denature the RNA and facilitate poly-A RNA binding to the beads, incubate the mixture at 65 °C for 5 min and then cool to 4 °C.
3. When the temperature has reached 4 °C, place the tubes at room temperature for 5 min to allow the RNA to bind to the beads.
4. Place the sample onto a magnetic stand to separate the poly-A RNA-bound beads from the solution. Remove and discard the supernatant and remove the sample from the magnetic stand.
5. Wash beads by adding 200 μL of Bead Washing Buffer and gently pipet up and down to mix. Place on magnetic stand at room temperature for 5 min.
6. Remove and discard all of the supernatant and remove sample from the magnetic stand. Add 50 μL Elution Buffer and gently pipet the entire volume up and down to mix.
7. Incubate the sample at 80 °C for 2 min and then cool to 25 °C. This elutes the mRNA and any nonspecifically bound contaminant rRNA from the beads.
8. At room temperature, add 50 μL of Bead Binding Buffer to allow the mRNA to rebind to the beads and reduce the amount of rRNA that bound nonspecifically. Gently mix the entire volume up and down and incubate at room temperature for 5 min.

9. Place the sample on the magnetic stand for 5 min. Remove and discard all of the supernatant and remove the sample from the magnetic stand.
10. Wash beads by adding 200 μL of Bead Washing Buffer and gently pipet up and down to mix. Place on magnetic stand at room temperature for 5 min.
11. Remove and discard all of the supernatant and remove sample from the magnetic stand.

    Add 19.5 μL of Fragment, Prime, Finish mix and gently pipet the entire volume up and down to mix. The Fragment, Prime, Finish mix contains random hexamers for reverse transcriptase priming and serves as the first strand cDNA synthesis reaction buffer.
12. Incubate the sample at 94 °C for 8 min and then cool to 4 °C. Place on magnetic stand and transfer 17 μL to a new tube.

3.9 cDNA Synthesis

1. Synthesize first strand cDNA. Add 8 μL of First Stand Synthesis Act D Mix and SuperScript II to the sample and gently pipet to mix. Incubate the sample at 25 °C for 10 min, 42 °C for 15 min, 70 °C for 15 min, and then cool to 4 °C. Proceed immediately to Second Strand cDNA synthesis.
2. Synthesize second strand cDNA. Add 5 μL of Resuspension Buffer to the sample, followed by 20 μL of Second Strand Marking Master Mix. Pipet gently to mix. Incubate at 16 °C for 1 h and then bring sample to room temperature.
3. Purify the sample using AMPure XP beads that have been equilibrated to room temperature for at least 30 min and mixed well. Add 90 μL of AMPure XP beads to the 50 μL of ds cDNA. Gently pipet up and down to mix and incubate at room temperature for 15 min.
4. Place the samples on a magnetic separator for 5 min, or until all beads are bound to the side of the tube and the solution has cleared. Remove and discard supernatant.
5. Add 200 of μL 80% ethanol without disturbing the beads. Incubate for 30 s and remove and discard all of the supernatant.
6. Repeat the 80% ethanol wash one time for a total of two washes.
7. Let the sample dry for approximately 15 min at room temperature and then remove from the magnetic stand. Add 20 μL of Resuspension Buffer and gently pipet up and down to resuspend.
8. Incubate at room temperature for 2 min. Place on magnetic stand for 5 min and transfer 17.5 μL supernatant to a new tube.

3.10 Adenylate 3′ Ends and Ligate Adapters

1. Add 12.5 μL of A-Tailing Mix to sample, gently pipet up and down to mix, and incubate at 37 °C for 30 min, 70 °C for 5 min, and then cool to 4 °C. Proceed immediately to Ligate Adapters.
2. Add 2.5 μL of Resuspension Buffer, 2.5 μL of Ligation Mix, and 2.5 μL of the appropriate RNA adapter index (*see* **Note 7**) to each sample. Gently pipet up and down 10 times to mix.
3. Incubate at 30 °C for 10 min.
4. Add 5 μL of Stop Ligation Buffer and gently pipet up and down to mix.
5. Purify the ligation reaction by adding 42 μL of well-mixed, room-temperature AMPure XP beads to the ligation reaction and mix by gently pipetting up and down 10 times. Incubate at room temperature for 15 min.
6. Place the samples on a magnetic separator for 5 min, or until all beads are bound to the side of the tube and the solution has cleared. Remove and discard supernatant.
7. Add 200 μL of 80% ethanol without disturbing the beads. Incubate for 30 s and remove and discard all of the supernatant.
8. Repeat the 80% ethanol wash one time for a total of two washes.
9. Let the sample dry for approximately 15 min at room temperature and then remove from the magnetic stand. Add 52.5 of μL Resuspension Buffer and gently pipet up and down to resuspend.
10. Incubate at room temperature for 2 min. Place on magnetic stand for 5 min or until liquid is clear. Transfer 50 μL of supernatant to a new tube being careful not to disturb the beads.
11. Vortex the AMPure XP beads well and add 50 μL to the sample. Mix well by gently pipetting up and down 10 times.
12. Incubate at room temperature for 15 min.
13. Place on magnetic stand for 5 min or until solution is clear, remove and discard supernatant.
14. Add 200 μL of 80% ethanol without disturbing the beads. Incubate for 30 s and remove and discard all of the supernatant.
15. Repeat the 80% ethanol wash one time for a total of two washes.
16. Allow the sample to dry for approximately 15 min at room temperature and then remove from the magnetic stand. Add 22.5 μL of Resuspension Buffer and gently pipet up and down to resuspend.

17. Incubate at room temperature for 2 min. Place on magnetic stand for 5 min or until liquid is clear. Transfer 20 μL of the supernatant to a new tube being careful not to disturb the beads.

3.11 Enrich DNA Fragments

1. Add 5 μL of PCR Primer Cocktail and 25 μL of PCR Master Mix to sample and gently pipet up and down to mix.
2. Perform PCR to enrich for DNA fragments with adapters on both ends. Incubate at 98 °C for 30 s and 15 cycles of 98 °C for 10 s, 60 °C for 30 s, 72 °C for 30 s, then followed with a final extension at 72 °C for 5 min.
3. Purify the PCR reaction by adding 47.5 μL of well-mixed AMPure XP beads to the ligation reaction and mix by gently pipetting up and down 10 times. Incubate at room temperature for 15 min.
4. Place the samples on a magnetic separator for 5 min, or until all beads are bound to the side of the tube and the solution has cleared. Remove and discard supernatant.
5. Add 200 μL of 80% ethanol without disturbing the beads. Incubate for 30 s and remove and discard all of the supernatant.
6. Repeat the 80% ethanol wash one time for a total of two washes.
7. Let the sample dry for approximately 15 min at room temperature and then remove from the magnetic stand. Add 32.5 μL of Resuspension Buffer and gently pipet up and down to resuspend.
8. Incubate at room temperature for 2 min. Place on magnetic stand for 5 min or until liquid is clear. Transfer 30 μL of supernatant to a new tube being careful not to disturb the beads.

3.12 Validate Library

1. Assess the size and distribution of the purified library on the 2100 Bioanalyzer using the DNA1000 assay.
2. Allow kit reagents to equilibrate to room temperature for 30 min before use. To prepare gel–dye mix, add 25 μL of DNA dye concentration (blue dot) to a DNA gel matrix vial (red dot), vortex well, spin down, and apply to a spin filter. Centrifuge at 2240 × *g* for 15 min. Protect from light and store at 4 °C. Use within 6 weeks of preparation.
3. Place a new chip on the priming station, load 9 μL prepared gel–dye mix into well labeled "G" (third row, black circle). Close the priming station and depress the plunger until it engages with the clip. After 60 s, release the clip, wait 5 s, and then slowly pull back the plunger to the 1 mL position.

4. Add 9 μL of prepared gel–dye mix into each of the wells marked G (rows 1 and 2).
5. Pipet 5 μL of the supplied marker (green dot) in all sample and ladder wells. Do not leave any wells empty.
6. Add 1 μL of DNA ladder (yellow dot) to the ladder well and 1 μL of purified sample library to each sample well.
7. Place the loaded chip into the supplied vortex mixer and mix for 1 min. The chip must be run within 5 min.
8. Read samples using the integrated DNA 1000 series algorithm. The final library product should be a prominent band at approximately 260 bp (Fig. 3).

3.13 Library Quantification

1. Dilute sample library to a final dilution of 1:1,000,000 in 10 mM Tris–HCl.
2. Prepare qPCR master mix by combining 12 μL of KAPA SYBR Fast qPCR Master Mix (2×) + Primer Premix (10×) with

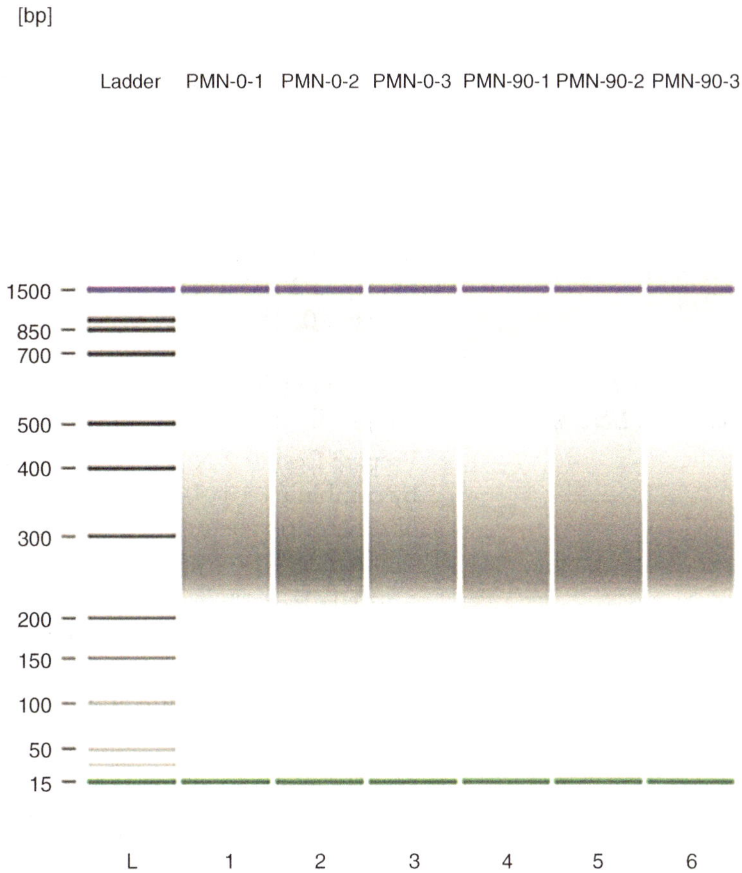

Fig. 3 Analysis of the distribution of TruSeq stranded mRNA sample library size using an Agilent 2100 Bioanalyzer. The final library should have an average molecular size distribution around 260 bp

4.0 μL of PCR-grade water. Prepare enough master mix for three replicates each of six standards, one no-template control, and diluted sample library.

3. Dispense 16 μL of qPCR master mix into each well of a 96-well PCR plate and add 4 μL of diluted standard, sample library, or dilution buffer (no-template control). Seal and transfer to qPCR instrument and perform the following cycling protocol: Incubate at 95 °C for 5 min then 35 cycles of 95 °C for 30 s, 60 °C for 45 s.

4. Use the instrument software to generate the standard curve and ensure that the average ΔC_q value between DNA standards is in the range of 3.1–3.6 and the calculated efficiency is in the range of 90–110% (i.e., PCR product has increased 1.8 to 2.2-fold per cycle and the slope of the standard curve it between −3.1 and −3.6).

5. Calculate the concentration of the sample library by converting the average C_q value for each sample library to a concentration (in pM) using the standard curve. Determine the average size of the sample library using the Bioanalyzer 2100 software region table tab to create a region on the size distribution (Fig. 4). The concentration of each sample library should be size adjusted by multiplying the calculated average concentration with the following factor: 452 (size of DNA standard)/ 312 (average fragment length of library in bp).

6. Multiply the size-adjusted concentration with the dilution factor to determine the stock library concentration. Dilute sample library to 2 nM in 10 mM Tris–HCL, pH 8.5 containing 0.1% Tween 20.

3.14 Denature and Dilute Libraries

1. For pooling libraries, combine an equal amount of each 2.0 nM sample library into one tube so that the volume is >15.0 μL.

2. Prepare fresh 0.2 N NaOH. Denature the sample library/pool by combining 10 μL of 0.2 N NaOH with 10 μL of 2 nM stock library/pool, vortex briefly, and incubate at room temperature 5 min.

3. Add 10 μL of 200 mM Tris–HCl, pH 7 and vortex briefly. Add 970 μL of prechilled HT1 to the denatured sample library/ pool, vortex briefly, and place sample on ice until ready to proceed to final dilution and sequencing. The denatured sample library/pool is now at 20 pM.

4. Prepare 2 nM of PhiX control library by combining 5 μL of 10 nM PhiX with 20 μL of RSB. Vortex briefly to mix. Store at −20 °C for up to 3 months.

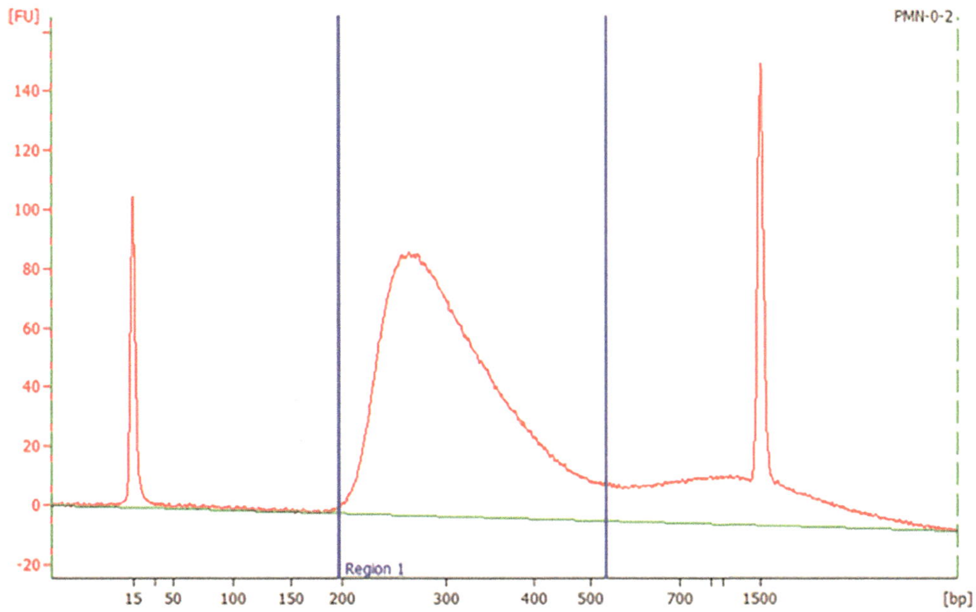

Fig. 4 Electropherogram profile used to determine average fragment size of the library using the region table feature of the Bioanalyzer 2100 software

5. Denature PhiX by combining 5 μL of 2 nM PhiX with 5 μL of freshly diluted 0.2 N NaOH. Vortex briefly and incubate at room temperature for 5 min.
6. Add 5 μL of 200 mM Tris–HCl, pH 7.0, vortex briefly and centrifuge.
7. Add 985 μL of prechilled HT1 to the tube of denature PhiX to achieve a final volume of 1.0 mL at 20 pM.
8. Dilute the 20 pM denatured PhiX to 1.8 pM by adding 117 μL of PhiX to 1183 μL of prechilled HT1. Invert to mix, and store on ice (can be aliquoted and stored at −20 °C for up to 3 months.)

3.15 Sequence Libraries on NextSeq 550

1. Thaw reagent cartridge (~ 1 h) in a room temperature water bath (do not submerge the cartridge). Invert several times to mix and inspect to ensure all reagents are thawed. Gently tap on bench to dislodge water from base and reduce air bubbles and store at 4° C until ready to use (within 1 week of thawing).
2. Remove flow cell from 4 °C storage and allow unwrapped package to equilibrate at room temperature for 30 min. Remove flow cell from package and clean glass surface with lint-free alcohol rub and dry with lint-free tissue (may not be necessary is surface is clean).
3. Dilute 20 pM denatured sample library/pool to 1.8 pM final concentration by combining 117 μL of sample with 1183 μL of prechilled HT1. Add 1.3 μL of denatured 1.8 pM PhiX control library to the 1.3 mL final volume.

4. Vortex gently to mix, pulse-centrifuge, and store on ice until ready to load into reagent cassette.
5. Clean the foil seal covering the library sample reservoir #10 with a low-lint tissue and pierce the seal using a clean 1.0 mL pipette tip. Load 1.3 mL of prepared library into the reservoir.
6. Create sample sheet using Illumina Experiment Manager software. Select appropriate instrument from the Instrument Selection menu and enter sample information following the prompts for workflow parameters, including read type (paired end) and number of cycles for reads1 (75 cycles) and read 2 (75 cycles).
7. Set up the sequencing run following the on-screen instructions on the sequencing instrument. Instructions for loading the flow cell, reagent cassette, and buffer cassette are provided.

4 Notes

1. The one-cycle target labeling kit contains reagents for cDNA synthesis, IVT labeling, sample cleanup and control reagents. Reagents are provided for 30 reactions.
2. Do not exceed 10^7 PMNs per Qiagen RNA mini-column. Numbers in excess do not increase RNA quantity and effectively reduce purity of final RNA. Larger quantities of total RNA can be obtained by pooling multiple preparations or, alternatively, using larger RNA purification columns.
3. The second purification of PMN total RNA on RNeasy columns is essential for obtaining high-quality RNA. Omission of this step may lead to incomplete cDNA synthesis and subsequent poor (5′:3′) ratios of labeled target cRNA.
4. The optimal excitation and emission spectra of the RiboGreen fluorophore should be empirically determined on the fluorometer to ensure that sample readings remain in the detection range.
5. Residual DNA from the PMN total RNA preparations will contribute to the overall signal obtained in the RiboGreen assay. We have found that the levels of contaminating DNA present in the RNA samples do not affect the overall yield of labeled target cRNA. Contaminating DNA should be removed for NGS experiments (Subheading 3.2). The concentration of cRNA will also be obtained in the determination of RNA quality (Subheading 3.4) and can be used to verify the RiboGreen results.
6. Unless stated otherwise, the incubation steps for all cDNA and target labeling reactions are performed in a thermocycler.

7. Refer to Illumina's Index Adapter Pooling Guide (Document # 1000000041074 v06) for recommendations on pooling multiple samples for sequencing.

Acknowledgments

This work was supported by the Intramural Program of the National Institutes of Allergy and Infectious Diseases, NIH.

References

1. Kasprisin DO, Harris MB (1977) The role of RNA metabolism in polymorphonuclear leukocyte phagocytosis. J Lab Clin Med 90:118–124
2. Chang FY, Shaio MF (1990) In vitro effect of actinomycin D on human neutrophil function. Microbiol Immunol 34:311–321
3. Kasprisin DO, Harris MB (1978) The role of protein synthesis in polymorphonuclear leukocyte phagocytosis II. Exp Hematol 6:585–589
4. Cline MJ (1966) Phagocytosis and synthesis of ribonucleic acid in human granulocytes. Nature 212:1431–1433
5. Cline MJ (1966) Ribonucleic acid biosynthesis in human leukocytes. Effects of phagocytosis on RNA metabolism. Blood 28:188–200
6. Jack RM, Fearon DT (1988) Selective synthesis of mRNA and proteins by human peripheral blood neutrophils. J Immunol 140:4286–4293
7. Newburger PE, Dai Q, Whitney C (1991) In vitro regulation of human phagocyte cytochrome b heavy and light chain gene expression by bacterial lipopolysaccharide and recombinant human cytokines. J Biol Chem 266:16171–16177
8. Newburger PE, Ezekowitz RA, Whitney C et al (1988) Induction of phagocyte cytochrome b heavy chain gene expression by interferon gamma. Proc Natl Acad Sci U S A 85:5215–5219
9. Itoh K, Okubo K, Utiyama H et al (1998) Expression profile of active genes in granulocytes. Blood 92:1432–1441
10. Subrahmanyam YV, Yamaga S, Prashar Y et al (2001) RNA expression patterns change dramatically in human neutrophils exposed to bacteria. Blood 97:2457–2468
11. Borjesson DL, Kobayashi SD, Whitney AR et al (2005) Insights into pathogen immune evasion mechanisms: *Anaplasma phagocytophilum* fails to induce an apoptosis differentiation program in human neutrophils. J Immunol 174:6364–6372
12. Kobayashi SD, Braughton KR, Palazzolo-Ballance AM et al (2010) Rapid neutrophil destruction following phagocytosis of *Staphylococcus aureus*. J Innate Immun 2:560–575
13. Kobayashi SD, Braughton KR, Whitney AR et al (2003) Bacterial pathogens modulate an apoptosis differentiation program in human neutrophils. Proc Natl Acad Sci U S A 100:10948–10953
14. Kobayashi SD, Voyich JM, Braughton KR et al (2003) Down-regulation of proinflammatory capacity during apoptosis in human polymorphonuclear leukocytes. J Immunol 170:3357–3368
15. Kobayashi SD, Voyich JM, Braughton KR et al (2004) Gene expression profiling provides insight into the pathophysiology of chronic granulomatous disease. J Immunol 172:636–643
16. Kobayashi SD, Voyich JM, Buhl CL et al (2002) Global changes in gene expression by human polymorphonuclear leukocytes during receptor-mediated phagocytosis: cell fate is regulated at the level of gene expression. Proc Natl Acad Sci U S A 99:6901–6906
17. Kobayashi SD, Voyich JM, Somerville GA et al (2003) An apoptosis-differentiation program in human polymorphonuclear leukocytes facilitates resolution of inflammation. J Leukoc Biol 73:315–322
18. Kobayashi SD, Voyich JM, Whitney AR et al (2005) Spontaneous neutrophil apoptosis and regulation of cell survival by granulocyte macrophage-colony stimulating factor. J Leukoc Biol 78:1408–1418
19. Holland SM, DeLeo FR, Elloumi HZ et al (2007) STAT3 mutations in the hyper-IgE syndrome. N Engl J Med 357:1608–1619
20. Theilgaard-Monch K, Jacobsen LC, Borup R et al (2005) The transcriptional program of

terminal granulocytic differentiation. Blood 105:1785–1796
21. Theilgaard-Monch K, Knudsen S, Follin P et al (2004) The transcriptional activation program of human neutrophils in skin lesions supports their important role in wound healing. J Immunol 172:7684–7693
22. Fessler MB, Malcolm KC, Duncan MW et al (2002) Lipopolysaccharide stimulation of the human neutrophil: an analysis of changes in gene transcription and protein expression by oligonucleotide microarrays and proteomics. Chest 121:75S–76S
23. Tsukahara Y, Lian Z, Zhang X et al (2003) Gene expression in human neutrophils during activation and priming by bacterial lipopolysaccharide. J Cell Biochem 89:848–861
24. Kluger Y, Tuck DP, Chang JT et al (2004) Lineage specificity of gene expression patterns. Proc Natl Acad Sci U S A 101:6508–6513
25. Zhang X, Kluger Y, Nakayama Y et al (2004) Gene expression in mature neutrophils: early responses to inflammatory stimuli. J Leukoc Biol 75:358–372
26. Kobayashi SD, DeLeo FR (2009) Role of neutrophils in innate immunity: a systems biology-level approach. Wiley Interdiscip Rev Syst Biol Med 1:309–333
27. Nauseef WM (2014) Isolation of human neutrophils from venous blood. Methods Mol Biol 1124:13–18
28. Siemsen DW, Malachowa N, Schepetkin IA et al (2014) Neutrophil isolation from nonhuman species. Methods Mol Biol 1124:19–37
29. Kobayashi SD, Sturdevant DE, DeLeo FR (2007) Genome-scale transcript analyses in human neutrophils. Methods Mol Biol 412:441–453

Part V

NADPH Oxidase and Production of Reactive Oxygen Species

Chapter 22

Measurement of Respiratory Burst Products, Released or Retained, During Activation of Professional Phagocytes

Claes Dahlgren, Halla Björnsdottir, Martina Sundqvist, Karin Christenson, and Johan Bylund

Abstract

Activation of professional phagocytes, potent microbial killers of our innate immune system, is associated with an increased cellular consumption of molecular oxygen (O_2). The O_2 molecules consumed are reduced by electrons delivered by a membrane localized NADPH-oxidase that initially generate one- and two electron reduced superoxide anions (O_2^-) and hydrogen peroxide (H_2O_2), respectively. These oxidants can then be processed into other highly reactive oxygen species (ROS) that can kill microbes, but that may also cause tissue destruction and drive other immune cells into apoptosis. The development of basic techniques to measure and quantify ROS generation by phagocytes is of great importance, and a large number of methods have been used for this purpose. A selection of methods (including chemiluminescence amplified by luminol or isoluminol, absorbance change following reduction of cytochrome c, and fluorescence increase upon oxidation of PHPA) are described in detail in this chapter with special emphasis on how to distinguish between ROS that are released extracellularly, and those that are retained within intracellular organelles. These techniques can be valuable tools in research spanning from basic phagocyte biology to diagnosis of diseases linked to the NADPH-oxidase and more clinically oriented research on innate immune mechanisms and inflammation.

Key words Reactive oxygen species, Superoxide anions, Hydrogen peroxide, Myeloperoxidase, Intracellular NADPH-oxidase activity, Plasma membrane NADPH-oxidase activity, Subcellular granules, Chemiluminescence, Cytochrome c reduction, PHPA oxidation

1 Introduction

Professional phagocytes of the innate immune system increase their consumption of oxygen (O_2) during phagocytosis of microbial intruders or upon interaction with certain inflammatory mediators. The electron transport system responsible for O_2 consumption, the NADPH-oxidase, is essential for protection against invading pathogens, as demonstrated by the susceptibility of individuals with a nonfunctional oxidase (e.g., patients suffering from chronic granulomatous disease; CGD) to bacterial and fungal infections (*see* **Note 1** [1, 2]). The NADPH-oxidase is of importance also for the ability

Mark T. Quinn and Frank R. DeLeo (eds.), *Neutrophil: Methods and Protocols*, Methods in Molecular Biology, vol. 2087, https://doi.org/10.1007/978-1-0716-0154-9_22,

to properly regulate inflammatory signaling, as demonstrated by the pronounced inflammatory symptoms associated with CGD [3]. Superoxide (O_2^-), the reduced oxygen metabolite generated by the NADPH-oxidase, serves as a precursor for formation of hydrogen peroxide (H_2O_2) and other secondary reactive oxygen species (ROS) that can be measured following activation. This induced ROS production is accompanied by a marked increase in cellular O_2 consumption and the process is therefore often referred to as the "respiratory burst" (*see* **Note 2**). Among the human phagocytes, neutrophils are especially competent to produce ROS, and the basic structure of the phagocyte NADPH-oxidase has been uncovered primarily through studies of neutrophils from patients with CGD (*see* **Note 1**). In resting phagocytes of healthy individuals, the cytosolic components of the oxidase are separated from the membrane bound components, and during phagocyte activation the cytosolic and membrane components assemble to form a functional multicomponent electron-transfer system which catalyzes the reduction of molecular O_2 at the expense of cytosolic NADPH [4]. Although the intracellular signals responsible for assembly of the active oxidase in many respescts remain unknown, electrons are ferried from the cytosol across a membrane and delivered to O_2 present in an intracellular compartment (e.g., a phagosome) or extracellularly [5, 4].

A logic link between phagocyte ROS production and microbial killing was established when it was demonstrated that human phagocytes generate O_2^- [6] and that this process was lacking in phagocytes isolated from highly immunocompromised CGD patients [7, 8]. Even though decades of research in this area have passed, a detailed mechanistical explanation for how ROS participate in the killing of microbes is still lacking ([9–11] *see* **Note 3**). It is for example not entirely established which particular ROS that kill microbes; O_2^- and H_2O_2 are not sufficiently reactive to account for the bactericidal effects (*see* **Notes 3** and **4**). However, these oxidants can be metabolized into other ROS with stronger antimicrobial effects, most notably those that are formed by the granule localized enzyme myeloperoxidase (MPO) (*see* **Note 5**). It has also been claimed that the actual ROS are not directly responsible for bacterial killing but the primary role of the NADPH-oxidase is instead to provide optimal conditions for other (non-ROS) killing molecules [12]. Clearly there is still much to learn regarding microbial killing by phagocytes and the role of the NADPH-oxidase in this process.

Assembly of the neutrophil NADPH-oxidase at cellular membranes can result in the release of ROS to the extracellular milieu or intracellularly into a phagosome. In addition, it is clear that neutrophils also are able to produce substantial amounts of intracellular ROS in the total absence of phagosome formation (reviewed in [5, 13]). In neutrophils, only a fraction of the flavocytochrome b (the membrane component of the oxidase) is localized in the plasma membrane, whereas 80–85% is found in membranes of the peroxidase-negative granules [14, 15], suggesting that NADPH-oxidase activation in granule membranes are central for the generation of intracellular ROS [16, 17]. Interestingly, the molecular make-up of the NADPH-oxidase complexes responsible for extracellular and intracellular ROS production appears to differ and specific defects in intracellular ROS production have recently been described for neutrophils derived from patients lacking the cytosolic NADPH-oxidase component $p40^{phox}$ (*see* **Note 6**). Neutrophils from such $p40^{phox}$ deficient CGD patients lack the ability to form intracellular ROS, whereas their capacity to form extracellular ROS seems rather normal ([13, 18, 19], *see* **Notes 1** and **6**). In addition, neutrophils from patients with the inflammatory syndrome SAPHO (synovitis, acne, pustulosis, hyperostosis, osteitis) has been reported to displaye severely decreased intracellular ROS production while extracellular ROS production was intact [20, 21]. However, this defect was not found in other SAPHO patients and is thus not a general feature of SAPHO [22].

Different agonists activate the two pools of NADPH-oxidase (in the plasma membrane or in the granules) differently, suggesting that the signaling as well as the molecular mechanisms for regulation differ depending on the localization of the oxidase [23–27]. Not much is known about the signals leading to the differential activation of the two pools of the oxidase, but as indicated above it is clear that $p40^{phox}$ is needed only for intracellular activation [18, 19]. In addition, the dependency of PI-3-kinase activity as well as the PKC isotype specificity differs between the two responses [25, 28–30].

As will be clear from this chapter, it is not trivial to distinguish between neutrophil ROS generated extracellularly from those retained within intracellular locations (phagosomes or granules), and it is even more difficult to define the biological consequences of ROS production at these distinct subcellular locations. Judging by the clinical picture of classical CGD patients (the neutrophils of which lack both extra- as well as intracellular ROS) and $p40^{phox}$ CGD patients (the neutrophils of which specifically lack intracellular ROS), it seems that intracellular ROS are critical for controlling inflammatory reactions, possibly by curbing inflammatory signaling ([5] *see* **Notes 1** and **6**). One defined cellular process where intracellular ROS play a specific role is during the formation of neutrophil extracellular traps (NETs; *see* **Note** 7), a spectacular form of cell death by which phagocytes capture and eliminate microbes by casting out DNA decorated with nuclear and granular proteins [31, 32]. In vitro, neutrophils from classical as well as $p40^{phox}$ CGD fail to form NETs in response to PMA [19, 32] and the intracellular ROS that drive NET formation need to be processed by MPO [33] which may take place inside a vesicle that results from fusion between MPO containing (azurophilic) and flavocytochrome b containing (specific) granules [34].

Table 1
Techniques used for measuring cellular production of different reactive oxygen species

Technique	Measuring principle	Cellular localization	Comment
Superoxide anion			
Photometry	SOD-inhibitable reduction of cytochrome c	Extracellular	Easy to follow kinetics of the response, provided that the stimulus is nonparticulate; H_2O_2 may interfere with the assay; low sensitivity.
Luminometry	Peroxidase-dependent isoluminol-amplified chemiluminescence	Extracellular	High sensitivity; easy to follow kinetics of the response; detects O_2^- despite the requirement for a peroxidase.
	Lucigenin-amplified chemiluminescence	Extracellular	High sensitivity, but less than the isoluminol system; easy to follow kinetics of the response.
Precipitation reaction	NBT reduction	Intracellular	Simple to count the number of positive cells microscopically, but laborious to make quantitative; should include SOD and catalase to remove extracellular ROS.
Hydrogen peroxide			
Fluorometry	Peroxidase-dependent oxidation of PHPA ± azide	Extracellular Intracellular	Fluorescence increases, making kinetics easy to follow; SOD is required for conversion of O_2^- to H_2O_2; low sensitivity NaN_3 inactivates MPO and catalase, thus allowing H_2O_2 generated intracellularly to leak out and be detected extracellularly
	Peroxidase-dependent oxidation of Scopoletin ± azide	Extracellular Intracellular	Fluorescence decreases, making kinetics more difficult to follow; SOD is required for conversion of O_2^- to H_2O_2; higher sensitivity than the PHPA system. See above regarding NaN_3
Nonidentified oxygen radical			
Precipitation reaction	DAB oxidation	Intracellular	Simple to count the number of positive cells microscopically, laborious to make quantitative; should include SOD and catalase to remove extracellular ROS
Fluorometry	Oxidation of 2,7-dichlorofluorescein or dihydrorhodamine 123	Intracellular	Many different oxidants can change the fluorescence of these substrates, making it difficult to use the technique quantitatively; difficult to follow kinetics. Should include SOD and catalase to remove extracellular ROS that may otherwise leak back into cells
Luminometry	Peroxidase-dependent luminol-amplified chemiluminescence	Intracellular	Possibly detects O_2^-; high sensitivity; easy to follow kinetics; dependent on endogenous MPO; should include SOD and catalase to remove extracellularly produced metabolites

The importance of ROS production for the antimicrobial and pathophysiological activity of phagocytes makes it essential to have access to adequate techniques for their analysis. Furthermore, the ability to properly distinguish between extra- and intracellular ROS production is critical for the capacity to discriminate classical CGD from the variant $p40^{phox}$ CGD. A number of different techniques to monitor ROS production from phagocytes (detecting O_2^-, H_2O_2, or undefined oxidants) have been developed over the years, and they all have their individual advantages and limitations (Table 1). Although there is a need for better methods to measure the different types of ROS and to determine the precise subcellular site for their generation [35], we currently have to rely on the techniques that are at hand. To properly evaluate data obtained using the available methods it is instrumental to have a solid understanding of where, how and what the different assays measure. In this chapter we describe a selection of basic, easy to perform techniques that are frequently used to follow cellular production of ROS (i.e., O_2^-, H_2O_2, or undefined oxidants), not only in the extracellular milieu, but also in intracellular compartments.

2 Materials

1. 5 M NaOH stock solution (can be kept at room temperature protected from light).
2. Krebs-Ringer phosphate (KRG) buffer: 120 mM NaCl, 5 mM KCl, 1.7 mM KH_2PO_4, 8.3 mM Na_2HPO_4, 10 mM glucose, 1 mM $CaCl_2$, and 1.5 mM $MgCl_2$, pH 7.3.
3. 50 mM 5-amino-2,3-dihydro-1,4-phthalazinedione (Luminol) stock solution: prepare stock solution by dissolving luminol powder in freshly prepared 0.1 M NaOH. Prepare 500 μM luminol working solution fresh on the day of the experiment by diluting the 50 mM luminol stock solution 1:100 in KRG buffer (*see* **Note 8**).
4. 50 mM 6-amino-2,3-dihydro-1,4-phthalazinedione (Isoluminol) stock solution: prepare stock solution by dissolving isoluminol powder in freshly prepared 0.1 M NaOH. Prepare 500 μM isoluminol working solution fresh on the day of the experiment by diluting the 50 mM isoluminol stock solution 1:100 in KRG buffer (*see* **Note 8**).
5. Horseradish peroxidase (HRP) solution: 80 U/mL HRP in physiological saline. Store in small aliquots at −20 °C.
6. *p*-Hydroxyphenylacetate (PHPA) solution: 10 mg/mL PHPA in physiological saline. Store in small aliquots at −70 °C.
7. Superoxide dismutase (SOD) solution: 5000 U/mL SOD in physiological saline. Store in small aliquots at −70 °C.

8. Catalase solution: wash 200,000 U/mL particulate catalase twice by suspending in H_2O and centrifugation at 15,000 × *g* for 10 min. Dissolve the final pellet in physiological saline. Store in small aliquots at −70 °C (*see* **Note 9**).
9. Sodium azide (NaN_3) solution: 10 mM NaN_3 in KRG. NaN_3 solution can be stored at 4°C for several months.
10. 1.2 mM Cytochrome C (cytC) solution: 15 mg/mL cytC in KRG. Prepare solution fresh on the day of the experiment.
11. 5 mM 2-(4-iodophenyl)-3-(4-nitrophenyl)-5-(2,4-disulfophenyl)-2-tetrazolium, monosodium salt (WST-1; Dojindo Laboratories, Kumamoto, Japan) solution: dissolve 33 mg/mL WST-1 in KRG. Prepare solution fresh on the day of the experiment.
12. Dihydrorhodamine 123 (DHR123) solution: 5 mg/mL DHR123 in dimethyl sulfoxide (DMSO). Store in small aliquots at −20 °C. Prepare 10 μg/mL working solution fresh on the day of the experiment.
13. Diphenyleneiodonium chloride (DPI) solution: 1.2 mM DPI, a widely used inhibitor of NADPH-oxidase activity, in DMSO. This stock solution can be kept at room temperature protected from light for very long time. Prepare the working solution fresh on the day of the experiment by diluting the stock 1/100 in phosphate buffered saline (37 °C) during heavy mixing. Adding DPI to any of the measuring systems described here, at concentrations between 1 and 10 μM [36], inhibits the NADPH-oxidase activity by 50–95% (*see* **Note 10**).
14. Latrunculin A solution: 100 μg/mL latrunculin A in DMSO. Store in small aliquots at −20 °C. Prepare fresh stock solutions containing 250 ng/mL latrunculin A on the day of the experiment and use 15–25 ng/mL solution as a working solution to inhibit actin polymerization (*see* **Note 11**).
15. Neutrophils purified from peripheral human blood by standard techniques (or any cell that can generate ROS). Store the cells on melting ice until use.
16. Chemiluminometer; preferentially equipped with a temperature-controlled sample holder.
17. Photometer, preferentially equipped with a temperature-controlled cuvette holder.
18. Fluorometer, preferentially with a temperature-controlled cuvette holder.
19. Flow cytometer.

3 Methods

A large number of techniques have been developed over the years to measure the cellular production of different ROS (Table 1). Ideally, techniques for measuring cellular production of ROS should: (1) be specific for a particular oxygen metabolite, (2) be sensitive, (3) not interfere with cellular function, (4) distinguish the localization of the ROS production, (5) not require specialized laboratory equipment that is expensive or complicated to handle, and (6) be easy to standardize. Nevertheless, no single technique has hitherto been found that satisfies all of these criteria [35]. However, new and more selective (?) redox probes that could be valuable tools to measure phagocyte NADPH-oxidase activities are possibly on their way [37]. The drawbacks of the various techniques differ, and more than one technique usually has to be included in the methodological repertoire of oxygen radical scientists. Described in detail below are three techniques (luminol/isoluminol-amplified chemiluminescence, cytochrome C reduction, and PHPA-oxidation) that we routinely use to study respiratory burst activity in human phagocytes. These methods serve as valuable tools in both basic and clinically oriented research involving phagocyte function. We also describe DHR123 fluorescence since it is a widely used method, despite the fact that it suffers from a variety of drawbacks making it unsuitable for quantitative and kinetic studies (*see* Subheading 3.2.3).

3.1 Detection of Extracellular ROS

It is generally assumed that the NADPH-oxidase is assembled and activated either in the plasma membrane or in intracellular membranes such as the phagosomal membrane [9, 38]. The ROS generated will then either be released from the cells (activation in the plasma membrane) or be retained inside the phagocyte (activation in internal membranes). The techniques described in this section are suitable for the specific detection of ROS produced extracellularly.

3.1.1 Extracellular O_2^- Detection by Isoluminol-Enhanced Chemiluminescence

There are several dyes that after being excited by ROS release energy in the form of light (i.e., chemiluminescence). Among these dyes, the membrane-permeable luminol (Fig. 1) is the most intensively characterized and most frequently used in the free radical research field (*see* **Note 12** and Subheading 3.2.1). Luminol has an amino group in position 5 of the phthalate ring, and changing the position of the amino group in that ring does not change the ability of the molecule to detect ROS. However, moving the amino group away from the first carbon atom in the aromatic ring makes the molecule more hydrophilic and less able to pass over biological membranes. Hence, isoluminol (Fig. 1) can be used to exclusively measure ROS *released* from activated cells [39, 40]. Luminol and

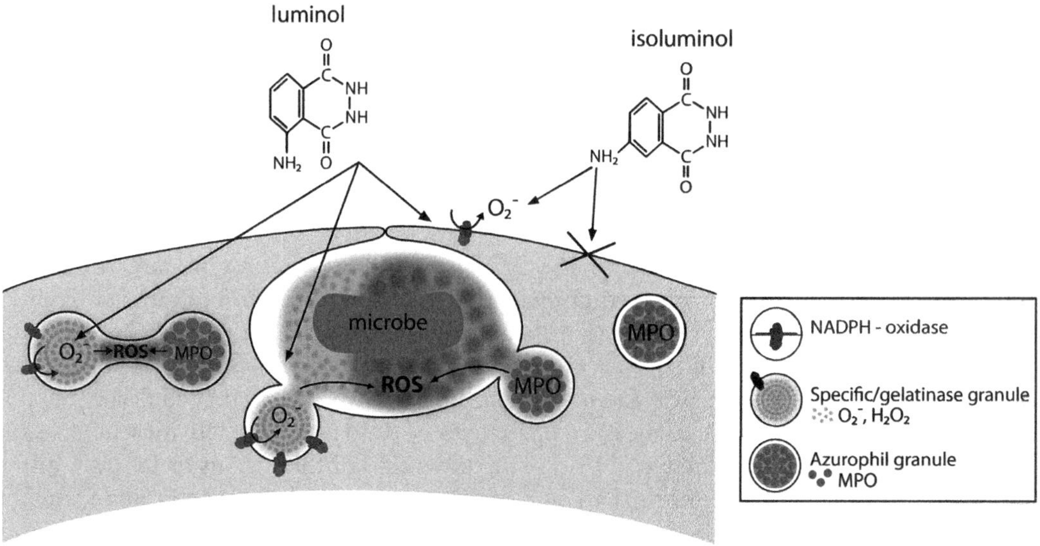

Fig. 1 Molecular structure and membrane permeability of luminol and isoluminol

isoluminol are activity amplifiers that decreases the ROS detection limits substantially, thus making these techniques very sensitive (*see* **Note 13**). The isoluminol system described here (as well as the luminol system described in Subheading 3.2.1) is a robust, highly sensitive technique that enables kinetic monitoring of phagocyte O_2^- production both in basic research, and for more clinically oriented investigations of phagocyte functions. While these methods provide relative measure of ROS, they do not quantify the exact amount of O_2^- present in the samples. There are some alternative substrates for the chemiluminescence reaction, but there are no strong arguments for using these in neutrophil respiratory burst measurements (*see* **Note 14**).

1. Add 650 μL of KRG (or 550 μL if latrunculin A is to be used) to Eppendorf tubes or luminometer tubes for each sample (*see* **Note 15**).
2. Add 100 μL of the 500 μM isoluminol solution.
3. Add 50 μL of HRP solution (*see* **Note 16**).
4. Add 100 μL of cells in KRG (10^2–5×10^6 cells, depending on the activity of the cells and the sensitivity of the luminometer; *see* **Note 17**).
5. Latrunculin A can be added (100 μL) if the response needs to be increased (*see* **Note 11**).
6. Equilibrate samples to the desired temperature.
7. Add 100 μL of a stimulus dissolved in KRG to activate the cells (*see* **Note 18**).

8. Measure chemiluminescence with an ordinary β-counter or one of the many luminometers present on the market (preferentially equipped with a temperature-controlled sample holder), including those adapted to microtiter plates (*see* **Notes 18** and **19**).

3.1.2 Spectrophotometric Determination of Extracellular O_2^- by Cytochrome C or WST-1

Various ROS can reduce a number of different substrates, and techniques that exploit the absorbance change of chromogenic substrates are commonly used to measure ROS production. Here we describe a simple and highly reproducible technique that relies on the O_2^--dependent reduction of the membrane-impermeable substrate cytochrome C (cytC). The same method can be used with other substrates, e.g., cytC can be replaced by the tetrazolium salt WST-1, without making any other changes. The reduction of cytC can be detected by a photometric change in absorbance at 550 nm, while WST-1 reduction is followed at 540 nm. As there is a one-to-one molar stoichiometry between the amount of O_2^- produced and the number of cytC molecules reduced, the actual amount of O_2^- produced can be quantified with this technique.

1. Prepare two cuvettes for each sample (i.e., a sample and a reference cuvette).
2. Add 700 μL of KRG to the sample cuvette and 690 μL of KRG to the reference cuvette (*see* **Note 20**).
3. Add 100 μL of cytC or WST-1 solution to both cuvettes.
4. Add 10 μL of SOD solution to the reference cuvette only (*see* **Note 20**).
5. Add 100 μL of cells ($\geq 5 \times 10^5$ cells) in KRG to both cuvettes.
6. Equilibrate both samples to the desired temperature in a temperature-controlled cuvette holder or water bath.
7. Add 100 μL of a stimulus dissolved in KRG to activate the cells (*see* **Note 21**).
8. Continuously monitor the change in absorbance at 550 nm (for cytC) in a spectrophotometer.
9. Determine the change in absorbance (ΔOD_{550} for cytC) for each sample by subtracting the absorbance of the reference cuvette containing SOD from that of the sample cuvette for each sample for each time point.
10. Calculate the molar amount of O_2^- generated per unit time and volume using the Beer-Lambert law with an extinction coefficient (ε) of 21.1 mM^{-1} cm^{-1} for reduced cytC at 550 nm. For a standard 1 mL assay containing 10^6 cells in a cuvette with a 1 cm light path, the amount of O_2^- generated can easily be calculated using Eq. (1) (*see* **Notes 22** and **23**):

$$\Delta OD \times 47.4 = \text{nmoles } O_2^-/10^6\text{cells/time unit} \qquad (1)$$

11. When comparing the two substrates, WST-1 generates a somewhat greater increase in absorbance, but this cannot be directly transferred to an increased sensitivity. With a diluted cell suspension, the sensitivity of the cytC reduction assay is actually somewhat greater than that of the WST-1 assay system (*see* **Note 24**).

3.1.3 Extracellular ROS Detection by Fluorescence

Membrane-impermeable PHPA is oxidized by H_2O_2 in the presence of HRP (which does not penetrate intact phagocytes), resulting in an increase in fluorescence. Extracellular production of H_2O_2 can subsequently be followed continuously using a fluorometer, including microtiter plate fluorometers.

1. Add 550 μL of KRG to Eppendorf tubes for each sample (*see* **Note 15**).
2. Add 100 μL of PHPA solution.
3. Add 50 μL of HRP solution.
4. Add 100 μL of SOD solution (*see* **Note 25**).
5. Add 100 μL of cells ($\geq 5 \times 10^5$ cells) in KRG.
6. Equilibrate samples to the desired temperature.
7. Add 100 μL of a stimulus dissolved in KRG to activate the cells (*see* **Note 18**).
8. The change in fluorescence emission at 400 nm with an excitation wavelength of 317 nm is measured.
9. Calculate the amount of H_2O_2 produced from a standard curve of the PHPA system, calibrated with different concentrations of H_2O_2 added to nonactivated cell samples (*see* **Note 26**).

3.2 Detection of Intracellular ROS

As mentioned above, the extracellular release of ROS represent only a fraction of the total ROS that neutrophils are capable of producing. Intracellular production of ROS in neutrophils is not restricted to phagosomes, indicating that a complete NADPH-oxidase can be assembled and activated also in flavocytochrome b-containing granule membranes (*see* **Note 27**). Thus, it is important to measure both intra- and extracellular ROS production by phagocytes treated with various activating stimuli. Here we describe some methods for selective measurement of intracellular ROS production by phagocytes treated with various activating stimuli.

3.2.1 Intracellular ROS Detection by Chemiluminescence

The membrane-permeable dye, luminol (Fig. 1), is excited by phagocyte-generated ROS, resulting in chemiluminescence. By adding membrane-impermeable enzymes (SOD and catalase) to reaction mixtures, extracellularly released O_2^- and H_2O_2 are removed, leaving ROS generated specifically in intracellular compartments to be measured (*see* **Note 13**).

1. Add 680 μL of KRG to Eppendorf tubes or luminometer tubes for each sample (*see* **Note 15**).
2. Add 100 μL of fresh 500 μM luminol solution.
3. Add 10 μL of SOD solution.
4. Add 10 μL of catalase solution.
5. Add 100 μL of cells in KRG (10^2–5×10^6 cells, depending on the activity of the cells and the sensitivity of the luminometer; *see* **Note 17**).
6. Equilibrate samples to the desired temperature.
7. Add 100 μL of a stimulus dissolved in KRG to activate the cells (*see* **Note 18**).
8. Measure chemiluminescence with a β-counter or luminometer, including those adapted to microtiter plates (*see* **Note 28**).

3.2.2 Intracellular H_2O_2 Detection by PHPA Fluorescence

The O_2^- generated inside an intracellular compartment cannot normally be determined extracellularly, simply because neutrophils lack the anion channels needed for O_2^- to pass biological membranes [41]. Similarly, H_2O_2 generated in an intracellular compartment cannot be measured extracellularly, but for other reasons. The H_2O_2 can pass biological membranes but it is rapidly consumed by endogenous peroxidases and catalase on its way from intracellular compartments to the plasma membrane. However, if endogenous MPO and catalase are inhibited (e.g., with NaN_3), the intracellularly produced H_2O_2 can leak out of the cell and be detected extracellularly by the PHPA technique described above (Subheading 3.1.3) [42, 43].

1. Prepare two Eppendorf tubes for each sample (i.e., a sample and a reference tube).
2. Add 500 μL of KRG to the sample tube and 600 μL of KRG to the reference tube (*see* **Note 15**).
3. Add 100 μL of PHPA solution.
4. Add 50 μL of HRP solution.
5. Add 100 μL of NaN_3 solution (final concentration 1 mM) to the reference sample only (*see* **Note 29**).
6. Add 100 μL of cells ($\geq 5 \times 10^5$ cells) in KRG.
7. Equilibrate samples to the desired temperature.
8. Add 100 μL of a stimulus dissolved in KRG to activate the cells (*see* **Note 18**).
9. Measure change in fluorescence emission at 400 nm with an excitation wavelength of 317 nm.

10. Calculate intracellular H_2O_2 generation by subtracting the fluorescence of samples lacking NaN_3 (extracellular H_2O_2) from that of samples containing NaN_3 (total H_2O_2).
11. Calculate the amount of H_2O_2 produced from a standard curve of the PHPA system, calibrated with different concentrations of H_2O_2 added to nonactivated cell samples (*see* **Note 26**).

3.2.3 Intracellular ROS Detection by DHR123

Several probes exist for the detection of cellular ROS by flow cytometry. One of the most widely used is dihydrorhodamine 123 (DHR123; *see* **Note 30**), which is a nonfluorescent molecule that readily diffuses across cell membranes. Once inside the cell, DHR123 can be oxidized by ROS to the highly fluorescent, cationic rhodamine 123, which can be measured by flow cytometry or fluorometrically. Using flow cytometry, the dye should theoretically measure mainly intracellular ROS, but extracellular H_2O_2 may also pass over the plasma membrane and react with the probe intracellularly [44], which complicates the interpretation of data (Fig. 2). Also, it is not clear which particular oxygen species that directly reacts with DHR123 to generate fluorescence [45], but the signal is partly dependent on cellular MPO (Fig. 2). Further, the technique is neither directly quantitative nor reliable for kinetic studies. Thus, DHR123 (and similar probes) is likely more useful as a generic redox sensitive probe (*see* **Note 31**), than for detailed studies of phagocyte respiratory burst products.

In the protocol described below, extracellular ROS are neutralized by the addition of SOD and catalase to the measuring system, and a comparison between the fluorescent signals obtained in the presence or absence of these antioxidants reveals that the relative contribution of extracellular ROS is quite significant (Fig. 2).

1. Add 100 μL of isolated neutrophils (5×10^5 to 10^6 cells/mL) in KRG to Eppendorf tubes or FACS tubes.
2. Add 1 μL of SOD solution (final concentration of 50 U/mL).
3. Add 1 μL of catalase solution (final concentration of 2000 U/mL).
4. Add 10 μL of fresh 10 μM DHR123 solution (*see* **Note 32**).
5. Incubate samples for 15 min at 37 °C.
6. Add 10 μL of stimulus dissolved in KRG to activate the cells (e.g., 50 nM PMA).
7. Incubate samples for an additional 15 min at 37 °C.
8. Add 300 μL ice cold KRG.
9. Keep tubes in darkness on ice and analyze directly by flow cytometry using an excitation/emission spectrum of 488/525 nm. Set up samples in duplicates, with one sample containing SOD and catalase. After gating away cellular debris,

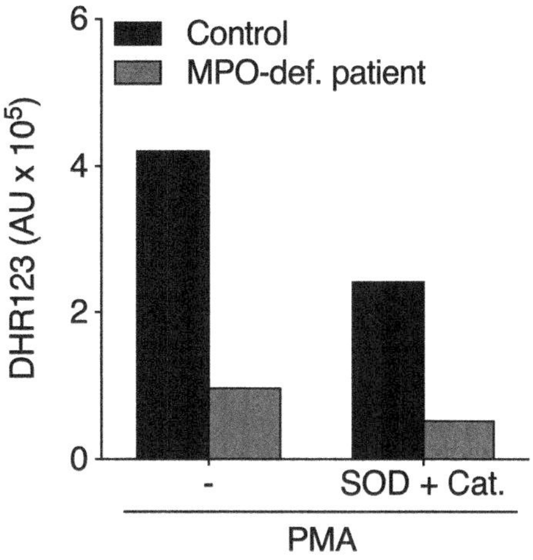

Fig. 2 Neutrophil derived ROS measured by DHR123. Human DHR123-stained neutrophils from one control (black) and one patient with total MPO-deficiency (grey) were stimulated with PMA for 15 min in the absence or presence of SOD and catalase (to scavenge extracellular ROS) before analyzed by flow cytometry. Neutrophils from the patient with MPO-deficiency displayed a considerably reduced fluorescence signal indicating that an intracellular peroxidase is required for DHR123 fluorescence to occur

record the geometric mean fluorescence intensity values of stimulated samples and compare to an unstimulated sample.

4 Notes

1. The dominating form of CGD is the result of mutations in the *CYBB* gene that encodes the membrane protein, gp91^{phox}. Such mutations are inherited in an X-linked recessive manner and thus mainly affect males. Less common are autosomal recessive mutations in the genes encoding p22^{phox} (that forms a membrane heterodimer with gp91^{phox}), or one of the proteins p47^{phox} or p67^{phox} that are cytosolic in resting cells. These represent cases of classical CGD, but mutations in additional genes may also result in other forms of the disease (*see* also **Note 6** below). CGD patients are not only overly susceptible to various infections, but they also commonly suffer from a variety of inflammatory complications indicative of a malfunction in the mechanisms that control inflammation. Importantly, the manifestations of CGD can be quite varied even with identical mutations, indicating that both genetic and environmental factors influence the progression and severity of clinical symptoms [46, 47].

2. The electrons transported by the NADPH-oxidase are generated by the hexose monophosphate shunt in the cytoplasm. The entire electron transporting enzyme is sometimes called NOX2 (NADPH-oxidase 2). This is not entirely correct since this name originally refers only to the larger of the two membrane-bound subunits ($gp91^{phox}$) encoded for by the *NOX2* gene (also called the *CYBB* gene).
3. The primary ROS produced by the NADPH-oxidase, O_2^-, is not very bactericidal per se, but its killing effect can be markedly increased by its transformation into H_2O_2, especially in the presence of MPO and halides [48], suggesting that the very reactive hypochlorous acid (HOCl) is responsible for microbial killing inside the phagosome [49, 50]. However, a number of observations challenge this view and we clearly do not yet have a complete picture of how NADPH-oxidase activity mediates microbial killing in the phagosome [13, 51–53].
4. Superoxide anion and H_2O_2 may be further reduced by a number of different cellular protection systems, including superoxide dismutase (SOD; catalyzes the reduction of O_2^- to H_2O_2), catalase (catalyzes the reduction of H_2O_2 to H_2O), and glutathione peroxidase (catalyzes the oxidation of glutathione by H_2O_2). The O_2^- and H_2O_2 can act as substrates for reactions with other cellular ROS (e.g., O_2^- reacts with nitric oxide to form the very reactive peroxynitrite molecule [54]) or cellular enzymes, such as MPO, that uses H_2O_2 in combination with chloride ions to generate the highly microbicidal HOCl [55].
5. Many of the various assays for detecting ROS are more or less dependent on the enzymatic activity of MPO (Table 1). This is important information to keep in mind since azide (NaN_3), a widely used preservative in antibodies and other biological reagents, is an MPO-inhibitor.
6. New genetic variants of CGD have recently been described [19, 56], and in one of these the cytoplasmic NADPH-oxidase subunit $p40^{phox}$ is affected [18, 19]. Neutrophils lacking $p40^{phox}$ secrete normal levels of ROS, whereas ROS production at intracellular sites is reduced or absent. Such cells are also unable to form neutrophil extracellular traps (NETs) when activated by PMA. In patients with this novel subtype of CGD ($p40^{phox}$ deficiency), the inflammatory manifestations typical for classical CGD are persistent, but the severity of infections seems milder.
7. Neutrophils may undergo a spectacular form of cell death in order to capture and eliminate microbes as an alternative strategy to classical phagocytosis [31]. This violent form of cell death, called NETosis, is distinct from necrosis and is a process

by which neutrophils release their nuclear material and the DNA becomes decorated with a large number of proteins originating from the different neutrophil storage granules. Originally, NET formation was described as an antibacterial defense strategy whereby neutrophils may entangle and kill extracellular microbes. Since then the regulatory potential of NETs as modulators of inflammation has become evident from a large number of studies of noninfectious diseases such as SLE, psoriasis, small vessel vasculitis, rheumatoid arthritis, gout, venous thrombosis, and cardiovascular diseases. In some of these conditions of aseptic inflammation, the autoantibodies used to diagnose patients are often directed toward components present in NETs [57–59].

8. Luminol and isoluminol stock solutions can be stored at room temperature, protected from light, for several months.
9. It should be noticed that many catalase batches contain rather high levels of endotoxin.
10. The most widely used pharmacological inhibitor of NADPH-oxidase activity, DPI, effectively blocks NADPH-oxidase-derived ROS formation but it is in fact a general inhibitor of flavoproteins and as such it is not specific for the phagocyte NADPH-oxidase. Accordingly, DPI also affects a number of other cellular systems, and it was in fact originally described as an inhibitor of mitochondria by blocking the flavin components of the electron transport chain [60]. Thus, the use of DPI for the purpose of claiming that the ROS detected in a cellular system originate from the activated NADPH-oxidase is not conclusive. There are, however, possibly new inhibitors on their way [61], and recently a small molecule named GSK2795039 was introduced and shown to be a quite specific inhibitor for flavocytochrome b, the membrane component of the phagocyte NADPH-oxidase [62].
11. Despite the fact that latrunculin A inhibits actin polymerization by a slightly different mechanism than cytochalasins, this drug can replace cytochalasin B and amplify and/or prolongate the NADPH-oxidase response induced by a large number of agonists for G-protein coupled receptors [63].
12. In order for luminol to react with oxidants in an intracellular compartment, it has to cross one or more biological membranes. Currently, very little is known about the diffusion properties of luminol or the diffusion limitations in different phagocyte membranes (plasma membrane, granule membranes, and/or phagolysosomal membranes), yet it is clear that the position of the amino group in the phthalate ring of the molecule is of prime importance (Fig. 1 and [40]).

13. The requirement for ROS in the luminol-dependent chemiluminescence reaction is shown by the absence of chemiluminescence obtained when using cells isolated from patients with CGD [40, 64–66]. In addition, neutrophils isolated from donors with a complete deficiency of MPO produce very low levels of chemiluminescence, despite a pronounced (in fact higher than normal) production of O_2^- and H_2O_2 (as measured with other techniques described here [33, 67]), demonstrating that MPO is critical in the chemiluminescence-producing reaction. The fact that addition of MPO to the extracellular fluid is not sufficient to fully regenerate the chemiluminescence activity of MPO-deficient cells to that of control cells suggests that intracellular MPO (stored in the azurophil granules) participates in the excitation of luminol by reacting with ROS generated in an intracellular compartment. This suggestion is further supported by the fact that part of the chemiluminescence response is insensitive to cell-impermeable extracellular scavengers of O_2^- and H_2O_2 [27]. Furthermore, we have recently shown that inhibition of phospholipase A2 (PLA2) can abrogate intracellular chemiluminescence specifically, without affecting intracellular ROS production or MPO activity [68]. This indicates that PLA2 activity is not needed for activation of the NADPH-oxidase per se, but rather contributes to the intracellular processing of ROS and facilitates fusion between MPO-containing (azurophil) granules and flavocytochrome b-containing (specific and/or gelatinase) granules. This study also highlights the value of combining more than one method for the measurements of ROS.
14. The luminol derivative, L-012, which is an even more sensitive chemiluminescence amplifier than luminol [69, 70], has been used in model systems of whole animals to measure ROS produced by inflammatory cells in vivo [71]. The basic measuring characteristics of L-012 are the same as for isoluminol (*see* Subheading 3.1.1), but the high light output makes it difficult to work with; it is e.g., necessary to decrease the exposure time in the photomultiplier. The background activity of L-012 is also very high, while the signal to noise ratio is basically the same as for luminol/isoluminol. Hence, there are no strong arguments for using L-012 instead of luminol/isoluminol as probe to measure phagocyte NADPH-oxidase activity.

    The chemiluminescence kit from National Diagnostics (Atlanta, GA, USA), called Diogenes Cellular Luminescence Enhancement System is a chemiluminescence enhancer of unknown origin that is "non-denaturing to living cells." Oxygen radicals generated by neutrophils after stimulation of cells with the chemoattractant fMLF can be followed using this kit and most of the activity is inhibited by SOD. The kit (used as suggested by the manufacturer) is however less sensitive than the HRP/isoluminol technique (*see* Subheading 3.1.1). Neutrophils activated by fMLF produce O_2^- mainly at the plasma membrane (extracellular release) but the enhancer manufactured by National Diagnostics can be used to measure also intracellular ROS formation in a way similar to the luminol/SOD/catalase system (Subheading 3.2.1). Given that this kit is less sensitive, much more expensive, and that it has a very limited shelf life as compared to the luminol/isoluminol systems described in this chapter, there are no strong arguments for using this kit to measure phagocyte NADPH-oxidase activity.

15. The methods described are for 1 mL assay volumes; however, they can easily be adapted to a microtiter plate format by adjusting the volumes accordingly. Based on a 200 μL assay volume for microtiter plate wells, the volumes for each step would then be reduced to a fifth of those used in the 1 mL assay. Microtiter plate assays are quite convenient and reduce sample and reagent requirements, as well as time required to measure signal readouts; *see* also **Note 18** below.
16. Although the light-generating reaction is peroxidase-dependent, the amount of H_2O_2 released from the cells does not affect the activity detected [40]. Peroxidases (MPO or HRP) catalyze the chemiluminescence reaction using mainly O_2^- and not H_2O_2, and samples can be analyzed in the presence and absence of SOD, in parallel, to directly determine the portion of the activity that reflects O_2^- production.
17. Future developments of new luminescent dyes that are more sensitive than luminol/isoluminol should allow measurements of respiratory burst activity in individual cells, or even measurements at the subcellular level.
18. The neutrophil respiratory burst can be activated by a number of different soluble and particulate stimuli, including chemoattractants, certain cytokines, phorbol esters, calcium ionophores, certain lectins, and various microorganisms (opsonized as well as unopsonized). All of these stimuli elicit a neutrophil chemiluminescence response in the presence of luminol [25, 26, 72–77]. Different stimuli trigger ROS production with markedly different kinetics, which is especially important to keep in mind when analyses are to be performed on microtiter plate-based assays; for slow ROS responses (induced by e.g., PMA stimulation as well as phagosomal production) it is often possible to add stimuli manually (with a multipipette) to wells of a microtiter plate. In contrast, in order to properly follow rapid responses (induced e.g., by

agonists for G-protein coupled receptors) a plate reader equipped with injectors is to prefer.

19. The relation between the amount of O_2^- produced and the amount of detected light or change in absorbance is dependent on the luminometer, but can easily be determined by a direct comparison with the SOD-inhibitable reduction of cytC [78] described in Subheading 3.1.2.

20. In order to specifically determine the amount of O_2^- produced, SOD is added to reference samples so that any part of the signal that is not inhibited by SOD (i.e., not due to O_2^-) can be subtracted.

21. The activity induced by soluble stimuli can easily be measured using cytC reduction, but when using chemoattractants the response induced in naive cells is usually very low. In order to increase the induced response, latrunculin A can be included in the measuring system (*see* **Note 11**). The cytC absorbance is stable, however when a particulate prey (e.g., bacteria) is used to activate the cells, a discontinuous assay system has to be used in which the samples are centrifuged (to remove the particles that could otherwise scatter the light) and O_2^- production is determined from the sample and reference supernatants.

22. The O_2^- production is often calculated and reported as the amount produced by 10^6 cells. If the number of cells in the sample differs from 10^6, then the equation will need to be adjusted by dividing by the actual number of cells in units of 10^6. For example, if 5×10^5 (0.5×10^6) or 1×10^5 (0.1×10^6) cells are in the sample, then the value obtained from Eq. (1) must be divided by 0.5 and 0.1, respectively, and so on.

23. If a microtiter plate format is used for the cytC assay, two parameters need to be changed in the formula: (1) the light path length will no longer be 1 cm and should be determined by measuring the distance from the bottom of the plate to the top of the sample liquid meniscus (usually about 0.75 cm) and (2) the assay volume is now 200 μL instead of 1 mL. As an example: O_2^- can be calculated for a 200 μL assay in a microtiter plate well with a 0.75 cm light path using Eq. (2):

$$\Delta OD_{550} \times 12.64 = \text{nmoles } O_2^-/\text{time unit} \quad (2)$$

This number can then be divided by the number of cells in the well to obtain O_2^- production per unit of cells (*see* **Note 22** above).

24. Accumulation of H_2O_2 in the measuring system (which can occur when a large proportion of the available cytC is reduced and the H_2O_2 consuming enzymes are inhibited) can result in reoxidation of the ferrous form of cytC, thus giving an underestimated result. For stoichiometric reasons, problematic levels

of substrate consumption are reached more rapidly in the cytC reduction assay than when WST-1 is used as substrate.

25. In order to reflect the release of ROS using PHPA (which specifically detects H_2O_2), SOD has to be added to the system to convert all O_2^- to H_2O_2.
26. The PHPA can be replaced by other fluorogenic substrates that, when oxidized by the peroxidase-H_2O_2 complex, changes its fluorescence properties. Scopoletin (7-hydroxycoumarin) is one such substrate that in its reduced form emits light at 460 nm when excited at 350 nm, whereas the oxidized form of Scopoletin loses its fluorescence [79]. The Scopoletin-assay system is more sensitive than the PHPA system but has the disadvantage that the fluorescence is lost in relation to H_2O_2 production, making the "measuring window" rather narrow. Amplex® Red (10-acetyl-3,7-dihydroxyphenoxazine) is another substrate that increases in fluorescence (excitation wavelength 590 nm; emission wavelength 645 nm) when oxidized [80]. The basic measuring characteristics of Amplex-red should be same as for PHPA, but the fluorescence output is higher. The background activity of Amplex-red is also higher, giving a signal to noise ratio that is basically the same as for PHPA.
27. Human neutrophils contain at least four types of granules that are mobilized (induced to fuse with the plasma membrane) hierarchically during in vivo extravasation of the cells from the blood stream to the tissue [81, 82]. Granule mobilization can also be induced by in vitro priming, e.g., short (ca. 20 min) treatment with TNFα. Apparently, the membranes of the mobilizable granules store most of the cellular content of flavocytochrome b and translocation of flavocytochrome b from the granules to the plasma membrane is associated with a reduction of the NADPH-oxidase activity in the granule fraction (as measured in a cell-free system using subcellular fractions from activated neutrophils) [16].
28. The light-generating reaction is dependent on the endogenous peroxidase MPO. It is possible for MPO to use O_2^- as well as H_2O_2 in the light-generating reaction, but the actual oxygen metabolite(s) measured intracellularly, has not been identified. It is not possible to directly quantify the amounts of O_2^- produced with this technique, and a change in luminescence could be due to either a change in O_2^-/H_2O_2 production, a change in the availability or enzyme activity of MPO [68], or a change in the basic conditions (protein content; pH; ionic composition) in the intracellular compartment in which the reaction takes place. It should also be noted that luminol can in some situations inhibit the extracellular release of O_2^- [83],

suggesting that this dye should only be used to measure ROS generated in intracellular compartments and that the molecular basis for any detected responses should be verified with another technique. Interestingly, while giving rise to chemiluminescence, luminol effectively scavenges the ROS that is being measured and luminol thus blocks PMA-triggered NET formation as effectively as a cell permeable MPO inhibitor [33].

29. Even though NaN_3 inhibits the major H_2O_2-consuming enzymes (i.e., catalase and MPO) in neutrophils, intracellularly generated H_2O_2 may be consumed or scavenged by other routes. Thus, the calculated amount of intracellularly generated H_2O_2 will most likely be an underestimation of what is really produced.

30. An alternative probe that we have used, with very similar results, is H_2DCFDA. In the protocol described, DHR123 can be exchanged for H_2DCFDA (Molecular Probes/Invitrogen) at a final concentration of 1–10 μM. The excitation/emission spectrum of H_2DCFDA is 492–495/517–527 nm.

31. The DHR123 is by no means specific for NADPH-oxidase derived ROS, as evidenced by the fact that non-phagocytes [84] give rise to DHR123 signals. The same is true also for CGD neutrophils, although in contrast to control neutrophils, these cells do not increase DHR123 signal after stimulation [85]. This lack of specificity for NADPH-oxidase derived ROS is shared by the probe H_2DCFDA (*see* **Note 30** above) which also gives a clear signal in CGD cells [66].

    A new rhodamine-based probe (R19-S) was recently introduced to specifically measure HOCl, a product generated by the MPO-H_2O_2-halide system, and the dye has been used to measure ROS following phagocytosis [86]. It should be noticed, however, that despite the fact that most of the activity should be present intracellularly (in the phagosome/phagolysosome), most of the fluorescence signal detected during uptake of zymosan particles, was recovered in cell supernatants. This suggests that the signal is the result of extracellularly produced H_2O_2 reacting with MPO released from the azurophil granules [86], and this is possibly due to limited membrane permeability of the probe.

32. For the protocol described, a final concentration of DHR123 between 1 and 10 μg/mL has been used with similar results.

Acknowledgments

This work was supported by the Swedish Research Council, the Swedish Society for Medical Research, the IngaBritt and Arne

Lundberg Research Foundation, the Swedish state under the LUA-ALF and TUA agreements, the Swedish Heart- and Lung Foundation, and the King Gustaf V Memorial Foundation. We thank Maria Hjulström and Hülya Çevik-Aras for performing chemiluminescence determinations with the National diagnostic kit and L-012, respectively.

References

1. Quie PG, White JG, Holmes B et al (1967) In vitro bactericidal capacity of human polymorphonuclear leukocytes: diminished activity in chronic granulomatous disease of childhood. J Clin Invest 46:668–679
2. Segal AW (2005) How neutrophils kill microbes. Annu Rev Immunol 23:197–223
3. Rieber N, Hector A, Kuijpers T et al (2012) Current concepts of hyperinflammation in chronic granulomatous disease. Clin Dev Immunol 2012:252460
4. Quinn MT, Gauss KA (2004) Structure and regulation of the neutrophil respiratory burst oxidase: comparison with nonphagocyte oxidases. J Leukoc Biol 76:760–781
5. Bylund J, Brown KL, Movitz C et al (2010) Intracellular generation of superoxide by the phagocyte NADPH oxidase: how, where, and what for? Free Radic Biol Med 49:1834–1845
6. Babior BM, Kipnes RS, Curnutte JT (1973) Biological defense mechanisms. The production by leukocytes of superoxide, a potential bactericidal agent. J Clin Invest 52:741–744
7. Baehner RL, Murrmann SK, Davis J et al (1975) The role of superoxide anion and hydrogen peroxide in phagocytosis-associated oxidative metabolic reactions. J Clin Invest 56:571–576
8. Johnston RB Jr, Keele BB Jr, Misra HP et al (1975) The role of superoxide anion generation in phagocytic bactericidal activity. Studies with normal and chronic granulomatous disease leukocytes. J Clin Invest 55:1357–1372
9. Hampton MB, Kettle AJ, Winterbourn CC (1998) Inside the neutrophil phagosome: oxidants, myeloperoxidase, and bacterial killing. Blood 92:3007–3017
10. Slauch JM (2011) How does the oxidative burst of macrophages kill bacteria? Still an open question. Mol Microbiol 80:580–583
11. Winterbourn CC, Kettle AJ (2013) Redox reactions and microbial killing in the neutrophil phagosome. Antioxid Redox Signal 18:642–660
12. Levine AP, Duchen MR, de Villiers S et al (2015) Alkalinity of neutrophil phagocytic vacuoles is modulated by HVCN1 and has consequences for myeloperoxidase activity. PLoS One 10:e0125906
13. Dahlgren C, Karlsson A, Bylund J (2019) Intracellular neutrophil oxidants: from laboratory curiosity to clinical reality. J Immonol 202 (11):3127–3134
14. Borregaard N, Heiple JM, Simons ER et al (1983) Subcellular localization of the b-cytochrome component of the human neutrophil microbicidal oxidase: translocation during activation. J Cell Biol 97:52–61
15. Hager M, Cowland JB, Borregaard N (2010) Neutrophil granules in health and disease. J Intern Med 268:25–34
16. Dahlgren C, Johansson A, Lundqvist H et al (1992) Activation of the oxygen-radical-generating system in granules of intact human neutrophils by a calcium ionophore (ionomycin). Biochim Biophys Acta 1137:182–188
17. Karlsson A, Dahlgren C (2002) Assembly and activation of the neutrophil NADPH oxidase in granule membranes. Antioxid Redox Signal 4:49–60
18. Matute JD, Arias AA, Wright NA et al (2009) A new genetic subgroup of chronic granulomatous disease with autosomal recessive mutations in p40 phox and selective defects in neutrophil NADPH oxidase activity. Blood 114:3309–3315
19. van de Geer A, Nieto-Patlan A, Kuhns DB et al (2018) Inherited p40phox deficiency differs from classic chronic granulomatous disease. J Clin Invest 128(9):3957–3975
20. Bjorkman L, Dahlgren C, Karlsson A et al (2008) Phagocyte-derived reactive oxygen species as suppressors of inflammatory disease. Arthritis Rheum 58:2931–2935
21. Ferguson PJ, Lokuta MA, El-Shanti HI et al (2008) Neutrophil dysfunction in a family with a SAPHO syndrome-like phenotype. Arthritis Rheum 58:3264–3269
22. Wekell P, Bjornsdottir H, Bjorkman L et al (2016) Neutrophils from patients with SAPHO syndrome show no signs of aberrant NADPH oxidase-dependent production of

intracellular reactive oxygen species. Rheumatology (Oxford) 55:1489–1498
23. Dahlgren C, Karlsson A (2002) Ionomycin-induced neutrophil NADPH oxidase activity is selectively inhibited by the serine protease inhibitor diisopropyl fluorophosphate. Antioxid Redox Signal 4:17–25
24. Karlsson A, Follin P, Leffler H et al (1998) Galectin-3 activates the NADPH-oxidase in exudated but not peripheral blood neutrophils. Blood 91:3430–3438
25. Karlsson A, Nixon JB, McPhail LC (2000) Phorbol myristate acetate induces neutrophil NADPH-oxidase activity by two separate signal transduction pathways: dependent or independent of phosphatidylinositol 3-kinase. J Leukoc Biol 67:396–404
26. Lock R, Dahlgren C, Linden M et al (1990) Neutrophil killing of two type 1 fimbria-bearing *Escherichia coli* strains: dependence on respiratory burst activation. Infect Immun 58:37–42
27. Serrander L, Larsson J, Lundqvist H et al (1999) Particles binding beta(2)-integrins mediate intracellular production of oxidative metabolites in human neutrophils independently of phagocytosis. Biochim Biophys Acta 1452:133–144
28. Brown GE, Stewart MQ, Liu H et al (2003) A novel assay system implicates PtdIns(3,4)P(2), PtdIns(3)P, and PKC delta in intracellular production of reactive oxygen species by the NADPH oxidase. Mol Cell 11:35–47
29. Kent JD, Sergeant S, Burns DJ et al (1996) Identification and regulation of protein kinase C-delta in human neutrophils. J Immunol 157:4641–4647
30. Sergeant S, McPhail LC (1997) Opsonized zymosan stimulates the redistribution of protein kinase C isoforms in human neutrophils. J Immunol 159:2877–2885
31. Brinkmann V, Reichard U, Goosmann C et al (2004) Neutrophil extracellular traps kill bacteria. Science 303:1532–1535
32. Brinkmann V (2018) Neutrophil extracellular traps in the second decade. J Innate Immun 10 (5–6):414–421
33. Bjornsdottir H, Welin A, Michaelsson E et al (2015) Neutrophil NET formation is regulated from the inside by myeloperoxidase-processed reactive oxygen species. Free Radic Biol Med 89:1024–1035
34. Bjornsdottir H, Welin A, Dahlgren C et al (2016) Quantification of heterotypic granule fusion in human neutrophils by imaging flow cytometry. Data Brief 6:386–393
35. Murphy MP, Holmgren A, Larsson NG et al (2011) Unraveling the biological roles of reactive oxygen species. Cell Metab 13:361–366
36. Hancock JT, Jones OT (1987) The inhibition by diphenyleneiodonium and its analogues of superoxide generation by macrophages. Biochem J 242:103–107
37. Kalyanaraman B, Hardy M, Podsiadly R et al (2017) Recent developments in detection of superoxide radical anion and hydrogen peroxide: opportunities, challenges, and implications in redox signaling. Arch Biochem Biophys 617:38–47
38. Kobayashi T, Robinson JM, Seguchi H (1998) Identification of intracellular sites of superoxide production in stimulated neutrophils. J Cell Sci 111(Pt 1):81–91
39. Dahlgren C, Karlsson A (1999) Respiratory burst in human neutrophils. J Immunol Methods 232:3–14
40. Lundqvist H, Dahlgren C (1996) Isoluminol-enhanced chemiluminescence: a sensitive method to study the release of superoxide anion from human neutrophils. Free Radic Biol Med 20:785–792
41. Halliwell B, Gutteridge JM (1985) The importance of free radicals and catalytic metal ions in human diseases. Mol Asp Med 8:89–193
42. Root RK, Metcalf J, Oshino N et al (1975) H2O2 release from human granulocytes during phagocytosis. I. Documentation, quantitation, and some regulating factors. J Clin Invest 55:945–955
43. Root RK, Metcalf JA (1977) H_2O_2 release from human granulocytes during phagocytosis. Relationship to superoxide anion formation and cellular catabolism of H_2O_2: studies with normal and cytochalasin B-treated cells. J Clin Invest 60:1266–1279
44. van Pelt LJ, van Zwieten R, Weening RS et al (1996) Limitations on the use of dihydrorhodamine 123 for flow cytometric analysis of the neutrophil respiratory burst. J Immunol Methods 191:187–196
45. Wardman P (2007) Fluorescent and luminescent probes for measurement of oxidative and nitrosative species in cells and tissues: progress, pitfalls, and prospects. Free Radic Biol Med 43:995–1022
46. Segal BH, Leto TL, Gallin JI et al (2000) Genetic, biochemical, and clinical features of chronic granulomatous disease. Medicine (Baltimore) 79:170–200
47. Segal BH, Veys P, Malech H et al (2011) Chronic granulomatous disease: lessons from a rare disorder. Biol Blood Marrow Transplant 17:S123–S131

48. Rosen H, Klebanoff SJ (1979) Bactericidal activity of a superoxide anion-generating system. A model for the polymorphonuclear leukocyte. J Exp Med 149:27–39
49. Weiss SJ, Klein R, Slivka A et al (1982) Chlorination of taurine by human neutrophils. Evidence for hypochlorous acid generation. J Clin Invest 70:598–607
50. Green JN, Kettle AJ, Winterbourn CC (2014) Protein chlorination in neutrophil phagosomes and correlation with bacterial killing. Free Radic Biol Med 77:49–56
51. Chapman AL, Hampton MB, Senthilmohan R et al (2002) Chlorination of bacterial and neutrophil proteins during phagocytosis and killing of *Staphylococcus aureus*. J Biol Chem 277:9757–9762
52. Kutter D, Devaquet P, Vanderstocken G et al (2000) Consequences of total and subtotal myeloperoxidase deficiency: risk or benefit ? Acta Haematol 104:10–15
53. Segal AW, Garcia RC, Harper AM et al (1983) Iodination by stimulated human neutrophils. Studies on its stoichiometry, subcellular localization and relevance to microbial killing. Biochem J 210:215–225
54. Beckman JS, Koppenol WH (1996) Nitric oxide, superoxide, and peroxynitrite: the good, the bad, and ugly. Am J Phys 271: C1424–C1437
55. Klebanoff SJ (2005) Myeloperoxidase: friend and foe. J Leukoc Biol 77:598–625
56. Arnadottir GA, Norddahl GL, Gudmundsdottir S et al (2018) A homozygous loss-of-function mutation leading to CYBC1 deficiency causes chronic granulomatous disease. Nat Commun 9:4447
57. Delgado-Rizo V, Martinez-Guzman MA, Iniguez-Gutierrez L et al (2017) Neutrophil extracellular traps and its implications in inflammation: an overview. Front Immunol 8:81
58. Jorch SK, Kubes P (2017) An emerging role for neutrophil extracellular traps in noninfectious disease. Nat Med 23:279–287
59. Van Avondt K, Hartl D (2018) Mechanisms and disease relevance of neutrophil extracellular trap formation. Eur J Clin Invest 48(Suppl 2):e12919
60. Holland PC, Sherratt HS (1972) Biochemical effects of the hypoglycaemic compound diphenyleneiodonnium. Catalysis of anion-hydroxyl ion exchange across the inner membrane of rat liver mitochondria and effects on oxygen uptake. Biochem J 129:39–54
61. Altenhofer S, Radermacher KA, Kleikers PW et al (2015) Evolution of NADPH oxidase inhibitors: selectivity and mechanisms for target engagement. Antioxid Redox Signal 23:406–427
62. Hirano K, Chen WS, Chueng AL et al (2015) Discovery of GSK2795039, a novel small molecule NADPH oxidase 2 inhibitor. Antioxid Redox Signal 23:358–374
63. Dahlgren C, Gabl M, Holdfeldt A et al (2016) Basic characteristics of the neutrophil receptors that recognize formylated peptides, a danger-associated molecular pattern generated by bacteria and mitochondria. Biochem Pharmacol 114:22–39
64. Bylund J, Campsall PA, Ma RC et al (2005) Burkholderia cenocepacia induces neutrophil necrosis in chronic granulomatous disease. J Immunol 174:3562–3569
65. Bylund J, MacDonald KL, Brown KL et al (2007) Enhanced inflammatory responses of chronic granulomatous disease leukocytes involve ROS-independent activation of NF-kappa B. Eur J Immunol 37:1087–1096
66. Sundqvist M, Christenson K, Bjornsdottir H et al (2017) Elevated mitochondrial reactive oxygen species and cellular redox imbalance in human NADPH-oxidase-deficient phagocytes. Front Immunol 8:1828
67. Dahlgren C, Stendahl O (1983) Role of myeloperoxidase in luminol-dependent chemiluminescence of polymorphonuclear leukocytes. Infect Immun 39:736–741
68. Bjornsdottir H, Granfeldt D, Welin A et al (2013) Inhibition of phospholipase A(2) abrogates intracellular processing of NADPH-oxidase derived reactive oxygen species in human neutrophils. Exp Cell Res 319:761–774
69. Imada I, Sato EF, Miyamoto M et al (1999) Analysis of reactive oxygen species generated by neutrophils using a chemiluminescence probe L-012. Anal Biochem 271:53–58
70. Nishinaka Y, Aramaki Y, Yoshida H et al (1993) A new sensitive chemiluminescence probe, L-012, for measuring the production of superoxide anion by cells. Biochem Biophys Res Commun 193:554–559
71. Kielland A, Blom T, Nandakumar KS et al (2009) In vivo imaging of reactive oxygen and nitrogen species in inflammation using the luminescent probe L-012. Free Radic Biol Med 47:760–766
72. Almkvist J, Dahlgren C, Leffler H et al (2002) Activation of the neutrophil nicotinamide adenine dinucleotide phosphate oxidase by galectin-1. J Immunol 168:4034–4041
73. Fu H, Belaaouaj AA, Dahlgren C et al (2003) Outer membrane protein a deficient

Escherichia coli activates neutrophils to produce superoxide and shows increased susceptibility to antibacterial peptides. Microbes Infect 5:781–788

74. Fu H, Karlsson J, Bylund J et al (2006) Ligand recognition and activation of formyl peptide receptors in neutrophils. J Leukoc Biol 79:247–256
75. Karlsson J, Fu H, Boulay F et al (2005) Neutrophil NADPH-oxidase activation by an annexin AI peptide is transduced by the formyl peptide receptor (FPR), whereas an inhibitory signal is generated independently of the FPR family receptors. J Leukoc Biol 78:762–771
76. Lundqvist H, Follin P, Khalfan L et al (1996) Phorbol myristate acetate-induced NADPH oxidase activity in human neutrophils: only half the story has been told. J Leukoc Biol 59:270–279
77. Thoren F, Romero A, Lindh M et al (2004) A hepatitis C virus-encoded, nonstructural protein (NS3) triggers dysfunction and apoptosis in lymphocytes: role of NADPH oxidase-derived oxygen radicals. J Leukoc Biol 76:1180–1186
78. Foyouzi-Youssefi R, Petersson F, Lew DP et al (1997) Chemoattractant-induced respiratory burst: increases in cytosolic Ca^{2+} concentrations are essential and synergize with a kinetically distinct second signal. Biochem J 322 (Pt 3):709–718
79. Boveris A, Martino E, Stoppani AO (1977) Evaluation of the horseradish peroxidase-scopoletin method for the measurement of hydrogen peroxide formation in biological systems. Anal Biochem 80:145–158
80. Mohanty JG, Jaffe JS, Schulman ES et al (1997) A highly sensitive fluorescent micro-assay of H_2O_2 release from activated human leukocytes using a dihydroxyphenoxazine derivative. J Immunol Methods 202:133–141
81. Faurschou M, Borregaard N (2003) Neutrophil granules and secretory vesicles in inflammation. Microbes Infect 5:1317–1327
82. Zarember KA, Kuhns DB (2011) Editorial: will the real neutrophil please stand up? J Leukoc Biol 90:1039–1041
83. Faldt J, Ridell M, Karlsson A et al (1999) The phagocyte chemiluminescence paradox: luminol can act as an inhibitor of neutrophil NADPH-oxidase activity. Luminescence 14:153–160
84. Leutner S, Schindowski K, Frolich L et al (2005) Enhanced ROS-generation in lymphocytes from Alzheimer's patients. Pharmacopsychiatry 38:312–315
85. Mauch L, Lun A, O'Gorman MR et al (2007) Chronic granulomatous disease (CGD) and complete myeloperoxidase deficiency both yield strongly reduced dihydrorhodamine 123 test signals but can be easily discerned in routine testing for CGD. Clin Chem 53:890–896
86. Albrett AM, Ashby LV, Dickerhof N et al (2018) Heterogeneity of hypochlorous acid production in individual neutrophil phagosomes revealed by a rhodamine-based probe. J Biol Chem 293:15715–15724

Chapter 23

Cell-Free NADPH Oxidase Activation Assays: A Triumph of Reductionism

Edgar Pick

Abstract

The superoxide ($O_2^{\cdot -}$)-generating NADPH oxidase complex of phagocytes comprises a membrane-associated heterodimeric flavocytochrome, known as cytochrome b_{558} (consisting of NOX2 and p22^{phox}) and four cytosolic regulatory proteins, p47^{phox}, p67^{phox}, p40^{phox}, and the small GTPase Rac. Under physiological conditions, in the resting phagocyte, $O_2^{\cdot -}$ generation is initiated by engagement of membrane receptors by a variety of stimuli, followed by signal transduction sequences leading to the translocation of the cytosolic components to the membrane and their association with the cytochrome, a process known as NADPH oxidase assembly. A consequent conformational change in NOX2 initiates the electron flow along a redox gradient, from NADPH to molecular oxygen (O_2), leading to the one-electron reduction of O_2 to $O_2^{\cdot -}$. Historically, methodological difficulties in the study of the assembled complex derived from stimulated cells, due to its lack of stability, led to the design of "cell-free" systems (also known as "broken cells" or in vitro systems). In a major paradigm shift, the cell-free systems have as their starting point NADPH oxidase components derived from resting (unstimulated) phagocytes, or as in the predominant method at present, recombinant proteins representing the components of the NADPH oxidase complex. In cell-free systems, membrane receptor stimulation and the signal transduction sequence are absent, the accent being placed on the actual process of assembly, all of which takes place in vitro. Thus, a mixture of the individual components of the NADPH oxidase is exposed in vitro to an activating agent, the most common being anionic amphiphiles, resulting in the formation of a complex between cytochrome b_{558} and the cytosolic components and $O_2^{\cdot -}$ generation in the presence of NADPH. Alternative activating pathways require posttranslational modification of oxidase components or modifying the phospholipid milieu surrounding cytochrome b_{558}. Activation is commonly quantified by measuring the primary product of the reaction, $O_2^{\cdot -}$, trapped immediately after its generation by an appropriate acceptor in a kinetic assay, permitting the calculation of rates of $O_2^{\cdot -}$ production, but numerous variations exist, based on the assessment of reaction products or the consumption of substrates. Cell-free assays played a paramount role in the identification and characterization of the components of the NADPH oxidase complex, the performance of structure–function studies, the deciphering of the mechanisms of assembly, the search for inhibitory drugs, and the diagnosis of various forms of chronic granulomatous disease (CGD).

Key words Reactive oxygen species, Superoxide, NADPH oxidase, Cytochrome b_{558}, NOX2, NOXes, Cytosolic components, p47^{phox}, p67^{phox}, Rac, Cell-free assays, Anionic amphiphile, Arachidonic acid, GTP, Kinetic assay, Superoxide dismutase, Prenylation, "Peptide walking"

Mark T. Quinn and Frank R. DeLeo (eds.), *Neutrophil: Methods and Protocols*, Methods in Molecular Biology, vol. 2087, https://doi.org/10.1007/978-1-0716-0154-9_23,

1 Introduction

Reductionism is a dirty word, and a kind of "holistier than thou" self-righteousness has become fashionable
—Richard Dawkins, The Extended Phenotype: The Gene as the Unit of Selection (1982).

1.1 The NOX2 NADPH Oxidase

Phagocytic cells (principally, neutrophils and macrophages) destroy engulfed bacterial, fungal, and protozoal pathogens by a number of effector mechanisms. Among these, reactive oxygen species (ROS) occupy a prominent position. ROS are derived from the primordial oxygen radical, superoxide ($O_2^{\cdot-}$), which is produced in response to appropriate engagement of membrane receptors by a tightly regulated enzyme complex, known as the $O_2^{\cdot-}$-generating phagocyte NADPH oxidase (briefly, "oxidase") (reviewed in [1]). The oxidase catalyzes the formation of $O_2^{\cdot-}$ by the NADPH-driven one-electron reduction of molecular oxygen (O_2). The functionally competent oxidase complex is composed of a membrane-associated flavocytochrome b_{558}, which is a heterodimer of two subunits (NOX2, also known as gp91^{phox}, and p22^{phox}), and four cytosolic components, p47^{phox}, p67^{phox}, p40^{phox}, and the small GTPase Rac1/2 (reviewed in [2–4]). An extensive review of the theoretical and methodological aspects of the NADPH oxidase complex was recently published in the form a book [5].

The only catalytic component of the oxidase is NOX2, a glycosylated protein of 570 amino acids, consisting of six transmembrane α-helices linked by three external and two cytosol-facing loops and a cytosolic segment, also known as the dehydrogenase region (DHR). NOX2 is harboring all redox stations carrying the flow of electrons from NADPH to O_2, namely, an NADPH-binding site and noncovalently bound FAD, both present in the DHR, and two hemes, bound to pairs of histidines present in the second and fifth transmembrane helices. In resting phagocytes, the components of the complex exist as distinct entities, oxidase activation being the consequence of the interaction of flavocytochrome b_{558} with cytosolic components, a process requiring translocation of the cytosolic components to the membrane environment of flavocytochrome b_{558}. This process involves a complex set of protein–protein and protein–lipid interactions leading to the formation of the assembled oxidase generating $O_2^{\cdot-}$ (reviewed in [6, 7]) (Fig. 1).

The dominant model for oxidase assembly envisages the decisive interaction being that of p67^{phox} with the DHR of NOX2, resulting in a conformational change in NOX2 that initiates the electron flow [9]. Because p67^{phox} does not possess an intrinsic membrane attachment domain of its own, it requires the assistance of p47^{phox} and Rac, to be brought in contact with NOX2 [10–12]. The roles of p47^{phox} and Rac in assisting the association of

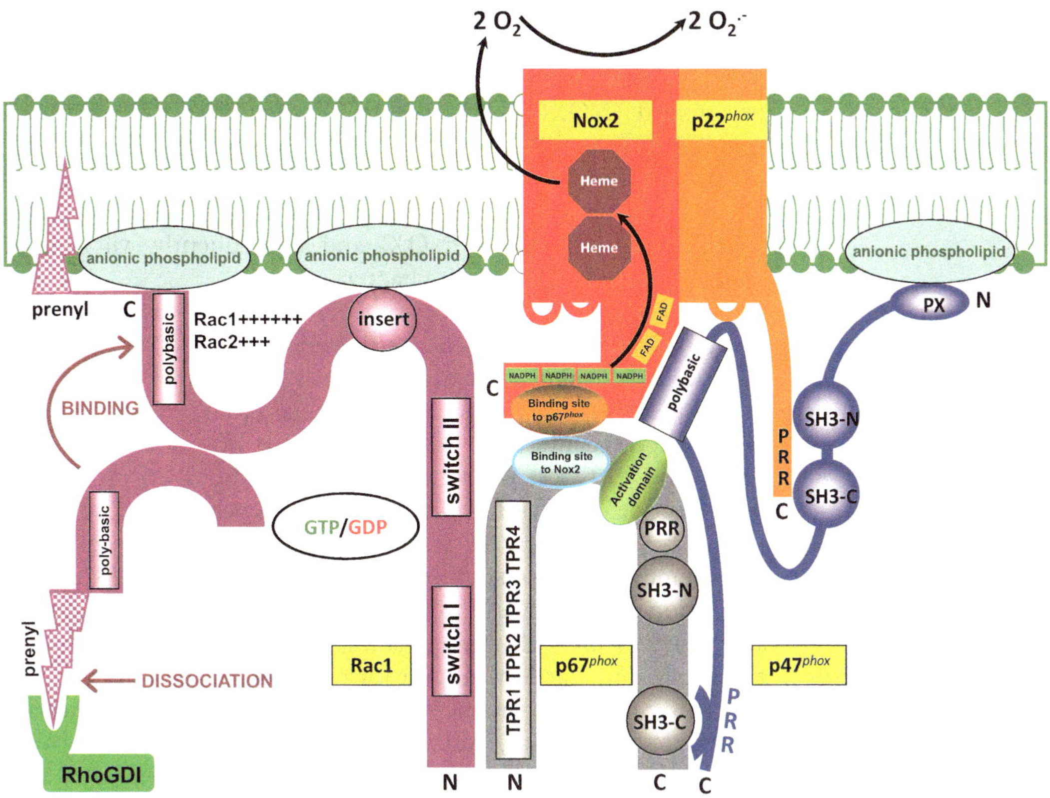

Fig. 1 Schematic representation of the assembled phagocyte NADPH oxidase. *TPR* tetratricopeptide repeat, *SH3* Src homology 3, *PRR* proline rich region, *PX phox* homology. (This is a modified version of Fig. 1, appearing in Ref. 8, by permission of Springer Science + Business Media)

p67*phox* with NOX2 are not interchangeable; under certain in vitro conditions, oxidase activation can take place in the absence of p47*phox* but not that of Rac [11, 13, 14]. These differences in the "assistance" provided to p67*phox* are of significance in the design and interpretation of cell-free assays and will be discussed in this chapter. At a more general level, oxidase assembly can be looked at as a "propagated wave" of conformational changes, a term inspired by early work on the chemical transmission of nerve impulses [15]. Thus, a conformational change induced in one oxidase component by a modifying process, is "propagated" to another component in which a conformational change is also induced. Phosphorylation of p47*phox* leads to binding to p22*phox* and the ensuing rapprochement and binding to NOX2, and GDP to GTP exchange in Rac leads to binding to p67*phox* and the induction of a conformational change in p67*phox*. Models of oxidase assembly were predominantly deducted from biochemical and molecular biology-based studies but these were recently backed by morphological methodologies applied to the intact phagocyte [16]. Discussing NOX2 oxidase assembly cannot be isolated from mentioning that

NOX2 is a member of a family of seven isoforms (NOX 1, 2, 3, 4, and 5 and Duox1 and 2) sharing major structural and functional characteristics and cytosolic homologs (the p47^{phox} homolog, NOX organizer 1 (NOXO1), and the p67^{phox} homolog, NOX activator 1 (NOXA1)) (reviewed in [17, 18]). However, NOX isoforms, other than NOX2, do not function as significant ROS generators in phagocytes and utilize activating pathways partially overlapping or different from those of NOX2 and, consequently, will only be discussed peripherally in this chapter.

The early literature on the assay of NADPH oxidase was rooted in the seminal finding by Bernard Babior that the primary ROS produced by phagocytizing leukocytes is $O_2^{\cdot-}$, which can be converted to H_2O_2 and can act as bactericidal agent [19–21]. To this contributed the awareness of the multiplicity of stimulants (microorganisms, particles, and soluble agents) causing ROS production by phagocytes [22] and the finding that, independent of the nature of the stimulant, oxidase activity could be demonstrated in the membrane fraction of stimulated but not resting cells [23]. Thus, subjection of phagocytes to a stimulant, followed by cell disruption, separation of the cell homogenate into a particulate (membrane) and a cytosolic fraction, and measurement of NADPH-dependent $O_2^{\cdot-}$ production by the membrane fraction, became the standard procedure for assessing oxidase activity [24]. In the course of work with the membrane fraction of stimulated phagocytes, two facts became firmly established: first, the physiological substrate of the enzyme was NADPH and not NADH [23], and, second, the enzyme required supplementation with FAD for optimal activity [25]. These basic facts appear rather straightforward today but 40 years ago in a publication by a superb biochemist, it was stated that the $O_2^{\cdot-}$-generating enzyme of leukocytes was present mainly in the cytosol, was NADH-specific, and, perhaps, was not a single enzyme [26].

Working with membranes from stimulated cells was technically difficult because the activity, as expressed in reaction rates, declined rapidly over time, was inactivated at 37 °C, and was sensitive to high salt concentrations and to harsh sonic disruption of the cells [24]. These difficulties prompted a search for an alternative procedure for the assay of the oxidase. An additional stimulus for such a search was the unexplained mechanism responsible for the majority of the autosomal recessive cases of chronic granulomatous disease (CGD). In the early 1980s, cytochrome b_{558} was the only known component of what turned out to be a multicomponent complex and it was not known that some components are located in the cytosol. The fact that most patients with autosomal recessive CGD possessed a normal cytochrome b_{558} (found to be missing in the X-linked form of the disease) was puzzling.

Useful historical perspectives of oxidase research "prehistory," written by some of the "founding fathers" of the field, are found in Refs. 3, 27, 28.

1.2 The "Phylogeny" of the Cell-Free System

As on many other occasions, the design of a system to allow activation of the oxidase in subcellular fractions derived from resting (unstimulated) phagocytes, has a complex parenthood. During the period preceding the development of the cell-free system, a number of studies were published linking ROS production by phagocytes to arachidonic acid metabolism. In some studies, a link between products of arachidonic acid oxidation, by the cyclo-oxygenase or lipoxygenase pathways, and ROS generation was examined but yielded negative results. However, going upstream by looking for the role of phospholipase A_2, the enzyme responsible for the production of arachidonic acid from phospholipids, in ROS generation, demonstrated a clear and probably causal relationship. Thus, the phospholipase A_2 inhibitor *p*-bromophenacyl bromide inhibited $O_2^{\cdot-}$ generation by human leukocytes [29] but this result was tempered by questions about the specificity of the inhibitor. Independently from this approach, some long chain unsaturated fatty acids and some medium chain saturated fatty acids were found to elicit an increase in cyanide-resistant oxygen consumption [30] and robust $O_2^{\cdot-}$ production by leukocytes in a dose dependent manner [31] and with an order of effectiveness arachidonic > linolenic > linoleic > oleic acids [32]. With remarkable foresight, Kakinuma suggested the need for both an anionic and a hydrophobic character of the fatty acids to enable the elicitation of ROS production by phagocytes [30]. The mechanism by which certain fatty acids elicited ROS production by leukocytes remained unsolved but was looked upon as being related to the stimulation of oxygen consumption and $O_2^{\cdot-}$ generation by detergents such as saponin [33], and deoxycholate and digitonin [34], a connection which surfaced again when the properties of oxidase activators in the cell-free system were being defined. These results were followed by definitive proof being presented of digitonin eliciting $O_2^{\cdot-}$ generation by leukocytes in a dose- and temperature-dependent manner [35]. This led us to initiate a systematic study intended to explore the link between the stimulation of $O_2^{\cdot-}$ production in guinea pig macrophages in response to the multiple agents [22] and the liberation of arachidonic acid by the action of phospholipase A_2. We found, indeed, that eight out of ten elicitors of $O_2^{\cdot-}$ production also caused arachidonic acid and thromboxane B_2 liberation [36]. Three procedures to inhibit phospholipase A_2 blocked $O_2^{\cdot-}$ production elicited by most stimulants, confirming the results of Smolen and Weissmann [29]. We also confirmed the ability of long-chain unsaturated but not saturated fatty acids to elicit $O_2^{\cdot-}$ production and showed that native arachidonic acid was as effective as the lipoxygenase-derived oxygenation product, 15-hydroxyeicosatetraenoic acid, and that the cyclooxygenase-derived product, prostaglandin E_2, was ineffective. An important point was also the finding of the need for the ionized form of arachidonic acid for activity; the methyl ester form was inactive

[36, 37]. We also found that the second product of the action of phospholipase A_2, lysophosphoglycerides, were inactive. A message of significance for the future design of cell-free oxidase activation systems is that arachidonic acid itself and not one of its oxygenation products acts as a messenger for ROS generation by phagocytes. This was proven by our results [36] and by several additional studies [38–40], which did not confirm earlier reports of inhibitors of lipoxygenase interfering with ROS production by leukocytes [41, 42].

1.3 Involvement of a PhospholipaseA2 in the Activation of the Oxidase in the Intact Cell—Controversies Galore

The involvement of phospholipase A_2, resulting in arachidonic acid liberation, in oxidase activation in vivo is a controversial and unsolved question. Suffice it to state that whether such mechanism is at work is irrelevant to the practical use of unsaturated fatty acids and anionic detergents as oxidase activators in the cell-free system, but a brief review of the rather confusing data is warranted. Inhibition of phospholipase A_2 did not affect oxidase activation by all stimulants. Thus, activation by phorbol ester and arachidonic acid in macrophages [36] and by phorbol ester in neutrophils [43] were found to be resistant to inhibition. However, diametrically opposite results were also described, such as ROS production by neutrophils elicited by arachidonic acid [44] and phorbol ester [45, 46] being blocked by the phospholipase A_2 inhibitors *p*-bromophenacyl bromide and mepacrine.

Further confusion was introduced by reports of the involvement of a cytosolic phospholipase A_2. The canonical model assumed that the phospholipase is translocated to the membrane and that the liberated arachidonic acid promoted the assembly of the oxidase complex. However, work with a human myeloid cell line suggested that a cytosolic phospholipase A_2 acts after the assembly of the complex [47], possibly by enhancing the affinity of the assembled oxidase for NADPH [48]. On the other hand, experiments with human monocytes led to the proposal that a cytosolic phospholipase A_2 was involved in the translocation of $p47^{phox}$ and $p67^{phox}$ to the membrane, a process mediated by arachidonic acid [49]. The marked difference in the mechanisms of action proposed for cytosolic phospholipase A_2 in neutrophils versus monocytes as well as the lack of later confirmatory studies suggest that caution should be exercised in relating to these studies.

1.4 Polytheism or Monotheism in the Activation of the NOX2 Oxidase

A question of paramount importance is whether stimulants of ROS production by phagocytes utilize a single or multiple pathways leading to the assembly of the oxidase complex. Data based on the use of inhibitors indicate that there are at least two transduction sequences: (a) Via activation of several isoforms of protein kinase C (PKC) (early work reviewed in Ref. 50; more recent work reviewed in Ref. 7), and (b) Via activation of phospholipase A_2, resulting in liberation of arachidonic acid (discussed above). The typical PKC-mediated pathway is activated following binding of the phorbol ester, phorbol 12-myristate 13-acetate (PMA) to receptors, and is susceptible to inhibition by PKC inhibitors [43, 46]. The target of PKC are several serines, located within the polybasic region at the C-terminus of p47^{phox}, the phosphorylation of which leads to the disruption of an intramolecular autoinhibitory bond between the polybasic region and the src homology 3 (SH3) tandem, leading to binding of p47^{phox} to p22^{phox} [51–54]. Interestingly, at this downstream point, the two pathways meet. Thus, the PKC-induced conformational change, resulting from the phosphorylation of serines, was found to be mimicked by that caused by the interaction of arachidonic acid and anionic detergents with p47^{phox}. The evidence for the latter is extensive and accumulated independently of the development of the cell-free system. Thus, it was found that exposure of p47^{phox} to arachidonic acid or the anionic amphiphile sodium dodecyl sulfate (SDS) caused a dose-dependent quenching of intrinsic tryptophan fluorescence, which could also be induced by PKC-mediated phosphorylation of p47^{phox} in vitro [55]. The fact that this was due to a conformational change was supported by the finding that the circular dichroism spectrum of p47^{phox} was also changed upon addition of SDS [55]. Further proof for arachidonic acid and anionic amphiphiles modifying the conformation of p47^{phox} was provided by changes in the exposure of cysteines [56, 57].

It, thus, appears that the two most investigated pathways leading to the activation of the oxidase (phospholipase A_2-derived arachidonic acid and PKC-mediated phosphorylation of p47^{phox}) have a common mechanism, the induction of a conformational change in p47^{phox}. This is also reflected in the fact that, in parallel to the design of the anionic amphiphile-activated cell-free system to be discussed in this chapter, there were a number of attempts to activate the oxidase in vitro by PKC. These originated in two groups. Tauber et al. described the activation of the oxidase by PMA in a system consisting of neutrophil plasma membrane, cytosol, phospholipid, ATP and NADPH [58, 59]. Purified PKC or the catalytic fragment of PKC were found to be able to replace the cytosol [60]. Babior et al. [61] also designed a PKC-based cell-free oxidase activation system. They first reported that the oxidase could be activated in a mixture of membrane, cytosol, PKC, phospholipids, and ATP, and supplemented with p47^{phox}; p47^{phox} prephosphorylated by PKC could replace PKC and native p47^{phox} [61]. Furthermore, the oxidase could be activated by prephosphorylated p47^{phox} in an assay mixture consisting of membrane, and recombinant p67^{phox} and Rac [62]. In a study intended to compare oxidase activation in vitro by PKC and anionic amphiphile, p47^{phox} was truncated to remove serines essential for PKC-mediated phosphorylation. This protein was unable to support PKC-mediated activation but was functional in the amphiphile-activated system, a result which was seen as supportive for a dichotomy of the activation mechanism [63]. In light of the overwhelming evidence that the final downstream result of both pathways is a conformational change in p47^{phox}, the design of in vitro assays based on this effect would have seemed the choice option. In reality, the canonical amphiphile-activated cell-free assay was designed prior to this fact being known, solely motivated by the belief that it reproduces the phospholipase A_2 pathway thought to be at work in vivo. The fact that PKC-based cell-free systems are hardly if ever used is the consequence of the pragmatic observation that the levels of $O_2^{\cdot-}$ production are considerably lower than those achieved by anionic amphiphiles [59, 63].

1.5 The Orphan Pathways—Other Phospholipases and Phosphatidic Acid

In parallel to the predominant interest in transduction mechanisms centered on phospholipase A_2 and PKC, considerable work was done on the involvement of phospholipases C and D and their products in the activation of the oxidase. Phospholipase C generates diacylglycerol from phospholipids and is linked to the PKC pathway by virtue of its known ability to activate PKC (reviewed in [64]). At the experimental level, exogenous phospholipase C was found to elicit $O_2^{\cdot-}$ production by phagocytes [22], which was, paradoxically, blocked by phospholipase A_2 inhibitors [36]. More accent was placed on the participation of phospholipase D in the mediation of oxidase activation. Phospholipase D generates phosphatidic acid by direct action on phospholipids but also serves as a source of diacylglycerol by the action of phosphatidate phosphohydrolase on phosphatidic acid (reviewed in [65]). A correlation of $O_2^{\cdot-}$ production with phosphatidic acid generation, via phospholipase D activation, in neutrophils stimulated by certain ligands was observed [66, 67]. The mechanism of phosphatidate-mediated activation is controversial and is of significance for the analysis of the mechanism of cell-free activation of the oxidase. Evidence was presented for phosphatidic acid acting directly on cytochrome b_{558} (in fact, on NOX2), as shown by the ability to activate the oxidase in vitro without the assistance of cytosolic factors [68]. Although this conclusion received strong support from later work from our laboratory (*see* Subheading 4.4.1), work from other groups indicated that phosphatidic acid activates the oxidase in vitro in a system requiring the presence of both the membrane and the cytosolic components [69, 70]. Phosphatidic acid was a weak activator but its action was enhanced by diacylglycerol in a synergistic manner [69]. It was next reported that phosphatidic acid acts by activating protein kinases to phosphorylate p47^{phox} [71] and p22^{phox} [72], representing a full-circle return to the PKC-mediated pathways.

2 Cell-Free NADPH Oxidase Activation Systems

2.1 The Advantages of Knowing Less

Chance favors the (un)prepared mind
A paraphrase of a saying of Louis Pasteur

The road to the design of the cell-free system started far away from its unexpected climax. The author's laboratory was interested in establishing more quantitative criteria for the reported enhancement of ROS production by "activated" macrophages, a term usually meaning macrophages exposed to the cytokine interferon γ. Pioneering work in this area originated in the group of Zanvil Cohn [73, 74] and we followed up on this by using cytokine-treated guinea pig peritoneal macrophages [75, 76]. We were eager to identify a reliable biochemical parameter related to the enhanced ROS production and, naïvely, assumed that this must be a change in the composition or function of the enzyme producing $O_2^{\cdot -}$. At that time, the only way to do this was to recover the membrane fraction from activated macrophages and measure $O_2^{\cdot -}$ production [23, 24]. This attempt did not yield results and we decided to go back to square one and learn more about the enzyme (which was thought to be composed of a single protein). As described in Subheading 1.2, the data in the literature on the involvement of phospholipase A_2 and its product, arachidonic acid, in the activation of the oxidase, as well as our own results [22, 36] led us to the hypothesis that arachidonic acid might activate the oxidase in membranes derived from resting (unstimulated) phagocytes. In retrospect, knowing very little about the oxidase was, probably, the "luck component" that allowed Yael Bromberg and the author to do experiments which investigators who were more familiar with the ability of fatty acids and anionic surface-active agents to activate the oxidase in intact cells did not "dare" to do.

2.2 The Canonical Cell-Free System

Live (Cell)—Free or Die
A paraphrase of the official motto of the U.S. state of New Hampshire, written by General John Stark, July 31, 1809.

In experiments performed in 1983 in the author's laboratory, it was first found that significant NADPH-dependent $O_2^{\cdot -}$ production was elicited in homogenates of unstimulated guinea pig peritoneal macrophages, freed of nuclei, exposed to the Na salt of arachidonic acid [77] (*see* **Note 1**). The reaction was strictly NADPH-dependent (not supported by NADH, NADP or NAD), exhibited peak activity at a concentration close to 100 μM arachidonic acid (lower and higher concentrations were less active) (*see* **Note 2**), was enhanced by supplementation with FAD, was resistant to sodium azide, and inhibited by Ca^{2+} (*see* Subheading 3.1). The K_m for NADPH was 49 μM and quantitative parameters were similar to those described for the oxidase recovered from stimulated phagocytes. In addition to arachidonic acid (C20:4), long chain fatty acids of lesser degrees of unsaturation, such as oleic (C18:1), linoleic (C18:2), and linolenic (C18:3) acids were also found to elicit $O_2^{\cdot -}$ production. The cyclooxygenase product prostaglandin E_2 was inactive and the lipoxygenase product 15-hydroperoxy-5,8,11,13-eicostetraenoic acid (15-HPETE) was equal in activating ability to C20:4. The methyl ester of arachidonic acid was inactive. None of the numerous stimulants, eliciting $O_2^{\cdot -}$ production in intact phagocytes were capable of acting as activators in the cell-free system.

It, thus, came as a surprise when membranes prepared by centrifugation of the homogenate did not respond by $O_2^{\cdot -}$ production when exposed to arachidonic acid under conditions identical to those which yielded a positive response with homogenates. In a radical paradigm changing experiment, the cytosolic fraction, removed following the sedimentation of the membranes, was added back to the membranes and the "reconstituted" homogenate exposed to arachidonic acid. This led to an almost full recovery of $O_2^{\cdot -}$ production, a result that was interpreted as demonstrating the dependence of oxidase activation on a cytosolic component. This result was unexpected and represents a good example of what Thomas Kuhn defined as an "anomaly" [78] (*see* **Note 3**). The popularity acquired by the cell-free system as a methodological advance obscured its true significance as a conceptual revolution that led to the discovery and characterization of the cytosolic factors and the elucidation of the molecular basis of the various forms of CGD.

The cell-free system developed in the author's laboratory was first presented in June 1983 at a conference at the Pasteur Institute in Paris. The "cell-free concept" was clearly ripe for discovery, as shown by the fact that a similar cell-free oxidase activating system was described simultaneously with but independently of the author's by Heyneman and Vercauteren [79] (*see* **Note 4**). Their paper, accepted 14 days before the acceptance of the author's paper, showed that horse leukocyte membranes combined with cytosol generated $O_2^{\cdot -}$ in response to sodium oleate (C18:1) in a reaction which was NADPH-dependent and resistant to cyanide. They were also the first to report that a homogenate from leukocytes of a patient with CGD did not generate $O_2^{\cdot -}$ in response to oleate. Their work was inspired, similar to ours, by reports on the elicitation of ROS generation in intact phagocytes by fatty acids [30–32], by their own work on oleate-induced $O_2^{\cdot -}$ production by horse leukocytes [80] (*see* **Note 5**), and by a description of the activation of membranes derived from resting cells by dialysis [81] (*see* **Note 6**). This was followed by the description by two groups of arachidonate induced cell-free oxidase activation in homogenates of human neutrophils, exhibiting the same dual requirement for both membranes and cytosol and properties identical to those shown with material derived from guinea pig macrophages and horse neutrophils [82, 83]. This proved that the fatty acid-elicited oxidase activation in vitro was a general phenomenon, as shown by the fact that identical findings were made in two types of phagocytic cells and in three species. The report by McPhail et al. [82] also contained the first indication that the freshly discovered cytosolic component(s) were translocated to the membrane upon stimulation of intact leukocytes; thus, low concentrations of whole cell stimulants endowed membranes isolated from such cells to respond to arachidonic acid in the absence of cytosolic components.

A methodological but also conceptual revolution was the introduction of the semirecombinant cell-free system. In this, the membrane is either used in the native form or represented by a purified and relipidated cytochrome b_{558} preparation, and the cytosol is replaced by a mixture of purified recombinant components [84]. This system permits introduction of strict quantification of the components participating in cell-free oxidase activation, the performance of dose–response assays, and control over the ratios among cytosolic components, among these and cytochrome b_{558}, and among the activating amphiphile and the membrane and cytosolic components.

Soon after the introduction of the cell-free system, the mechanism by which fatty acids cause oxidase activation became a subject of intense investigation. A systematic study of the fatty acid specificity of cell-free oxidase activation revealed that various fatty acids differ considerably in the concentration required for the induction of maximal oxidase activity but that bell-shaped dose–response curves are typical [85]. This report was also among the first to explore the ability of saturated fatty acids and the *trans* form of some fatty acids to serve as activators. A detailed discussion of the intricacies of the mechanism of action of fatty acids appears in Subheading 3.1.

A major breakthrough in deciphering this mechanism was the finding that SDS or lithium dodecyl sulfate (LiDS) can replace fatty acids as activators in the cell-free system, with very similar characteristics and dose–response curves [86]. Other anionic detergents, such as sodium cholate, sodium deoxycholate, digitonin, and saponin, containing fused aromatic rings, were inactive. Sodium dodecyl sulfonate, like SDS, consists of an aliphatic hydrophobic moiety and an anionic polar head and, thus, could serve as an activator [86]. Cell-free oxidase activation by fatty acids and some detergents was clearly one and the same and the term "anionic amphiphiles" became the accepted term for all activators belonging to both chemical categories.

2.3 How Do Anionic Amphiphiles Work

Yet another "beyond pragmatism" value of the cell-free system is that it provided novel insights into the molecular basis of oxidase activation. The predominant explanation for the oxidase-activating ability of anionic amphiphiles is that their main target is p47^{phox}. As discussed in Subheading 1.4, there is ample experimental proof for anionic amphiphiles causing a conformational change in p47^{phox}, mimicking in vitro the phosphorylation of critical serines within the polybasic region, occurring in vivo [55–57]. Some caution is, however, required concerning this "single target" interpretation because of the existence of a body of evidence for a direct effect of anionic amphiphiles on cytochrome b_{558}, leading to conformational changes which might contribute significantly to oxidase activation [87–91]. Finally, there is also evidence for p67^{phox} and Rac serving as possible targets. Thus, arachidonic acid was found to elicit GDP to GTP exchange on Rac and to promote the direct interaction of the p67^{phox}-Rac-GTP complex with the DHR of NOX2 [92]. This latter effect could be due to an effect on the p67^{phox}-Rac-GTP complex or to an effect on NOX2. Complex and yet partially understood effects of arachidonic acid on p67^{phox} were also described. Using thiol accessibility as a measure of conformational changes, it was shown that arachidonic acid influences the secondary structure of the p47^{phox}–p67^{phox} complex by an effect on both partners [93].

2.4 Support for and Opposition to the Cell-Free Paradigm

Support for the proposal that the cell-free system is, at least in part, an in vitro reflection of oxidase assembly in vivo, was provided by the authors of the first descriptions of the system, who reported that a homogenate from leukocytes of one patient with CGD did not respond to oleic acid [79] and that patients with the X-linked form of CGD possess a defective membrane component but a normal cytosolic component(s) [83, 94]. This observation was followed by the reciprocal finding that most CGD patients with the autosomal recessive mode of inheritance have a normal membrane component but a defective cytosolic component(s) [95–97]. These clinical correlates were essential for "legitimizing" the cell-free system as reflecting the activity of the $O_2^{\cdot -}$-forming oxidase in intact phagocytes. Further support for the usefulness of the cell-free system was provided by the unending list of its applications. Among the most significant were its role in the identification and characterization of p47^{phox} and p67^{phox} [98, 99], the finding that Rac1 [100] and Rac2 [101] are essential for oxidase activation, and providing experimental support for the thesis that cytochrome b_{558} is the only catalytic component of the oxidase complex [102, 103], by demonstrating the ability of purified cytochrome b_{558} to produce $O_2^{\cdot -}$ in a cell-free system in the presence [104] and, unexpectedly, in the absence [105, 106] of cytosolic components.

It is of interest to mention that the cell-free system as a method and its theoretical implications related to the mechanism of oxidase assembly were not universally accepted. The original paper describing the ability of SDS to replace arachidonate as an activator [86] was rejected by a major journal principally on the basis that it represented a mere extension of observations made in the past on the effect of detergents on whole cells (*see* **Note 7**). The reviewer referred to the reports on the ability of saponin [33], deoxycholate [34] and digitonin [34, 35] to elicit $O_2^{\cdot-}$ production by intact phagocytes, ignoring or overlooking the statement made in the paper that these detergents were inactive in the cell-free system [86]. Skepticism toward the cell-free system representing in vitro an event occurring in vivo went on even after the publication of the papers from the author's and other laboratories (*see* [107] and **Note 8**). In another publication, the ability of fatty acids to elicit $O_2^{\cdot-}$ production by whole neutrophils was linked to their ability to bind to the cells in a saturable and reversible manner and their property of activating the oxidase was related to the secondary activation of a phospholipase C rather than to a direct effect on the enzyme [108].

However, as knowledge and use of the cell-free system expanded, criticism subsided gradually and positive reverberations became predominant as, to use the phrase of Thomas Kuhn, *"the initially anomalous has become the anticipated"* (*see* [21, 109–112] and **Note 9**). In addition to the original papers [77, 79, 82, 83, 86], two brief reviews provide useful descriptions of the history and evolution of cell-free systems but are lacking the rendition of newer developments [113, 114].

3 A Few Yet Unanswered Questions

3.1 What Makes Fatty Acids Tick

The qualities required for a fatty acid to elicit ROS production in intact phagocytes were shown not to be automatically applicable to their ability to act as oxidase activators in the cell-free system. Nevertheless, there are many similarities and the large volume of work in cell-based systems should not be ignored. Thus, in intact cells, long chain unsaturated fatty acids were the most potent activators [30, 32] but saturated fatty acids with 10, 12, 14, and 16 carbons were also capable of activation [31, 115, 116]. A single study described the ability of very long chain fatty acids (22–34 carbons) to elicit moderate $O_2^{\cdot-}$ production by neutrophils, their ability decreasing with increasing chain length [117]. In the cell-free system, long chain unsaturated fatty acids were the most active [77, 85] but certain saturated fatty acids were reported to be capable of activation by two groups of investigators [116, 118, 119]. Medium chain length saturated fatty acids (12, 14, 16 carbons) were active, whereas shorter (8 and 10 carbons) and longer (18 and 20 carbons) chain fatty acids were inactive [118], although a different group reported lack of activation by saturated fatty acids of all chain lengths [85].

The complexity of the mechanism of oxidase activation by fatty acids is illustrated by the recent finding that the *trans* form of arachidonic acid is incapable of cell-free oxidase activation and acts as an inhibitor of oxidase activation induced by the *cis* form [120]. The inhibition appears to be due to major "denaturating" conformational changes in p67^{phox} caused by the *trans* form of arachidonic acid. In contrast to the inability of the *trans* form of arachidonic acid to activate the oxidase, an earlier report describes activation of the oxidase by the *trans* forms of oleic and linoleic acids [85, 116]. Finally, nitroarachidonic acid was reported to prevent oxidase assembly, based on whole cell studies; its effect in the cell-free system has not yet been investigated [121].

An important methodological aspect in cell-free oxidase activation is the role of calcium ions. It was amply demonstrated that only the ionized form of fatty acids activate the oxidase in the cell-free system; methyl esters are inactive [77, 85]. The presence of Ca^{2+} in the assay buffer was found not to be required for activation [77, 83]; on the contrary, it acted as an inhibitor [77] (*see* **Note 10**). The effect of Ca^{2+} on the ability of fatty acids to activate the oxidase was studied in depth on whole cells and related to the Krafft point [115] (*see* **Note 11**). Fatty acids were divided in three groups: Group A comprised fatty acids with a high Krafft point, which were inactive even in the absence of Ca^{2+} (stearic acid, C18:0); group B comprised acids with medium Krafft points, active only in the absence of Ca^{2+} (myristic acid, C14:0), and group C comprised fatty acids with low Krafft points, capable of activation even in the presence of Ca^{2+} (oleic acid, C18:1, and linoleic acid, C18:2). To this latter category belong the fatty acids which were the most effective in the cell-free system because their Krafft points were below the temperature at which cell-free assays are usually run (close to 24 °C).

The relationship of the Krafft point to the critical micellar concentration (CMC) (*see* **Note 11**) raised the question of the connection between optimal activating concentrations of fatty acids and anionic detergents in the cell-free system and the CMC of the activators. The optimal activating concentration of anionic amphiphiles was commonly found to be close to 100 μM [77, 82, 83, 85, 86]; however, concentrations of 1.5 mM [79] or varying between 20 and 800 μM were also reported (reviewed in [93]). These differences are likely to be the result of the use of different sources of oxidase components, as poignantly illustrated by an optimum of 200 μM arachidonate, when the membranes were of bovine leukocytes, and 600 μM, when membranes originated in yeast expressing cytochrome b_{558} [120]. An additional influencing factor is the lipid composition of the membranes used in the assay [120] and the lipids used in reconstituting solubilized membrane [122].

The CMC of the long-chain unsaturated fatty acids active in the cell-free system are as follows: arachidonate, 73 μM; linolenate, 150 μM; linoleate, 93 μM, and oleate, 86 μM [123]. The CMC of SDS is 8.1–8.4 mM and that of the detergent of choice in the author's laboratory, LiDS), is 7–10 mM [124, 125]. Thus, the optimal activating concentrations of the fatty acids in the cell-free system and the concentrations found to cause a conformational changes in p47*phox* [55, 57] and cytochrome b_{558} [87–90] were in the range of their CMC. In sharp contrast, this correlation was absent in the case of anionic detergents, which activated the oxidase at concentrations 70–100 times lower than their CMC values. It can, thus, be concluded that the CMC of an anionic amphiphile is not an indicator of the optimal activating concentration in the cell-free system.

In all considerations related to the degree at which the cell-free system is representative of events occurring in the intact cell, the issue of the concentration of free fatty acids likely to exist in phagocytes upon stimulation by agents initiating ROS production is raised (discussed in Ref. 126). Unfortunately, the available information is scarce; resting leukocytes were reported to contain 0.5–1 μM free arachidonate [127] but no information was found on intracellular levels upon stimulation of cells by agents activating the oxidase.

3.2 Getting in and Getting Out—How Do Substrates and Products Find Their Way in the Cell-Free System

Activation of the oxidase in both the intact cell and the cell-free system consists of two processes: the assembly of the oxidase complex from its components and the catalytic phase, namely, the flow of electrons from NADPH to O_2. In the whole cell, the process of assembly appears straightforward and can be visualized as the translocation of the cytosolic components to the membrane vicinity of cytochrome b_{558}, culminating in the interaction of p67*phox* with NOX2, as discussed in Subheading 1.1. This interaction initiates the catalytic phase, the consequence of a yet poorly understood conformational change in the DHR of NOX2. The location of the NADPH- and FAD-binding sites in the DHR allow easy contact with both substrates present in the cytosol. Less well understood are the electron transfer from FAD to the membrane-located two hemes and the precise mechanism of O_2 reduction to $O_2^{\cdot-}$ by the hemes (ligation or "outer sphere"). This ideal "geography" does not exist in the cell-free systems, a fact most pronounced in the now dominant use of the semirecombinant system, consisting of purified recombinant cytosolic components and some form of membrane or purified cytochrome b_{558}. Other differences from the whole cell situation are the absence of p40*phox* and the fact that p47*phox* and p67*phox* are added as separate entities and not in the form of the preformed complex present in the cytosol of resting phagocytes [128].

In most of the original cell-free assays, membranes were used in their native form, prepared by plain centrifugation of a cell homogenate freed of nuclei and debris [77, 82, 83]. In the case of human neutrophils, a "whole" membrane fraction contains the plasma membranes, as well as the specific and azurophil granules. Most of

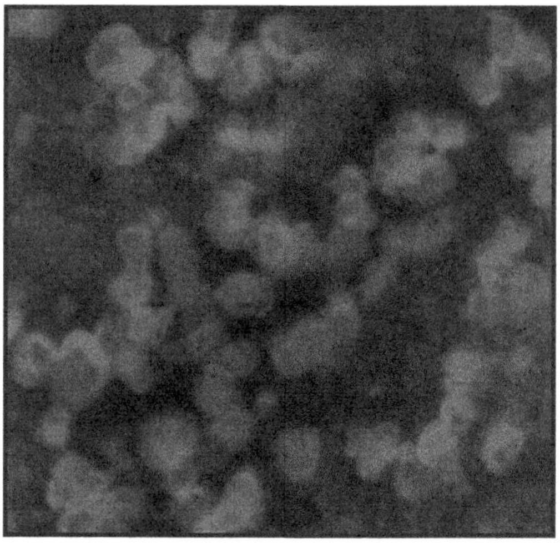

Fig. 2 Electron microscopy (negative staining)-derived image of guinea pig macrophage membrane liposomes obtained by solubilization of membranes by octyl glucoside followed by dialysis to remove the detergent. The liposomes have a diameter of 314 nm

cytochrome b_{558} is found in the specific granules, with a lesser amount present in the plasma membranes [129] and since the only membrane component participating in $O_2^{\cdot-}$ generation is cytochrome b_{558} [104], only the cytochrome b_{558}-containing fractions are relevant. An analysis of the subcellular compartmentalization of membranes in the neutrophil, which produce $O_2^{\cdot-}$ in a cell-free system, revealed that both plasma membranes and specific granules were involved [130]. Thus, the cell-free system reproduces accurately the in vivo situation, as shown by the finding that in intact neutrophils $O_2^{\cdot-}$ production occurs both at the level of the plasma membrane and in an intracellular compartment corresponding to granules [131]. To the best of my knowledge, the subcellular origin of macrophage/monocyte membranes participating in cell-free activation has not been ascertained.

In order to enable better reproducibility and more accurate measurement of cytochrome b_{558} content, the use of solubilized membranes was introduced [132]. This consisted of solubilizing the membranes with the aid of the nonionic detergent octyl glucoside, at a concentration just above its CMC of 25 mM. Following removal of the detergent by dialysis, liposomes are formed, which exhibit $O_2^{\cdot-}$ generating ability superior to that of native membranes (Fig. 2). Such liposomes contained the "natural" phagocyte lipids but membrane material freed of endogenous lipid could be reconstituted ("relipidated") with nonphagocyte natural or synthetic phospholipids, yielding membrane material with excellent activity in the cell-free system [122]. Finally, purified cytochrome

b_{558} was relipidated to generate liposomes and these replaced total solubilized membrane in the cell-free assay [104].

The fact that $O_2^{\cdot-}$ production by the anionic amphiphile-activated cell-free system is similar in amplitude to that of stimulated whole cells and membranes derived from stimulated cells indicates that the components of the cell-free system are accessible to NADPH, amphiphilic activator, and oxygen and that the cytosolic components can interact with cytochrome b_{558} in the membrane. The mechanics of cell-free systems based on membrane liposomes are poorly understood. If membrane liposomes are "outside out," the access of cytosolic components and NADPH to the inner face of the membrane is impaired but the liberation of $O_2^{\cdot-}$ to the assay buffer is assured. If the membrane liposomes are "inside out," there is access of cytosolic components and NADPH to the inner face of the membrane but contact of $O_2^{\cdot-}$ with the assay buffer is impaired. Two possible solutions to this quagmire were suggested: (a) The process of sonic disruption, commonly used to prepare cell homogenates, causes an artificial permeabilization of the membrane allowing access to the interior of the liposome [133], or (b) The anionic amphiphile serving as activator enhances membrane permeability. This later possibility is made unlikely by the existence of several cell-free systems activated in the absence of anionic amphiphiles [11, 12, 133–138].

4 More About the Cell-Free System

4.1 Alternative Sources for Membranes

The difficulty of obtaining human and animal phagocytes in sufficient amounts, combined with the wish to reduce the use of animals in research, led to increased use of cell lines as a source of membranes for the performance of cell-free assays. It is beyond the purpose of this chapter to list all the cell types in use and only a few examples will be given. A popular cell line is the human myeloid leukemia cell PLB-985 [139]. This can be differentiated by a number of maturation inducing agents to express cytochrome b_{558}, although levels are considerably lower than those of neutrophils and macrophages. A variation of PLB-985 was constructed which mimics cells of patients with the x-linked form of CGD, lacking NOX2 (X-CGD PLB-985) [140]. Such a cell allows transfection with mutants of NOX2, making it a very useful tool in studying the functional consequences of NOX2 mutations. Wild type and X-CGD PLB-985 cells, transfected with NOX2 mutants, were used as a source of membranes for cell-free assays upon supplementation with cytosolic components and arachidonate, as activator [141, 142]. Cell lines also proved useful when cell-free systems were utilized to test compounds for oxidase inhibitory ability for potential application as therapeutic agents. A typical example is testing of a novel small molecule for a direct inhibitory effect of NOX2 by assessing its ability to interfere with cell-free NADPH oxidase in two systems [143]. In the first, membranes from differentiated PLB-985, expressing NOX2 and p22^{phox}, were used, combined with recombinant cytosolic components and with SDS, serving as activator. In the second, baby hamster kidney cells (BHK) [144] were transduced with a modified baculovirus into which NOX2 and p22^{phox} cDNAs were cloned, resulting in cells expressing the NOX2-p22^{phox} heterodimer. Membranes originating from these cells were used in a cell-free system as done with PLB-985 cells.

A radically different approach was to attempt the expression of cytochrome b_{558} in the yeast *Pichia pastoris*, leading to the successful production of the NOX2-p22^{phox} heterodimer, which was reconstituted in liposomes with an optimized phospholipid composition [145]. Such liposomes were functional in a cell-free system when combined with recombinant cytosolic components and activated by arachidonate, although arachidonate dose-response profiles were different from those of neutrophil membranes [120]. Attempts to express cytochrome b_{558} mediated by baculovirus in insect cells resulted in either a nonfunctional protein or low yields and did not reach the stage of application in the cell-free system [145–147].

4.2 Variations on the Theme of the Semirecombinant Cell-Free System—Focus on Cytosol

Variations of the semirecombinant amphiphile-dependent cell-free system were introduced focused on the cytosolic components. One development was the idea to fuse cytosolic components in the hope that the chimeric constructs will support a more active and more stable oxidase. The first of these was a [p47^{phox}–p67^{phox}] chimera consisting of p47^{phox}(1–286) fused to p67^{phox}(1–210), based on the expectation that p47^{phox} truncated at residue 286 will be active because of the absence of autoinhibition and the p67^{phox} moiety comprises the four tetratricopeptide repeat (TPR) motifs, essential for binding Rac, and the "activation domain" [134]. A significant observation, to receive much support in the future, was that the chimera combined with Rac, was capable of activation in the absence of an amphiphilic activator if the membrane was replaced by purified cytochrome b_{558} relipidated with anionic phospholipids. A distinctive feature of the [p47^{phox}–p67^{phox}] chimera-based system was the generation of a remarkably stable complex (half-life of 184 min) [148], which could be further enhanced by chemical crosslinking of the chimera with cytochrome b_{558} [149]. An interesting practical application of this stable and activator-independent cell-free system was its use as a $O_2^{\cdot -}$-generating device for the study of oxidative damage to cells [150]. A second chimera-driven cell-free system was based on the construction of a [p67^{phox}-Rac1] fusion protein consisting of p67^{phox}(1–210) and full-length Rac1 (1–192) [12, 151–153]. For functionality, the chimera was subjected to GDP to GTP exchange or a Q61L mutation was introduced in the Rac moiety to assure that it was constitutively in the GTP-bound form. A cell-free assay in which the [p67^{phox}-Rac1] chimera replaced the individual components exhibited high activity and improved stability and was only partially dependent on p47^{phox} but an amphiphilic activator was required [151, 152]. The [p67^{phox}-Rac1] chimera could be prenylated at the Rac moiety, a modification which endowed it with the ability to activate the oxidase in a cell-free system in the absence of both p47^{phox} and an amphiphilic activator [12, 135].

The newest development in this area was the introduction of tripartite chimeras (trimeras) [137, 138]. The prototype is [p47^{phox}(1–286)-p67^{phox}(1–212)-Rac1(1–192)] and comprises most of the domains present in individual components essential for oxidase activation. Similar to [p47^{phox}–p67^{phox}] chimeras, the p47^{phox} moiety was truncated at residue 286, relieving autoinhibition, and the phox homology (PX) domain was conserved. The p67^{phox} moiety comprised the four TPR motifs, essential for binding Rac, and the "activation domain." The Rac1 moiety was represented by full-length Rac1 and the trimera was either subjected to GDP to GTP exchange or constructed with a Q61L mutation in the Rac moiety to assure a constitutive GTP-bound form. The trimera supported amphiphile-dependent oxidase activation in the cell-free system with a lower EC_{50} than combined individual components and formed a more stable complex. A finding of high relevance was that supplementation of the membrane liposomes with anionic phospholipids made cell-free activation amphiphile-independent, suggesting that electrostatic interaction between the polybasic tail region of the Rac1 moiety and the negatively charged inner aspect of the membrane was sufficient for oxidase assembly and activation [137]. The trimera could be prenylated at the Rac moiety yielding a molecule which was capable of cell-free oxidase activation of native (unmodified) membrane liposomes in the absence of an amphiphile activator, demonstrating the importance of hydrophobic interactions (between the prenyl tail of the Rac moiety and the membrane) in oxidase assembly [138]. Of special significance for the relationship of cell-free activation to events in the whole cell is the recent description of the ability of the trimera transfected into COS cells expressing NOX2 and p22^{phox} to elicit spontaneous $O_2^{\cdot-}$ production in the absence of any whole cell stimulant [154]. The trimera was found to bind to the cell membrane, indicating that it became prenylated in the cell. Of interest for the mechanism of action of amphiphilic activators is the finding that arachidonic acid does not significantly modify the circular dichroism spectrum and the intrinsic tryptophan fluorescence spectrum of the trimera suggesting that it is expressed in an "active" conformation [154].

Finally, cell-free oxidase activation can also be supported by replacing Rac by a complex of Rac and GDP dissociation inhibitor for Rho (RhoGDI). Before the easy availability of recombinant Rac, [Rac-RhoGDI] purified from cytosol was used [155]. This was replaced by [Rac-RhoGDI] generated from recombinant Rac1 prenylated in vitro [135] and recombinant Rho-GDI and was used principally in the study of the mechanism of dissociation of the complex, an essential step preceding the translocation of Rac to the membrane ([156, 157], reviewed in Ref. 158).

4.3 Unconventional Cell-Free Systems and Permeabilized Cells

The canonical cell-free system is not capable of detecting the participation of p40^{phox} in oxidase activation [159]. To enable this, a system was designed based on permeabilization of neutrophils by streptolysin-O, resulting in the formation of cytosol-free "cores" [160]. These are used as a source of membranes, with the maintenance of membrane morphology and preservation of intracellular granules. Upon supplementation with cytosol, ATP, GTP and NADPH, a cell-free-like system is generated which responds to stimulants normally acting on intact cells, such as PMA, by $O_2^{\cdot -}$ production. Using this system, a role for p40^{phox} in oxidase activation in human neutrophils could be demonstrated [161]. Two unconventional permeabilized phagocyte cell assays were also published. In the first, neutrophil "cytoplasts" were prepared free of nuclei and granules and with an "outside-out" plasma membrane and shown to respond with ROS production to stimulants, such as PMA and zymosan [162]. The second described the property of permeabilized leukocytes to respond to stimulants and, at the same time, allow access of exogenous NADPH to the enzyme, exhibiting a much lower K_m claimed to be closer to the in vivo reality than the values measured in the cell-free systems [163].

4.4 A Cell-Free System Without Cytosolic Components

4.4.1 All Missing

A methodological variation of the canonical cell-free assays was the design of a system in which activation was performed in the absence of cytosolic activators. The logic behind this was the evidence that all redox stations are located in NOX2 [102, 103] and the hypothesis that a conformational change in NOX2 initiates the electron flow from NADPH to O_2. An important first description of oxidase activation by phosphatidic acid in the absence of cytosolic components, with the catalytic component having the characteristics of cytochrome b_{558}, was overlooked by most investigators [68]. In work published 5 years later, macrophage membrane-derived purified cytochrome b_{558} was relipidated with a mixture of crude soybean phosphatidylcholine (14–23% pure) and pure phosphatidic acid. The cytochrome was found to generate $O_2^{\cdot -}$ in vitro in the presence of FAD, NADPH, and a low amount of anionic amphiphile, in the absence of cytosolic components [105]. The level of $O_2^{\cdot -}$ production was about four times lower than that found in the canonical cytosol-dependent cell-free system. The proposed mechanism of action was that the specific milieu of cytochrome b_{558}, in which anionic phospholipids were dominant, facilitated electron transport from NADPH to FAD [106]. The discovery of a cytosol-independent oxidase activation process provided definitive functional proof for the presence of all redox stations on cytochrome b_{558}. The cytosol-independent cell-free assay was used successfully for the elucidation of the molecular basis in some cases of the X91^{+} form of CGD [164].

4.4.2 Only p47^{phox} Missing

A particular situation is cell-free activation in the absence of p47^{phox}. This was described independently by two groups [13, 14]. p47^{phox}-independent activation required purified and appropriately relipidated cytochrome b_{558} and high molar ratios of p67^{phox} and Rac1 relative to cytochrome b_{558}. Significantly, p67^{phox} truncated to remove both SH3 domains, supported oxidase activation both in the presence and absence of p47^{phox} [13]. These results contributed to establishing the concept of p67^{phox} as a NOX activator (NOXA) and p47^{phox}, as a NOX organizer (NOXO). Cell-free systems involving prenylated Rac, discussed in Subheading 4.5.2, are also functional in the absence of p47^{phox}.

4.5 Cell-Free Activation in the Absence of an Activator

4.5.1 Relieving Autoinhibition

The elucidation of one of the mechanisms by which anionic amphiphiles induce oxidase activation (severing the intramolecular bond in p47^{phox} between the $(SH3)_2$ tandem and the polybasic C-terminus) led to the design of a new type of amphiphile-independent cell-free system. In this, truncation of p47^{phox} at residue 286, which removes the polybasic C-terminus [133], or engineered mutations in p47^{phox}, which cause unmasking of $(SH3)_2$ [136], make the system amphiphile-independent. The need for amphiphile is circumvented because, in both cases, spontaneous interaction between the $(SH3)_2$ of p47^{phox} and the proline-rich region at the C-terminus of p22^{phox} is made possible. Surprisingly, amphiphile-independent activation involving p47^{phox} truncated at residue 286, required p67^{phox} to be truncated, too, at residue 242. Also, truncated p67^{phox} combined with full-length p47^{phox} failed to activate in the absence of amphiphile, suggesting that the amphiphile might also have an effect on p67^{phox} [133].

4.5.2 A Lipid Anchor

We developed a conceptually different amphiphile-independent cell-free system based on the use of prenylated Rac1, which binds to phagocyte membranes with high affinity and serves as a carrier for p67^{phox}, leading to oxidase activation in the absence of amphiphile and without the need for p47^{phox} [11, 165]. Later variations of this system are represented by a prenylated [p67^{phox}-Rac1-GTP] chimera, which activates the oxidase in the absence of amphiphile and of any other component [12, 135, 153], a tripartite chimera, consisting of the functional domains of p47^{phox}, p67^{phox}, and full-length prenylated Rac1-GTP [138], and prenylated Rac-GDP, in conjunction with a guanine nucleotide exchange factor (GEF) for Rac and GTP or ATP [166]. The essential difference between amphiphile-dependent and amphiphile-independent, prenylated Rac-dependent cell-free systems is poignantly illustrated by the inhibitory effects of a peptide, corresponding to the C-terminus of Rac1, and of the polybasic neomycin sulfate on the amphiphile-dependent activation [12, 165], as opposed to specific inhibition of the amphiphile-independent activation by phospholipid liposomes [11] and RhoGDI [11, 12].

4.5.3 "Reversed" Activation—Making the Membrane Anionic

The plasma membrane of mammalian cells contains 15–20% anionic phospholipids, a fact of considerable importance in leukocyte function (reviewed in Ref. 167). Yet another group of cell-free systems was developed, based on the rationale of artificially enriching phagocyte membranes with anionic phospholipids. This is expected to result in an increase in the negative charge at the cytosolic aspect of the membrane and should promote the binding of the cationic cytosolic components Rac and p47^{phox} (or positively charged regions in chimeras resulting from the fusion of cytosolic components) to the membrane and, possibly, decrease the electrostatic repulsion of the positively charged cytochrome b_{558}. The importance of the negative charge of phospholipids in which cytochrome b_{558} was reconstituted for cell-free oxidase activation was first shown by the group of Lambeth [10] and was followed by several studies supporting this thesis. Thus, a combination of a [p67^{phox}(1–210)–p47^{phox}(1–286)] chimera and Rac1-GTP activated phagocyte membranes enriched in anionic phospholipids, in the absence of amphiphile [134]. The tripartite chimera, [p47^{phox}(1–286)–p67^{phox}(1–212)-Rac1(192)], was also shown to be a potent oxidase activator in the absence of anionic amphiphile, provided that the membrane is enriched with one of the anionic phospholipids, phosphatidic acid (PA), phosphatidylglycerol (PG), phosphatidylserine (PS), or phosphatidylinositol (PI) [137]. Also, enrichment of phagocyte membrane with the anionic phospholipids PG or PA enables oxidase activation by p67^{phox} combined with [Rac1(GTP)-RhoGDI] complexes [156], and supplementation of membranes with phosphatidylinositol 3,4,5-triphosphate promotes oxidase activation by p67^{phox} and [Rac1(GDP)-RhoGDI] complexes in conjunction with GTP and a GEF [157], both in the absence of an anionic amphiphile and p47^{phox}. The fine mechanism behind this form of "spontaneous" activation is not explained by simple electrostatic attraction between the membrane and the cytosolic components and their chimeric variations because the overall charges of [p67^{phox}(1–210)–p47^{phox}(1–286)] and [p47^{phox}(1–286)–p67^{phox}(1–212)-Rac1(192)] chimeras are close to neutral and, thus, it is likely that particular positively charged domains in the cytosolic proteins and their chimeric constructs are participating in the interaction. An example of the major effect of electrostatics on cell-free activation is illustrated in Table 1.

Table 1
Amphiphile-independent cell-free oxidase activation by enrichment of membrane with anionic phospholipids

	NADPH oxidase activity (mol $O_2^{\cdot-}$/s/mol cytochrome b_{558} heme)			
Cytosolic activator(s)	Membrane + phosphatidic acid (PA)	Membrane + phosphatidyl-glycerol (PG)	Membrane + phosphatidyl-serine (PS)	Membrane + phosphatidyl-inositol (PI)
No activator (membrane only)	11.70 ± 1.11	2.47 ± 0.13	1.75 ± 0.08	1.87 ± 0.04
p47*phox* + p67*phox*	32.60 ± 2.60	12.37 ± 1.28	6.81 ± 0.58	2.91 ± 0.19
p67*phox* + Rac1	95.41 ± 3.64	34.29 ± 5.07	12.80 ± 1.20	5.63 ± 0.40
p47*phox* + p67*phox* + Rac1	112.83 ± 12.85	69.95 ± 8.46	48.57 ± 7.73	22.50 ± 3.22
[p67*phox*(1–212)-Rac1 (1–192)] chimera[a]	62.46 ± 4.81	12.92 ± 0.55	5.62 ± 0.57	3.11 ± 0.06
[p67*phox*(1–212)-Rac1 (1–192)] chimera + p47*phox*	80.03 ± 4.99	44.87 ± 2.43	22.23 ± 0.42	6.12 ± 0.13
[p47*phox*(1–286)-p67*phox*(1–210)] chimera[b]	19.76 ± 0.86	3.21 ± 0.29	1.90 ± 0.35	2.27 ± 0.15
[p47*phox*(1–286)-p67*phox*(1–210)] chimera + Rac1	93.68 ± 2.97	68.94 ± 3.06	48.65 ± 1.46	30.25 ± 3.39
[p47*phox*(1–286)-p67*phox*(1–212)-Rac1 (1–192)] chimera[c]	92.05 ± 2.38	50.76 ± 1.01	41.58 ± 1.72	37.80 ± 3.19

Assay mixtures contained solubilized macrophage membrane relipidated with PA, PG, PS, or PI, corresponding to 5 nM heme, and cytosolic activator(s), at 300 nM. p47*phox*, p67*phox*, and Rac1 were full-length proteins. Rac1 and the chimeras were nonprenylated. Rac1 and the chimeras were exchanged to the GTPase-resistant GTP analog, guanylyl imidodiphosphate (GMPPNP). The final concentrations of membrane phospholipid in the assays were 12 μM endogenous membrane phospholipid and 80 μM PA, PG, PS, or PI. Activation was in the absence of an anionic amphiphile. Methodology was as described in Ref. 137. The results represent means ± S.E. derived from three experiments

[a]Ref. 152
[b]Ref. 134
[c]Ref. 137

4.6 Beginning and End

All cell-free systems are reductionist constructions. The systems, in most of their variations, are missing all or part of the initiating transduction mechanism from membrane receptors to the enzyme and also lack the "termination" process occurring in the intact phagocyte. In vivo, NADPH oxidase activity is transient and $O_2^{\cdot-}$ production is regulated by the balance between assembly and disassembly of the complex (reviewed in [168]). Earlier studies in intact cells concluded that the turnover of cytosolic components in the complex [169] and in particular, p67*phox* and Rac [170], is very high, indicating a continuous exchange of bound for free cytosolic components. More recent studies indicate a stable accumulation of the key oxidase activator, p67*phox*, at the phagosomal membrane [171] but p47*phox* and Rac exhibited only a transient presence and detached from the phagosome prior to the end of ROS production [172]. Clearly, cell-free systems are not the methodology of choice for the assessment of the stability of the oxidase complex and the apparent termination of activity, when occurring in short-term assays, is due either to the exhaustion of NADPH or to the consumption of the reagents serving as $O_2^{\cdot -}$ traps. The brief duration of most contemporary cell-free assays also assures that the reaction components are unlikely to be inactivated in the course of the assay. In the past, it was thought that part of the $O_2^{\cdot -}$ generated in the system, which has escaped the intrinsic trap meant to bind the radical, might be dismutated to H_2O_2 and inactivate one or more of the oxidase components. Direct evidence for an inhibitory effect of ROS and H_2O_2, in particular, on p67*phox* and the membrane, prior to the assembly of the oxidase was presented [173]. The significance of this finding was emphasized by the detection of H_2O_2, derived from $O_2^{\cdot -}$ produced by the oxidase, at the cytosolic aspect of the phagocyte plasma membrane [174]. To prevent autooxidation by H_2O_2, catalase was added to the reaction, in order to degrade any H_2O_2 which might have been produced [175] but such supplementation is unnecessary in brief kinetic assays and in the presence of sufficient $O_2^{\cdot -}$-trapping reagent. In spite of the existing limitations, the stability of the assembled oxidase complex was studied in cell-free systems, too, and was found to be significantly increased by chemical cross-linking of membrane and unidentified cytosolic components [176], by chimerization of p47*phox* with p67*phox* [134, 148, 149] or with Rac1 [151], and, most pronouncedly, by using a tripartite chimera consisting of functional domains of p47*phox*, p67*phox*, and Rac1, as the activator [137]. It, thus, appears that procedures replacing the natural association/dissociation cycles between cytosolic components and between cytosolic and membrane components by covalent bonds, enhance the half-life of the oxidase complex.

4.7 What Are We Measuring in Cell-Free Assays

All cell-free systems are designed to mimic oxidase activation in vivo under in vitro conditions, starting from the equivalent of the state of the enzyme in resting cells. Enzyme activity is expressed as the reaction rate, based on the quantification of a reaction product or on the consumption of a reaction substrate. A comprehensive recent review dealing with the detection of $O_2^{\cdot -}$ and H_2O_2 production by NADPH oxidases, with emphasis on whole cells, serves as a useful introduction to this section [177].

The most commonly used techniques are the following:

1. *Reduction of cytochrome c by $O_2^{\cdot-}$.* This method is by far the most reliable, easy to perform, and convertible to a high throughput assay. It was first described in the landmark paper by Babior et al. [19] on the production of $O_2^{\cdot-}$ by phagocytosing leukocytes. The specificity of cytochrome *c* reduction is checked by its elimination in the presence of superoxide dismutase (SOD). This assay is also used in less common situations, such as in NOX1-based cell-free systems [178], and for assessing the cytosol-independent diaphorase activity of the DHR of NOX4 [179].
2. *Reduction of iodonitrotetrazolium violet (INT).* This method was introduced with the claim that INT is reduced by electrons originating in $FADH_2$ bound to the DHR of NOX2, by two-electron reduction and, thus, is measuring a step before the reduction of the two hemes and the generation of $O_2^{\cdot-}$ [180]. Most of the INT reduction was described as being SOD-resistant and not being dependent on p47^{phox} (however, *see* **Note 12**). INT reduction is useful for measuring the spontaneous or cytosolic factors-dependent "activity" of the DHR of NOX2 and other NOXes [181, 182].
3. *Reduction of nitrotetrazolium blue (NBT).* This method is used almost exclusively for measuring NADPH-dependent diaphorase activities of the DHR of NOX2 and other NOXes, in the presence or absence of cytosolic activators [181, 183, 184].
4. *Reduction of WST-1.* 4-[3-(4-Idophenyl)-2-(4-nitrophenyl)-2*H*-5-tetrazolio]-1,3-benzene disulfonate (WST-1) has recently become an alternative to cytochrome *c*, based on the claim of more efficient reduction by $O_2^{\cdot-}$ and a lesser chance of $O_2^{\cdot-}$ to be dismutated to H_2O_2 [185]. It was successfully applied to use in the cell-free system [143].
5. *Other artificial electron acceptors.* These include dichloroindophenol, potassium ferricyanide, and cytochrome b_5. Together with INT and NBT, they are used for measuring the constitutive diaphorase activities of the DHR of NOX4 [179, 182].
6. *Measuring the production of H_2O_2.* Quantification of the primordial ROS generated in the cell-free system, $O_2^{\cdot-}$, should, in most cases, by the default choice. On rare occasions, H_2O_2, derived by nonenzymatic dismutation of $O_2^{\cdot-}$, is measured in NOX2 cell-free systems. Unlike NOX2, NOX4 produces mainly H_2O_2 [186] and cell-free systems centered on NOX4 are based on the quantification of H_2O_2 [179, 187]. Most methodologies are based on horseradish peroxidase (HRP)-dependent oxidation of probes such as phenol red [188], assessed by colorimetry, or *N*-acetyl-3,7,-dihydroxyphenoxazine (Amplex Red), utilizing a fluorescence method [189]. Care must be exercised when using the Amplex Red

method because NADPH can oxidize Amplex Red directly [190], requiring the use of adequate controls and a lower concentration of NADPH [187]. Assay buffers containing HRP should not contain NaN_3, an inhibitor of peroxidases. Oxidase assay buffers intended for quantifying $O_2^{\cdot-}$ production, routinely contain 2 m*M* NaN_3 (*see* Subheading 5.1.5). An additional method is based on the oxidation of the nonfluorescent 2′7′-dichlorodihydrofluorescein (OxyBURST Green), resulting in a fluorescent product [191]. The product reacting with the reduced fluorochrome is probably H_2O_2. This latter method was used successfully in at least one cell-free system [143].

7. *Chemiluminescence assay for measuring* $O_2^{\cdot-}$. On some occasions, an enhanced sensitivity is required for the detection of $O_2^{\cdot-}$ in cell-free assays. For this purpose, bis-*N*-methyl acridinium nitrate [lucigenin) is used as the chemiluminescent detector. Although its validity as a specific probe was claimed [192], doubts were expressed about its specificity (reviewed in [177]). Its use in a canonical amphiphile-dependent cell-free system is illustrated in [193].
8. *NADPH consumption*. This is a simple technique, easily applicable to cell-free assays [194]. Its principal use is in situations in which a compound added to the reaction interferes with the detection reagent (*see* Ref. 119). As is the case for all substrate consumption assays, it has the disadvantage that the product of the reaction is presumed but not ascertained. However, when applied to the semirecombinant type of cell-free assay [84], there is an almost absolute certainty that NADPH is used exclusively for $O_2^{\cdot-}$ production. It is highly recommended to determine the concentration of NADPH solution used in the assay by its absorbance at 340 nm [195].
9. *Oxygen consumption*. This assay is rarely used at present in cell-free systems because of the cumbersome equipment required. It was popular in the early history of the cell-free system in order to establish the stoichiometry between oxygen consumption and $O_2^{\cdot-}$ production [118, 132]. Recent developments comprise the introduction of microplate-based oxygen consumption instruments (such as made by Seahorse Bioscience) [196]. To the best of this author's knowledge, these were not yet applied to cell-free assay but the potential exists.

    It is important to emphasize that an important element in the choice of the best methodology to be used in cell-free systems is the ability to run kinetic assays in which reaction velocities are recorded as V_{max} (reaction product or substrate consumption units/time unit). It is also preferable to assess the formation of the primary product of the oxidase, namely, $O_2^{\cdot-}$.

4.8 The "Two-Step" Assay

A modification of the cell-free system is the "two-step" assay, the purpose of which is to separate the assembly and catalytic phases in oxidase function. In the "two-step" assay, the components of the oxidase are first mixed in a small volume (usually 1/10–20 of the final reaction volume) and exposed to the activating amphiphile at a concentration resulting in maximal activation. This "first-step" mixture does not contain NADPH and the $O_2^{\cdot-}$ detection reagent. After incubation for a determined time interval (leading to the assembly of the oxidase complex), the mixture is supplemented with assay buffer containing the $O_2^{\cdot-}$ detection reagent and NADPH but no amphiphile, which results in the dilution of the amphiphile to a nonactivating concentration, and $O_2^{\cdot-}$ generation is measured (this second step represents the catalytic phase). The "two-step" assay is useful for testing the mechanism by which inhibitors affect oxidase activation. The ideal inhibitor interferes with the assembly or with the binding of a substrate (such as NADPH) but does not target the ROS detection reagents (*see* Subheading 6.7). Examples of the use of the "two-step" assay can be found in Refs. 119, 137, 197, 198.

4.9 A Few Leftovers: Eosinophils, Temperature, and Thermodynamics

The vast majority of cell-free assays utilize membranes from neutrophils and macrophages. On rare occasions, eosinophils were the origin of components. Eosinophils were reported to exhibit higher rates of ROS production than neutrophils [199, 200]. This was also reflected in higher activity in the cell-free system, a fact attributed to higher concentrations of, at the time, unidentified cytosolic components [200, 201].

Temperature has a major effect on the course of cell-free reactions. This was first examined by Ligeti et al. [202], who showed, by using a "two-step" cell-free assay, that oxidase assembly was temperature-dependent with full assembly being achieved in 5 min at 25 °C but only after 30 min, at 0 °C. An in-depth study of the effect of temperature on oxidase assembly in the cell-free system concluded that temperature effects were mediated by the thickness and fluidity (viscosity) of the lipid bilayer and not by changes in protein structure [203]. These characteristics of the membrane lipid environment affect the catalytic properties of NOX2 itself and not only the oxidase assembly. A major influence on viscosity was exerted by the sterol content of the membrane [204]. ROS production increased with increasing temperature, not supporting the claim (*see* Ref. 205) that maximal activity is reached at 37 °C. A thermodynamically constrained mathematical model for the kinetics of NADPH oxidase activation, based on cell-free and whole cells studies was recently published [206]. It concluded that the apparent K_m values for NADPH and O_2 were independent of temperature, but the apparent V_{max} increased with raising temperature (*see* **Note 13**).

4.10 Uses of Cell-Free Systems

Cell-free assays are extensively used in both basic research and clinical medicine. The tremendous expansion of the field of non-phagocytic NOXes has provided further impetus to their use and to the development of variations of the assay adapted to specific isoforms. However, the dominant applications of the cell-free system remain focused on NOX2, the main reason for this being the much more vigorous generation of ROS by this best known and most studied isoform, not requiring amplified detection means. At the time of the writing of this chapter, the original descriptions of the arachidonate and SDS-activated cell-free systems [77, 86] have accumulated a total of 750 citations and many authors cite later applications and variations of the original method. The principal uses of cell-free assays are the following:

1. The identification, quantification and functional assessment of oxidase components. At present this refers almost exclusively to components produced by recombinant technology and less commonly to those purified from cells. Although cell-free assays, if properly performed, are among the most sensitive techniques for the detection of oxidase components, obtaining quantitative data requires basing these on careful dose-response experiments with highly purified components (*see* Subheading 6.6.3). Thus, cell-free assays are mainly used to assess the functional competence of recombinant oxidase components.
2. Structure–function studies on recombinant oxidase components, subjected to mutagenesis, truncations, deletions, chimerization, and posttranslational modifications, such as prenylation. This is, at present, one of the most popular and rewarding applications, due to the very high sensitivity of the system, enabling detection of the effect of minor structural modifications on function. A few examples from the work of our group are: the effect of mutagenesis of Rac1 on its function [207], and the effect of mutagenesis of p67^{phox} and Rac1 moieties in [p67^{phox}-Rac1] chimeras [12, 135, 152], and of p47^{phox}, p67^{phox}, and Rac1 moieties in [p47^{phox}–p67^{phox}-Rac1] trimeras [137, 138] on their activity in the cell-free system.
3. The availability of numerous variations of the cell-free system has opened the way to novel applications which were not possible when only the canonical assay was available. Some examples are: the cell-free system in the absence of cytosolic activators [105, 106], used in the diagnosis of some forms of CGD [164]; the p47^{phox}-independent [13, 14] and the amphiphile- and p47^{phox}-independent variations [11, 165], allowing focusing on the interaction of cytochrome b_{558} with p67^{phox}, and a system dependent on the participation of GEFs [166].

4. The cell-free assay is also a sensitive, simple, and fast method to detect NADPH oxidase components by a complementation approach. The assay was most successfully applied to the study of complete or partial NADPH oxidase assembly. Mixtures of all or selected oxidase components are prepared, and oxidase assembly is induced in vitro under conditions mimicking those found optimal in the cell-free assay. The assembled and isolated components are separated by FPLC gel filtration. We found Superose 12 10/300 GL or Superdex 200 Increase10/300 GL columns most appropriate (both from GE Healthcare). In a typical experiment, macrophage membrane liposomes were mixed with recombinant p67^{phox} and prenylated Ra1Q61L and injected in the gel filtration column. The column eluate was passed through an absorbance monitor and fractions of the eluate were collected. Membrane liposomes eluted with the exclusion volume and were identified by an absorbance peak at 413 nm (cytochrome b_{558} heme). Samples from the collected fractions were analyzed by four types of cell-free assay: (1) Assembled oxidase was identified in the exclusion volume by $O_2^{\cdot -}$ production upon addition of NADPH only; (2) Cytochrome b_{558} was detected in the exclusion volume by $O_2^{\cdot -}$ production on addition of p67^{phox}, prenylated RacQ61L, and NADPH; (3) Free p67^{phox} was detected at the expected retention volume by $O_2^{\cdot -}$ production on addition of membrane liposomes, prenylated Rac1Q61L, and NADPH, and (4) Free Rac1Q61L was detected at the expected retention volume by $O_2^{\cdot -}$ production on addition of membrane liposomes, p67^{phox}, and NADPH. Binding of single oxidase components to the membrane liposomes, not resulting in an assembled oxidase, could be shown by complementation with the missing component (not bound), and NADPH. Examples of the use of this methodology for the study of oxidase assembly with p67^{phox} and prenylated Rac [11] and with [(p67^{phox}(1–212)-Rac1(1–192)] chimeras [135, 152, 153] were published. It is regrettable that this fruitful methodology is too little known and used.
5. Investigating the mechanism of action of oxidase inhibitors (*see* Subheading 6.7). With the ever increasing interest in the development of NOX inhibitors [143, 208–211] cell-free assays became one of the key methods in the search for such compounds, predominantly in the form of high throughput screening (HTS).
6. Diagnosis of the various forms of CGD and follow up on the success of therapeutic approaches applied to CGD patients. One of the first indicators of the importance of the cell-free system was its application to distinguishing between CGD

caused by mutations in cytochrome b_{558} [79, 83] and in the cytosolic components [95–99].

7. Following the discovery of NOX isoforms other than NOX2, there was much interest in the development of cell-free systems applicable to nonphagocytic NOXes. Most of the progress was made with NOX4 [179, 182, 186, 187, 212].

   Cell-free assays are characterized by simplicity, speed, rigorous but easily understandable expression of results, and good repeatability. They are well suited for work with multiple-well plates and are consequently ideal for HTS.

4.11 Methodological Reductionism Triumphant

Methodological reductionism is the concept that complex biological events should, if possible, be studied at the most elementary level, preferably down to that of interacting molecules [213]. Complex processes, such as activation of an oxidative burst in the intact phagocyte, are deconstructed to its component parts. The cell-free system is one of the best examples for the successful application of methodological reductionism. Soon after its discovery it was described in the following terms by the late Bernard Babior, a "founding father" of NADPH oxidase research: *"What was really needed to achieve an understanding of oxidase activation at the molecular level was a cell-free oxidase activating system that can be taken apart and analyzed component by component using biochemical techniques* [21]. The overwhelming majority of the results obtained employing cell-free methodology overlapped those obtained by working with whole phagocytes or cell lines transfected with oxidase components or whole organisms (knockout or natural disease). Occasionally, results obtained in the cell-free system differ from those obtained in whole cells, providing proof for the Aristotelian statement that "*the whole is more than the sum of its parts.*" Examples for such discrepancies are the findings that truncation of p67^{phox} at residue 246 (which removes both SH3 domains) [214] or deletion of the N-terminal SH3 domain [214, 215] led to the elimination of $O_2^{\cdot-}$ production by stimulated cells transfected with the p67^{phox} mutants but both mutants were found to be fully capable of supporting both amphiphile-dependent and -independent $O_2^{\cdot-}$ production in a cell-free system [214, 216].

In this chapter we describe the basic methodology for performing the different versions of the cell-free assay, commonly used for studying the NOX2 oxidase, and shall deal with theoretical considerations, interpretation of results, possible problems and their solutions, available alternatives, and the multiple applications of this method.

5 Materials (*See* Note 14)

5.1 Chemicals and Reagents

5.1.1 Preparation of Phagocyte Membranes

1. Paraffin oil, weight/mL = 0.85 (highly liquid; e.g., Merck). This was used for eliciting a sterile peritoneal exudate as a source of macrophages for the preparation of membranes.
2. Earle's balanced-salt solution: 6.8 g of NaCl, 0.4 g of KCl, 0.125 g of $NaH_2PO_4 \cdot H_2O$, 0.2 g of $MgSO_4 \cdot 7H_2O$, 1 g of glucose, 0.2 g of $CaCl_2$ (anhydrous), 1.25 g of $NaHCO_3$, and H_2O up to 1 L.
3. Sonication buffer: 8 mM potassium, sodium phosphate buffer, pH 7.0 (made from 61 parts of K_2HPO_4 and 39 parts of NaH_2PO_4 stock solutions), 131 mM NaCl, 340 mM sucrose, 2 mM NaN_3, 5 mM $MgCl_2$, 1 mM ethylene glycol-bis(-β-aminoethyl ether)-*N*,*N*,*N*′,*N*′-tetraacetic acid (EGTA), 1 mM dithiothreitol (DTT), 1 mM 4-(2-aminoethyl)-benzene-sulfonyl fluoride hydrochloride (AEBSF) (*see* **Note 15**), and 0.021 mM leupeptin hemisulfate. It is best to add DTT (200 mM), AEBSF (100 mM), and leupeptin (2.1 mM) from concentrated stock solutions (stock concentrations are listed in parentheses) just before using the buffer.
4. 3.5 M KCl solution in 20 mM Tris–HCl buffer, pH 7.5, for mixing with sonication buffer to reach a final concentration of 1 M KCl. It is used to wash macrophage membranes, to remove nonintegral membrane-attached proteins.
5. Octyl-β-D-glucopyranoside (octyl glucoside).
6. Solubilization buffer: 120 mM sodium phosphate buffer, pH 7.4, 1 mM $MgCl_2$, 1 mM EGTA, 2 mM NaN_3, 10 μM flavin adenine dinucleotide, disodium salt (FAD), and 20% v/v glycerol. Add 40 mM octyl glucoside (from powder), 1 mM DTT, 1 mM AEBSF, and 0.021 mM leupeptin from concentrated stock solutions just before using the buffer. Unused buffer can be divided in aliquots of 25–50 mL and stored frozen at −20 °C. The same basic buffer, not supplemented with octyl glucoside, FAD, and AEBSF serves for dialysis of solubilized membranes leading to the formation of membrane liposomes.
7. Sodium dithionite.
8. Phospholipids: 3-*sn*-phosphatidic acid (PA) (sodium salt, from egg yolk, 98%) and L-α-phosphatidyl-DL-glycerol (PG) (ammonium salt, synthetic, 99%). Dissolve phospholipids at 5 mM in solubilization buffer containing 40 mM octyl glucoside (but lacking protease inhibitors, DTT, and FAD). Dispense into aliquots of 0.5 mL and store at −75 °C.

5.1.2 Expression of Recombinant Cytosolic Components

All recombinant proteins used in the cell-free assays have an N-terminal 6His tag. The basic procedure for expression in *E. coli* and subsequent purification of the proteins is described in Ref. 138.

1. *E. coli* competent cells (Rosetta 2(DE3)pLysS; Novagen).
2. pET-30a expression vector (Novagen).
3. LB Broth.
4. Isopropyl β-D-1-thiogalactpyranoside (IPTG).
5. 10% Triton X-100 solution prepared in H_2O.
6. Protease inhibitor mixture cOmplete, EDTA-free (e.g., Sigma-Aldrich).

5.1.3 Purification of Recombinant Oxidase Cytosolic Components

1. Imidazole solution. Prepare a 2 M solution in H_2O, bringing it to pH 7.4 with 3 M HCl.
2. Ni Sepharose 6 Fast Flow (GE Healthcare).
3. *E. coli* lysis buffer also serving as binding buffer for metal affinity chromatography on Ni Sepharose: 20 mM sodium phosphate buffer, pH 7.4, 0.5 M NaCl, and 20 mM imidazole.
4. Washing buffer for metal affinity chromatography on Ni Sepharose: 20 mM sodium phosphate buffer, pH 7.4, 0.5 M NaCl, and 40 mM imidazole.
5. Elution buffer for metal affinity chromatography on Ni Sepharose: 20 mM sodium phosphate buffer, pH 7.4, 0.5 M NaCl, and 300–500 mM imidazole.
6. Buffer used in FPLC gel filtration experiments (phosphate buffered saline, PBS): 137 mM NaCl, 2.7 mM KCl, 4.3 Na_2PO_4, 1.4 mM KH_2PO_4, and 2 mM NaN_3, pH 7.3.
7. Gel filtration markers kit for protein molecular weights 12,000–200,000 Da.
8. "Protein assay dye reagent concentrate" for measuring protein concentration by the Bradford assay [217] (Bio-Rad).
9. 2 mg/mL bovine gamma globulin standard for Bradford assay.
10. NuPage 12% Bis-Tris electrophoresis gels, 1 mm gel thickness (Invitrogen).
11. NuPage MOPS SDS running buffer (Invitrogen).
12. NuPage LDS sample buffer (4×), and NuPage reducing agent (10×) (Invitrogen).
13. Precision Plus SDS PAGE protein standards, unstained (10–250 kDa) (e.g., Bio-Rad).
14. "Instant Blue" protein gel stain (Expedeon).

5.1.4 Reagents Required for Nucleotide Exchange on Rac (See ***Note 16****)*

1. Buffer for diluting recombinant Rac1 for performing nucleotide exchange: 50 mM Tris–HCl, pH 7.5, 150 mM NaCl, 4 mM $MgCl_2$, and 2 mM DTT. PBS is not compatible with high concentrations of $MgCl_2$.
2. 10 mM guanylylimidodiphosphate, trisodium salt hydrate (GMPPNP) stock solution: prepare in H_2O, aliquot, and store –75 °C.
3. 10 mM guanosine-5'-*O*-(3-(thio)triphosphate, tetralithium salt (GTPγS) stock solution: prepare in H_2O, aliquot, and store –75 °C.
4. 0.5 M ethylenedinitrilotetraacetic acid, disodium salt dihydrate (EDTA) stock solution: prepare in H_2O. In order to dissolve EDTA, the solution has to be brought to pH 8.0 with 10 M NaOH.
5. 1 M $MgCl_2$ stock solution: prepare in H_2O.
6. Recombinant rat geranylgeranyl transferase I, GST-Fusion, His-Tag, made in *E. coli* (Sigma-Aldrich, Calbiochem brand). 1 pmol enzyme will transfer ≥3 pmol geranylgeranyl to RhoA in 10 min at 37 °C, pH 7.2.
7. 1 mg/mL geranylgeranyl pyrophosphate: prepare in methanol/10 mM aqueous NH_4OH (7:3).
8. Prenylation buffer: 50 mM Tris–HCl buffer, pH 7.7, 0.1 mM $ZnCl_2$, 5 mM $MgCl_2$, and 2 mM DTT.
9. 10% (v/v) Triton X-114 solution: prepare in H_2O.

5.1.5 Cell-Free Assays

1. Cytochrome *c*, from equine heart, 95% (e.g., Sigma-Aldrich), (*see* **Note 17**).
2. 10 mM *p*-iodonitrotetrazolium violet (INT) solution: prepare in ethanol and keep frozen at –20 °C, in the dark.
3. 5 mM β-nicotinamide adenine dinucleotide 2′-phosphate reduced, tetrasodium salt, hydrate, ≥93% (NADPH): prepare in H_2O, divide into 1 mL aliquots, and store at –20 °C. Avoid frequent thawing and freezing (*see* **Note 18**).
4. Arachidonic acid (>95%, product number 10931, or ≥98.5%, product number A3611, both from Sigma Aldrich). A more convenient alternative is arachidonic acid sodium salt, ≥98.5%, product number SML 1395 (Sigma Aldrich).
5. 10 mM lithium dodecyl sulfate (LiDS) stock solution: prepare in H_2O and store at 4 °C for unlimited periods, provided that evaporation is prevented. Unlike SDS, LiDS does not precipitate out of aqueous solutions at 4 °C.
6. Superoxide dismutase (SOD), from bovine erythrocytes: prepare 10,000 U/mL stock solution in H_2O, aliquot in amounts of 100 μL, and store at –20 °C.

7. NADPH oxidase assay buffer: 65 mM sodium phosphate buffer, pH 7.0 (made from 61 parts K_2HPO_4 and 39 parts NaH_2PO_4 stock solutions), 1 mM EGTA, 10 μM FAD, 1 mM $MgCl_2$, 2 mM NaN_3, and 0.2 mM cytochrome *c* (*see* **Note 19**). The conductivity of this buffer is 7.7 mS/cm. When reduction of INT is measured, cytochrome *c* is replaced by 100 μM INT. When the assay is based on NADPH consumption, the buffer does not contain cytochrome *c* or INT. When the concentration of LiDS to be used in a large number of assays is known, this can be dissolved in the buffer. Assay buffer with and without LiDS can be divided into batches of 100 mL and stored at –20 °C for unlimited periods of time, in the dark, to prevent damage to FAD. The rationale for the components of the buffer is discussed in Subheading 6.6.2.

5.2 Disposable Plasticware

1. 96-well microplates, polystyrene, flat bottom, clear, with a well volume of 382 μL and a maximal height of 10.9 mm (*see* **Note 20**). When the wells in these plates are filled with a 0.21 mL reaction volume, the vertical light path is 0.575 cm. Plates intended for use in ELISA assays of medium or high hydrophilic protein-binding capacity are not recommended for use in cell-free assays. For the performance of the NADPH consumption assay, 96-well microplates permitting passage of UV light are used (*see* **Note 20**).
2. For the preparation of dilutions and the storage of recombinant proteins and membrane liposomes, "Protein LoBind" 1.5 mL conical tubes (Eppendorf), or "Non Stick Surface" 1.5 mL conical tubes (Labcon), made of polypropylene are recommended, to reduce binding of the proteins to the tube wall. Glass and polystyrene tubes should not be used.
3. For batch metal affinity purification of 6His-tagged recombinant proteins, disposable centrifuge columns (polypropylene, 10 mL capacity), with polyethylene bottom filter (30 μm pore size), and 30 mL column extenders (Pierce, Thermo Fisher Scientific), were found very useful for both binding to gel and elution from gel.
4. Centrifugal concentrators, 10,000 molecular weight cutoff, 4- and 15-mL (Amicon Ultra), for the concentration of all recombinant cytosolic components and buffer exchange.
5. Disposable Rapid-Flow Filter Unit, 0.45 μm pore size, aPES membrane, 250 or 500 mL capacity for filtration of all solutions used the metal affinity purification procedures (Nalgene).

5.3 Small and Medium-Sized Equipment for General Use and for Preparation of Recombinant Oxidase Components

1. Manual and electronic single-channel pipettors (range from 0.5 to 1000 μL). Dispensing mode is very useful for adding small equal amounts of reagents to 96-well plates.
2. Multipette Plus (manual) or Multipette *stream* (electronic) pipettors (Eppendorf) and various Combitips (Eppendorf), for distributing equal amounts of reagents in 96-well plates.
3. Multichannel (8-channel) manual (30–300 μL range) or electronic (15–300 μL range) pipettors.
4. Rotating tube mixer Rotamix RM1 (ELMI).
5. XCell SureLock Mini-Cell for SDS-PAGE of mini-gels (Invitrogen).
6. Electrophoresis power supply.
7. Thermomixer comfort, rotary mixer, and heater/cooler (Eppendorf).
8. Tabletop microcentrifuge.
9. FPLC gel filtration columns HiLoad 10/60 Superdex 75 prep grade (fractionation range: 3–70 kDa), for purification of p47^{phox}, p67^{phox} (1–212) and Rac, and HiLoad 10/60 Superdex 200 prep grade (fractionation range: 10–600 kDa), for purification of p67^{phox} (1–526) (GE Healthcare). Columns are fitted with a coolant jacket.

5.4 Large Equipment of General Use and for Preparation of Phagocyte Membranes and Recombinant Oxidase Components

1. "Warm room" set at 37 °C, containing shaking platform for culturing transformed *E. coli* cultures before induction by IPTG.
2. Innova 4230 or C24KC refrigerated incubator shaker (New Brunswick Scientific). These incubators are suitable for growth of *E. coli* at 18 °C, for expression of recombinant oxidase components by IPTG induction.
3. Refrigerated low-speed centrifuge (up to 7000 × *g*), with a swing-out rotor.
4. Refrigerated high-speed centrifuge (up to 48,000 × *g*), with fixed angle rotor.
5. Ultracentrifuge and fixed angle rotor.
6. High intensity ultrasonic processor (500-W) fitted with exchangeable 13 mm diameter tip or microprobe.
7. Optical microscope (with 10× and 40× magnification objective lenses and phase contrast capability).
8. Spectrophotometer (double-beam) UV/visible.
9. Akta Basic 10 chromatography system, to be used for FPLC gel filtration of recombinant oxidase components (GE Healthcare).
10. Coolant circulator for gel filtration columns (Fryka).

11. Compressed Helium gas tank for bubbling through FPLC buffer.

5.5 Large Equipment Specifically Intended for Performance of Cell-Free Assays

1. Tunable microplate absorbance readers. For reading of cell-free assays based on the reduction of cytochrome *c*, INT, or NBT, we used in the past a SpectraMax 340 reader, fitted with software SoftMax Pro, Version 5.2. Recently, we are using a VersaMax reader, fitted with software SoftMax Pro, Version 6.5.1. For the NADPH consumption assay, we are using a SpectraMax 190 reader (all instruments are manufactured by Molecular Devices) (*see* **Note 21**).
2. Vibrating platform shaker for 96-well plates.

6 Methods

Cell-free assays are used in an almost limitless variety of forms and applications. In the original forms of the assay, cytochrome b_{558} was represented by a total phagocyte membrane preparation and the cytosolic components, by total cytosol [77, 79, 82, 83]. In the variants developed later, a more sophisticated membrane preparation is utilized or the membrane is altogether replaced by purified relipidated cytochrome b_{558}. In all the assays to be described, we use a modified membrane preparation, originating from guinea pig peritoneal exudate macrophages, and the cytosol is replaced by purified recombinant proteins (p47^{phox}, p67^{phox} and Rac). Although a number of anionic amphiphiles can act as activators in cell-free systems, we shall limit our description to LiDS as the prototype activator. We describe one method for the detection of $O_2^{\cdot-}$ based on its trapping by oxidized cytochrome *c* [19], one method to assess the diaphorase activity by the reduction of INT [180], and one method, based on the consumption of NADPH [194].

6.1 Cytochrome b$_{558}$ (the Membrane Component)

6.1.1 Membrane Preparation

We here describe the preparation of membranes from elicited guinea pig peritoneal macrophages [86, 132] (*see* **Note 22**). Considering the paucity of granules in macrophages, differential centrifugation is not required to obtain granule-free pure plasma membrane preparations in these cells. Instead, we prepare of a "total" membrane fraction, defined by its sedimentation at 160,000 × *g*. The use of an uncharacterized membrane preparation is made possible by the fact that, in the cell-free assays to be described, the amounts of membrane added to the assay are based strictly on the cytochrome b_{558} content.

1. Inject guinea pigs (male or female, weighing 300–500 g) intraperitoneally with sterile light paraffin oil (15 mL per animal).

2. After 4 day, sacrifice the animals by CO_2 inhalation, and collect the peritoneal content via a 3-cm-long incision in the abdominal wall and repeated introduction and aspiration of 50-mL volumes of ice-cold Earle's balanced-salt solution.
3. Pass the collected peritoneal lavage through a 180 μm pore size nylon mesh sheet and collect in ice-cooled Erlenmeyer flasks. (*see* **Note 23**).
4. Centrifuge fluid for 20 min at 940 × *g* and 4 °C, to sediment the cells.
5. Repeat the procedure once more and suspend the cell sediment in ice-cold distilled H_2O (20 mL to a cellular pellet derived from 400 mL lavage fluid) to lyse red cells.
6. After 3 min, add an equal volume of ice-cold 0.29 M NaCl solution in water (20 mL), resulting in an isotonic NaCl concentration, mix well, and recentrifuge at 940 × *g*, as above. If necessary, the lysis procedure can be repeated once more.
7. Resuspend the cell pellet in Earle's solution (10 mL per animal), and count cells after diluting the suspension 1:10 in 1% v/v acetic acid in H_2O. The expected cell harvest varies from 1 to 2×10^8 cells per animal.
8. Pellet the cells at 940 × *g*, and resuspend in sonication buffer, at a concentration of 10^8 cells/mL, in 16 × 100 mm polypropylene tubes (4 mL/tube).
9. Sonicate samples (400-W ultrasonic processor), keeping tubes immersed in ice-water, with the microprobe lowered into the cell suspension 2/3 of its entire depth. Submit cells to three cycles of sonic disruption, at an amplitude of 10%, each cycle consisting of sonication for 9 s, followed by a 1-s rest, repeated three times. Check for quality of cell disruption by phase contrast microscopy at 40× magnification (~90% cell disruption is expected).
10. Centrifuge the cell homogenate for 10 min at 3000 × *g* and 4 °C in a swinging-bucket rotor to remove unbroken cells, aggregates, and nuclei. Collect the supernatant (postnuclear supernatant).
11. Centrifuge the supernatant for 2 h at 160,000 × *g* and 4 °C (ultracentrifuge is required). The supernatant from this step is collected and represents the cytosol. If an ultracentrifuge is not available, centrifuge for 4 h at 48,000 × *g* in a high-speed centrifuge.
12. Resuspend the membrane pellet in sonication buffer supplemented with 1 M KCl (*see* **Note 24**) at a volume identical to the original volume of the homogenate. Resuspend directly in the ultracentrifuge tubes (kept in ice-water) by adding buffer with 1 M KCl in small aliquots (1–2 mL) and very briefly and

gently sonicate after each addition, using a microprobe, until all the membrane is suspended (*see* **Note 25**).

13. Centrifuge the mixture for 2 h at 160,000 × *g*.
14. Discard the supernatant and freeze the membrane pellet at −75 °C. The membranes can be kept frozen indefinitely for future use. We found no reason for flash freezing or keeping membranes at a lower temperature.

6.1.2 Preparation of Membrane Liposomes

Although membrane preparations obtained as described above are adequate for use, we routinely use solubilized membrane preparations consisting of liposomes as our source of cytochrome b_{558} in cell-free assays. Liposomes are obtained by solubilizing membranes with octyl glucoside and then removing detergent by dialysis [122, 132].

1. Suspend frozen membranes in ice-cold solubilization buffer at a concentration of 5 × 10^8 cell equivalents/mL using the original ultracentrifuge tubes as containers.
2. Stir suspension with a small magnetic bar by placing the tube in a beaker containing ice-water. Continue solubilization until no intact membrane fragments remain. This might take 3–6 h, and it is important to keep the tubes ice-cooled throughout this period.
3. Centrifuge the solution for 1 h at 48,000 × *g* and 4 °C in a fixed angle rotor.
4. Transfer supernatant containing the solubilized membrane into a fresh tube (appears as a pale-yellow opalescent solution). Discard pellet.
5. Place solution in dialysis tubing with a 25,000 molecular weight cut-off (*see* **Note 26**) and dialyze against a 100-fold excess of detergent-free solubilization buffer for 18–24 h at 4°°C (*see* **Note 27**). The dialyzed detergent-free solubilized membrane consists of liposomes of about 300 nm, in diameter (Fig. 2).
6. Measure the concentration of cytochrome b_{558} as described below.
7. Supplement the preparation with 10 μM FAD, divide into aliquots of 1–1.5 mL, and store at −75 °C (*see* **Note 28**). The preparation is now ready to be used in all forms of cell-free assays (*see* **Note 29**).

6.1.3 Quantification of Cytochrome b558 Content

The results of cell-free assays should be expressed in turnover values (mol $O_2^{\cdot -}$ produced per time unit [s] per mol cytochrome b_{558} heme) (*see* **Note 30**). Thus, it is essential that the cytochrome b_{558} content of membranes is known. Cytochrome b_{558} content is expressed by heme concentration.

1. Dilute membrane liposome preparation in solubilization buffer without octyl glucoside and FAD (1:5 or 1:10 dilution). Place diluted samples in spectrophotometer cuvette (*see* **Note 31**).
2. Place cuvette in the sample compartment of a double-beam spectrophotometer, and place a cuvette containing the buffer used for diluting the membrane preparation in the reference compartment.
3. Run an absorption spectrum scan, from 400 to 600 nm, using a band-width of 1 nm, a scanning interval of 1 nm, and a scanning speed of 100 nm/min (*see* **Note 32**). Save this as "oxidized spectrum" in the computer linked to the spectrophotometer.
4. Add a few grains of sodium dithionite, rapidly mix, and run a spectral scan again. Save this as "reduced spectrum."
5. Subtract the "oxidized spectrum" from the "reduced spectrum," and save the resulting "reduced *minus* oxidized spectrum."
6. Detect and record absorbance values of major cytochrome b_{558} peaks, located at 558/559, 529, and 426/427 nm. Detect and record absorbance at the "valley," at 410/411 nm.
7. Calculate heme content, based on the Δ extinction coefficient of the 426/427 nm peak/411 nm valley pair, using $\Delta\varepsilon_{426/427\text{–}411\,\text{nm}} = 200\ \text{mM}^{-1}\ \text{cm}^{-1}$ [218] (*see* **Note 33**).
8. Normally, we obtain a concentration of cytochrome b_{558} heme of 400 pmol/10^8 cell membrane equivalents. Thus, the membrane liposome suspension of 5×10^8 cell membrane equivalents/mL has a heme concentration of 2 μM (*see* **Note 34**). A typical "reduced *minus* oxidized spectrum" is shown in Fig. 3.

6.2 Preparing Recombinant Cytosolic Components

All recombinant cytosolic components are expressed in *E. coli* as N-terminal 6His-tagged proteins. The methodology is described in Ref. 138. The key steps are briefly summarized below:

6.2.1 Expression of p^{47phox}, p^{67phox}, and Rac1

1. Transform *E. coli* competent cells (Rosetta 2(DE3)pLysS, Novagen) with the expression vector pET-30a-6His Kan^R, carrying cDNA encoding each of the three cytosolic components, following a standard protocol, as described in pET System Manual, 11th edition, Novagen.
2. Induce bacteria with 0.4 mM IPTG and grow at 18 °C for 18 h in a refrigerated incubator shaker. Induction at 18 °C is important for maximizing the recovery of the recombinant proteins in the soluble fraction after disruption of the bacteria.
3. Test by SDS-PAGE for successful induction of the desired protein by comparing bacterial cell extracts before and after induction by IPTG.

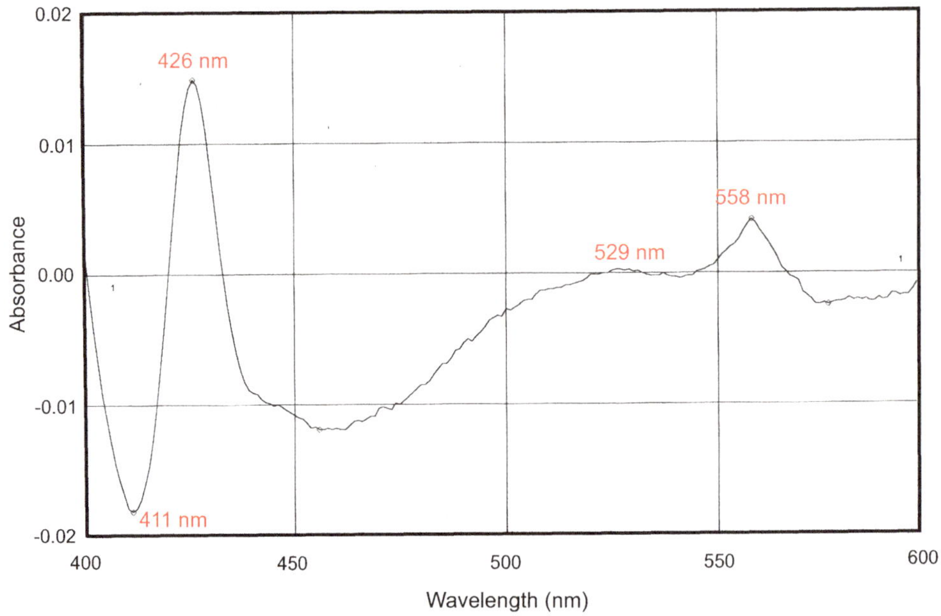

Fig. 3 Typical reduced *minus* oxidized spectrum of cytochrome b_{558} in total guinea pig macrophage membrane liposomes obtained by reduction with sodium dithionite. The absorption peaks at 558 nm (α band), 529 nm (β band), and 426 nm (γ or Soret band) are pointed out, as well as the valley at 411 nm. The concentration of heme was calculated based on the absorption at 426 nm *minus* absorption at 411 nm and found to be 1655 pmol/mL (1.655 μM)

4. Sediment bacteria by centrifugation at 3500 × *g* and resuspend in lysis buffer supplemented with protease inhibitor mixture cOmplete, EDTA-free (one tablet per 50 mL buffer). For 1 L of original bacterial culture, use 40–50 mL lysis buffer.
5. Freeze the bacterial suspension at −75 °C. Frozen suspensions can be kept frozen till processed further. This freezing step is essential for successful recovery of soluble recombinant protein.
6. Allow suspension to thaw slowly and add Triton X-100 to a final concentration of 1% v/v.
7. Subject the suspension (~50 mL) to sonic disruption in a glass beaker immersed in an ice/water mixture using a 500-W ultrasonic processor with 13 mm tip probe, for two to three 5 min intervals, at an amplitude of 20%, and alternating cycles of sonication for 2 s and 2 s of rest.
8. Centrifuge at 48,000 × *g* for 30 min and decant supernatant containing the soluble protein.
9. Measure protein concentration by the Bradford assay [217], modified for use with 96-well microplates (*see* Bio-Rad Technical Bulletin 1177 EG and **Note 35**).

6.2.2 Two-Stage Purification of Recombinant p47phox, p67phox, and Rac1

Purification by metal affinity chromatography on Ni sepharose

1. Mix 3 mL washed packed Ni sepharose beads with soluble fraction derived from sonic disruption of bacteria from 1 L culture (~40 mL). Incubate for 1 h at room temperature with top/bottom rotation using a rotating tube mixer at 10 RPM.
2. Transfer content into a centrifuge column with bottom filter and allow the fluid to flow by gravity into collecting tube.
3. Wash beads twice with 15 mL volumes of binding buffer and twice with 15 mL volumes of washing buffer.
4. Add 10 mL/column elution buffer, seal bottom and top apertures, and incubate for 30 min at room temperature with top/bottom rotation using a rotating tube mixer at 10 RPM.
5. Allow the eluate to run by gravity into a collecting tube and repeat procedure from one to four times (labeled as eluates 1, 2, 3, and 4).
6. Measure protein concentrations in all eluates and analyze by SDS-PAGE, for purity. The purity requirements for the performance of routine cell-free assays are met by purification on Ni Sepharose. When cell-free assays are used for purposes requiring strict quantification, such as dose responses or the search for inhibitors, higher degrees of purity are required (*see* **Note 36**).

Purification by FPLC Gel Filtration

1. Proceed to purification by gel filtration on HiLoad 16/60 Superdex 200 column, for p67*phox*(1–526), and on HiLoad 16/60 Superdex 75 column, for p67*phox*(1–212), p47*phox*(1–390), and Rac. Pool eluates with significant protein concentration from Ni Sepharose and concentrate to volumes of 2.5 mL per one column purification (=2% of column volume), using Amicon Ultra 15 mL (10 kDa molecular-weight-cut-off) centrifugal concentrators. Centrifuge concentrates at 12,000 × *g* for 30 min in a microcentrifuge at 4 °C and use supernatant for gel filtration.
2. Inject material into a column refrigerated by a coolant circulator and perform gel filtration using an Akta Basic 10 chromatography system with refrigerated PBS, degassed by bubbling of Helium, as the running buffer, at a flow rate at 1 mL/min. Record absorbance at 280 nm and collect 2 mL fractions. Analyze fractions by SDS-PAGE and pool those of highest purity.
3. Supplement purified recombinant proteins with 30% v/v glycerol, divide in small aliquots in "Protein LoBind" or "Non Stick Surface" tubes and store at −75 °C. Avoid repeated

thawing/freezing. In this state, they are active for an unlimited time period (*see* **Note 37**).

6.2.3 The Preference for Rac1

Rac2 is the predominant form of Rac in neutrophils [101], whereas monocytes and macrophages use Rac1 in oxidase activation [100, 155, 219] (reviewed in [158]). The oxidase can be activated in the cell-free system by both Rac1 and Rac2, in the nonprenylated form. However, one should be aware that this is an artifact since nonprenylated Rac does not exist, as such, in vivo. Translocation to the membrane of nonprenylated Rac is dependent exclusively on the net positive charge of the polybasic domain at the C-terminus. Because Rac1 contains six contiguous basic residues in this domain, whereas Rac2 contains only three that are only partially contiguous, nonprenylated Rac1 is much more active in the cell-free system than Rac2 [10, 138, 220]. Native, recombinant Rac1 contains exclusively GDP [221, 222]. Thus, before use in cell-free assays, Rac1 is subjected to nucleotide exchange to the nonhydrolyzable GTP analogs GMPPNP or GTPγS (for choosing between the two analogs *see* **Note 38**). Recently, we are using predominantly the Rac1 mutant Q61L, which is constitutively in the GTP-bound form [223].

6.2.4 Nucleotide Exchange on Rac

The procedure described here is for exchange to GMPPNP, but the same procedure is used for exchange to GTPγS. It is based on the removal of bound endogenous GDP by chelation of Mg^{2+} and replacement of the bulk of GDP by the GTP analog.

1. Decide on the size of the batch of recombinant Rac that is to be to subject to guanine nucleotide exchange. For use in cell-free assays, we normally perform exchange on aliquots of 10–20 nmol of Rac in Tris–HCl buffer, pH 7.5. Place in 1.5- or 2-mL polypropylene tube.
2. Add GMPPNP stock solution in a quantity representing a ten-fold molar excess over the amount of Rac. For example, if you intend to perform exchange on 10 nmol of Rac, add 100 nmol of GMPPNP (10 μL from the 10 mM stock solution).
3. Add EDTA solution to a final concentration of 12.5 mM. Incubate for 30 min at 30 °C in a rotary mixer set at 600 movements/min.
4. Stabilize the exchanged state of Rac by adding $MgCl_2$ solution to a final concentration of 25 mM.
5. Store exchanged protein in polypropylene tubes, frozen at −75 °C (*see* **Notes 39** and **40**).

6.2.5 Prenylation of Rac In Vitro

In the past, 6His-tagged Rac1 was cloned into the baculovirus genome, and this was used to infect cultures of Sf9 cells [11]. In this procedure, prenylated Rac was expressed in the cell membrane and thus had to be purified following membrane solubilization. This procedure was replaced by a much simpler method using in vitro enzymatic prenylation of nonprenylated recombinant Rac1 [135].

1. Add 10 nmol of nonprenylated Rac (*see* **Note 41**) to a 1.5 mL or 2 mL polypropylene tube.
2. Add 10 μL (20 nmol) of geranylgeranyl pyrophosphate stock solution, 10 μL (10 U) of geranylgeranyl transferase I stock solution, and prenylation buffer containing $ZnCl_2$ (*see* **Note 42**) to a final volume of 0.9 mL.
3. Incubate for 45 min at 37 °C in a rotary mixer (Thermomixer Comfort, Eppendorf) set at 600 rpm (*see* **Note 43**).
4. Add 60 μL of 70 mM octyl glucoside in H_2O (final octyl glucoside concentration is 4.375 mM), and reincubate for 45 min under the same conditions as above.
5. Sonicate the protein in a 400-W ultrasonic processor fitted with a cup horn filled with ice-water, for five cycles of 10 s each at 50% amplitude.
6. Add 0.24 mL of glycerol to bring the final volume to 1.2 mL. The final concentrations of components are 8.33 μM Rac, 3.5 mM octyl glucoside, and 20% v/v glycerol (this does not take into account the glycerol carried into the reaction by nonprenylated Rac itself).
7. Prenylated Rac can be stored at −75 °C. After thawing it should be centrifuged for 15 min at 10,000 × *g* to check for the presence of aggregates. If sediment is found, use the supernatant after measuring its protein content anew.

6.2.6 Checking the Degree of Rac Prenylation

Prenylation in vitro is a very reliable methodology provided that a trusted source of geranylgeranyl transferase I is available. Nevertheless, it is recommended that until enough experience is acquired, the degree of prenylation should be checked [224].

1. Remove an aliquot from the completed prenylation mixture before adding glycerol. Since the final detection method is based on SDS-PAGE, one has to remove sufficient protein to make detection easy. Normally, one third of a 10 nmol Rac prenylation mixture is used for confirming prenylation (about 3 nmol Rac). The procedure is best performed in 1.5 mL conical microcentrifuge tubes.
2. Add prenylation buffer up to a total volume of 0.9 mL.
3. Add 0.1 mL of 10% v/v Triton X-114 (1% final concentration).

4. Place the tube in ice-water for 30 min, vortexing the tube every 5 min.
5. Heat the mixture at 37 °C for 10 min in a heating block (Thermomix Comfort) or a water bath. Keep the tubes stationary (do not mix). This will cause the solution to become cloudy due to aggregation of Triton X-114 above its cloud point. Amphiphilic proteins, such as prenylated Rac, will associate with the detergent aggregates, whereas nonprenylated Rac will remain in the aqueous phase.
6. Centrifuge the mixture at 10,000–12,000 × *g* in a table top microcentrifuge at room temperature. This will result in phase separation with the upper (aqueous) phase containing nonprenylated Rac and the lower (detergent-enriched) phase containing prenylated Rac. Transfer upper phase into a fresh tube.
7. Add prenylation buffer to the lower phase to make the total volume equal to that of the upper phase and mix well.
8. Take equal sample volumes from the two phases and subject to SDS-PAGE.
9. Compare intensity of the Rac bands (21 kDa) in the two phases (*see* **Note 44**).

6.3 An Overview of Cell-Free Assay Design

For the proper application of cell-free assays, it is essential to recall a number of theoretical considerations, as outlined below.

1. $O_2^{\cdot -}$ is generated by the NOX2 component of cytochrome b_{558}, found in the membrane, and results are to be related to the amount of NOX2 heme present in the reaction.
2. All cytosolic components must be present at saturating quantities in relation to cytochrome b_{558}. These quantities are determined by dose-response experiments in which the concentration of one or all cytosolic components is varied in the presence of a constant amount of cytochrome b_{558} (membrane).
3. The amphiphile-independent cell-free system is a very useful variant of the canonical system, with specific applications in situations in which the emphasis is on interaction between NOX2 and p67*phox*. In spite of the fact that it was described two decades ago [11] and is technically simple, it has not gained wide acceptance.
4. Normally, amphiphile-dependent cell-free assays are performed with nonprenylated Rac1. However, identical results are obtained with prenylated Rac1, provided that p47*phox* and p67*phox* are present in the reaction. In the absence of p47*phox*, amphiphile exerts a paradoxical inhibitory effect in cell-free assays containing prenylated Rac and p67*phox* [165].

5. All cell-free assays comprise two stages: (a) the stage of oxidase complex assembly, in the course of which cytosolic components are translocated to the membrane, leading to the induction of conformational change in NOX2, and (b) the catalytic stage, initiated by the addition of NADPH, expressed in the production of $O_2^{\cdot -}$. In some forms of cell-free assay ("two-step" assay), the two stages are separated by the interruption of assembly just before the initiation of catalysis. Since the length of time required for full assembly is not always known, assembly might merge with the catalytic stage, although an effort is usually made to bring assembly to completion before the addition of NADPH.
6. Cell-free oxidase assembly is time- and temperature-dependent (*see* **Note 45**).
7. Oxidase activation in cell-free systems is reduced by an increase in the ionic strength of the assay buffer (*see* **Note 46**).
8. Kinetic models of anionic amphiphile-induced oxidase assembly have been proposed both before [197, 225] and after [88, 206, 226, 227] the identification of all the components of the oxidase. These models are frequently contradictory and it should be understood that in vitro studies yield results that might differ from those derived from whole cell studies [16, 171, 172, 228].
9. To the best of my knowledge, the only description of the electron flow in the canonical semirecombinant cell-free system, based on experimental data and not on modeling [206], is still the two decades-old paper of Koshkin et al. [229].

6.4 The Canonical Amphiphile-Dependent Cell-Free Assay—"Don't Leave Home Without It"

6.4.1 Cytochrome c Reduction

We describe here the basic methodology for performing cell-free oxidase activation in what is called the "semirecombinant" system. This is a modification of the original amphiphile-activated (membrane+cytosol) system [77, 79, 82, 83, 86]. Measuring ROS production by phagocytes by an end-point assay, using a microplate spectrophotometer, became the standard procedure about four decades ago [230, 231], and a kinetic assay was introduced a decade later, stimulated by the enhanced capabilities of the available instruments [232]. Kinetic assays in microtiter plates soon became the routine procedure for the performance of cell-free assays. This required adjustment of the assay from the 1–3 mL volumes, used in standard spectrophotometers, to 100–300 μL volumes used with 96-well microtiter plates (96-well plates). We describe a kinetic cell-free oxidase activation assay performed in 96-well plates in which the reaction components comprise solubilized macrophage membrane liposomes, recombinant $p47^{phox}$, $p67^{phox}$ and nonprenylated Rac1.

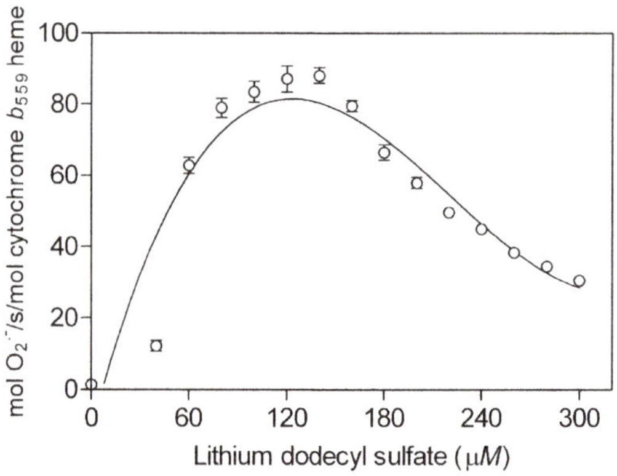

Fig. 4 Dose–response curve of lithium dodecyl sulfate (LiDS) in the amphiphile-dependent cell-free system. Assay mixtures consisting of solubilized macrophage membrane liposomes (5 nM cytochrome b_{558} heme) and recombinant p47^{phox} (100 nM), p67^{phox} (100 nM), and nonprenylated Rac1-GMPPNP (100 nM) were incubated with varied concentrations of LiDS as indicated. $O_2^{\cdot-}$ production was initiated by the addition of NADPH and measured by the kinetic cytochrome *c* reduction assay for 5 min. Results represent means ± S.E. of three experiments (Reproduced from Ref. 8 by permission of Springer Science + Business Media)

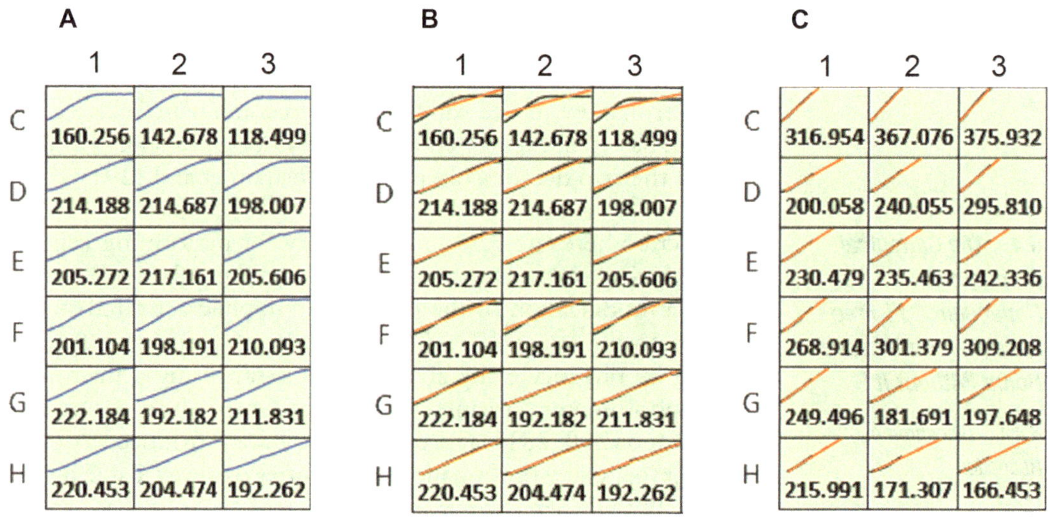

Fig. 5 Typical screen image of measuring the increase in absorbance at 550 nm in a cell-free assay using a VersaMax plate reader and SoftMax Pro 6.5.1 software. The settings are: first data point set to zero, results expressed as V_{max} (milli–Abs$_{550nm}$ per min), lag time zero, end time 300 s. (**a**) Raw data with end time of 300 s (blue data points). (**b**) An attempt to obtain linear curves (reduced data) is unsuccessful (black data points, orange curves) due to the presence of plateaus in many of the wells. (**c**) By reducing end time to 110 s, (reduced data) all curves become linear (black data point, orange curves) and V_{max} values are adjusted automatically

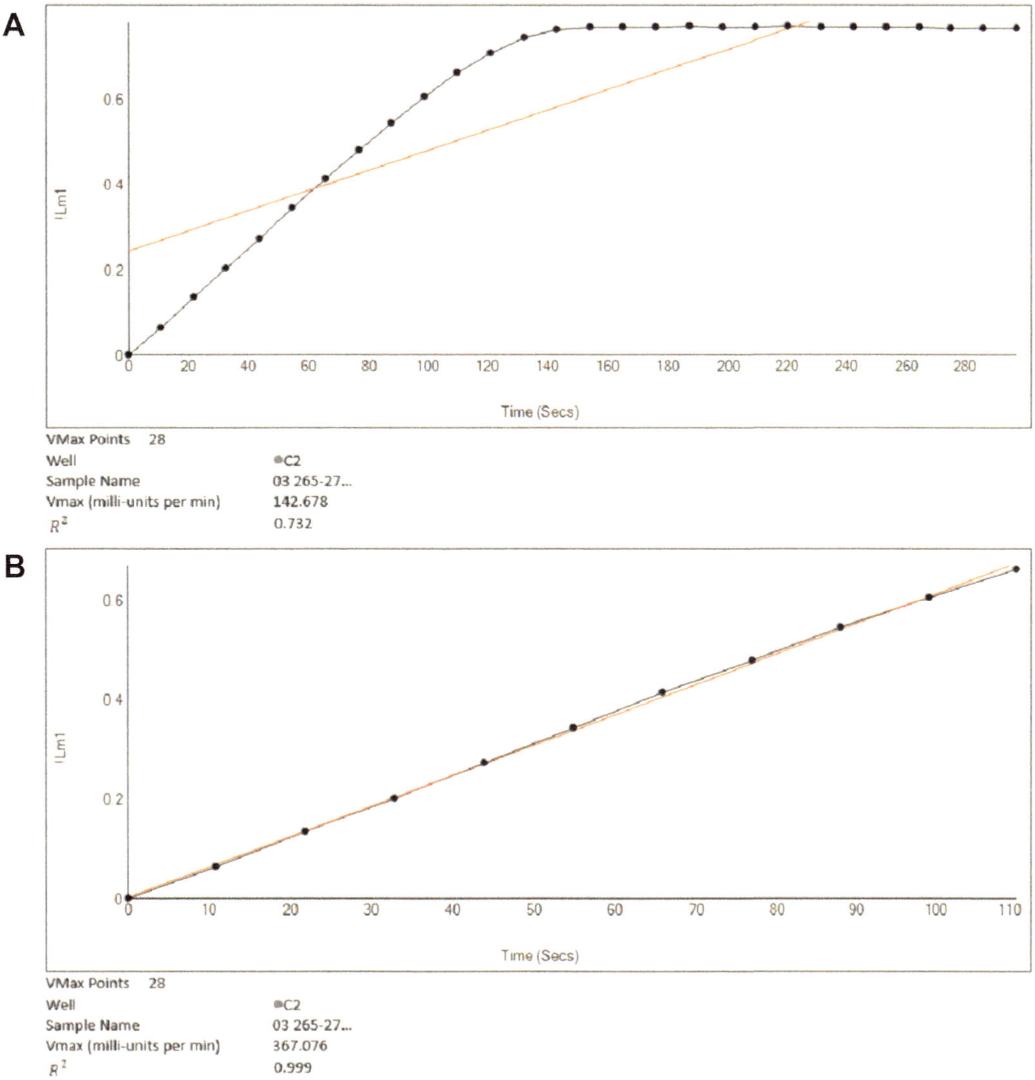

Fig. 6 Focusing on well C2 in the experiment illustrated in Fig. 5. (**a**) An attempt to obtain a linear curve in well C2. Black data points represent the original recording and the orange curve, the unsuccessful attempt to linearize, due to the presence of a plateau reached at 140 s. The V_{max} value of 142.678 units is incorrect, as also shown by the R^2 value of 0.732. (**b**) A linear curve is obtained by reducing end time to 110 s. The orange linear curve overlaps perfectly the original black data points. The V_{max} increases to 360.076 units and the R^2 value is 0.999

1. Add 20 μL of solubilized membrane liposomes (50 nM cytochrome b_{558} heme) to the wells of a 96-well plate. This is intended to result in a final concentration of 5 nM cytochrome b_{558} heme in 200 μL (the total volume of the reaction before addition of NADPH) and equals 1 pmol cytochrome b_{558} heme/well.

2. Add 20 μL of a mixture of p47^{phox}, p67^{phox}, and nonprenylated Rac1-GMPPNP (or Rac1 Q61L mutant), each at a concentration tenfold higher than that desired as the final concentration in 200 μL. As an example, if a final concentration of 100 nM is to be achieved for all three components, add 20 μL of a solution containing 1 μM of each component (*see* **Note 47**). Make all dilutions of membrane and cytosolic components in oxidase assay buffer without LiDS. Dispensing 10 or 20 μL aliquots of membrane or cytosolic components to the wells is best performed with single-channel electronic pipettors, in the dispensing mode, or with Multipette or multichannel pipettors.
3. Add 160 μL/well of assay buffer containing an optimized concentration of LiDS. We typically use a digital 8-channel pipette. For this protocol, the final concentration of LiDS causing maximal activation is 120–130 μM (Fig. 4) (*see* **Note 48**). Because the amphiphile is diluted 1.25-fold by the volumes of membrane and cytosolic components previously added to the wells, the concentration of the amphiphile in the assay buffer has to be adjusted accordingly. An alternative procedure, to be used when the optimal activating concentration of LiDS is being explored, is to add 150 μL assay buffer and immediately add 10 μL of a LiDS stock solution, made to result in the desired final concentration.
4. Place the plate on an orbital shaker and mix contents for 90 s at 500–600 movements/min and room temperature (*see* **Note 49**).
5. Dispense 10 μL of NADPH/well using a multichannel pipettor, as fast as possible. This results in a final concentration of 238 μM NADPH in a total volume of 210 μL per well, which is above the K_m for NADPH of the oxidase in the cell-free system [77].
6. Transfer the plate quickly to the microplate spectrophotometer, set at "temperature off" on the control panel (this means reading at ambient temperature, corresponding to 23–24 °C) and mix contents for 5 s before the first read. The instrument should be set to record increase in absorbance at 550 nm over a time period of 5 min. Include blank wells containing 200 μL assay buffer to which 10 μL of 5 mM NADPH were added simultaneously with its addition to the sample wells (*see* **Note 50** for details of microplate reader settings).
7. Results in the kinetic read type are expressed as "V_{max} = increase in Abs_{550nm}/min." The software of the instrument calculates these values by dividing the Δmilli–Abs_{550nm} over time, by the number of minutes elapsed. Thus, it is essential for the increase in absorbance curve to be linear. The curves turn nonlinear whenever one or more components of the reaction

is/are exhausted. Although every effort is made to prevent this from occurring, by choosing the right amounts of enzyme (cytochrome b_{558} in the membrane), cytochrome *c*, and NADPH (*see* **Note 51**), it occasionally happens. In this case, the linear portion of the curve is chosen, and values are recalculated as Δmilli–Abs_{550nm}/min for the revised time interval (shorter than 5 min) (*see* Figs. 5 and 6). Contemporary microplate readers are fitted with the appropriate software, allowing fast and simple recalculation of the slopes after selecting the linear segment. Δmilli–Abs_{550nm}/min values are transformed to nmol cytochrome *c* reduced per min per well content of 210 μL, based on the extinction coefficient $\Delta E_{550} = 21\ mM^{-1}\ cm^{-1}$ for reduced minus oxidized cytochrome *c* (=nmol $O_2^{\cdot -}$/min/well) (*see* **Note 52**).

8. Express the final results as "turnover": the amount of $O_2^{\cdot -}$ produced per time unit per mol membrane cytochrome b_{558} heme (mol $O_2^{\cdot -}$/s/mol cytochrome b_{558} heme; *see* *y*-axis of graphs in most figures in this chapter). This is easily calculated by knowing the nmol $O_2^{\cdot -}$/min/well values and the amount of cytochrome b_{559} heme per well (1 pmol, when 20 μL of membrane, at a concentration of 50 nM of cytochrome b_{558} heme, are added per well). For each experimental condition, perform the assay in triplicate wells and make the software calculate mean values and standard deviations (*see* **Note 53**).
9. It is essential to include SOD control wells in cell-free assays to assure that the reduction of cytochrome *c* is indeed due to $O_2^{\cdot -}$. This requires that parallel SOD-containing wells are included for every group of wells in which $O_2^{\cdot -}$ production is detected. Use a large excess of 100 U SOD/mL by adding 10 μL/well of a 2000 U/mL of SOD solution before the addition of NADPH. Addition of SOD is expected to prevent cytochrome *c* reduction by 95% or more (*see* **Note 54**).
10. A number of additional control reactions are requirements for the proper execution of cell-free assays, and no assay is complete without the inclusion of reactions wells in which one of the following components is omitted (i.e., anionic amphiphile, NADPH, membrane, and all or each of the individual cytosolic components).

6.4.2 INT Reduction

Cytochrome *c* reduction can be replaced by INT reduction. The method is identical to that described at Subheading 6.4.1, with the exception that the assay buffer contains 100 μM INT instead of cytochrome *c* (*see* **Note 55** for converting increase in absorbance at 490 nm data to $O_2^{\cdot -}$ values).

Three problems are related to the use of the INT technique:

1. The first is whether INT is reduced by electrons originating from reduced NOX2-bound FAD ($FADH_2$) [180]. In our

hands, however, when using the canonical semirecombinant amphiphile-activated cell-free system, at least 80% of INT reduction is SOD-sensitive and thus mediated by $O_2^{\cdot -}$. "True" INT reduction, representing diaphorase activity, only occurs when transfer of electrons from $FADH_2$ to hemes and to O_2 is prevented by anaerobiosis. This technique is very rarely used (*see* Ref. 142) and is not a realistic alternative for routine application.

2. Second, SDS and LiDS react with INT, forming an unidentified material which absorbs at 490 nm. We have no experience with using arachidonate to replace the anionic detergents.
3. Third, in our hands, INT is about 50% less effective than cytochrome *c* in detecting $O_2^{\cdot -}$ production in the canonical cell-free assay, under strictly identical conditions. The reason for this is unclear.

INT reduction should, therefore, be used predominantly for measuring the diaphorase activity of the DHR of NOX2 [181, 183, 184] and of other NOXes [179, 182, 212].

6.4.3 NADPH Consumption

The method is similar to that described at Subheading 6.4.1, with the exception that the assay buffer contains no electron acceptor and a negative slope, corresponding to the conversion of NADPH to $NADP^+$, is recorded [194] (*see* **Note 56** for converting *decrease* in absorbance at 340 nm data to $O_2^{\cdot -}$ values). As in the cytochrome *c* and INT assays, the catalytic phase of the reaction is initiated by the addition of NADPH to the wells. A number of issues are to be taken in consideration:

1. The technique is useful when there is evidence for interference by a component of the cell-free reaction with electron acceptors, as illustrated in Refs. 180, for INT, and 119, 233, for cytochrome *c*, or in the presence of a reducing agent.
2. It is ideal for use with amphiphile-dependent semirecombinant cell-free systems, in which the presence of contaminating NADPH reductases is unlikely. Even in their presence, the absolute dependence on an amphiphile activator makes the assay applicable.
3. The sensitivity of the assay is comparable to that based on cytochrome *c* reduction.
4. There is a requirement for microplates allowing passage of UV light.

6.5 Amphiphile-Independent Cell-Free Assays

The ability to activate the oxidase in vitro in the absence of an anionic amphiphile was first reported by Hata et al. [133], based on C-terminal truncation of both p47*phox* and p67*phox*. Amphiphile-independent systems were also described by Ebisu et al. [134], using a chimeric construct consisting of truncated p67*phox* and p47*phox*, and by Peng et al. [136]), who prevented the establishment of intramolecular bonds in p47*phox*, by mutagenesis. The two latter groups and we [138] also observed that acidification of the membrane phospholipid environment made the presence of an anionic amphiphile unnecessary.

A conceptually distinct situation, in which oxidase activation can be achieved in the absence of amphiphile and of p47*phox*, is represented by a cell-free system consisting of membrane liposomes, p67*phox*, and prenylated Rac [11]. We proposed that proper targeting of p67*phox* to the membrane in conjunction with the induction of a conformational change in p67*phox* by Rac is sufficient for the initiation of electron flow in NOX2 [11, 12, 135]. Variations of this system include activation by combinations of p67*phox*, prenylated Rac, GTP, and a Rac GEF [166], and amphiphile-independent oxidase activation by p67*phox* and prenylated [Rac-RhoGDI] complexes [156, 157].

We describe two methods for amphiphile-independent cell-free oxidase activation. One assay is based on the use of prenylated Rac and does not require the participation of p47*phox*; the other makes use of our ability to modify the charge of phospholipids in phagocyte membranes and works with nonprenylated Rac.

6.5.1 Amphiphile-Independent Cell-Free Oxidase Activation in Mixtures of Membrane, p67phox, and Prenylated Rac1

The amphiphile-independent cell-free system is useful for investigating the role of Rac and Rac-p67*phox* interaction in oxidase assembly. This particular aspect of assembly is more difficult to explore in the presence of p47*phox*, which has not only an assembly-initiating function, but also a role in the stabilization of the assembled complex [153]. Other situations in which the amphiphile-independent cell-free system is the assay of choice are when the effects of regulators of Rac are to be explored in vitro. One example is provided by Rac GEF-dependent oxidase activation in a cell-free system consisting of membrane, p67*phox*, prenylated Rac1-GDP, GTP, and a Rac GEF, such as Trio or Tiam1 [166]. Another example is the ability of [prenylated Rac1-RhoGDI] complexes in conjunction with p67*phox*, to activate the oxidase when added to phagocyte membrane liposomes enriched in anionic phospholipids [156] or specific phosphoinositides, in the presence of GTP and a GEF [157], in the absence of amphiphile.

Applications of amphiphile-independent cell-free assays also comprise the study of inhibitors (proteins, peptides, phospholipids, nucleotides, detergents, drugs) on the various stages of oxidase assembly. An example is the study of the effect of the amphiphilic activator LiDS on oxidase activation by p67*phox* and prenylated Rac1-GMPPNP, in the absence of p47*phox*. We found LiDS to exert a marked dose-dependent inhibitory effect, in the 25–200 μM concentration range, which was relieved by the presence of p47*phox* [165]. Further examples are the distinct effects of a

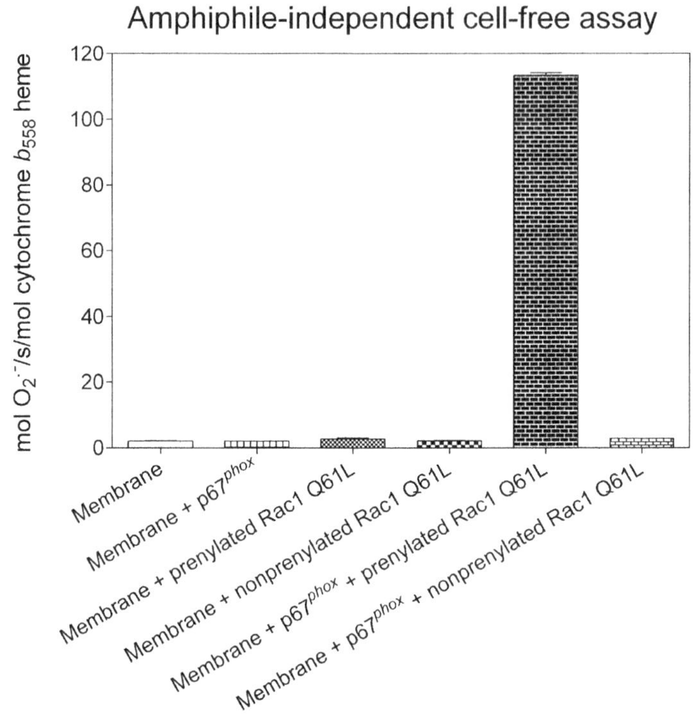

Fig. 7 Typical amphiphile-independent cell-free assay. The complete reaction mixture contained solubilized macrophage membrane liposomes (5 nM cytochrome b_{558} heme), recombinant p67^{phox} (300 nM), and recombinant Rac1 Q61L prenylated in vitro (300 nM). The contents were incubated without amphiphile for 5 min at room temperature. $O_2^{\cdot-}$ production was initiated by the addition of NADPH (238 μM) and measured by the kinetic cytochrome *c* reduction assay for 5 min. The compositions of the incomplete assay mixtures are indicated on the *x*-axis. Results represent means ± SE of three experiments (Reproduced from Ref. 8 by permission of Springer Science + Business Media)

number of compounds (GTP and GDP, a C-terminal Rac1 peptide, RhoGDI, the p21-binding domain of p21-activated kinase (PBD of PAK), and neomycin sulfate) on amphiphile-dependent and -independent cell-free oxidase activation, reflecting the existence of different pathways of assembly [165].

1. Subject Rac1 to nucleotide exchange with GMPPNP or use Rac1 Q61L mutant. It is preferable to perform nucleotide exchange before prenylation. This will reduce possible loss of prenylated protein during exchange by binding to surfaces due to hydrophobicity.
2. Prenylate Rac1-GMPPNP, as described before.
3. Add 20 μL/well of solubilized membrane liposomes (50 nM cytochrome b_{558} heme) to the wells of a 96-well plate. This is intended to result in a final concentration of cytochrome b_{558}

heme of 5 nM in 200 μL (the total volume of the reaction, before the addition of NADPH) and equals 1 pmol cytochrome b_{558} heme/well.

4. Add 20 μL of a mixture of p67^{phox} and prenylated Rac1-GMPPNP, each at a concentration ten-fold higher that that desired as the final concentration in 200 μL. If the requirements of the experiment are to add each component separately, add 10 μL of each component from a 20-fold concentrated stock solution. All dilutions of membrane and cytosolic components are made in assay buffer without LiDS (*see* **Note 57**).
5. Add 160 μL/well of assay buffer without LiDS, using a multichannel pipette. Place the plate on an orbital shaker and mix for 5 min at 500–600 movements/min and room temperature (*see* **Note 58**).
6. Dispense 10 μL of NADPH solution/well using an electronic pipettor, in the dispensing mode, or a multichannel pipettor. This results in a final concentration of 238 μM NADPH in a total volume of 210 μL per well.
7. Record activity and convert to turnover values as described for the amphiphile-dependent system (*see* Subheading 6.4). An example of such an assay, with the required control mixtures, is illustrated in Fig. 7.

6.5.2 Amphiphile-Independent Cell-Free Oxidase Activation in Mixtures of Negatively Charged Membrane, p^{47phox}, p^{67phox}, and Nonprenylated Rac1

Preparing Membrane Liposomes Enriched In Anionic Phospholipids

1. Dilute solubilized macrophage membrane in solubilization buffer containing 40 mM octyl glucoside to a concentration of cytochrome b_{558} heme of 1.2 nmol/mL.
2. Add PA or PG, both at a concentration of 5 mM, at a ratio of one part membrane and four parts phospholipids (v/v). This results in a final concentration of 240 pmol/mL cytochrome b_{558} heme and 4 mM anionic phospholipids.
3. Dialyze the membrane-phospholipid mixture (*see* **Note 26**) against a 100-fold excess of detergent-free solubilization buffer (lacking AEBSF and FAD) for 18–24 h at 4 °C (*see* **Note 27**).
4. Measure the concentration of cytochrome b_{558}, and supplement the preparation with 10 μM FAD.
5. Divide into aliquots of 1–1.5 mL, and store at −75 °C in polypropylene tubes.

Amphiphile-Independent Cell-Free Oxidase Activation with Anionic Membrane Liposomes.

This cell-free assay is a hybrid between the canonical amphiphile-dependent system (from which it borrowed the anionic charge requirement and the fact that Rac is nonprenylated) and the amphiphile-independent assay (based on the use of prenylated Rac).

1. Add 20 μL/well of membrane liposomes enriched in PA or PG (50 nM cytochrome b_{558} heme and about 0.8 mM anionic phospholipid) to the wells of a 96-well plate. This is intended to result in a final concentration of cytochrome b_{558} heme of 5 nM and close to 80 μM anionic phospholipid in 200 μL (the total volume of the reaction, before the addition of NADPH) and equals 1 pmol cytochrome b_{558} heme/well.
2. Add 20 μL of a mixture of p47^{phox}, p67^{phox}, and nonprenylated Rac1-GMPPNP or Rac1 Q61L mutant, each at a concentration ten-fold higher that that desired as the final concentration in 200 μL (*see* **Note 59**). All dilutions of membrane and cytosolic components are made in assay buffer without LiDS. Concentrations of p47^{phox}, p67^{phox}, and nonprenylated Rac required for reaching maximal activation in this system are higher than those customary in the canonical amphiphile-dependent assay.
3. Add 160 μL/well of assay buffer without LiDS. Place the plate on an orbital shaker and mix for 90 s at 500–600 movements/min and room temperature.
4. Dispense 10 μL of NADPH solution/well. This results in a final concentration of 238 μM NADPH in a total volume of 210 μL per well.
5. Record activity and convert to turnover values as described for the amphiphile-dependent system.

6.6 "Sense and Sensibility" (Sensitivity) in Cell-Free Assays

Here, we discuss a number of methodological issues related to the proper way of performing cell-free assays. Emphasis will be placed on untested or unproven assumptions and some "sacred cows" will be questioned.

6.6.1 LiDS, SDS, or Arachidonate?

1. Unless the purpose of performing the cell-free assay is to explore the oxidase activating capabilities of arachidonic acid itself, arachidonic acid isomers, arachidonic acid oxidation products, or related compounds, such as nitroarachidonic acid, there are few reasons justifying the use of arachidonic acid as an activator.
2. Arachidonic acid activates the oxidase only in its ionized salt form. Stock solutions are tedious to prepare starting from the acid form and it is preferable to work with the sodium salt. However, all forms of arachidonic acid are very sensitive to oxidation by air and the oxidase activating activity of undefined oxidation products is unknown. Thus, we recommend using LiDS or SDS as standard amphiphilic activators.

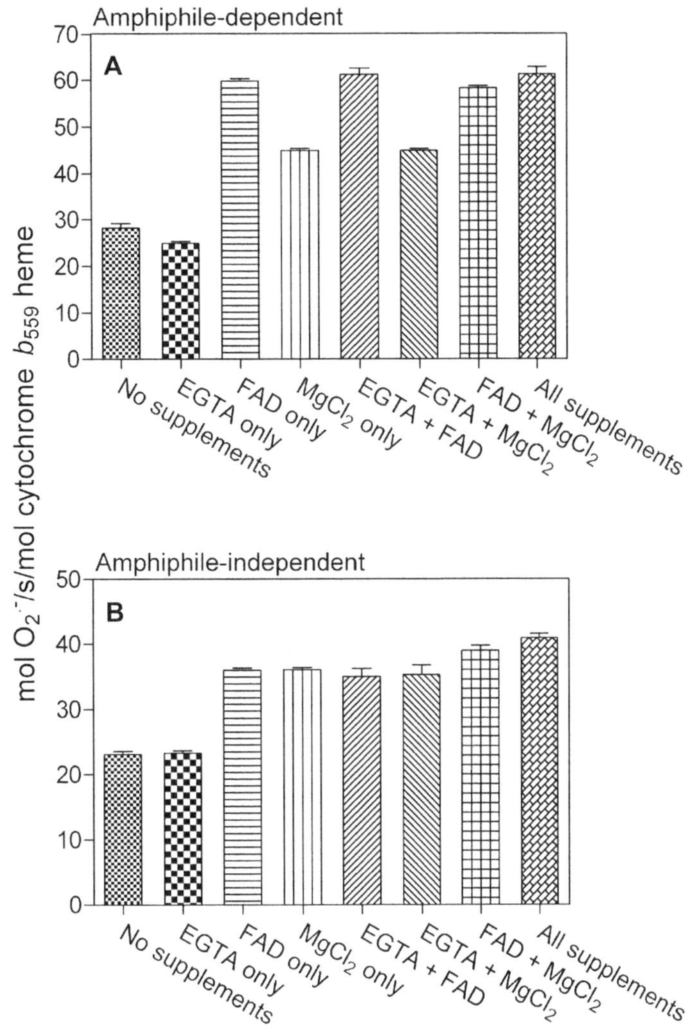

Fig. 8 Which supplements to the cell-free NADPH oxidase assay buffer are essential? Cell-free assays were performed in the canonical amphiphile-dependent system (**a**) and in the amphiphile-independent system, based on the use of prenylated Rac1 (**b**). The basic assay buffer was supplemented with our without 1 m*M* EGTA, 10 μM FAD, or 1 mM $MgCl_2$, or combinations of two or all three of these, as shown on the *x*-axis of panels **a** and **b**. (**a**) Amphiphile-dependent cell-free systems consisting of solubilized macrophage membrane liposomes (5 nM cytochrome b_{559} heme) and recombinant p47^{phox} (30 n*M*), p67^{phox} (30 nM), and nonprenylated Rac1-GMPPNP (30 nM) were incubated with 130 μM LiDS, as described. (**b**) Amphiphile-independent cell-free systems consisting of solubilized macrophage membrane liposomes (5 nM cytochrome b_{558} heme), recombinant p67^{phox} (300 nM), and recombinant Rac1-GMPPNP prenylated in vitro (300 nM*)* were incubated without amphiphile, as described. In both panels **a** and **b**, $O_2^{\cdot-}$ production was initiated by the addition of NADPH and measured by the kinetic cytochrome *c* reduction assay for 5 min. Results

3. SDS and LiDS are equally good activators, but concentrated solutions of SDS must be kept at room temperature, whereas LiDS solutions can be stored at 4 °C, minimizing evaporation. For the reasons listed above, LiDS and SDS yield more reproducible results than arachidonate. We find no basis for the claim that arachidinate is to be preferred because it represents a more "physiologic" form of activation.

6.6.2 "To Supplement or not to Supplement—That is the Question"

1. Calcium: in early experiments, we found that activation was reduced by Ca^{2+} and moderately enhanced by the Ca^{2+} chelator EGTA [77] (*see* **Note 60**). We reinvestigated the necessity of Ca^{2+} chelation in the LiDS-activated and amphiphile-independent systems by examining the effect of EGTA, alone or in association with other supplements. As seen in Fig. 8, when using high purity salts for preparing the assay buffer, the presence of Ca^{2+} is unlikely and, consequently, EGTA had no enhancing effect on oxidase activation in both the amphiphile-dependent and -independent systems.
2. FAD: a flavin requirement was observed in the oxidase isolated from stimulated phagocytes [25], and, early in the development of the arachidonate-activated cell-free system, it was found that exogenous FAD enhanced activation [77]. The most likely explanation is that NOX2 lost the noncovalently bound FAD during preparation of membranes, leading to a need to reflavinate cytochrome b_{558}. Here, we compared cell-free oxidase activation in the presence and absence of 10 μM FAD in the assay buffer by using solubilized membrane liposomes, which are routinely supplemented with FAD. As apparent in Fig. 8, FAD enhanced both amphiphile-dependent (Fig. 8a) and amphiphile-independent (Fig. 8b) oxidase activation, the effect being more pronounced on the amphiphile-dependent activation (*see* **Note 61**).
3. Magnesium: a requirement for Mg^{2+} was described early in cell-free studies, and it was suggested that the metal interacted with a saturable oxidase component at a K_m of about 1 mM [225]. The identity of this component was not established at the time, but after the discovery of the involvement of Rac in oxidase assembly, it became common belief that the requirement for millimolar concentrations of Mg^{2+} was related to its role in preventing the dissociation of GTP from Rac [234]. As shown in Fig. 8, supplementation of the assay buffer with 1 mM Mg^{2+} enhanced oxidase activation in both the amphiphile-dependent (Fig. 8a) and -independent (Fig. 8b)

Fig. 8 (continued) illustrated represent means ± SE of three experiments (Reproduced from Ref. 8 by permission of Springer Science + Business Media)

systems. Higher concentrations of Mg^{2+} (up to 5 mM) were not more effective than 1 mM (results not shown). Combining supplementation with FAD with that with Mg^{2+} did not result in an additive or synergistic effect; activities were identical to those found with FAD alone. Also, combining supplementation with FAD or Mg^{2+} with EGTA, or adding all three supplements, had no additive or synergistic effect. The almost identical ability of FAD and Mg^{2+} to improve assembly and the lack of an additive or cooperative effect suggest that they act by the same mechanism, most likely related to the stability of the NOX2-FAD bond and not to that of the Rac-GTP bond (*see* **Note 62**).

6.6.3 "Measure for Measure"—The Intricacies of Dose—Response Studies with Cytosolic Oxidase Components

1. Most cell-free oxidase activation assays follow the principle of a constant amount of membrane and variable amounts of cytosolic components. This leaves open the issue of quantitative relationships among cytosolic components (*see* **Note 63**).
2. A problem we frequently encountered when performing cell-free assays was determining the optimal methodology for relating activity turnover values to the amounts of cytosolic proteins added to a constant amount of membrane. Figure 9 summarizes the two main approaches used in our laboratory. In these experiments, the concentration of the membrane (cytochrome b_{558}) was constant. The concentrations of cytosolic components were either varied all in parallel or individually, in which case the other components were added at the maximal concentration in the range studied. Assays were run either in the amphiphile-dependent system (Fig. 9a), or in the amphiphile-independent system (Fig. 9b). In the amphiphile-dependent system, the concentration of LiDS was kept constant at 130 μM because the optimal activating concentration of LiDS did not vary with the concentration of cytosolic components within the 0–1 μM range when using purified recombinant cytosolic components.
3. It is apparent that when all components are varied in parallel, the dose–response curve has a sigmoidal shape, whereas when a single component is varied in the presence of an excess of the other component(s), the curves are hyperbolic. The highest levels of activation are seen when the concentrations of Rac1 and $p47^{phox}$ (amphiphile-dependent system) and Rac1 (amphiphile-independent system) are varied individually, in the presence of an excess of the other component(s); the lowest activities are found when $p67^{phox}$ is varied individually (*see* **Note 64**).

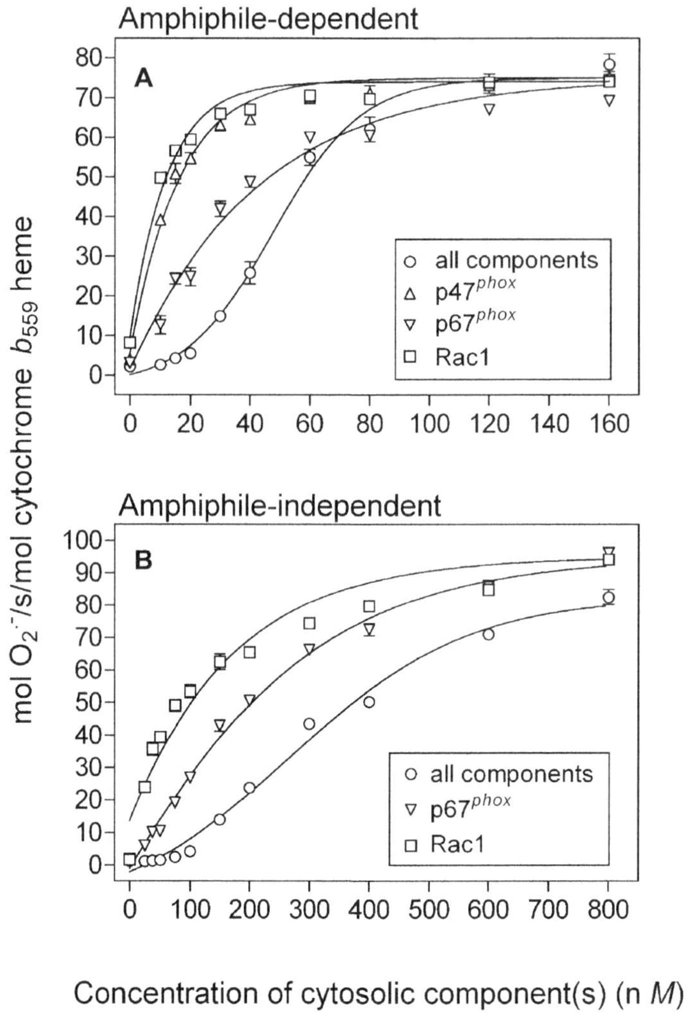

Fig. 9 The effect of concentration ratios among cytosolic components on the nature of the dose–response curves in cell-free assays. (**a**) Four types of amphiphile-dependent cell-free assays, consisting of various combinations of cytosolic components, were performed. All four consisted of solubilized macrophage membrane liposomes (5 nM cytochrome b_{558} heme), recombinant p47^{phox} (varied from 0 to 160 nM), and recombinant p67^{phox} (varied from 0 to 160 nM), and recombinant nonprenylated Rac1-GMPPNP (varied from 0 to 160 nM). The four combinations of components were as follows: (1) All three cytosolic components were present at equal concentrations (varied from 0 to 160 nM); (2) p47^{phox} was varied from 0 to 160 nM, whereas p67^{phox} and Rac1 were both present at a concentration of 160 nM; (3) p67^{phox} was varied from 0 to 160 nM, whereas p47^{phox} and Rac1were both present at a concentration of 160 nM, and (4) Rac1 was varied from 0 to 160 nM, whereas p47^{phox} and p67^{phox} were both present at a concentration of 160 nM. In all cases, the components were incubated with 130 μM LiDS, as described. (**b**) Three types of amphiphile-independent cell-free assays, consisting of various combinations of cytosolic components, were performed. All three consisted of solubilized macrophage

6.6.4 To Exchange or to Add?

1. In the early period of the use of cell-free assays, it was reported repeatedly that the addition of GTP or nonhydrolyzable GTP analogs (GTPγS or GMPPNP) was an absolute requirement for oxidase activity. Many of these observations were made in cell-free systems consisting of membrane and total cytosol before the identification of Rac as the small GTPase involved in oxidase activation [198, 202, 235, 236].
2. With the advent of the semirecombinant systems, which involved the use of recombinant Rac1 or Rac2, the "habit" of supplementing the assay buffer with GTP analogs persisted when native Rac (Rac-GDP, not exchanged to GTP) was present in the reaction. The assumed explanation for this was that added GTP analogs were bound to Rac-GDP in a nucleotide exchange reaction taking place simultaneously with oxidase assembly (*see* **Note 65**).
3. Because the concentration of Mg^{2+} in the assay buffer is prohibitive for spontaneous nucleotide exchange, the ability of prenylated Rac to take up GTP from the medium points to the intervention of a GEF. In a semirecombinant cell-free system, GEF can originate only in the membrane but its presence, identity, and quantity are unknown parameters in the vast majority of cases and will depend on the animal species and nature of the phagocyte serving as the source for the membrane (reviewed in Ref. 158).
4. Another common assumption is that native Rac (Rac-GDP) is inactive in cell-free systems (however, *see* Ref. 221). We have shown in the past that this is true only below a certain quantitative threshold and when this is exceeded, significant activity can be achieved. Thus, in the canonical amphiphile-dependent cell-free system, the differences in V_{max} between Rac1-GDP and Rac1-GTPγS were marked at 20 nM Rac but minimal, at 200 nM Rac [207], reflecting the difference in affinity for $p67^{phox}$.

Fig. 9 (continued) membrane liposomes (5 nM cytochrome b_{558} heme), recombinant $p67^{phox}$ (from 0 to 800 nM), and recombinant Rac1-GMPPNP prenylated in vitro (from 0 to 800 n*M*). The three combinations of components were: (1) The two cytosolic components were present at equal concentrations (varied from 0 to 800 nM); (2) $p67^{phox}$ was varied from 0 to 800 nM, whereas Rac1 was present at a concentration of 800 nM, and (3) Rac1 was varied from 0 to 800 nM, whereas $p67^{phox}$ was present at a concentration of 800 nM. The components were incubated in the absence of an anionic amphiphile. In all assays, $O_2^{\cdot -}$ production was initiated by the addition of NADPH and measured by the kinetic cytochrome *c* reduction assay for 5 min. Results illustrated in both panels represent means ± SE of three experiments. (Reproduced from Ref. 8 by permission of Springer Science + Business Media)

5. Because of contradictory and controversial results, we recommend that one should not rely on "in assay" nucleotide exchange, achieved by the addition of GTP analogs to the assay buffer, and always perform quantifiable nucleotide exchange on both nonprenylated and prenylated Rac, before their use in the assays. Following this advice will prevent inconsistent and poorly reproducible results, due to lack of conversion of Rac from the GDP- to the GTP-bound form.

6.7 Use of the Cell-Free System for the Discovery of Oxidase Inhibitors

Cell-free systems are ideally suited for identifying potential oxidase inhibitors and for investigating their mechanism of action. The search for oxidase inhibitors received enormous impetus by the accumulating evidence for the involvement of nonphagocytic NOXes in the pathogenesis of a wide variety of diseases (reviewed in Refs. 208–211). So far, cell-free assays appropriate for measuring the activity of nonphagocytic NOXes are few and their use is not widespread. Thus, the cell-free assay is mostly applied to NOX2-based situations, whether in phagocytes or other cells. A central place is taken by synthetic peptide analogs of oxidase components (reviewed in [187, 237–240]. Peptide analogs are used for two purposes: (1) As a mean of locating functional domains in individual oxidase components, and (2) To identify peptides with the potential of being used as therapeutic agents to dampen ROS production in disease situations in which excessive ROS production represents a primary or secondary pathogenic mechanism.

To achieve the first goal, arrays of overlapping peptides "covering" part of or the whole sequence of an oxidase component were tested for an effect on cell-free activation, a methodology that became known as "peptide walking." This was applied to Rac1 [241], $p47^{phox}$ [242], $p67^{phox}$ [216], $p22^{phox}$ [243], and NOX2 [244]. The second goal yielded rather disappointing results, with only one peptide, corresponding to residues 86–94 in the cytosolic loop B of NOX2, found to inhibit oxidase activation in whole cells and organs and in an animal model, thus exhibiting a therapeutic potential [178, 245]. Nevertheless, the cell-free system continues to be used as the essential method for identifying potential small molecule inhibitors and for elucidating the mechanism of inhibition (*see* a few examples in Refs. 143, 246, 247). An intrinsic advantage of the assay is its applicability to HTS, using 96-well or 384-well plates.

We shall briefly summarize some of the critical issues to be considered when using the cell-free system for the identification of peptide or other small molecule oxidase inhibitors.

1. When assessing the significance of inhibition results, it is recommended to run dose-response studies in a routine manner. These should be performed within a concentration range to enable the calculation of IC_{50} values. It is essential that the

A

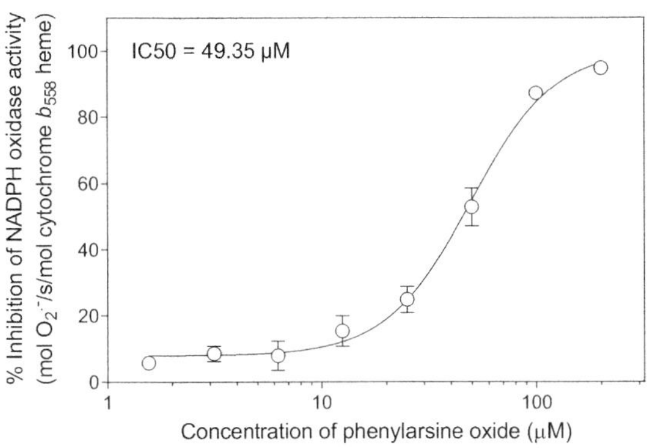

B

	% inhibition of NADPH oxidase
log(agonist) vs. response -- Variable slope (four parameters)	
Best-fit values	
Bottom	7.935
Top	100.4
LogEC50	1.693
HillSlope	2.216
EC50	49.35
Span	92.47
Std. Error	
Bottom	2.017
Top	5.160
LogEC50	0.03775
HillSlope	0.3674
Span	6.029
95% CI (asymptotic)	
Bottom	3.727 to 12.14
Top	89.64 to 111.2
LogEC50	1.615 to 1.772
HillSlope	1.450 to 2.983
EC50	41.17 to 59.16
Span	79.89 to 105.0
Goodness of Fit	
Degrees of Freedom	20
R squared	0.9756
Adjusted R squared	0.9720
Sum of Squares	698.9
Sy.x	5.911
Number of points	
# of X values	24
# Y values analyzed	24

Fig. 10 A typical inhibition of NADPH oxidase activity curve by phenylarsine oxide, as investigated by its ability to interfere with cell-free activation in a canonical amphiphile-dependent cell-free assay. This consisted of macrophage membrane liposomes (5 nM cytochrome b_{558} heme), p47^{phox}, p67^{phox}, and Rac1Q61L (100 nM, each), and LiDS (120 μM). Phenylarsine oxide (1.5–200 μM) was added to the assay buffer as the first component of the reaction, followed by membrane, cytosolic components, and LiDS. After incubation for 3 min at 24 °C, NADPH was added (238 μM). Control mixtures contained the corresponding concentrations of the organic solvent used to dissolve phenylarsine oxide. Results represent means ± SE of three experiments and were analyzed and plotted by GraphPad Prism, Version 8 (GraphPad Software)

compound does not exert a nonspecific inhibitory effect on the actual measurement of $O_2^{\cdot-}$ production. This can be easily tested by adding the peptide to a xanthine–xanthine oxidase $O_2^{\cdot-}$-generating system.

2. Ideally, peptide inhibitors are expected to interfere with oxidase activation in the cell-free system by competing with the intact oxidase component, from which the peptide was derived, for interaction with another component of the oxidase complex. To test such an assumption, kinetic studies are required in order to demonstrate that inhibition is competitive. This was found to be the case with some peptides [220, 242] but, occasionally, what appeared as competition [248] did not withstand kinetic analysis [249]. In yet another study, a compound thought to compete with NADPH for binding to NOX 2 turned out, based on careful kinetic analysis, to compete with p67^{phox} [247].
3. Most Inhibitors active in cell-free systems are expected to interfere with the process of oxidase assembly. Such peptides inhibit only when added before the initiation of assembly and are inactive when added after the completion of assembly. The preferential inhibition upon peptide addition before assembly

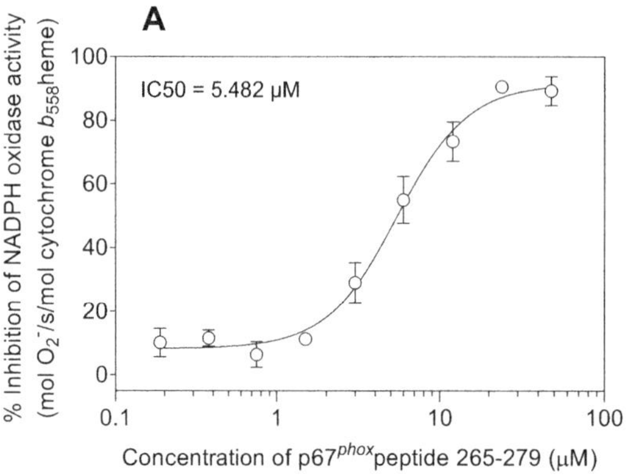

B

	% Inhibition of NADPH oxidase activity
log(inhibitor) vs. response – Variable slope (four parameters)	
Best-fit values	
Bottom	8.222
Top	91.46
LogIC50	0.7390
HillSlope	1.986
IC50	5.482
Span	83.24
Std. Error	
Bottom	2.798
Top	4.350
LogIC50	0.04913
HillSlope	0.4053
Span	5.712
95% CI (profile likelihood)	
Bottom	2.258 to 13.56
Top	83.45 to 102.4
LogIC50	0.6377 to 0.8535
HillSlope	1.340 to 2.983
IC50	4.342 to 7.136
Goodness of Fit	
Degrees of Freedom	23
R squared	0.9559
Sum of Squares	1385
Sy.x	7.761
Number of points	
# of X values	27
# Y values analyzed	27

Fig. 11 A typical inhibition of NADPH oxidase activity curve by a p67^{phox} peptide corresponding to residues 265–279, as investigated by its ability to interfere with cell-free activation in a canonical amphiphile-dependent cell-free assay. This consisted of macrophage membrane liposomes (5 nM cytochrome b_{558} heme), p47^{phox}, p67^{phox}, and Rac1Q61L (100 nM, each), and LiDS (120 μM). The peptide (0.187–48 μM) was added to the assay buffer as the first component of the reaction, followed by membrane, cytosolic components, and LiDS. After incubation for 3 min at 24 °C, NADPH was added (238 μM). Control mixtures contained the corresponding concentrations of the mixture of organic solvent (75%) and H_2O (25%) used to dissolve the peptide. Results represent means ± SE of three experiments and were analyzed and plotted by GraphPad Prism, Version 8 (GraphPad Software)

is, however, not universal; some peptides and small molecule inhibitors were also found to inhibit when added after assembly, raising the possibility that they may be capable of dissociating assembled complexes [165, 244]. On rarer occasions, peptides interfere with the catalytic (redox) function of NOX2, such as by competing for the binding of cofactors, or by another mechanism.

4. It is essential to control the sequence specificity of the inhibitory action of oxidase-analog peptides. This involves testing of scrambled and retropeptides and unrelated peptides or small molecule compounds of similar size, charge, or hydrophobicity. Charge and hydrophobicity are important parameters in protein-protein and protein–lipid interactions and there are numerous examples of situations in which what was expected to be sequence-specific inhibition by peptides, turned out to be sequence-independent (*see* Refs. 165, 243, 244, 250). On the other hand, lack of sequence specificity should not be an automatic disqualifier of the inhibitory peptide for possible therapeutic applications.

5. An important methodological consideration is the type of cell-free assay used for testing inhibitors. Thus, the amphiphile-dependent assay (with nonprenylated Rac1) is to be used when a charge effect is involved [167], as shown by the inhibitory effects of a positively charged C-terminal Rac1 peptide or the cationic neomycin sulfate. The very same compounds were inactive when tested in an amphiphile-independent assay (with prenylated Rac1), in which hydrophobic binding of Rac1 to the membrane is predominant. In the latter situation, RhoGDI and phosphatidylcholine vesicles were found to prevent activation but were inactive when tested in an amphiphile-dependent assay (with nonprenylated Rac1). Examples for charge-related and hydrophobicity-related oxidase inhibition are described in Refs. 11, 12, 165.
6. Occasionally, peptides expected to be inhibitory on the basis of the fact they correspond to domains of previously known functional significance, were found to be inactive. Thus, peptides corresponding to the switch I region in Rac1 [241], the proline-rich region in p22^{phox} [243], and the "activation domain" in p67^{phox} [216], were not inhibitory in the cell-free assay.
7. Lack of effect on oxidase activation under cell-free conditions of a compound found active in whole cells might mean that the inhibitor acts upstream of the oxidase, most likely by interfering with phagocyte receptor-ligand interaction or with a transductional step linking the receptor to the oxidase.

Figure 10 illustrates a typical dose–response curve of the small molecule oxidase inhibitor phenylarsine oxide. Its inhibitory action was found to be NOX2-specific [251, 252] and its putative target is the ^{369}Cys-Gly-Cys^{371} triad in the DHR of NOX2, with a role in the stabilization of the binding of p67^{phox} to NOX2 [233, 253, 254].

Figure 11 illustrates a typical dose–response curve of oxidase inhibition by a p67^{phox} peptide, corresponding to residues 265–279. This is one of three overlapping peptides, within the sequence 259–279 in the N-terminal SH3 region of p67^{phox}, identified by "peptide walking," as inhibitors of the oxidase in a cell-free system [216].

6.8 Epilogue

We have been in the "cell-free business" for a long time. The message we would like to leave with you is as follows: Like good wine, the assay improved by aging. Running cell-free assays is not only useful, simple, reproducible, and economical but, more than anything else, it is real fun. There is nothing like the feeling of taking out four proteins from the freezer and having a $O_2^{\cdot-}$ producing system on your desk in less than an hour. There is only one warning message: Cell-free assays are addictive; you might never go back to your beloved phagocytes again.

7 Notes

1. It was established that only the ionized form of arachidonic acid and other long chain unsaturated fatty acids were capable of oxidase activation, a fact also supported by the ability of non-fatty acid anionic amphiphiles to do the same [86]. When arachidonic acid is added as a solution in an organic solvent (ethanol), it is probably converted to the ionized form by reacting with components of the assay buffer. Due to uncertainty concerning the concentration of the ionized form, it is recommended to add arachidonic acid prepared as a Na salt, as described in [36].
2. It is probably true that there is no universal optimum concentration of arachidonate or other fatty acid for the induction of oxidase activation in the cell-free system. The data in the literature vary widely and it is a futile endeavor to try understanding the reasons in each particular case. The targets of the activating fatty acids are p47^{phox} [55, 57] and cytochrome b_{558} [87–91] and the optimal concentration is likely to be dependent on the concentration of the target molecules and their physical form in the assay (such as membranes or membrane liposomes).
3. The discovery of what was thought initially to be a single cytosolic component is a fitting illustration of the process of scientific discovery described by Thomas Kuhn as follows: "...awareness of anomaly opens a period in which conceptual categories are adjusted until the initially anomalous has become the anticipated. At this point the discovery has been completed" [78].
4. The almost simultaneous description by several groups of the cell-free system is yet another example of what Thomas Kuhn has so aptly described: "The very fact that a significant scientific novelty so often emerges simultaneously from several laboratories is an index both to the traditional nature of normal science and the completeness with which that traditional pursuit prepares the way for its own change" [78].
5. An important fact established by the experiments described in this paper [80] was that following elicitation of ROS production in leukocytes by oleate, an active and relatively stable oxidase could be recovered in the plasma membrane fraction. This exhibited a clear preference for NADPH and the expected stoichiometry of $2O_2$ + NADPH = $2O_2^{\cdot-}$ + NADP. Membranes derived from oleate-stimulated leukocytes of a CGD

patient failed to show oxidase activity. This modest but significant paper was essential for "legitimizing" fatty acid elicited ROS production by leukocytes as equal to that elicited by more conventional means, such as phagocytosis.

6. Present inspection of the paper by DeChatelet et al. [81] reveals that it is unlikely that the enzyme activated by dialysis was the NADPH oxidase. This is shown by NADH and NADPH serving as equally efficient substrates and by the ability to activate membranes derived from leukocytes of a CGD patient.

7. One of the reviewers of our manuscript describing cell-free activation of the oxidase by SDS recommended rejection of the manuscript and offered the following comments: "*the manuscript contains some interesting material. . .I express some disappointment that the authors do not approach the literature more thoroughly and attempt to relate effects seen here to those observed earlier with numerous detergents on whole cells.*"

8. The authors of a publication reexamining the relationship of the release of arachidonate to ROS production by macrophages [107] stated, "*. . .activation of the oxidase in disrupted neutrophil preparations with high concentrations of arachidonic acid alone* [citing Refs. 77, 79, 82, 83] *appears to reflect nonphysiologic changes in the lipophilic environment of the enzyme, mimicking the action of detergents such as digitonin* [citing [34]], *SDS* [citing erroneously [86]], *and saponin.*"

9. The cell-free assay was hailed as "*a major advance that has capped years of work on this problem*" [21] and a "*groundbreaking discovery of a method to activate the respiratory burst oxidase*" [109]. It was judged as having "*revolutionized our understanding of phagocyte superoxide generation*" [110]. It drew the comments that "*many of the recent advances in our understanding of the components of the oxidase have been the result of studies using a broken cell system, first developed by Bromberg and Pick and Heyneman and Vercauteren and extended by other investigators*" [111], and "*use of the cell-free system made possible the discovery if cytoplasmic proteins required for normal phagocyte oxidase activity*" [112]).

10. In order to eliminate calcium from the buffer used in the cell-free assay, the calcium chelator ethylene glycol-bis(2-aminoethylether)-*N,N,N′,N′*-tetraacetic acid (EGTA) was added at a concentration of 1 mM [77, 83].

11. The Krafft point is the narrow temperature range above which the solubility of a detergent rises sharply. At this temperature the solubility of the detergent becomes equal to the critical micellar concentration (CMC). A typical micelle of an amphiphilic detergent in aqueous solution is an aggregate with the

hydrophilic head regions in contact with the surrounding aqueous solvent and the hydrophobic tail regions oriented toward the micelle center. In colloidal chemistry, the CMC is defined as the concentration of detergent above which micelles form at a certain temperature.

12. The experience of our laboratory with the use of the INT reduction assay in the canonical cell-free assay is that about 80% of the reduction is SOD-sensitive and, thus, due to reduction by $O_2^{\cdot -}$.
13. This study must be approached with caution. The authors have no hands-on experience with the NADPH oxidase and the cell-free system. The main problem is, however, that the model is based on the uncritical acceptance of what was published over a period of five decades, comprising results that could not be reproduced or proven to be incorrect.
14. Subheadings 5 and 6 in this chapter are biased toward semi-recombinant cell-free systems using membranes derived from guinea pig peritoneal exudate macrophages and human recombinant cytosolic components. The rationale behind this combination is the ease of preparing large amounts of membranes with high cytochrome b_{558} content, excellent activity in cell-free assays, ease of solubilization, and years-long stability when kept frozen at −75 °C.
15. AEBSF was found to be an inhibitor of oxidase assembly in the cell-free assay [255], with an IC_{50} of 0.87 mM. However, the concentrations of AEBSF carried over into the assays were 250 to 350 times lower.
16. Currently, we prefer to replace the use of Rac1 exchanged to GMPPNP or GTPγS with the Rac1 mutant Q61L, which is constitutively in the GTP-bound form [223].
17. In the author's laboratory cytochrome *c* made by Sigma-Aldrich, product number C2506, is used. This is prepared from equine heart using trichloroacetic acid and has a purity of ≥95% and a reduced cytochrome c content of ≤5%, and was proven as very satisfactory over a long period of use. Alternatives are product numbers C2867 (purity of ≥99%) and C7752 (prepared using acetic acid, purity of ≥95%). We have no direct experience with these products and found that purity exceeding 95% is unnecessary and pricing is prohibitive. We measure the *total* concentration of cytochrome *c* in the assay buffer (before supplementation with FAD) by performing an absorbance wavelength scan (400–600 nm) on the native and sodium dithionite-reduced solution, determining the difference in absorbance at 550 nm between the reduced and oxidized samples and calculating the concentration by applying

the extinction coefficient, $\Delta E_{550} = 21\ mM^{-1}\ cm^{-1}$ for reduced minus oxidized cytochrome *c*.

18. In the author's laboratory NADPH made by Sigma-Aldrich, product number N1630 is used. As a note of caution, it is recommended to check the actual concentration of the 5 mM stock solutions by measuring absorbance at 340 nm, and using the extinction coefficient, $E_{340} = 6.22\ mM^{-1}\ cm^{-1}$, as described in Ref. 195.

19. This buffer was first described by us for use in the arachidonate-activated rudimentary cell-free system, based on the use of membrane and total cytosol [77] and was modified later [207].

20. In the author's laboratory, 96-well plates (flat-bottom, clear, Cat. No. 655101, Greiner Bio-One) are used for cytochrome *c*, INT, and NBT reduction assays, and 96-well plates (UV—Star, flat-bottom, Cat. No. 655801, Greiner Bio-One), for NADPH consumption assays.

21. For reading cytochrome *c* reduction, a narrow bandwidth, centered on 550 nm, improves the sensitivity of the assay. SpectraMax 340 (wavelength range 340–750 nm) has a bandwidth of 5 nm; VersaMax (wavelength range 340–850 nm) and SpectraMax 190 (wavelength range 190–850) have a bandwidth of 2 nm.

22. In this chapter we illustrate a typical cell-free assay by using membranes of elicited guinea pig macrophages, this being the routine source of phagocytes in our laboratory. The use of the admittedly more common human blood neutrophils as a source of membranes is described in numerous publications, such as Refs. [82, 83, 96, 97].

23. Analysis of the cellular content of the lavage revealed that it consisted of more than 90% macrophages at 4–5 days following the injection of paraffin oil.

24. The 1 M KCl concentration is achieved by mixing 2.5 volumes of sonication buffer with 1 volume of 3.5 M KCl in 20 mM Tris–HCl, pH 7.5.

25. Washing of membranes with 1 M KCl is intended to remove loosely attached cytosolic components, with special emphasis on the removal of membrane-bound Rac. It was reported that omission of the KCl wash step resulted in an apparent decrease in the dependence of cell-free oxidase activation on added Rac, which was due to the presence of membrane-bound Rac [256].

26. A dialysis membrane with a molecular weight cutoff of 25,000 is chosen because octyl glucoside is used for the solubilization of membranes at a concentration of 40 mM, which is well above its CMC of 25 mM. The micellar molecular weight of

octyl glucoside is 8000 and a large pore size membrane facilitates diffusion of the detergent from the dialysis bag to the surrounding buffer.

27. The dialysis buffer used for removing octyl glucoside and generating the membrane liposomes is identical to the solubilization buffer but contains no octyl glucoside, AEBSF, leupeptin hemisulfate, and FAD. The addition of serine protease inhibitors, especially AEBSF, to membrane preparations is avoided because of the reported likelihood of an interaction with cytochrome b_{558} [255]. FAD is omitted because it interferes with the spectroscopic measurement of cytochrome b_{558} content by sodium dithionite reduction.
28. Membrane liposomes can be kept frozen at −75 °C for long periods. We found minimal decreases in the cytochrome b_{558} heme content of frozen membranes over time. Keeping the material in small aliquots is recommended but thawing and refreezing for up to ten times was not found to cause significant damage. When membrane liposomes are used after long periods of storage or have an uncertain data record, it is recommended to determine the cytochrome b_{558} content again (*see* Subheading 6.1.3).
29. Routine use of solubilized membranes in the form of liposomes has the built-in advantage of providing a ready-made source of a membrane preparation which elutes as a single well-defined peak by gel filtration on Superose12 or Superdex 200 FPLC columns (in the exclusion volume). This allows the detection of translocation of cytosolic components to the membrane in an in vitro model of oxidase assembly (*see* Subheading 4.10 and Refs. 11, 135, 152, 153).
30. A relevant discussion of the importance of expressing results of cell-free assays in turnover values is found in [257, 258].
31. We use disposable 10 × 4 × 45 mm cuvettes (Sarstedt).
32. We use a Uvikon 943 double-beam spectrophotometer (Kontron Instruments), but any instrument with similar characteristics (scanning ability, narrow band-width, and basic data analysis capabilities) should be suitable.
33. Quite a number of alternative *Δ* extinction coefficients for calculating the cytochrome b_{558} heme concentration have been published. Most of these are centered on the 558/559 nm peak. Although this peak is more specific for cytochrome b_{558}, the fact that it is much lower than the 426/427 nm peak increases the chances of error, and we prefer to base our calculations on the 426/427 nm peak. We found no evidence for the presence of significant amounts of *b*-type cytochromes, other than b_{558}, in macrophage membranes [259].

34. The high concentration of cytochrome b_{558} heme (2 μM) might raise the question of whether freeing the solubilized membrane of octyl glucoside by dialysis is a necessary step in preparing membranes intended for the performance of cell-free assays. In these, the concentration of cytochrome b_{558} heme is 5–10 nM, and one might assume that just diluting the membrane 200–400-fold is sufficient for reducing the concentration of octyl glucoside well below the CMC, without dialysis. It is our experience, however, that this is not the case. Solubilized membrane preparations, dialyzed free of detergent, are clearly superior in cell-free assays to material prepared by dilution.
35. It is recommended to use the same protein standard throughout in order to compare protein concentrations over long time periods. No external protein standard reflects the true protein concentration of the recombinant proteins but what is important is to maintain the same level of "error." We use bovine gamma globulin, as the standard.
36. When a cell-free assay is focused on the characterization of a specific cytosolic component (such as studying the effect of mutations), we may opt for preparing the particular component at a high degree of purity and combining it with the other components with a lower degree of purity. This applies particularly to the performance of dose responses of one component in the presence of fixed amounts of the complementing components, present at a concentration equal to the highest concentration in the dose response (*see* Subheading 6.6.3).
37. Highly purified cytosolic components have the tendency to self-aggregate, even in the presence of 30% glycerol. Repeated thawing–freezing is likely to promote aggregation. Aggregated protein is inappropriate for cell-free assays and its presence distorts the results of protein concentration measurements. It is, thus, highly recommended to subject frozen proteins, after thawing, to centrifugation at 12,000 × *g* at 4 °C, for 30 min, in a table top microcentrifuge accommodating 1.5 mL conical tubes. Measure protein concentration in the supernatants and keep the supernatants at 4 °C during the performance of the assay. The material can be refrozen at −75 °C for further storage but recentrifugation and rechecking protein concentration are recommended when used in a fresh assay.
38. In the Rho GTPase folklore, it is commonly believed that GMPPNP is more resistant to the intrinsic GTPase activity than GTPγS but that the affinity for GTPγS is higher than that for GMPPNP. We are now employing exclusively GMPPNP for generating the GTP-bound form of Rac1 or Rac2.

39. Rac exchanged to GMPPNP was found, in our hands, to be stable for several weeks, and we found no reason to perform the exchange a short time before using the protein in cell-free assays. However, when larger amounts of Rac are subjected to exchange, it is wise to divide it in smaller aliquots, in order to avoid repeated thawing and freezing.
40. It is obvious that, when using this methodology, unbound GMPPNP is not removed and is present in the Rac preparations at a concentration roughly ten-fold higher than that of the protein. This means that free GMPPNP is transferred to the cell-free assays, a fact which is to be taken into consideration. When removal of unbound nucleotide is desired to eliminate possible unwanted effects of free GMPPNP, the nucleotide-exchanged Rac is subjected to buffer exchange by centrifugal ultrafiltration, using 4 mL centrifugal filter units fitted with 10,000-molecular-weight-cut-off membranes. Three volumes of 4 mL each are filtered, using the buffer in which Rac is found, supplemented with 20% v/v glycerol, and the sample is reconstituted to its original volume. Protein concentration is measured again, to check for some unavoidable loss in the course of ultrafiltration.
41. Prenylation can be performed after and before nucleotide exchange to a nonhydrolyzable GTP analog, but we prefer to prenylate after nucleotide exchange.
42. $ZnCl_2$ is an essential cofactor of geranylgeranyl transferase I.
43. This method can be applied to larger amounts of Rac, provided that nonprenylated Rac is sufficiently concentrated. However, the heated rotary mixer (Thermomixer Comfort) accommodates tubes with maximal volumes of 1.5 or 2 mL, which are also convenient for storage.
44. Two additional bands of 48 and 43 kDa are visible in the aqueous phase. These represent the α and β subunits of the enzyme geranylgeranyl transferase I.
45. This was first shown by Ligeti et al. [202]. Thus, at 25 °C, full assembly was achieved after 5 min of exposure to the amphiphile, whereas at close to 0 °C, the process took 30 min. Note, however, that, at both temperatures, the same level of assembly was ultimately achieved.
46. This observation was first made by Pilloud (Dagher) et al. [197] in a system composed of membrane and whole cytosol, and was thought to be related to the effect of salt on the micellar state of the anionic amphiphile activator. More recent work in a semirecombinant system demonstrated that increasing ionic strength prevents binding of nonprenylated Rac1 (via its polybasic tail) to negatively charged phospholipids in the membrane [10]. No effect of ionic strength was seen on Rac2,

which possesses a lesser positive charge at its C-terminus. These facts are of practical importance for the design of an optimal assay buffer and are also of relevance for the mechanism of amphiphile-independent cell-free activation of membranes enriched in anionic phospholipids [136, 137, 165].

47. If the requirements of the experiment are to add each component separately or in groupings of two, add 10 μL of each component from a 20-fold concentrated stock solution or 20 μL of a ten-fold concentrated mixture of two components and 10 μL of a 20-fold concentrated stock of the third component. This will bring the total volume of the added components to 30 μL and requires the reduction of the volume of added membrane preparation to 10 μL and the proportional increase in the concentration of the membrane stock preparation to 20-fold.

48. The optimization of amphiphile concentration was first discussed in 1989 [197] (a paper well worth reading even today) and reviewed recently [260]. Because anionic amphiphiles act on both cytosolic [55, 59] and membrane [87–91] components, optimization might be a complex issue when setting up radically new conditions and/or when nonpurified components are used, some of which might bind the anionic amphiphile. It is a much simpler procedure when using purified recombinant cytosolic components.

49. Optimization of the time required for the amphiphile-dependent assembly of the oxidase complex is discussed in [202] and an excellent in-depth study, comprising stopped flow measurements, was recently published [227]. We found 90 s to be sufficient in the overwhelming majority of cases. However, it is important to point out that prolonging the time of assembly for up to 5 min might be advantageous and, in most cases, is not damaging. Thus, when in the preliminary stages of a project involving cell-free activation, exploring longer assembly times is recommended. Once the minimal time assuring full activation is found, this can be used routinely, but adding to it a "safety time supplement" is a wise move.

50. We currently use a VersaMax plate reader fitted with Softmax Pro 6.5.1. software. Settings are: Read Mode: Absorbance, Read Type: Kinetic, Shake Settings: Before First Read, 5 s, Time: 5 min, Interval: 0.11 min, Reads: 28, Min Interval: 0.05 min, Read Area: Multiple Cell, Plate Type: 96 well standard clear bottom, Reduction Settings: Optical Density, First Data Point Set To Zero; Wavelength Combination:!Lm1 (set at 550 nm), Kinetic Reduction: V_{max} (milli-units per min), V_{max} Points: 28 of 28, Lag Time: 0, End Time = 300 s.

51. Turnovers in the amphiphile-activated cell-free system rarely exceed 100 mol $O_2^{\cdot-}$/s/mol cytochrome b_{558} heme. This corresponds to 100 pmol $O_2^{\cdot-}$/s/pmol cytochrome b_{558} heme. Thus, in each well containing 1 pmol cytochrome b_{558} heme, 100 pmol $O_2^{\cdot-}$ are produced per second, which means 6 nmol per min, and 30 nmol per 5 min. The total amount of cytochrome *c* present in the well is 40 nmol, which is sufficient for binding all the $O_2^{\cdot-}$ produced in 5 min. A total of 48 nmol NADPH are added to the well, which based on a stoichiometry of 1 mol NADPH supporting the production of 2 mol of $O_2^{\cdot-}$, is more than sufficient for the production of 30 nmol of $O_2^{\cdot-}$. When measuring high activity oxidase preparations, "bending" of the curve nevertheless occurs, in spite of apparently sufficient total amounts of NADPH and cytochrome *c* at the start of the reaction, this being due to the presence of lesser and lesser amounts of NADPH and oxidized cytochrome *c* as we approach the end of the reaction. Our group is using single sources of cytochrome *c* and NADPH as components of the assay buffer. Because of variations in the degree of purity, the amount of water, and the proportion of oxidized and reduced material, it is wise to verify the actual amount of the two compounds present in the assay buffer (*see* Subheading 5.1.5). In the presence of sufficient cytochrome *c*, no need was found for the addition of catalase [175] to prevent reoxidation of cytochrome *c* by H_2O_2 originating in the spontaneous dismutation of $O_2^{\cdot-}$ that escaped scavenging by cytochrome *c*. However, if catalase is to be added, the assay buffer should be modified not to contain NaN_3 because of its inhibitory effect on catalase.

52. The extinction coefficient for the absorbance at 550 nm of reduced minus oxidized cytochrome *c*, as applied to a 1-cm path length, must be modified for the vertical path length of the microplate wells. This varies with the dimension and shape of the wells and the volume of the reaction mixture present in the well. Some advanced microplate spectrophotometers have a "PathCheck" sensor, allowing the normalization of absorbance values to a 1-cm path length. This allows the use of the canonical extinction coefficient to calculate the concentration of reduced cytochrome *c* in the well, without the need for any correction. This has only to be translated into the total amount of reduced cytochrome *c* in 210 μL, which permits the calculation of the turnover. With instruments not having the "PathCheck" option, the length of the vertical path length in the well must be determined by other means. Once it is known, the following equation will allow the direct calculation of nmol $O_2^{\cdot-}$ per min per well: nmol $O_2^{\cdot-}$/min/well = Δmilli–Abs_{550nm}/min × 0.047619 × reaction volume (in mL)/path

length (in cm). As an example, when then the reaction volume is 0.21 mL and the path length is found to be 0.575 cm, nmol $O_2^{\cdot -}$/min/well = Δmilli–Abs$_{550nm}$/min × 0.047619 × 0.21/ 0.575 = Δmilli–Abs$_{550nm}$/min × 0.017391.

53. The proper way to express results of cell-free oxidase activation assays is as turnover values. Unless there is a compelling reason for not doing so, oxidase activities should be related to the heme content of cytochrome b_{558} present in the membrane and not to cell number equivalents, total membrane protein, or the protein concentration of one or the other of the cytosolic components. The, unfortunately, common habit of expressing cell-free assay results as % change relative to a "basal" value can be thoroughly misleading in the absence of the information on the turnover corresponding to that basal value. This is critical when the effect of inhibitors on oxidase activity is expressed. Thus, a 50% inhibition is meaningless when the basal turnover value is 2 mol $O_2^{\cdot -}$/s/mol cytochrome b_{558} heme but is potentially meaningful when the value is 80 mol $O_2^{\cdot -}$/s/mol cytochrome b_{558} heme [257, 258].

54. From the advent of semirecombinant cell-free assays and the increasing rarity of the use of whole cytosol or partially purified cytosolic fractions, the need for the SOD control has been drastically reduced. In our laboratory, where semirecombinant cell-free assays are performed routinely, we have not encountered a single occasion of nonspecific cytochrome *c* reduction. When a reducing agent is carried over into the reaction, cytochrome *c* reduction occurs practically at time zero and will be reflected in the absence of the typical kinetics. The presence of a cytochrome *c* reductase in the membrane preparation remains possible but will act independently of the presence of amphiphile and cytosolic components. SOD controls should be used when cell-free assays are utilized as a diagnostic means on unpurified biologic material and in novel experimental situations. Yet another control for the specificity of cytochrome *c* reduction by $O_2^{\cdot -}$, rarely applied at present, is the use of acetylated cytochrome *c* as the $O_2^{\cdot -}$ trap [175]. Acetylation of lysine residues in cytochrome *c* decreases direct electron transfer from reductases, while maintaining the ability of $O_2^{\cdot -}$ to reduce cytochrome *c* [261].

55. The extinction coefficient for the absorbance of reduced INT at 490 nm relevant to a 1 cm path length has to be modified for the vertical path length of the microplate wells. When the microplate spectrophotometer does not have a "PathCheck" sensor, one has to know the total reaction volume per well and the length of the vertical path length. Once these are known, the following equation will allow the direct calculation of nmol INT reduced per min per well: nmol reduced INT/min/

well = Δmilli–Abs$_{490nm}$/min × 0.095328 × reaction volume (in mL)/path length (in cm). If the results are to be expressed in $O_2^{\cdot-}$ equivalents, the values are to be multiplied by 2, to account for the fact that reduction of INT is a two-electron reaction. As an example, when then the reaction volume is 0.21 mL and the path length is found to be 0.575 cm, nmol reduced INT/min/well = Δmilli–Abs$_{490nm}$/min × 0.095238 × 0.21/0.575 = Δmilli–Abs$_{490nm}$/min × 0.034782 (0.069564, for $O_2^{\cdot-}$ equivalent).

56. For the NADPH consumption test it is recommended to use 96-well plates for work at UV wavelengths (UV-Star, flat-bottom, Greiner Bio-One) and a microplate spectrophotometer capable to measure absorbance at 340 nm (we prefer to use the SpectraAmax 190 (wavelength range 190–850 nm, Molecular Devices) but VersaMax (wavelength range 340–850 nm, Molecular Devices) is also adequate. The minimum absorbance limit of the microplate reader has to be adjusted to a negative value, to allow the recording of negative absorbance kinetics relative to the blank represented by assay buffer with NADPH. The extinction coefficient for the absorbance of reduced NADPH at 340 nm relevant to a 1 cm path length [195] has to be modified for the vertical path length of the microplate wells. When the microplate spectrophotometer does not have a "PathCheck" sensor, one has to know the total reaction volume per well and the length of the vertical path length. Once these are known, the following equation will allow the direct calculation of nmol reduced NADPH consumed per min per well: nmol reduced NADPH consumed/min/well = ΔmAbs$_{340nm}$/min × 0.160771 × reaction volume (in mL)/path length (in cm). If the results are to be expressed in $O_2^{\cdot-}$ equivalents, the values are to be multiplied by 2, to account for the fact that one molecule of NADPH donates two electrons and yields two molecules of $O_2^{\cdot-}$. Note also that the consumption of NADPH will generate negative rate values when related to the reduced NADPH blank value. As an example, when then the reaction volume is 0.21 mL and the path length is found to be 0.575 cm, nmol reduced NADPH consumed/min/well = Δmilli–Abs$_{340nm}$/min × 0.160771 × 0.21/0.575 = Δmilli–Abs$_{340nm}$/min × 0.058716 (0.117432, for $O_2^{\cdot-}$ equivalent).

57. The concentrations of p67^{phox} and prenylated Rac required for reaching maximal activation, in the absence of amphiphile and p47^{phox}, are usually higher than those necessary for amphiphile-dependent activation in the presence of identical amounts of membrane. We routinely start with a concentration of 300 nM for both p67^{phox} and prenylated Rac.

58. In the amphiphile-independent cell-free system, we found on most occasions that prolonging the incubation to 5 min or longer is required for obtaining maximal oxidase activities. Optimization of the length of the assembly time is recommended.
59. The presence of p47^{phox} is not an absolute requirement for oxidase activity to be detected in this assay, but turnover values are higher in its presence.
60. This effect is thought to be due to an elevation of the Krafft point of the fatty acids in the presence of Ca^{2+} [115].
61. These results support the proposal that the stability of the FAD-Nox2 bond is enhanced by anionic amphiphile-induced changes in Nox2 and by the process of assembly with the cytosolic components [262].
62. In the experiments described in Fig. 8, both nonprenylated and prenylated Rac1 were exchanged to GMPPNP and the exchange stabilized by 25 mM $MgCl_2$. Thus, it is unlikely that significant dissociation of GMPPNP from Rac can occur upon dilution of exchanged Rac preparations in assay buffer during the 90-s time interval of oxidase assembly. However, assay buffer is also used for intermediary dilution steps of recombinant cytosolic components, and it is possible that the diluted proteins are, sometimes, kept in assay buffer for longer periods of time. In the particular case of Rac, it is recommended to make such dilutions in the buffer used for diluting Rac in preparation for nucleotide exchange, which contains 4 mM $MgCl_2$.
63. There is good evidence for equimolar and simultaneous translocation of the cytosolic components in neutrophils stimulated by two elicitors of a respiratory burst and it was also found that this translocation corresponded temporally with the generation of $O_2^{\cdot-}$ [228].
64. It is of interest that sigmoidal dose–response curves were described in the early period of the use of cell-free assays, when total cytosol was used and its amount was related to activity in the presence of a constant amount of amphiphile [197, 225].
65. This assumption was examined by Heyworth et al. [263] and Fuchs et al. [256], who showed that exchange with GTP added to the assay buffer only takes place with prenylated Rac and that preloading with GTP by Mg^{2+} chelation is required for non-prenylated Rac to work in the amphiphile-dependent cell-free assay.

Acknowledgments

The research described in this chapter was supported by the Julius Friedrich Cohnheim-Minerva Center for Phagocyte Research, the Ela Kodesz Institute of Host Defense against Infectious Diseases, Israel Science Foundation Grants 428/01, 19/05, 49/09, 300/13, and 144/17, the Roberts-Guthman Chair in Immunopharmacology, the Walter J. Levy Benevolent Trust, the David Roberts Fund, and the Joseph and Shulamit Salomon Fund. It is important to point out that the cell-free system was discovered almost simultaneously by several investigators: R.A. Heyneman and R.E. Vercauteren, in Belgium, and Linda McPhail and John Curnutte, in the USA. All these investigators, independently, contributed greatly to the birth of the "cell-free" paradigm. Edgar Pick would like to thank the many postdoctoral fellows, students, and assistants, who over a period of close to four decades, were responsible for this considerable body of work. Among the many, Ms. Yael Bromberg, deserves special thanks for her dedication and perseverance up to and after the moment when we first saw cytochrome *c* being reduced by a homogenate of resting macrophages. Thanks go to my fellow scientists, too many to name, who provided materials and invaluable advice, for making this work possible. There is no greater satisfaction than the realization of the fact that, on so many occasions, what started as collaboration (called "networking," these days) or competition, evolved into long-lasting friendships. Finally, no hard feelings are left toward the reviewers who rejected one of our papers describing the cell-free system and toward those who labeled the cell-free system an "*in vitro artefact*," of no relevance to what happens in vivo. Time is a great healer.

References

1. Nauseef WM (2007) How human neutrophils kill and degrade microbes: an integrated view. Immunol Rev 219:88–102
2. Quinn MT, Gauss KA (2004) Structure and regulation of the neutrophil respiratory burst oxidase: comparison with nonphagocyte oxidases. J Leukoc Biol 76:760–781
3. Cross AR, Segal AW (2004) The NADPH oxidase of phagocytes—prototype of the NOX electron transport chain systems. Biochim Biophys Acta 1657:1–22
4. Sumimoto H (2008) Structure, regulation and evolution of Nox-family NADPH oxidases that produce reactive oxygen species. FEBS J 275:3249–3277
5. Knaus UG, Leto TL (eds) (2019) NADPH oxidases: methods and protocols. Springer Science + Business Media, LLC, New York
6. Nauseef WM (2004) Assembly of the phagocyte NADPH oxidase. Histochem Cell Biol 122:277–291
7. Groemping Y, Rittinger K (2005) Activation and assembly of the NADPH oxidase: a structural perspective. Biochem J 386:401–416
8. Pick E (2014) Cell-free NADPH oxidase assays: "in vitro veritas". In: Quinn MT, DeLeo FR (eds) Neutrophil methods and protocols, 2nd edn. Springer Science + Business Media, LLC, New York
9. Han C-H, Freeman JLR, Lee T et al (1998) Regulation of the neutrophil respiratory burst oxidase. Identification of an activation domain. J Biol Chem 273:16663–16668
10. Kreck ML, Freeman JL, Lambeth JD (1996) Membrane association of Rac is required for

high activity of the respiratory burst oxidase. Biochemistry 35:15683–15692
11. Gorzalczany Y, Sigal N, Itan M et al (2000) Targeting of Rac1 to the phagocyte membrane is sufficient for the induction of NADPH oxidase assembly. J Biol Chem 275:40073–40081
12. Sarfstein R, Gorzalczany Y, Mizrahi A et al (2004) Dual role of Rac in the assembly of NADPH oxidase, tethering to the membrane and activation of $p67^{phox}$. A study based on mutagenesis of $p67^{phox}$-Rac1 chimeras. J Biol Chem 279:16007–16016
13. Freeman JL, Lambeth JD (1996) NADPH oxidase activity is independent of $p47^{phox}$ in vitro. J Biol Chem 271:22578–22582
14. Koshkin V, Lotan O, Pick E (1996) The cytosolic component $p47^{phox}$ is not a *sine qua non* participant in the activation of NADPH oxidase but is required for optimal superoxide production. J Biol Chem 271:30326–30329
15. Dale H (1936) Some recent extensions of the chemical transmission of the effects of nerve impulses. Nobel lecture, December 12, 1936. From Nobel Lectures Physiology or Medicine, 1922–1941. Elsevier Publishing Company, Amsterdam
16. Ziegler CS, Bouchab L, Tramier M et al (2019) Quantitative live-cell imaging and 3D modeling reveal critical functional features in the cytosolic complex of phagocyte NADPH oxidase. J Biol Chem 294:3824–3836
17. Leto TL, Geiszt M (2006) Role of Nox family NADPH oxidases in host defense. Antioxid Redox Signal 8:1549–1561
18. Bedard K, Krause K-H (2007) The NOX family of ROS-generating NADPH oxidases: physiology and pathophysiology. Physiol Rev 87:255–313
19. Babior BM, Kipnes RS, Curnutte JT (1973) Biological defense mechanisms. The production by leukocytes of superoxide, a potential bactericidal agent. J Clin Invest 52:741–744
20. Cheson BD, Curnutte JT, Babior BM (1977) The oxidative killing mechanism of the neutrophil. Prog Clin Immunol 3:1–65
21. Babior BM (1984) Oxidants from phagocytes: agents of defense and destruction. Blood 64:959–966
22. Pick E, Keisari Y (1981) Superoxide anion production and hydrogen peroxide production by chemically elicited macrophages—induction by multiple nonphagocytic stimuli. Cell Immunol 59:301–318
23. Babior BM, Curnutte JT, McMurrich BJ (1976) The particulate superoxide-forming system from human neutrophils: properties of the system and further evidence supporting its participation in the respiratory burst. J Clin Invest 58:989–996
24. Markert M, Andrews PC, Babior BM (1984) Measurement of $O_2^{\cdot-}$ production by human neutrophils. The preparation and assay of NADPH oxidase-containing particles from human neutrophils. Meth Enzymol 105:358–365
25. Babior BM, Kipnes RS (1977) Superoxide-forming enzyme from human neutrophils: evidence for a flavin requirement. Blood 50:517–524
26. Badwey JA, Curnutte JT, Karnovsky ML (1979) The enzyme of granulocytes that produces superoxide and peroxide – an elusive pimpernel. New Engl J Med 300:1157–1160
27. Rossi F (1986) The $O_2^{\cdot-}$-forming NADPH oxidase of the phagocytes: nature, mechanism of activation and function. Biochim Biophys Acta 853:65–89
28. Vignais PV (2002) The superoxide-generating NADPH oxidase: structural aspects and activation mechanism. Cell Mol Life Sci 59:1248–1459
29. Smolen JE, Weissmann G (1980) Effects of imdomethacin, 5,8,11,14-eicosatetraynoic acid, and bromophenacyl bromide on lysosomal enzyme release and superoxide anion generation by human polymorphonuclear leukocytes. Biochem Pharmacol 29:533–538
30. Kakinuma K (1974) Effects of fatty acids on the oxidative metabolism of leukocytes. Biochim Biophys Acta 348:76–85
31. Kakinuma K, Minakami S (1978) Effects of fatty acids on superoxide radical generation in leukocytes. Biochim Biophys Acta 538:50–59
32. Badwey JA, Curnutte JT, Karnovsky ML (1981) *cis*-polyunsaturated fatty acids induce high levels if superoxide production by human neuttrphils. J Biol Chem 256:12640–12643
33. Zatti M, Rossi F (1967) Relationship between glycolysis and respiration in surfactant-treated leucocytes. Biochim Biophys Acta 148:553–555
34. Graham RC, Karnovsky MJ, Shafer AW et al (1967) Metabolic and morphological observations on the effect of surface-active agents on leukocytes. J Cell Biol 32:629–647
35. Cohen HJ, Chovaniec ME (1978) Superoxide generation by digitonin-stimulated Guinea pig granulocytes. A basis for a continuous assay for monitoring superoxide production and for the study of the activation of the generating system. J Clin Invest 61:1081–1087

36. Bromberg Y, Pick E (1983) Unsaturated fatty acids as second messengers of superoxide generation by macrophages. Cell Immunol 79:243–252
37. Steinbeck MJ, Robinson JM, Karnovsky MJ (1991) Activation of the neutrophil NADPH-oxidase by free fatty acids requires the ionized carboxyl group and partitioning into membrane lipid. J Leukoc Biol 49:360–368
38. Ingraham LM, Weening RS, Clarke MF et al (1982) Relation of respiratory burst and arachidonate metabolism during phagocytosis by Guinea pig alveolar macrophages. J Lab Clin Med 99:908–916
39. Ward PA, Sulavik MC, Johnson KJ (1985) Activated rat neutrophils. Correlation of arachidonate products with enzyme secretion but not with $O_2^{\cdot-}$ generation. Am J Pathol 120:112–120
40. Bomalaski JS, Baker DG, Brophy L et al (1989) A phospholipase A_2-activating protein (PLAP) stimulates human neutrophil aggregation and release of lysosomal enzymes, superoxide, and eicosanoids. J Immunol 142:3957–3962
41. Bokoch GM, Reed PW (1979) Inhibition of the neutrophil oxidative response to a chemotactic peptide by inhibitors of arachidonic acid oxygenation. Biochem Biophys Res Commun 90:481–487
42. Bokoch GM, Reed PW (1980) Stimulation of arachidonic acid metabolism in the polymorphonuclear leukocyte by an *N*-formylated peptide. Comparison with ionophore A23187. J Biol Chem 265:10223–10226
43. Maridonneau-Parini I, Tringale S, Tauber AI (1986) Identification of distinct activation pathways of the human neutrophil NADPH-oxidase. J Immunol 137:2925–2929
44. Maridonneau-Parini I, Tauber AI (1986) Activation of NADPH-oxidase by arachidonic acid involves phospholipase A_2 in intact human neutrophils but not in the cell-free system. Biochem Biophys Res Commun 138:1099–1105
45. Henderson LM, Chappell JB, Jones OTG (1989) Superoxide generation is inhibited by phospholipase A_2 inhibitors. Role of phospholipase A_2 in the activation of the NADPH oxidase. Biochem J 264:249–255
46. Sakata A, Ida E, Tominaga M, Onoue K (1987) Arachidonic acid acts as an intracellular activator of NADPH-oxidase in Fcγ receptor-mediated superoxide generation in macrophages. J Immunol 138:43543–44359
47. Dana R, Malech HL, Levy R (1994) The requirement for phospholipase A_2 for activation of the assembled oxidase in human neutrophils. Biochem J 297:217–223
48. Dana R, Leto TL, Malech HL et al (1998) Essential requirement of cytosolic phospholipase A_2 for activation of the phagocyte NADPH oxidase. J Biol Chem 273:441–445
49. Zhao X, Bey EA, Wientjes FB et al (2002) Cytosolic phospholipase A_2 ($cPLA_2$) regulation of human monocyte NADPH oxidase activity. $cPLA_2$ affects translocation but not phosphorylation of $p67^{phox}$ and $p47^{phox}$. J Biol Chem 277:25385–25392
50. Tauber AI (1987) Protein kinase C and the activation of the human neutrophil NADPH oxidase. Blood 69:711–720
51. Leto TL, Adams AG, de Mendez I (1994) Assembly of the phagocyte NADPH oxidase: binding of Src homology 3 domains to proline-rich targets. Proc Natl Acad Sci U S A 91:10650–10654
52. Sumimoto H, Kage Y, Nunoi H et al (1994) Role of Src homology 3 domains in assembly and activation of the phagocyte NADPH oxidase. Proc Natl Acad Sci U S A 91:5345–5349
53. Groemping Y, Lapouge K, Smerdon SJ et al (2003) Molecular basis of phosphorylation-induced activation of the NADPH oxidase. Cell 113:343–355
54. Marcoux J, Man P, Petit-Hartlein I et al (2010) $p47^{phox}$ molecular activation for assembly of the neutrophil NADPH oxidase complex. J Biol Chem 285:28980–28990
55. Swain SD, Helgerson SL, Davis AR et al (1997) Analysis of activation-induced conformational changes in $p47^{phox}$ using tryptophan fluorescence spectroscopy. J Biol Chem 272:29502–29510
56. Park J-W, Babior BM (1998) Activation of the leukocyte NADPH oxidase subunit $p47^{phox}$ by protein kinase C. A phosphorylation-dependent change in the conformation of the C-terminal end of $p47^{phox}$. Biochemistry 36:7474–7480
57. Park H-S, Park J-W (1998) Fluorescent labeling of the leukocyte NADPH oxidase subunit $p47^{phox}$: evidence for amphiphile-induced conformational changes. Arch Biochem Biophys 360:165–172
58. Cox JA, Jeng AY, Sharkey NA et al (1985) Activation of the human neutrophil nicotinamide dinucleotide phosphate (NADPH)-oxidase by protein kinase C. J Clin Invest 76:1932–1938
59. Cox J, Jeng AY, Blumberg PM et al (1987) Comparison of subcellular activation of the human neutrophil NADPH-oxidase by

arachidonic acid, sodium dodecyl sulfate (SDS) and phorbol myristate acetate (PMA). J Immunol 138:1884–1888

60. Tauber AI, Cox JA, Curnutte JT et al (1989) Activation of human neutrophil NADPH oxidase in vitro by the catalytic fragment of protein kinase C. Biochem Biophys Res Commun 158:884–890
61. Park J-W, Hoyal CR, El Benna J et al (1997) Kinase-dependent activation of the leukocyte NADPH oxidase in a cell-free system. Phosphorylation of membranes and $p47^{phox}$ during oxidase activation. J Biol Chem 272:11035–11043
62. Rossetti Lopes L, Hoyal CR, Knaus UG et al (1999) Activation of the leukocyte NADPH oxidase by protein kinase C in a partially recombinant cell-free system. J Biol Chem 274:15533–15537
63. Park J-W, Scott KE, Babior BM (1998) Activation of the leukocyte NADPH oxidase in a cell-free system: phosphorylation vs. amphiphiles. Exp Hematol 26:37–44
64. Nishizuka Y (1992) Intracellular signaling by hydrolysis of phospholipids and activation of protein kinase C. Science 259:607–614
65. Jenkins GM, Frohman MA (2005) Phospholipase D: a lipid centric review. Cell Mol Life Sci 62:2305–2316
66. Korchak HM, Vosshall LB, Hainess KA et al (1988) Activation of the human neutrophil by calcium-mobilized ligands. II correlation of calcium, diacylglycerol and phosphatidic acid generation with superoxide anion generation. J Biol Chem 263:11098–11105
67. Rossi F, Grzeskowiak M, Della-Bianca V et al (1990) Phosphatidic acid and not diacylglycerol generated by phospholipase D is functionally linked to the activation of the NADPH oxidase by FMLP in human neutrophils. Biochem Biophys Res Commun 168:320–327
68. Bellavite P, Corso F, Dusi S et al (1988) Activation of NADPH-dependent superoxide production in plasma membrane extracts of pig neutrophils by phosphatidic acid. J Biol Chem 263:8210–8214
69. Qualliotine-Mann D, Agwu DE, Ellenburg MD et al (1993) Phosphatidic acid and diacylglycerol synergize in a cell-free system for activation of NADPH oxidase from human neutrophils. J Biol Chem 268:23843–23849
70. Erickson RW, Langel-Peveri P, Traynor-Kaplan AE et al (1999) Activation of human neutrophil NADPH oxidase by phosphatidic acid or diacylglycerol in a cell-free system. Activity of diacylglycerol is dependent on its conversion to phosphatidic acid. J Biol Chem 274:22243–22250
71. McPhail LC, Qualliotine-Mann D, Waite KA (1995) Cell-free activation of neutrophil NADPH oxidase by a phosphatidic acid-regulated protein kinase. Proc Natl Acad Sci U S A 92:7931–7935
72. Regler DS, Waite KA, Wallin R et al (1999) A phosphatidic acid-activated protein kinase and conventional protein kinase C isoforms phosphorylate $p22^{phox}$, an NADPH oxidase component. J Biol Chem 274:36601–36608
73. Johnston RB, Godzik CA, Cohn ZA (1978) Increased superoxide anion production by immunologically activated and chemically elicited macrophages. J Exp Med 148:115–127
74. Murray HW, Cohn ZA (1980) Macrophage oxygen-dependent antimicrobial activity. III. Enhanced oxidative metabolism as an expression of macrophage activation. J Exp Med 152:1596–1609
75. Freund M, Pick E (1985) The mechanism of action of lymphokines VIII. Lymphokine-enhanced spontaneous hydrogen peroxide production by macrophages. Immunology 54:35–45
76. Freund M, Pick E (1986) The mechanism of action of lymphokines IX. The enzymatic basis of hydrogen peroxide production by lymphokine activated macrophages. J Immunol 137:1312–1318
77. Bromberg Y, Pick E (1984) Unsaturated fatty acids stimulate NADPH- dependent superoxide generation by cell-free system in macrophages. Cell Immunol 88:213–221
78. Kuhn TS (1962) The structure of scientific revolutions. The University of Chicago Press, Chicago
79. Heyneman RA, Vercauteren RE (1984) Activation of a NADPH oxidase from horse poymorphonuclear leukocytes in a cell-free system. J Leukoc Biol 36:751–759
80. Heyneman RA (1983) Subcellular localization and properties of the NADPH oxidase from equine polymorphonuclear leukocytes. Enzyme 29:198–207
81. Dechatelet LR, McCall C, Shirley PS (1980) Activation by dialysis of NADPH oxidase (s) from human neutrophils. J Reticuloendothel Soc 28:533–545
82. McPhail LC, Shirley PS, Clayton CC et al (1985) Activation of the respiratory burst enzyme from human neutrophils in a cell-free system. J Clin Invest 75:1735–1739
83. Curnutte JT (1985) Activation of human neutrophil nicotinamide adenine dinucleotide phosphate reduced (triphosphopyridine

nucleotide, reduced) oxidase by arachidonic acid in a cell-free system. J Clin Invest 75:1740–1743

84. Abo A, Boyhan A, West I et al (1992) Reconstitution of neutrophil NADPH oxidase activity in the cell-free system by four components: p67^{phox}, p47^{phox}, p21rac1, and cytochrome b_{-245}. J Biol Chem 267:16767–16770
85. Seifert R, Schultz G (1987) Fatty acid-induced activation of NADPH oxidase in plasma membranes of human neutrophils depends on neutrophil cytosol and is potentiated by stable guanine nucleotides. Eur J Biochem 162:563–569
86. Bromberg Y, Pick E (1985) Activation of NADPH-dependent superoxide production in a cell-free system by sodium dodecyl sulfate. J Biol Chem 260:13539–13545
87. Doussière J, Gaillard J, Vignais PV (1996) Electron transfer across the O_2^- generating flavocytochrome of neutrophils. Evidence for transition from low-spin state to a high-spin state of the heme iron component. Biochemistry 35:13400–13410
88. Doussiere J, Bouzidi F, Poinas A et al (1999) Kinetic study of the activation of the neutrophil NADPH oxidase by arachidonic acid. Antagonistic effects of arachidonic acid and phenylarsine oxide. Biochemistry 38:16394–16406
89. Foubert TR, Burritt JB, Taylor RM et al (2002) Structural changes are induced in human neutrophil cytochrome *b* by NADPH oxidase activators, LDS, SDS and arachidonate: intermolecular resonance energy transfer between trisulfopyrenyl-wheat germ agglutinin and cytochrome b_{558}. Biochim Biophys Acta 1567:221–231
90. Taylor RM, Foubert TR, Burritt JB et al (2004) Anionic amphiphile and phospholipid-induced conformational changes in human neutrophil flavocytochrome b observed by fluorescence resonance energy transfer. Biochim Biophys Acta 1663:201–213
91. Taylor RM, Riesselman MH, Lord CI et al (2012) Anionic lipid-induced conformational changes in human phagocyte flavocytochrome b precede assembly and activation of the NADPH oxidase complex. Arch Biochem Biophys 521:24–31
92. Matono R, Miyano K, Kiyohara T et al (2014) Arachidonic acid induces direct interaction of the p67^{phox}-Rac complex with the phagocyte oxidase Nox2 leading to superoxide production. J Biol Chem 289:24874–24884
93. Bizouarn T, Karimi G, Masoud R et al (2016) Exploring the arachidonic acid-induced structural changes in phagocyte NADPH oxidase p47^{phox} and p67^{phox} via thiol accessibility and SCRD spectroscopy. FEBS J 283:2896–2910
94. Curnutte JT, Kuver R, Scott PJ (1987) Activation of neutrophil NADPH oxidase in a cell-free system. Partial purification of the components and characterization of the activation process. J Biol Chem 262:5563–5569
95. Curnutte JT, Babior BM (1987) Chronic granulomatous disease. Adv Hum Genet 16:229–297
96. Curnutte JT, Berkow RL, Roberts RL et al (1988) Chronic granulomatous disease due to a defect in the cytosolic factor required for nicotinamide dinucleotide phosphate oxidase activation. J Clin Invest 81:606–610
97. Curnutte PJ, Scott PJ, Babior BM (1989) Functional defect in neutrophil cytosols from two patients with autosomal recessive cytochrome-positive chronic granulomatous disease. J Clin Invest 83:1236–1240
98. Volpp BD, Nauseef WM, Clark RA (1988) Two cytosolic neutrophil oxidase components absent in autosomal chronic granulomatous disease. Science 242:1295–1297
99. Nunoi H, Rotrosen D, Gallin JI et al (1988) Two forms of autosomal chronic granulomatous disease lack distinct neutrophil cytosol factors. Science 242:1298–1301
100. Abo A, Pick E, Hall A et al (1991) Activation of the NADPH oxidase involves the small GTP-binding protein p21^{rac1}. Nature 353:668–670
101. Knaus UG, Heyworth PG, Evans T et al (1991) Regulation of phagocyte oxygen radical production by the GTP-binding protein Rac2. Science 254:1512–1515
102. Rotrosen D, Yeung CL, Leto TL et al (1992) Cytochrome b_{558}: the flavin-binding component of the phagocyte NADPH oxidase. Science 256:1459–1462
103. Segal AW, West I, Wientjes F et al (1992) Cytochrome b-245 is a flavocytochrome containing FAD and the NADPH-binding site of the microbicidal oxidase of phagocytes. Biochem J 284:781–788
104. Knoller S, Shpungin S, Pick E (1991) The membrane –associated component of the amphiphile-activated cytosol-dependent superoxide-forming NADPH oxidase of macrophages is identical to cytochrome b_{559}. J Biol Chem 266:2795–2804
105. Koshkin V, Pick E (1993) Generation of superoxide by purified and relipidatred

cytochrome b_{559} in the absence of cytosolic activators. FEBS Lett 327:57–62
106. Koshkin V, Pick E (1994) Superoxide production by cytochrome b_{559}. Mechanism of cytosol-independent activation. FEBS Lett 338:285–289
107. Tsunawaki S, Nathan CF (1986) Release of arachidonate and reduction of oxygen. Independent metabolic bursts of the mouse peritoneal macrophage. J Biol Chem 261:11563–11570
108. Badwey JA, Curnutte JT, Robinson JM et al (1984) Effects of fatty acids on release of superoxide and on change of shape by human neutrophils. Reversibility by albumin. J Biol Chem 259:7870–7877
109. Babior BM, Woodman RC (1990) Chronic granulomatous disease. Semin Hematol 27:247–259
110. Nunoi H, Matsuda I (1992) Molecular basis of chronic granulomatous disease – analysis of the cytosol factors for NADPH oxidase. Rinsho Byori 40:380–384
111. Nauseef WM, McCormick S, Renee J et al (1993) Functional domains in an arginine-rich carboxyl-terminal region of p47phox. J Biol Chem 268:23646–23651
112. Nauseef WM (2014) Identification and quantification of superoxide anion: essential steps in the elucidation of the "respiratory burst". J Immunol 193:5357–5358
113. Bromberg Y, Sha'ag D, Shpungin S et al (1986) Activation of the superoxide forming NADPH oxidase in a macrophage-derived cell-free system by fatty acids and detergents. In: Zor U, Naor Z, Cohen F (eds) Advances in prostaglandin, thromboxane and leukotriene research, vol 16. Raven Press, New York, p 153
114. Dagher M-C, Pick E (2007) Opening the black box: learning from cell-free systems on the phagocyte NADPH oxidase. Biochimie 89:1123–1132
115. Yamaguchi T, Kaneda M, Kakinuma K (1986) Effect of saturated and unsaturated fatty acids on the oxidative metabolism of human neutrophils. The role of calcium ion in the extracellular medium. Biochim Biophys Acta 861:440–446
116. Tanaka T, Kanegasaki S, Makino R et al (1987) Saturated and trans-unsaturated fatty acids elicit high levels of superoxide generation in intact and cell-free preparations of neutrophils. Biochem Biophys Res Commun 14:696–612
117. Hardy SJ, Ferrante A, Poulos A et al (1994) Effect of exogenous fatty acids with greater than 22 carbon atoms (very long chain fatty acids) on superoxide production by human neutrophils. J Immunol 153:1754–1761
118. Tanaka T, Makino R, Iizuka T et al (1988) Activation by saturated and monounsaturated fatty acids of the O_2^--generating system in a cell-free preparation from neutrophils. J Biol Chem 263:13670–13676
119. Nishida S, Yoshida LS, Shimoyama T et al (2005) Fungal metabolite gliotoxin targets flavocytochrome b_{558} in the activation of the human neutrophil NADPH oxidase. Infect Immun 73:235–244
120. Souabni H, Thoma V, Bizouarn T et al (2012) *trans* arachidonic acid acid isomers inhibit NADPH-oxidase activity by direct interaction with enzyme components. Biochim Biophys Acta 1818:2314–2324
121. Gonzales-Perilli L, Alvarez MN, Prolo C et al (2013) Nitroarachidonic acid prevents NADPH oxidase assembly and superoxide radical production in activated macrophages. Free Rad Biol Med 58:126–133
122. Shpungin S, Dotan I, Abo A, Pick E (1989) Activation of the superoxide forming NADPH oxidase in a cell-free system by sodium dodecyl sulfate. Absolute lipid dependence of the solubilized enzyme. J Biol Chem 264:9195–9203
123. Glick J, Santoyo G, Casey PJ (1996) Arachidonate and related unsaturated fatty acids selectively inactivate the guanine nucleotide-binding regulatory protein G_z. J Biol Chem 271:2949–2954
124. Anlansson EAG, Wall SN, Almgren M et al (1976) Theory of the kinetics if micellar equilibria and quantitative interpretation of chemical relaxation studies of micellar solutions of ionic surfactants. J Phys Chem 80:905–922
125. Cifuentes A, Bernal JL, Diez-Masa JC (1997) Determination of critical micelle concentration values using capillary electrophoresis instrumentation. Anal Chem 69:4271–4274
126. Brash AR (2001) Arachidonic acid as a bioactive molecule. J Clin Invest 107:1339–1345
127. Chilton FH, Fonteh AN, Surette ME et al (1996) Control of arachidonate levels within inflammatory cells. Biochim Biophys Acta 1299:1–15
128. Lapouge K, Smith SJM, Groemping Y et al (2002) Architecture of the p40-p47-p67^{phox} complex in the resting state of the NADPH oxidase. A central role for p67^{phox}. J Biol Chem 277:10121–10128
129. Borregaard N, Heiple JM, Simons ER et al (1983) Subcellular localization of the *b*-cytochrome component of the human neutrophil

microbicidal oxidase: translocation during activation. J Cell Biol 97:52–61

130. Clark RA, Leidal KG, Pearson DW et al (1987) NADPH oxidase of human neutrophils. Subcellular localization and characterization of an arachidonate-activatable superoxide-generating system. J Biol Chem 262:4065–4074

131. Lundquist H, Follin P, Khalfan L et al (1996) Phorbol myristate acetate-induced NADPH oxidase activity in human neutrophils: only half the story has been told. J Leukoc Biol 59:270–279

132. Pick E, Bromberg Y, Shpungin S et al (1987) Activation of the superoxide forming NADPH oxidase by sodium dodecyl sulfate. Characterization of the membrane-associated component. J Biol Chem 262:16476–16483

133. Hata K, Ito T, Takeshige K et al (1998) Anionic amphiphile-independent activation of the phagocyte NADPH oxidase in a cell-free system by p47*phox* and p67*phox*, both in C terminally truncated forms. Implications for regulatory Src himology 3 domain-mediated interactions. J Biol Chem 273:4232–4236

134. Ebisu K, Nagasawa T, Watanabe K et al (2001) Fused p47*phox* and p67*phox* truncations efficiently reconstitute NADPH oxidase with higher activity than the individual components. J Biol Chem 276:24498–24505

135. Gorzalczany Y, Alloul N, Sigal N et al (2002) A prenylated p67*phox*-Rac1 chimera elicits NADPH-dependent superoxide production by phagocyte membranes in the absence of an activator and of p47*phox*. Conversion of a pagan NADPH oxidase to monotheism. J Biol Chem 277:18605–18610

136. Peng G, Huang J, Boyd M et al (2003) Properties of phagocyte NADPH oxidase p47*phox* mutants with unmasked SH3 (Src homology 3) domains: full reconstitution of oxidase activity in a semi-recombinant cell-free system lacking arachidonic acid. Biochem J 373:221–229

137. Berdichevsky Y, Mizrahi A, Ugolev Y et al (2007) Tripartite chimeras comprising functional domains derived from the three cytosolic components p47*phox*, p67*phox* and Rac1 elicit activator-independent superoxide production by phagocyte membranes. Role of membrane lipid charge and of specific residues in the chimeras. J Biol Chem 282:22122–22139

138. Mizrahi A, Berdichevsky Y, Casey PJ et al (2010) A prenylated p47*phox*-p67*phox*-Rac1 chimera is a quintessential NADPH oxidase activator. Membrane association and functional capacity. J Biol Chem 285:25485–25499

139. Tucker KA, Lilly MB, Heck L et al (1987) Characterizaton of a new human diploid cell line (PLB-985) with granulocytic and monocytic differentiating capacity. Blood 70:372–378

140. Zhen L, King AAJ, Xiao Y et al (1993) Gene targeting of X chromosome-linked chronic granulomatous disease locus in a human leukemia cell line and rescue by expression of recombinant gp91phox. Proc Natl Acad Sci U S A 90:9832–9836

141. Li XJ, Grunwald D, Mathieu J et al (2005) Crucial roles of two potential cytosolic regions of Nox2, ^{191}ISSTKTIRRS^{200} and ^{483}DESQANHFAVHHDEEKD^{599}, on NADPH oxidase activation. J Biol Chem 280:14962–14973

142. Picciocchi A, Debeurme F, Beaumel S et al (2011) Role of putative second transmembrane region of Nox2 protein in the structural stability and electron transfer of the phagocytic NADPH oxidase. J Biol Chem 286:28357–28369

143. Hirano K, Chen WS, Chueng ALW et al (2015) Discovery of GSK2795039, a novel small molecule NADPH oxidase 2 inhibitor. Antioxid Redox Signal 23:358–374

144. Macpherson IA, Stoker MGP (1962) Polyoma transformation of hamster cell clones—an investigation of genetic factors affecting cell competence. Virology 16:147–151

145. Ostuni MA, Lamanuzzi LB, Bizouarn T et al (2010) Expression of functional mammal flavocytochrome b_{558} in yeast: comparison with improved insect cell system. Biochim Biophys Acta 2798:1179–1188

146. Katkin JP, Malech HL, Leto TL (1992) Baculovirus mediated expression of human phagocytic cell oxidase cytochrome b_{558} in sf9 insect cells. Inflammation 16:393–410

147. Rotrosen D, Yeung CL, Katkin JP (1993) Production of recombinant cytochrome b_{558} allows the reconstitution of the phagocyte NADPH oxidase solely from recombinant proteins. J Biol Chem 268:14256–14260

148. Miyano K, Fukuda H, Ebisu K et al (2003) Remarkable stabilization of neutrophil NADPH oxidase using RacQ61L and a p67*phox*-p47*phox* fusion protein. Biochemistry 42:184–190

149. Miyano K, Kitahara H, Ohmi S et al (2004) Inactivation of neutrophil NADPH oxidase upon dilution and its prevention by cross-link and fusion of phox proteins. Arch Biochem Biophys 431:129–137

150. Tamura M, Nagasawa T, Tange T et al (2005) A new type of O_2^-- generating tool for oxidative stress studies by remodeling NADPH oxidase. J Biotechnol 120:421–429
151. Miyano K, Ogasawara S, Han C-H et al (2001) A fusion protein between Rac and p67^{phox} (1-210) reconstitutes NADPH oxidase with higher activity and stability than individual components. Biochemistry 40:14089–14097
152. Alloul N, Gorzalczany Y, Itan M et al (2001) Activation of the superoxide-generating NADPH oxidase by chimeric proteins consisting of segments of the cytosolic component p67^{phox} and the small GTPase Rac1. Biochemistry 40:14557–14566
153. Mizrahi A, Berdichevsky Y, Ugolev Y et al (2006) Assembly of the phagocyte NADPH oxidase complex: chimeric constructs derived from the cytosolic components as tools for exploring structure - function relationships. J Leukoc Biol 79:881–895
154. Massoud R, Sarfaty X, Erard M et al (2017) Conversion of Nox2 into a constitutive enzyme *in vitro* and in living cells, after its binding with a chimera of the regulatory subunits. Free Rad Biol Med 113:470–477
155. Pick E, Gorzalczany Y, Engel S (1993) Role of the rac1 p21-GDP-dissociation inhibitor for *rho* heterodimer in the activation of the superoxide-forming NADPH oxidase of macrophages. Eur J Biochem 217:441–455
156. Ugolev Y, Molshanski-Mor S, Weinbaum C, Pick E (2006) Liposomes comprising anionic but not neutral phospholipids cause dissociation of [Rac(1 or 2)-RhoGDI] complexes and support amphiphile-independent NADPH oxidase activation by such complexes. J Biol Chem 281:19204–19219
157. Ugolev Y, Berdichevsky Y, Weinbaum C et al (2008) Dissociation of Rac1(GDP)-RhoGDI complexes by the cooperative action of anionic liposomes containing phosphatidylinositol 3,4,5-triphosphate, Rac guanine nucleotide exchange factor, and GTP. J Biol Chem 283:22257–22271
158. Pick E (2014) Role of the rho GTPase Rac in the activation of the phagocyte NADPH oxidase. Outsourcing a key task. Small GTPases 5:e27952
159. Cross AR, Erickson RW, Curnutte JT (1999) The mechanism of activation of NADPH oxidase in the cell-free system: the activation process is primarily catalytic and not through the formation of a stoichiometric complex. Biochem J 341:251–255
160. Brown GE, Stewart MQ, Liu H et al (2003) A novel assay system implicates PtdIns(3,4)P_2, PtdIns(3)P. and PKCδ in intracellular production of reactive oxygen species by the NADPH oxidase. Mol Cell 11:35–47
161. Bissonnette SA, Glazier CM, Stewart MQ et al (2008) Phosphatidylinositol 3-phosphate-dependent and -independent functions of p40^{phox} in activation of the neutrophil NADPH oxidase. J Biol Chem 283:2108–2119
162. Roos D, Voetman AA, Meerhof LJ (1983) Functional activity of enucleated human polymorphonuclear leukocytes. J Cell Biol 97:368–377
163. Bauldry SA, Nasrallah VN, Bass DA (1992) Activation of NADPH oxidase in human neutrophils permeabilized with *Staphilococcus aureus* α-toxin. A lower Km when the enzyme is activated in situ. J Biol Chem 267:323–330
164. Leusen JHW, Meischl C, Eppink MHM et al (2000) Four novel mutations in the gene encoding gp91^{phox} of human NADPH oxidase: consequences for oxidase activity. Blood 95:666–673
165. Sigal N, Gorzalczany Y, Pick E (2003) Two pathways of activation of the superoxide-generating NADPH oxidase of phagocytes *in vitro*—distinctive effects of inhibitors. Inflammation 27:147–159
166. Mizrahi A, Molshanski-Mor S, Weinbaum C et al (2005) Activation of the phagocyte NADPH oxidase by Rac guanine nucleotide exchange factors in conjunction with ATP and nucleoside diphosphate kinase. J Biol Chem 280:3802–3811
167. Pick E (2010) Editorial: when charge is in charge - "Millikan" for leukocyte biologists. J Leukoc Biol 87:537–540
168. DeCoursey TE, Ligeti E (2005) Regulation and termination of NADPH oxidase activity. Cell Mol Life Sci 62:2173–2193
169. Akard LP, English D, Gabig TG (1988) Rapid deactivation of NADPH oxidase in neutrophils: continuous replacement by newly activated enzyme sustains the respiratory burst. Blood 72:322–327
170. van Bruggen R, Anthony E, Fernandez-Borja M et al (2004) Continuous translocation of Rac2 and the NADPH oxidase component p67^{phox} during phagocytosis. J Biol Chem 279:9097–9102
171. Tlili A, Erard M, Faure M-C et al (2012) Stable accumulation of p67^{phox} at the phagosomal membrane and ROS production within the phagosome. J Leukoc Biol 91:83–95

172. Faure MC, Sulpice J-C, Delattre M et al (2013) The recruitment of p47*phox* and Rac2G12V at the phagosome is transient and phosphatidylserine dependent. Biol Cell 105:501–518
173. Ostuni MA, Gelinotte M, Bizouarn T et al (2010) Targeting NADPH-oxidase by reactive oxygen species reveals an initial sensitive step in the assembly process. Free Rad Biol Med 49:900–907
174. Enyedi B, Zana M, Donko A et al (2013) Spatial and temporal analysis of NADPH oxidase-generated hydrogen peroxide by novel fluorescent reporter proteins. Antioxid Redox Signal 19:523–534
175. Yamaguchi T, Kaneda M, Kakinuma K (1983) Essential requirement of magnesium ion for optimal activity of the NADPH oxidase of guinea pig polymorphonuclear leukocytes. Biochem Biophys Res Comm 115:261–267
176. Tamura M, Takeshita M, Curnutte JT et al (1992) Stabilization of human neutrophil NADPH oxidase activated in a cell-free system by cytosolic proteins and by 1-ethyl-3-(3-dimethylaminopropyl) carbodiimide. J Biol Chem 267:7529–7538
177. Nauseef WM (2014) Detection of superoxide anion and hydrogen peroxide production by cellular NADPH oxidases. Biochim Biophys Acta 1840:757–767
178. Csányi G, Cifuentes-Pagano E, Al Gouleh I et al (2011) Nox2 B-loop peptide, Nox2ds, specifically inhibits the NADPH oxidase Nox2. Free Rad Biol Med 51:1116–1125
179. Nisimoto Y, Jackson HM, Ogawa H et al (2010) Constitutive NADPH-dependent electron transferase activity of the Nox4 dehydrogenase domain. Biochemistry 49:2433–2442
180. Cross AR, Yarchover JL, Curnutte JT (1994) The superoxide-generating system of human neutrophils possesses a novel diaphorase activity. Evidence for distinct regulation of electron flow within NADPH oxidase by p47*phox* and p67*phox*. J Biol Chem 269:21448–21454
181. Marques B, Liguori L, Paclet M-H et al (2007) Liposome-mediated cellular delivery of active gp91*phox*. PLoS One 2(9):e856
182. Nguyen MVC, Zhang L, Lhomme S et al (2012) Recombinant Nox4 cytosolic domain produced by a cell or cell-free base systems exhibits constitutive diaphorase activity. Biochem Biophys Res Commun 419:453–458
183. Han C-H, Nisimoto Y, Lee S-H et al (2001) Characterization of the flavoprotein domain of gp91phox which has NADPH diaphorase activity. J Biochem 129:513–520
184. Nisimoto Y, Ogawa H, Miyano K et al (2004) Activation of the flavoprotein domain of gp91*phox* upon interaction with N-terminal p67*phox* (1-210) and the Rac complex. Biochemistry 43:9567–9575
185. Tan AS, Berridge MV (2000) Superoxide produced by activated neutrophils efficiently reduces the tetrazolium salt, WST-1 to produce a soluble formazan: a simple colorimetric assay for measuring respiratory burst activation and for screening anti-inflammatory agents. J Immunol Methods 238:59–68
186. Takac I, Schroder K, Zhang L et al (2011) The E-loop is involved in hydrogen peroxide formation by the NADPH oxidase Nox4. J Biol Chem 286:13304–13313
187. Csányi G, Pagano P (2013) Strategies aimed at Nox4 oxidase inhibition employing peptides from Nox4 B-loop and C-terminus and p22*phox* N-terminus: an elusive target. Int J Hypertens 2013:842827. https://doi.org/10.1155/2013/842827
188. Pick E, Keisari Y (1980) A simple colorimetric method for the measurement of hydrogen peroxide produced by cells in culture. J Immunol Methods 38:161–170
189. Zhou M, Diwu Z, Panchuk-Voloshina N et al (1997) A stable nonfluorescent derivative of resorufin for the fluorometric determination of trace hydrogen peroxide: applications in detecting the activity of phagocyte NADPH oxidase and other oxidases. Anal Biochem 253:162–168
190. Votyakova TV, Reynolds IJ (2004) Detection of hydrogen peroxide with Amplex Red: interference by NADPH and reduced glutathione auto-oxidation. Arch Biochem Biophys 431:138–144
191. Rhyan TC, Weil GJ, Newburger PE et al (1990) Measurement of superoxide release in the phagovacuoles of immune complex-stimulated human neutrophils. J Immunol Methods 130:223–233
192. Li Y, Zhu H, Kuppusamy P et al (1998) Validation of lucigenin (bis-*N*-methylacridinium) as a chemilumigenic probe for detecting superoxide anion radical production by enzymatic and cellular systems. J Biol Chem 273:2015–2023
193. Yu L, Quinn MT, Cross AR et al (1998) Gp91*phox* is the heme binding subunit of the superoxide-generating NADPH oxidase. Proc Natl Acad Sci U S A 95:7993–7998

194. Sha'ag D (1989) Sodium dodecyl sulphate dependent NADPH oxidation: an alternative method for assaying NADPH-oxidase in a cell-free system. J Biochem Biophys Meth 19:121–128
195. Horecker BL, Kornberg A (1948) The extinction coefficients of the reduced band of pyridine nucleotides. J Biol Chem 175:385–390
196. Gerencser AA, Neilson A, Choi SW et al (2009) Quantitative microplate-based respirometry with correction for oxygen diffusion. Anal Chem 81:6868–6878
197. Pilloud M-C, Doussiere J, Vignais PV (1989) Parameters of activation of the membrane-bound $O_2^{\cdot -}$ generating oxidase from bovine neutrophils in a cell-free system. Biochem Biophys Res Commun 159:783–790
198. Aharoni I, Pick E (1990) Activation of the superoxide-generating NADPH oxidase of macrophages by sodium dodecyl sulfate in a soluble cell-free system: evidence for involvement of a G protein. J Leukoc Biol 48:107–115
199. Petreccia DC, Nauseef WM, Clark RA (1987) Respiratory burst of normal human eosinophils. J Leukoc Biol 41:283–288
200. Someya A, Nagaoka I, Iwabuchi K et al (1991) Comparison of O_2^{-}-producing activity of guinea-pig eosinophils and neutrophils in a cell-free system. Comp Biochem Physiol 100B:25–30
201. Bolscher BGJM, Koenderman L, Tool ATJ et al (1990) NADPH:O_2 oxidoreductase of human eosinophils in the cell-free system. FEBS Lett 268:269–273
202. Ligeti E, Doussiere J, Vignais PV (1988) Activation of the $O_2^{\cdot -}$generating oxidase in plasma membrane from bovine polymorphonuclear neutrophils by arachidonic acid, a cytosolic factor of protein nature, and nonhydrolyzable analogues of GTP. Biochemistry 27:193–200
203. Souabni H, Wien F, Bizouarn T et al (2017) The physicochemical properties of membranes correlate with the NADPH oxidase activity. Biochim Biophys Acta 1861:3520–3530
204. Massoud R, Bizouarn T, Houée-Levin C (2014) Cholesterol: a modulator of the phagocyte NADPH oxidase—a cell-free study. Redox Biol 3:16–24
205. Morgan D, Cherny VV, Murphy R et al (2003) Temperature dependence of NADPH oxidase in human eosinophils. J Physiol 550(2):447–458
206. Tomar N, Sadri S, Cowley AW Jr et al (2019) A thermodynamically-constrained mathematical model for the kinetics and regulation of NADPH oxidase 2 complex mediated electron transfer and superoxide production. Free Rad Biol Med 134:581–597
207. Toporik A, Gorzalczany Y, Hirschberg M et al (1998) Mutational analysis of novel effector domains in Rac1 involved in the activation of nicotinamide adenine dinucleotide phosphate (reduced) oxidase. Biochemistry 37:7147–7156
208. Lambeth JD, Krause K-H, Clark RA (2008) NOX enzymes as novel targets for drug development. Semin Immunopathol 30:339–363
209. Jaquet V, Scapozza L, Clark RA et al (2009) Small-molecule NOX inhibitors: ROS-generating NADPH oxidases as therapeutic targets. Antioxid Redox Signal 11:2535–2552
210. Diebold BA, Smith SME, Li Y et al (2015) NOX2 as a target for drug development: indications, possible complications, and progress. Antioxid Redox Signal 23:375–405
211. Altenhöfer S, Radermacher KA, Kleikers PWM et al (2015) Evolution of NADPH oxidase inhibitors: selectivity and mechanisms for target engagement. Antioxid Redox Signal 23:406–427
212. Jackson HM, Kawahara T, Nisimoto Y et al (2010) Nox4 B-loop creates an interface between the transmembrane and dehydrogenase domains. J Biol Chem 285:10281–10290
213. Bechtel W, Richardson R (1993) Discovering complexity: decomposition and localization as strategies in scientific research. Princeton University Press, Princeton
214. de Mendez I, Garrett MC, Adams AC et al (1994) Role of $p67^{phox}$ SH3 domains in assembly of the NADPH oxidase system. J Biol Chem 269:16326–16332
215. Maehara Y, Miyano K, Sumimoto H (2009) Role of the first SH3 domain of $p67^{phox}$ in activation of superoxide-producing NADPH oxidases. Biochem Biophys Res Commun 379:589–593
216. Zahavi A (2013) Elucidation of the domain (s) in the cytosolic NADPH oxidase component $p67^{phox}$ involved in binding to flavocytochrome b_{558}. Ph.D. Thesis, Tel Aviv University
217. Bradford MM (1976) A rapid and sensitive method for the quantitation of microgram quantities of protein utilizing the principle of protein-dye binding. Anal Biochem 72:248–254
218. Light D, Walsh C, O'Callaghan AM et al (1981) Characteristics of the cofactor

requirements for the superoxide-generating NADPH oxidase of human polymorphonuclear leukocytes. Biochemistry 20:1468–1476
219. Zhao X, Carnevale KA, Cathcart MK (2003) Human monocytes use Rac1, not Rac2, in the NADPH oxidase complex. J Biol Chem 278:40788–40792
220. Kreck ML, Uhlinger DJ, Tyagi SR et al (1994) Participation of the small molecular weight GTP-binding protein Rac1 in cell-free activation and assembly of the respiratory burst oxidase. Inhibition by a carboxyl-terminal Rac peptide. J Biol Chem 269:4161–4168
221. Bromberg Y, Shani E, Joseph G et al (1994) The GDP-bound form of the small G protein rac1 p21 is a potent activator of the superoxide forming NADPH oxidase of macrophages. J Biol Chem 269:7055–7058
222. Sigal N, Gorzalczany Y, Sarfstein R et al (2003) The guanine nucleotide exchange factor Trio activates the phagocyte NADPH oxidase in the absence of GDP to GTP exchange—"The emperor's new clothes". J Biol Chem 278:4854–4861
223. Xu X, Wang Y, Barry DC, Chanock SJ et al (1997) Guanine nucleotide binding properties of Rac2 mutant proteins and analysis of the responsiveness to guanine nucleotide dissociation stimulator. Biochemistry 36:626–632
224. Bordier C (1981) Phase separation of integral membrane proteins in triton X-114 solution. J Biol Chem 256:1604–1607
225. Babior BM, Kuver R, Curnutte JT (1988) Kinetics of activation of the respiratory burst oxidase in a fully soluble system from human neutrophils. J Biol Chem 263:1713–1718
226. Cross AR, Erickson RW, Curnutte JT (1999) Simultaneous presence of p47^{phox} and flavocytochrome b_{-245} are required for activation of NADPH oxidase by anionic amphiphiles. Evidence for an intermediate state of oxidase activation. J Biol Chem 274:15519–15525
227. Karimi G, Houée Levin C et al (2014) Assembly of phagocyte NADPH oxidase: a concerted process? Biochim Biophys Acta 1840:3277–3283
228. Quinn MT, Evans T, Loetterle LR et al (1993) Translocation of Rac correlates with NADPH oxidase activation. Evidence for equimolar translocation of oxidase components. J Biol Chem 268:20983–20987
229. Koshkin V, Lotan O, Pick E (1997) Electron transfer in the superoxide-generating NADPH oxidase complex reconstituted *in vitro*. Biochim Biophys Acta 1319:139–146
230. Pick E, Mizel D (1981) Rapid microassays for the measurement of superoxide and hydrogen peroxide production by macrophages in culture using an automatic enzyme immunoassasy reader. J Immunol Methods 46:211–226
231. Pick E (1986) Methods for studying the oxidative metabolism of macrophages. Microassays for O_2^- and H_2O_2 production and NBT reduction using an enzyme immunoassay microplate reader. Meth Enzymol 132:407–421
232. Mayo L, Curnutte JT (1990) Kinetic microplate assay for superoxide production by neutrophils and other phagocytic cells. Meth Enzymol 186:567–575
233. Bechor E, Dahan I, Fradin T et al (2015) The dehydrogenase region of the NADPH oxidase component Nox2 acts as a protein disulfide isomerase (PDI) resembling PDIA3 with a role in the binding of the activator protein p67^{phox}. Front Chem 3:3
234. Knaus UG, Heyworth PG, Kinsella BT et al (1992) Purification and characterization of Rac2. A cytosolic GTP-binding protein that regulates human neutrophil NADPH oxidase. J Biol Chem 267:23575–23582
235. Seifert R, Rosenthal W, Schultz G (1986) Guanine nucleotides stimulate NADPH oxidase in membranes of human neutrophils. FEBS Lett 105:161–165
236. Gabig TG, English D, Akard LP et al (1987) Regulation of neutrophil NADPH oxidase activation in a cell-free system by guanine nucleotides and fluoride. Evidence for participation of a pertussis and cholera toxin-insensitive G protein. J Biol Chem 262:1685–1690
237. El-Benna J, Dang PM-C, Périanin A (2010) Peptide-based inhibitors of the phagocyte NADPH oxidase. Biochem Pharmacol 80:778–785
238. Dahan I, Pick E (2012) Strategies for identifying synthetic peptides to act as inhibitors of NADPH oxidases, or "all that you did and did not want to know about Nox inhibitory peptides". Cell Mol Life Sci 69:2283–2305
239. El-Benna J, Dang PM-C, Périanin A (2012) Towards specific NADPH oxidase inhibition by small synthetic peptides. Cell Mol Life Sci 69:2307–2314
240. Cifuentes-Pagano ME, Meijles DN, Pagano PJ (2015) Nox inhibitors and therapies: rational design of peptidic and small molecule inhibitors. Curr Pharm Des 21:6023–6035
241. Joseph G, Pick E (1995) "Peptide walking" is a novel method of mapping functional

domains in proteins. Its application to the Rac1-dependent activation of NADPH oxidase. J Biol Chem 270:29079–29082

242. Morozov I, Lotan O, Joseph G et al (1998) Mapping of functional domains in p47^{phox} involved in the activation of NADPH oxidase by “peptide walking”. J Biol Chem 273:153435–115444

243. Dahan I, Issaeva I, Gorzalczany Y et al (2002) Mapping of functional domains in the p22^{phox} subunit of flavocytochrome b_{559} participating in the assembly of the NADPH oxidase complex by “peptide walking”. J Biol Chem 277:8421–8432

244. Dahan I, Molshanski-Mor S, Pick E (2012) Inhibition of NADPH oxidase activation by peptides mapping within the dehydrogenase region of Nox2 - a "peptide walking" study. J Leuk Biol 91:501–515

245. Rey FE, Cifuentes ME, Kiarash A et al (2001) Novel competitive inhibitor of NADPH oxidase assembly attenuates vascular O_2^- and systolic blood pressure in mice. Circ Res 89:408–414

246. Bosco E, Marchioni F, Kumar S et al (2012) Rational design of small molecule inhibitors targeting the Rac GTPase - p67phox signaling axis in inflammation. Chem Biol 19:228–242

247. Lejal N, Truchet S, Bechor E et al (2018) Turning off NADPH oxidase-2 by impeding p67^{phox} activation in infected mouse macrophages reduced viral entry and inflammation. Biochim Biophys Acta 1862:1263–1275

248. Rotrosen D, Kleinberg ME, Nunoi H et al (1990) Evidence for a functional cytoplasmic domain of phagocyte oxidase cytochrome b_{558}. J Biol Chem 265:8745–8750

249. Uhlinger DJ, Tyagi SR, Lambeth JD (1995) On the mechanism of inhibition of the neutrophil respiratory burst oxidase by a peptide from the C-terminus of the large subunit of cytochrome b_{558}. Biochemistry 34:524–527

250. Joseph G, Gorzalczany Y, Koshkin V et al (1994) Inhibition of NADPH oxidase activation by synthetic peptides mapping within the carboxy-terminal domain of small GTP-binding proteins. Lack of amino acid sequence specificity and importance of the polybasic motif. J Biol Chem 269:29024–29031

251. Le Cabec V, Maridonneau-Parini I (1995) Complete and reversible inhibition of NADPH oxidase in human neutrophils by phenylarsine oxide at a step distal to membrane translocation of the enzyme subunits. J. Biol Chem 270:2067–2073

252. Doussiere J, Poinas A, Blais C et al (1998) Phenylarsine oxide as an inhibitor of the activation of the neutrophil NADPH oxidase. Identification of the β subunit of the flavocytochrome *b* component of the NADPH oxidase as a target site for phenylarsine oxide by photoaffinity labeling and photoinactivation. Eur J Biochem 251:649–658

253. Dahan I, Smith SME, Pick E (2015) A Cys-Gly-Cys triad in the dehydrogenase region of Nox2 plays a key role in the interaction with p67^{phox}. J Leukoc Biol 98:859–874

254. Fradin T, Bechor E, Berdichevsky Y et al (2018) Binding of p67^{phox} to Nox2 is stabilized by disulfide bonds between cysteines in the ^{369}Cys-Gly-Cys^{371} triad in Nox2 and in p67^{phox}. J Leukoc Biol 104:1023–1039

255. Diatchuk V, Lotan O, Koshkin V et al (1997) Inhibition of NADPH oxidase activation by 4-(2-aminoethyl)-benzenesulfonyl fluoride and related compounds. J Biol Chem 272:13292–13301

256. Fuchs A, Dagher M-C, Jouan A et al (1994) Activation of the O_2^--generating NADPH oxidase in a semi-recombinant cell-free system. Assessment of the function of Rac in the activation process. Eur J Biochem 226:587–595

257. Pick E (2015) Absolute and relative activity values in assessing the effect of NADPH oxidase inhibitors. Antioxid Redox Signal 23:1250–1251

258. Jaquet V, Rutter AR (2015) Response to Pick. Antioxid Redox Signal 23:1251–1253

259. Pick E, Gadba R (1988) Certain lymphoid cells contain the membrane-associated component of the phagocyte-specific NADPH oxidase. J Immunol 140:1611–1617

260. Souabni H, Ezzine A, Bizouarn T et al (2017) Functional assembly of soluble and membrane recombinant proteins of mammalian NADPH oxidase complex. In: Lacapere J-J (ed) Membrane protein structure and function characterization: methods and protocols. Springer Science + Business Media, LLC, New York

261. Azzi A, Montecucco C, Richter C (1975) The use of acetylated ferricytochrome *c* for the detection of superoxide radicals produced in biological membranes. Biochem Biophys Res Commun 65:597–603

262. Hashida S, Yuzawa S, Suzuki NN et al (2004) Binding of FAD to cytochrome b_{558} is facilitated during activation of the phagocyte NADPH oxidase, leading to superoxide production. J Biol Chem 279:26378–26386

263. Heyworth PG, Knaus UG, Xu X et al (1993) Requirement for posttranslational processing of Rac GTP-binding proteins for activation of human neutrophil NADPH oxidase. Mol Biol Cell 4:261–269

Part VI

Analysis of Neutrophil Extracellular Traps

Chapter 24

Immunofluorescent Detection of NET Components in Paraffin-Embedded Tissue

Ulrike Abu-Abed and Volker Brinkmann

Abstract

Neutrophil extracellular traps (NETs) consist of decondensed chromatin fibers studded with granular and cytoplasmic proteins and peptides that are released by stimulated neutrophil granulocytes. If present in abundance (e.g., in large thrombi), NETs are depicted in H&E-stained tissue sections as pale bluish areas. Since no NET-specific antibodies exist, to unambiguously identify even small amounts of NETs in tissue, it is essential to demonstrate colocalization of nuclear and granular/cytoplasmic NET components which in unstimulated neutrophils are clearly separated. This requires good tissue preservation and a very defined immunolocalization, which can be achieved by using 2–3 μm thick sections of paraffin-embedded tissue. It provides sufficiently good tissue preservation for subcellular localization of two or more NET components, thereby allow to differentiate stimulated from unstimulated neutrophils and to clearly identify NETs. In this chapter, we will provide protocols for antigen retrieval and immunofluorescent labeling of NET components in paraffin-embedded tissue with commercially available antibodies.

Key words Neutrophil extracellular traps, Neutrophil granulocytes, Paraffin-embedded tissue, Antigen retrieval, Multicolor immunofluorescence, Slide digitalization, Confocal scanning microscopy

1 Introduction

Initially, NETs were regarded as a means of the innate immune system to counteract invading microorganisms [1]. Gradually it became clear that NETs can also have pathological potential including autoimmunity, infertility, coagulation, thrombosis, sepsis, neurodegeneration, and cancer (as recently reviewed [2]). The presence or absence of NETs in diseased tissue may have prognostic value, both positive or negative, and thus protocols to detect NETs or their absence in recent or archival tissue samples may promote further insight into NET-related pathomechanisms.

Several pathways to NET generation have been described including suicidal (or lytic) and vital NETosis (reviewed in [3]). Many molecular details have been revealed for suicidal NETosis: following stimulation of neutrophils, the multimeric NADPH

Mark T. Quinn and Frank R. DeLeo (eds.), *Neutrophil: Methods and Protocols*, Methods in Molecular Biology, vol. 2087, https://doi.org/10.1007/978-1-0716-0154-9_24,

oxidase complex assembles and produces reactive oxygen species (ROS) which activate the Raf-MEK-ERK pathway and trigger the release of Neutrophil Elastase (NE) from azurophilic granules [4–6]. NE translocates to the nucleus where it cleaves histones. Concurrently, peptidylarginine deiminase type 4 catalyzes the conversion of histone arginines to citrullines [7–9]. The nuclei of neutrophil granulocytes lose their characteristic lobate structure and start swelling until the nuclear envelope disintegrates. Karyoplasm, cytoplasm, and granular remnants mingle, and finally the cell membrane bursts and NETs are released [5].

This series of events results in distinct morphological modifications of neutrophils which consequently cannot be identified in standard histological stainings like H&E since their morphological hallmarks, that is, lobulated nuclei and granules, are lost in the process. Both neutrophils undergoing NETosis and NETs are fragile structures, so good tissue preservation is required for unambiguous detection of NETs in tissue, which rules out the use of cryo sections. To date, no antibodies exist that exclusively react with epitopes on NETs and do not stain unstimulated neutrophils or other tissue components. Thus, standard immunohistochemistry methods using a single primary antibody followed by an enzyme-based detection system cannot be used for NET localization. Instead, single enzyme-based stainings using several antibodies against NET components on consecutive sections have been applied to visualize NETs and fibrin meshworks in inflammatory lesions [10].

NETs or NETting neutrophils are characterized by colocalization of nuclear and granular (or cytoplasmic) proteins which in unstimulated neutrophils are clearly separated. Enzyme-based detection systems do not provide the required subcellular resolution and do not readily allow simultaneous localization of two antigens [11]. In contrast, immunofluorescent detection of two or more antigens in sections of formaldehyde-fixed paraffin embedded tissue sections is feasible on a routine basis, and can be combined with additional fluorochromes (e.g., DNA-intercalating dyes or microorganisms expressing fluorescent fusion proteins).

Paraffin embedding involves dehydration and heating of formaldehyde-fixed tissue which leads to formation of intra- and intermolecular methylene bridges that can mask epitopes [12, 13]. For antigen retrieval, rehydrated sections are heated in a suitable buffer [14, 15]. The optimal pH and temperature depends on the respective antigen–antibody pair. We provide protocols for successful antigen retrieval and antibody combinations that allow identification of NETs in human and murine tissue.

2 Materials

2.1 Tissue Fixation, Dehydration, Paraffin Embedding, Sectioning

1. Tris-buffered saline (TBS).
2. Paraformaldehyde solution: Heat TBS to ~60 °C. Add 2% (w/v) paraformaldehyde and dissolve while stirring gently. Avoid heating to >65 °C. Paraformaldehyde solution can be stored at −20 °C.
3. Automatic tissue processor if available (e.g., Leica TP 1020).
4. Ethanol.
5. Xylene (Dimethylbenzene).
6. Paraffin wax, melting point 56–58 °C, and paraffin mounting station.
7. Microtome (e.g., Microm 355S).
8. Glass slides, slide racks, jars.

2.2 Antigen Retrieval

1. HIER buffer: Tris–EDTA HIER solution, pH 9 (e.g., ScyTek).
2. Hot plate with temperature sensor.

2.3 Immunofluorescence Labeling

1. Moist chamber (plastic box with tightly fitting lid and moist filter paper and small tray for slides inside). Parafilm.
2. PAP pen (e.g., ImmEdge, Vectorlab).
3. Tris-buffered saline (TBS).
4. Blocking buffer: Add 1% BSA, 2% normal donkey serum, 5% cold water fish gelatin, 0.05% Tween 20, and 0.025% Triton X-100 to TBS.
5. Primary antibodies for detecting NET components both in human and murine tissue: rabbit anti-neutrophil elastase (ELANE; e.g., Atlas) and chicken anti-histone 2B (e.g., Abcam).
6. Secondary antibodies cross-absorbed against serum components of multiple species: donkey anti-rabbit labeled with Alexa Fluor 488 or, alternatively, Cy5. Donkey anti-chicken labeled with Cy3 (e.g., Jackson Immuno Research).
7. 2 μg/mL Hoechst 33342 DNA stain (e.g., Abcam).
8. 0.2 M Tris–HCl: Dissolve 5.8 g Tris base to 240 mL deionized H_2O and stir until dissolved. Adjust to pH 8.5 with HCl.
9. 8% Mowiol (polyvinyl alcohol)–glycerol mounting medium: Mix 40 g Mowiol with 120 g (105 mL) glycerol. Stir 10 min. Add 120 mL deionized H_2O, stir 10 min. Add 240 mL 0.2 M Tris–HCl to the Mowiol/glycerol mixture. While stirring, heat to 50 °C for about 2 h until the solution becomes clear. Store aliquots at −20 °C. Before using Mowiol, thawed aliquots should be centrifuged (4000 × *g*, 15 min).
10. Cover slips 60 mm × 24 mm, #1 (e.g., Menzel).

3 Methods

3.1 Fixation, Dehydration, Paraffin Embedding, Sectioning

1. Dissect fresh tissue into pieces not exceeding 20 mm × 30 mm × 3 mm. Immerse specimen into 2% freshly prepared or thawed paraformaldehyde solution in TBS, using 20 × tissue volume (*see* **Note 1**).
2. Fix at room temperature overnight (not longer than 20 h). Place samples into embedding cassettes, transfer to TBS.
3. Using an automatic tissue processor, dehydrate samples in a graded ethanol series (70%, 80%, 90%, 96%, 100%, 100%), each step should be 1 h.
4. Clear specimens twice with 100% xylene, each step should be 1 h.
5. Infiltrate in paraffin wax, two times for 1 h each step at 60 °C.
6. Mount specimens in embedding molds, using the cassette bottom as a cover, let paraffin solidify, and remove embedding molds.
7. Using a microtome, prepare 2–3 μm sections, let them float on a water bath at 37 °C.
8. Pick sections from the water surface with superfrost glass slides.
9. Keep sections on slides overnight at 40 °C to let tissue firmly adhere to the glass.

3.2 Rehydration, Heat-Induced Epitope Retrieval (HIER), Immunofluorescent Labeling

1. Place tissue sections on slides into racks.
2. Place rack into jars with the medium used for paraffin embedding, in reverse order, 5 min each step, until sections are fully hydrated in distilled H_2O.
3. Fill a jar with HIER buffer. Place the jar into a water bath on a temperature-controlled hot plate. Heat to 70 °C.
4. Transfer slide rack from the TBS to the HIER jar, incubate for 60–120 min at 70 °C.
5. Remove jar from hot plate and let buffer with slides cool down to room temperature. Rinse slides in rack 3× with deionized H_2O and transfer the slides to the TBS jar.
6. Take slides from rack one by one and carefully remove liquid from the slides around the sections, leaving the sections hydrated. Create liquid barrier around the sections with a PAP pen, and cover the sections with blocking buffer. Depending on the size of the sections, 100–200 μL should be sufficient. Transfer slides to the moist chamber and incubate with blocking buffer to prevent nonspecific binding (30 min, room temperature).

3.3 Immunofluorescence Labeling

1. Dilute both primary antibodies in blocking buffer to create 1 μg/mL antibody solutions. The combination of a rabbit antibody against ELANE and a chicken antibody against Histone 2B can be used both for human and murine tissue.
2. Remove blocking buffer from the sections and replace with the primary antibody mixture. Transfer slides to the moist chamber, seal with Parafilm, and incubate over night at room temperature.
3. Remove antibody solution and wash sections, 3 × 5 min with TBS.
4. Prepare working solution of secondary antibodies in blocking buffer without Triton X100. It is essential to use antibodies that have been preabsorbed against serum proteins from multiple species (*see* **Note 2**). For DNA counterstain, use Hoechst 33342 stain (*see* **Note 1**).
5. Cover sections with working solution of secondary antibodies. Transfer slides to the moist chamber, seal with Parafilm, and incubate 1 h at room temperature.
6. Remove antibody solution, wash sections three times for 5 min with TBS and one time with deionized water.
7. Cover sections with Mowiol solution and carefully position cover glass avoiding bubble formation (*see* **Note 3**). Let solidify.
8. Perform microscopic analysis with either a wide field microscope with appropriate band pass filters (for lenses up to 20× magnification) or a confocal microscope for detailed analysis (*see* **Note 3**). For an initial screen, we routinely digitize the entire section at 10× magnification with a slide scanner. These data can also be used to quantify the percentage of NET area in the tissue.
9. Tissue areas rich in NETs can be recognized with wide field fluorescence microscopy up to a primary magnification of 20×. Under these conditions, the section thickness roughly matches the focal depth of the objective. NET-containing tissue areas can rather easily be localized because the IgY antibody used to detect H2A stains areas of decondensed chromatin stronger than condensed chromatin in nuclei (Fig. 1c, d) (*see* **Notes 4** and **5**).
10. While wide field microscopy is useful for localization and quantification of NET-containing tissue areas, detailed analysis requires higher resolution and the use of confocal microscopes or deconvolution. In Figs. 2 and 3, which are maximum projections of confocal stacks, the diverging localization of NE, which is either highly concentrated in granules or diffusely spread in NETs, can easily be differentiated.

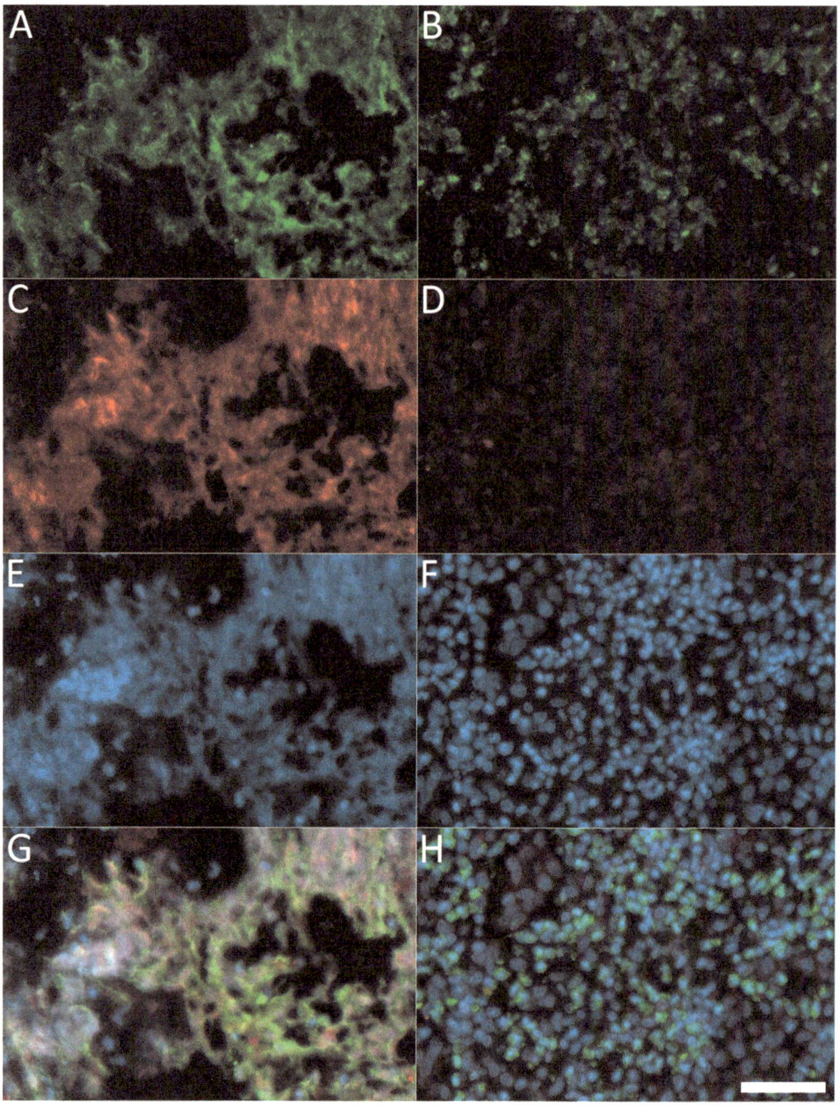

Fig. 1 Widefield fluorescence microscopy of a paraffin section of a human appendicitis sample. (Panels **a**, **c**, **e**, and **g**) depict a tissue area with NETs, while (panels **b**, **d**, **f**, and **h**) show a different area of the same section which is rich in neutrophils, but without NET formation. Staining is against NE (**a**, **b**, green), H2B (**c**, **d**, red), and DNA (**e**, **f**, blue). Note that H2B staining is considerably weaker in non-NETotic neutrophils and epithelial cells compared to areas rich in NETs. (**g** and **h**) represent the overlay of all three channels. Widefield microscopy, 20× objective, bar represents 25 μm

4 Notes

1. Tissue from experimental animals should be fixed as quickly as possible, preferably by perfusion. We use paraformaldehyde solutions in TBS, either freshly prepared or frozen, and keep

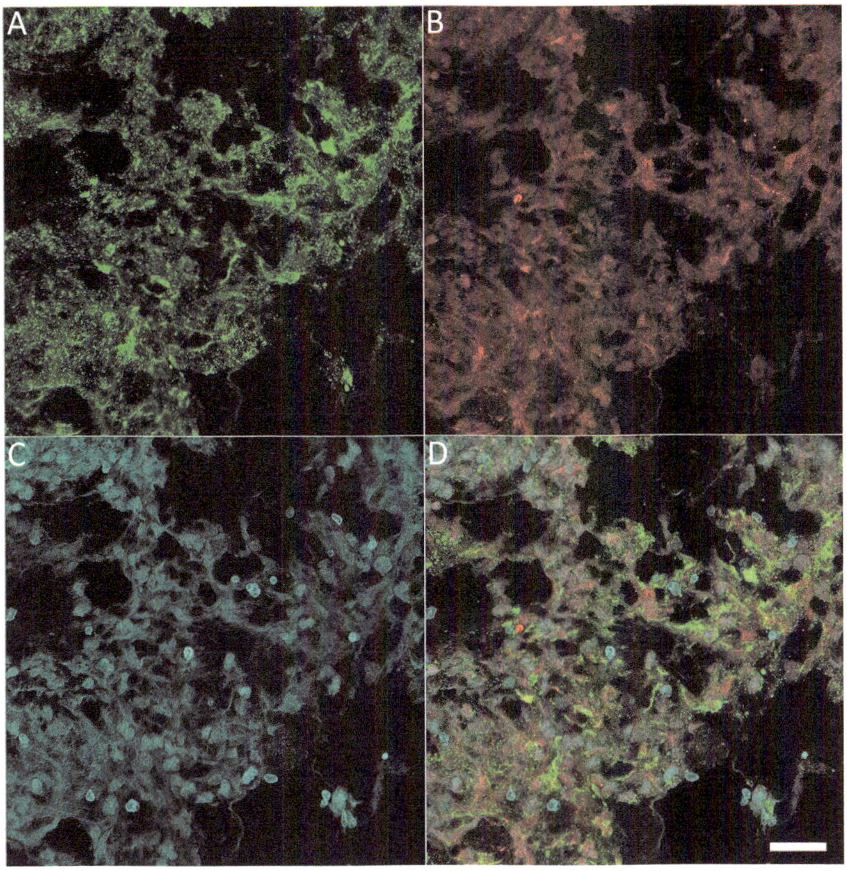

Fig. 2 Confocal fluorescence microscopy of NET components in a human appendicitis sample. Staining identical as in Fig. 1 (green—NE, red—H2B, blue—DNA). NE is contained in granules in many cells but is also present extracellularly as fibrous material overlapping with H2B and DNA to form NETs. Confocal microscopy, Z-stack presented as maximum projection. Bar represents 25 μm

the fixation time shorter than 20 h. In contrast, human tissue is mostly fixed in commercial formalin preparations which contain methanol, formic acid and additional aldehydes and ketones. Often, the time between excision and beginning of fixation is unknown as is the duration of fixation. Thus, tissue preservation can be suboptimal due to autolysis, and often samples are over fixed resulting in increased autofluorescence and epitope masking. This can partly be reverted by increasing duration or temperature of antigen retrieval. Tissue autofluorescence mainly occurs in the bluish/greenish part of the spectrum. While usually the Hoechst 33342 signal is sufficiently strong to be detected over a background of autofluorescence, antibody staining with secondary antibodies coupled to green-emitting fluorochromes like Alexa Fluor 488 can be unsatisfactory due to greenish tissue autofluorescence. This can be

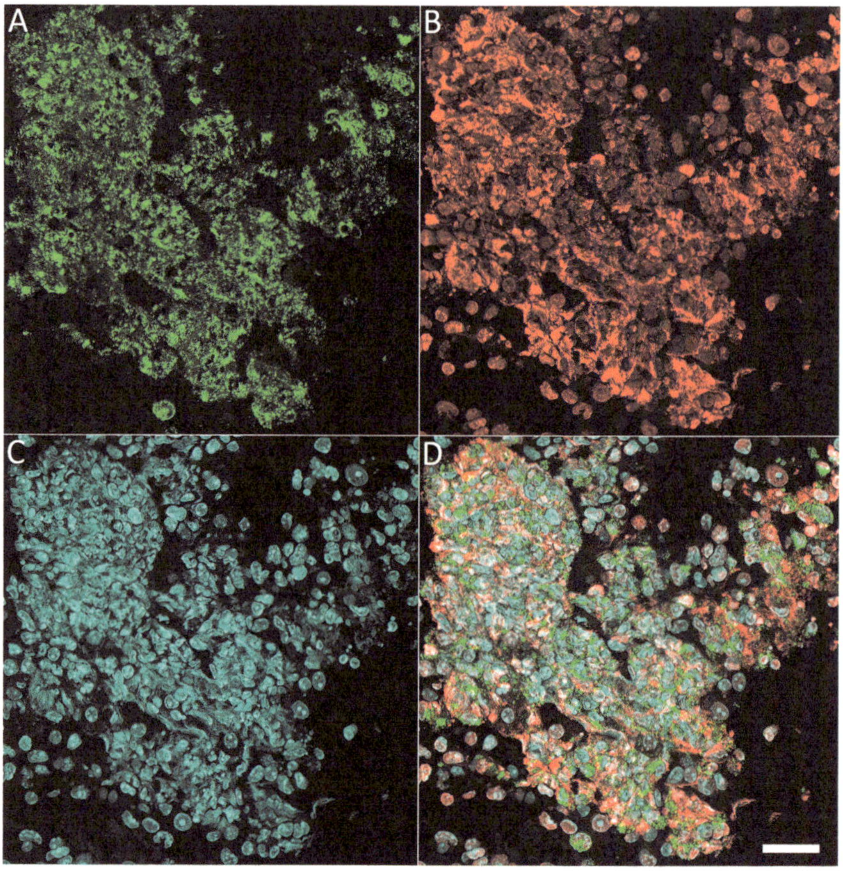

Fig. 3 Confocal fluorescence microscopy of NET components in a mouse lung infected with Mycobacterium tuberculosis. Staining identical as in Figs. 1 and 2 (green—NE, red—H2B, blue—DNA). NETs are clearly visible as whitish extracellular fibers resulting from overlapping signals of all three channels. Confocal microscopy, Z-stack presented as maximum projection. Bar represents 25 μm

overcome by using far-red fluorophores like Alexa Fluor 633. Since the human eye is not very sensitive for emissions beyond 600 nm, the signal can only be detected by a b/w camera or a confocal microscope.

2. For simultaneous staining with two primary antibodies raised in different hosts, it is absolutely essential to use secondary antibodies that have been preabsorbed against serum components of the respective hosts to avoid unspecific binding which would result in false-positive colocalization of the antigens. The quality of the set of secondary antibodies should be tested by omitting one of the primary antibodies at a time: the respective secondary antibody should not bind at all. For long-term storage, we dilute antibody conjugates to 50% by adding glycerol and store the aliquots at −20 °C.

3. Generally, we mount fluorescent specimens with Mowiol. Care should be taken to avoid bubble formation while the cover glass is applied. For high resolution imaging, the Mowiol layer between section and cover glass should be minimal to allow proper imaging of the entire section even with the limited focal depth of immersion objectives. This is achieved by gently pressing the cover glass down, so Mowiol will form a seal around the edge of the cover glass. This seal will harden after a couple of hours, while the Mowiol inside below the cover glass will remain fluid for weeks. If stored below 10 °C, these samples will be usable for months without great loss of fluorescence intensity. Since the Mowiol preparation we use mainly consists of glycerol, we prefer objectives that can be used with glycerol as immersion medium to match the refractive index.
4. Probably binding of the relatively large IgY (180 kDa compared to 150 kDa for IgG) is reduced in compacted chromatin. While NET-rich areas can be rather spacious as in this example of appendicitis, in dense tissues NETs can be significantly less extended and comprise just a few neutrophils as in myocarditis [16]. To clearly depict colocalization of nuclear and granular NET components in dense tissue, confocal microscopy or deconvolution of widefield images are often indispensable. Confocal microscopy also allows the use of high-resolution objectives which provide greatly improved subcellular resolution (Figs. 2 and 3).
5. Staining of NETs with DNA-intercalating dyes is rather weak compared to the signal intensity obtained in condensed nuclei (Fig. 1e, f). Delicate DNA staining of small NET areas can easily be missed in tissue areas of high cell density due to the high DNA content of neighboring condensed nuclei. These areas are better visualized by antibody staining, since many histone-binding antibodies stain decondensed chromatin as is present in NETs stronger than nuclear chromatin. One example is the IgY antibody used in this protocol to detect Histone 2B. It is essential to use narrow band fluorescence filters to avoid bleed-through fluorescence resulting in false positive overlapping signals.

Acknowledgments

The funding source of this work is the Max Planck Society. We thank Arturo Zychlinsky for critically reading the manuscript.

References

1. Brinkmann V, Reichard U, Goosmann C et al (2004) Neutrophil extracellular traps kill bacteria. Science 303:1532–1535
2. Papayannopoulos V (2018) Neutrophil extracellular traps in immunity and disease. Nat Rev Immunol 18:134–147
3. Brinkmann V (2018) Neutrophil extracellular traps in the second decade. J Innate Immun 10:414–421
4. Hakkim A, Fuchs TA, Martinez NE et al (2011) Activation of the Raf-MEK-ERK pathway is required for neutrophil extracellular trap formation. Nat Chem Biol 7:75–77
5. Fuchs TA, Abed U, Goosmann C et al (2007) Novel cell death program leads to neutrophil extracellular traps. J Cell Biol 176:231–241
6. Metzler KD, Fuchs TA, Nauseef WM et al (2011) Myeloperoxidase is required for neutrophil extracellular trap formation: implications for innate immunity. Blood 117:953–959
7. Papayannopoulos V, Metzler KD, Hakkim A et al (2010) Neutrophil elastase and myeloperoxidase regulate the formation of neutrophil extracellular traps. J Cell Biol 191:677–691
8. Neeli I, Khan SN, Radic M (2008) Histone deimination as a response to inflammatory stimuli in neutrophils. J Immunol 180:1895–1902
9. Wang Y, Li M, Stadler S et al (2009) Histone hypercitrullination mediates chromatin decondensation and neutrophil extracellular trap formation. J Cell Biol 184:205–213
10. Shiogama K, Onouchi T, Mizutani Y et al (2016) Visualization of neutrophil extracellular traps and fibrin meshwork in human fibrinopurulent inflammatory lesions: I. light microscopic study. Acta Histochem Cytochem 49:109–116
11. Robertson D, Savage K, Reis-Filho JS et al (2008) Multiple immunofluorescence labelling of formalin-fixed paraffin-embedded (FFPE) tissue. BMC Cell Biol 9:13
12. Rait VK, Xu L, O'Leary TJ et al (2004) Modeling formalin fixation and antigen retrieval with bovine pancreatic RNase A II. Interrelationship of cross-linking, immunoreactivity, and heat treatment. Lab Investig 84:300–306
13. Yamashita S, Okada Y (2005) Mechanisms of heat-induced antigen retrieval: analyses in vitro employing SDS-PAGE and immunohistochemistry. J Histochem Cytochem 53:13–21
14. Cattoretti G, Pileri S, Parravicini C et al (1993) Antigen unmasking on formalin-fixed, paraffin-embedded tissue sections. J Pathol 171:83–98
15. Shi SR, Imam SA, Young L et al (1995) Antigen retrieval immunohistochemistry under the influence of pH using monoclonal antibodies. J Histochem Cytochem 43:193–201
16. Weckbach LT, Grabmaier U, Uhl A et al (2019) Midkine drives cardiac inflammation by promoting neutrophil trafficking and NETosis in myocarditis. J Exp Med 216:350–368

Chapter 25

Detection, Visualization, and Quantification of Neutrophil Extracellular Traps (NETs) and NET Markers

Nicole de Buhr and Maren von Köckritz-Blickwede

Abstract

Neutrophil extracellular traps (NETs) have been identified as a key player in the pathogenesis of infection and inflammation in human and animals. On the one hand, NETs have been characterized as fundamental to the innate immune defense against different pathogens since they are able to entrap and immobilize invading pathogens. On the other hand, NETs have been shown to contribute to several diseases, based on their detrimental consequences. This chapter describes methods to detect NETs and NET markers in blood-derived isolated neutrophils of human, pigs, and horses in vitro, as well as NETs and NET marker detection in body fluids from in vivo studies. To avoid nonspecific background in NET-formation, a well-established isolation method for the neutrophils from fresh blood is needed. After stimulation of neutrophils to release NETs, NETs are stained with different antibodies to confirm the presence of extracellular DNA extrusion consisting of histone–DNA complexes, as well as granule components (e.g., myeloperoxidase or elastase). Furthermore, specific methods to quantify NETs and NET markers in the cerebrospinal fluid (CSF) and bronchoalveolar lavage fluid (BALF) are described in detail. In addition to immunofluorescence microscopy, quantification of NET markers from in vivo experiments in various body fluids is described (e.g., nuclease activity, free extracellular DNA, or cationic host defense peptides, such as the porcine PR-39 in BALF and CSF).

Key words Neutrophil extracellular traps, Immunofluorescence microscopy, Extracellular DNA, Myeloperoxidase, Elastase, Histone–DNA complexes, PicoGreen, Porcine, Human and equine neutrophils

1 Introduction

Since the original discovery in 2004, it has been well established that neutrophils can release neutrophil extracellular traps (NETs) as web-like fibers consisting of DNA and granular proteins [1]. Those fibers can mediate entrapment and immobilization of several pathogens and thereby contribute to efficient host defense against extracellular pathogens (e.g., *Staphylococcus aureus* [1–4]). On the other hand, excessive and dysregulated NET formation may lead to detrimental consequences (e.g., autoimmune diseases, thrombosis, metabolic disorders, lung diseases and fibrosis, cancer, and other

Mark T. Quinn and Frank R. DeLeo (eds.), *Neutrophil: Methods and Protocols*, Methods in Molecular Biology, vol. 2087,
https://doi.org/10.1007/978-1-0716-0154-9_25,

diseases [5]). If NET-formation and its subsequent elimination are finely balanced, NETs can likely contribute to a protective immune response against pathogenic infections. The role of NETs is evident during health and disease phenotypes [6], and NETs have different functions during infections with different pathogens. Therefore, detailed pathogen-specific studies for evaluation of the role of NETs are needed. For a better understanding of the specific contribution of NETs to health and disease and for a detailed analysis of the mechanisms involved in NET-formation, specific NET visualization and quantification methods are necessary.

Here, we describe methods to detect and visualize NETs and NET markers in vitro from isolated blood-derived neutrophils and from body fluids (e.g., CSF and BALF). The key visualization method for detection, as well as quantification of NETs, is immunofluorescence microscopy [7]. Since the major backbone of NETs is DNA [1, 8], different DNA intercalating dyes, such as propidium iodide, SYTOX Orange, etc., are often used to visualize NETs. Furthermore, fluorescence-based quantification of extracellular DNA labeled with DNA-intercalating dyes is routinely used to quantify NET-formation. However, it is important to note that cationic molecules can block the binding of the DNA-intercalating dyes to NETs and thereby hamper its visualization and valid quantification [9]. One example of such cationic molecules is host defense peptides (e.g., human cathelicidin LL-37), which can associate with NETs. Therefore, antibody-based techniques are highly recommended as the first choice for visualizing and quantifying NETs [7]. For an illustration of the differences, see visualization of NET-staining in Fig. 1.

For quantification of NET fluorescent images, most investigators use hand-counting of NETs per image or per neutrophil count. This approach has the advantage of specific counting of NETs based on extracellular extrusions and allows one to differentiate simple necrotic cell death from NET formation. On the other hand, the results may be biased by the observer, as only specific areas of the sample are visualized and counted. Furthermore, this approach does not allow for rapid screening of a large number of samples. Several authors have described automated systems to quantify NET-formation, and an overview of examples with advantages and disadvantages is provided in Table 1 [10–14]. However, one problem with automated systems is that automated NET analysis techniques must be adjusted for different animal species. The highly variable core morphology in, for example, bovine and equine neutrophils results in problems with adjusting automated techniques established for human neutrophils.

The simple use of DNA-intercalating dye harbors the risk of detecting not only NET-forming cells, but also necrotic cells. Furthermore, as mentioned above, associated proteins or peptides may block the staining. Therefore, antibody-based techniques are

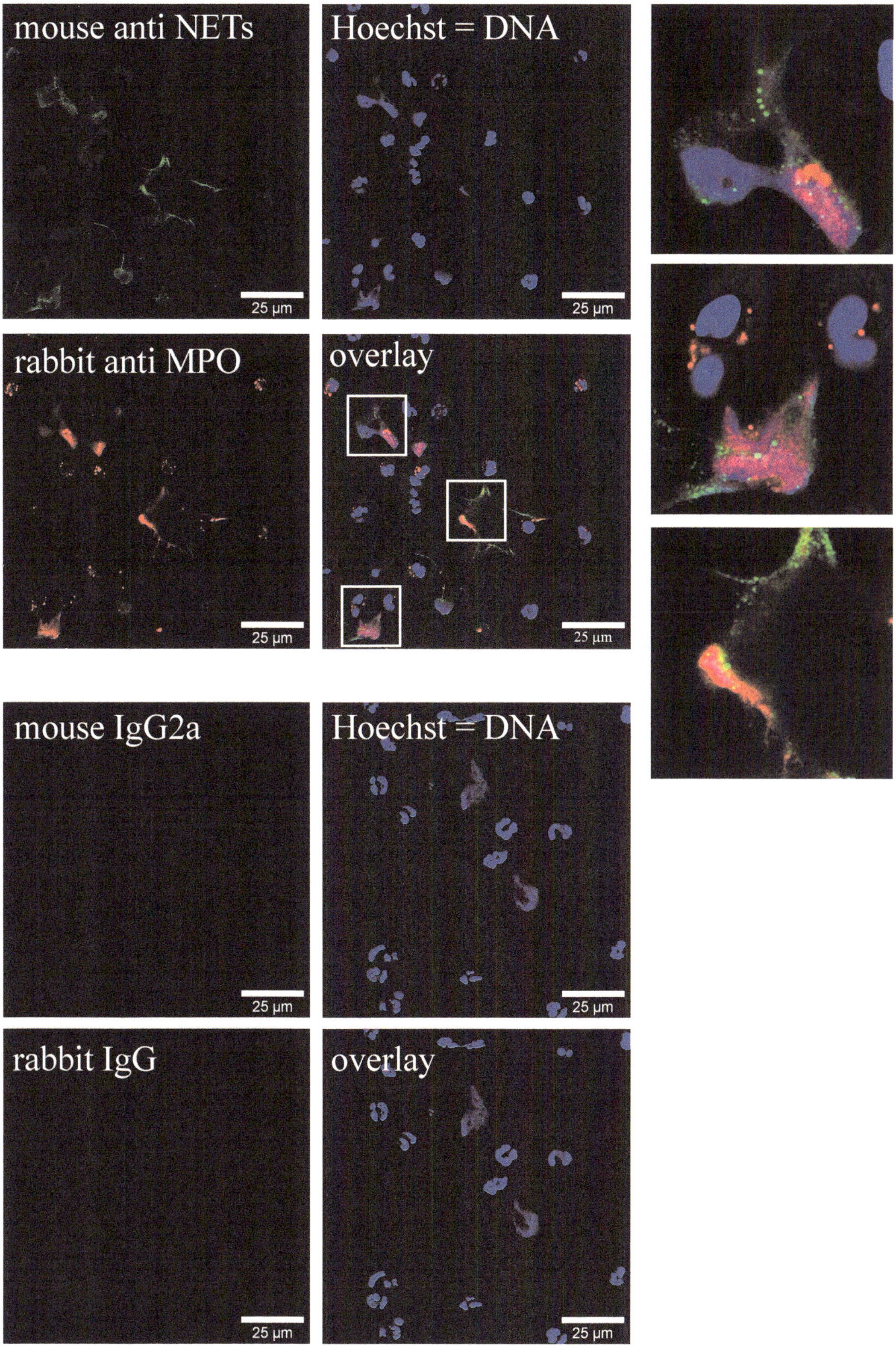

Fig. 1 Equine NETs Neutrophil extracellular traps visualized using a combination of DNA-intercalating dye Hoechst (blue), antibody against DNA–histone 1 complexes/NETs (green), and antibody against granule protein elastase (red). Pay attention to the more complex NET image by using an antibody against NETs or elastase compared to single staining by the DNA-intercalating dye Hoechst

Table 1
Examples for automated methods to quantify NETs

Software name and reference	Staining technique	Characteristics	Problems
NETQuant (Matlab) [10]	MPO or elastase and DAPI	NETs defined by three criteria (NET area, colocalization of DNA and MPO or elastase, and circularity of nucleus)	Programming of software can be altered by everybody and may lead to major changes (thus changes need to be published each time when tool is modified)
DANA (Image J) [11]	DAPI	DNA area	Large clusters of cells must be manually excluded; not very specific
Computational NET detection [12]	Image stream system (Sigma) and confocal microscopy; NET-marker: DAPI, MPO and histone H1	Very specific analysis based on a learning machine which can differentiate cell morphologies	Very complex to understand and learn
Flow cytometry [13]	H3cit antibody, MPO	H3cit as key marker for NET formation	H3cit indpendent pathways are not detected; late stage NETosis cells might be lost
GreenGlo™ dye [14]	DNA-intercalating dyes	Two different excitation/ emission spectra of the dye for nucleic DNA versus extracellular DNA	May be phototoxic over time; cannot differentiate between necrotic and NETotic cells

commonly used and are more specific, especially when used in combination with detection of DNA-histone-complexes and associated proteins [e.g., elastase or myeloperoxidase (MPO)]. The use of citrullinated histones (H3cit) or activation of peptidyl-arginine-deiminase (PAD-4) as NET markers is controversial [15, 16]. However, H3cit is often used especially during in vivo experiments as a NET-marker in combination with elastase or MPO as a neutrophil-specific marker [17–19].

In addition to immunofluorescence microscopy, the detection of NET-specific markers (e.g., free DNA, IL-17, and PR-39) can be used and facilitates quantification. Free DNA is described as the most prominent marker for NETs [20–22]. As NETs consist of a DNA-backbone, many studies use PicoGreen to detect free DNA and quantify by these NETs. Overall, consideration of numerous controls, such as pathogen background and stimulus background is essential.

IL-17 release was detected during NET-formation [23], and antimicrobial peptides of the cathelicidin family (e.g., porcine PR-39 [24] or human LL-37 [9]) have been shown to be stored in neutrophil granules and to be present in high amounts embedded in porcine or human NETs. Thus, those markers can be used as an additional quantification method in body fluids to enable unbiased quantification by biochemical assays or ELISAs for specific proteins. Since the host releases nucleases (e.g., DNase 1), which function in clearing NETs to avoid their detrimental effects [25], the quantification of nuclease protein amount or activity can also be used as an additional indirect marker for the formation of NETs in body fluids.

2 Materials

2.1 Isolation Granulocytes (Human, Porcine, Equine)

2.1.1 Plasticware

1. Sterile 50 ml polypropylene tubes (e.g., Falcon tubes).
2. Sterile tubes and tips (different sizes).
3. Sterile glass Pasteur pipettes (equine and porcine neutrophil preparation).
4. Sterile plastic Pasteur pipettes (human neutrophil preparation).
5. 10 ml Li-Heparin tubes (*see* **Note 1**).

2.1.2 Reagents

1. 0.4% trypan blue solution.
2. Endotoxin-free H_2O (*see* **Note 2**).
3. Endotoxin-free 1× PBS (*see* **Note 2**).
4. RPMI medium (without phenol red if later fluorescent analysis is planned).
5. Diff Quick/HAEMA staining solution.

Isolation Human Granulocytes

1. Fresh venous blood from healthy donor should be collected slowly in Li-Heparin tubes (*see* **Note 1**) and transported as fast as possible without cooling to the laboratory.
2. PolymorphPrep (1.113 g/ml).

Isolation Porcine and Equine Granulocytes

1. Fresh venous blood from a healthy animal should be taken slowly and as stress-free as possible in Li-Heparin tubes (*see* **Note 1**) and transported as fast as possible without cooling to the laboratory.
2. Ice-cooled, sterile 0.2% and 1.6% NaCl solutions prepared with endotoxin-free H_2O.
3. Biocoll (1.077 g/ml).

2.2 NET Induction

2.2.1 Plasticware

1. 48-well plate (for suspension culture).
2. 8 mm glass coverslips (thickness adjusted to the microscope used).
3. Parafilm.

2.2.2 Reagents

1. 0.01% poly-L-lysine solution: sterile filtered, suitable for cell culture.
2. Endotoxin-free 1× PBS.
3. 25 nM phorbol 12-myristate 13-actetate (PMA) solution: dissolve PMA in DMSO (*see* **Note 3**).
4. 10 mM methyl-β-cyclodextrin: always prepare fresh in RPMI (MW = 1331 g/mol). Prepare per well 100 μl (20 mM), because of a 1:2 dilution by cell suspension (*see* **Note 3**).
5. RPMI 1640 medium with L-glutamine (without phenol-red if later fluorescent analysis is planned).
6. 16% paraformaldehyde (PFA) solution.

2.3 NET-Staining for Immunofluorescence Microscopy

2.3.1 In Vitro Samples

Reagents

1. Permeabilization solution: 0.5% Triton X-100 in 1× PBS (always freshly prepared).
2. Blocking buffer: 3% normal donkey serum, 3% cold water fish gelatin, 1% bovine serum albumin (BSA), and 0.5% Tween 20 in 1× PBS. Prepare fresh and store for a maximum of 1 week at 4 °C.
3. 1× PBS.
4. Antibodies (*see* Table 2).
5. 50 mg/ml bisbenzimide H 33342 trihydrochloride (Hoechst 33342) stock solution: prepare in distilled H_2O and use at 1:1000 dilution (*see* **Note 4**).
6. Prolong Gold (Invitrogen).

2.3.2 Immunofluorescence Analysis of In Vivo Samples

Instruments and Reagents

1. Glass bottom plate (96-well) (*see* **Note 5**).
2. Hanks' balanced-salt solution (HBSS).
3. RPMI 1640 medium with L-glutamine but without phenol red.
4. 0.01% poly-L-lysine solution: sterile filtered, suitable for cell culture.
5. 16% paraformaldehyde (PFA) solution.
6. ProLong Gold with DAPI (Invitrogen).

Table 2
Antibodies used for visualization of NETs

Primary antibody	Host/isotype	Company	Concentration	Order number	Dilution	Species tested
Mouse anti-DNA-Histone1 monoclonal IgG2a	Mouse/ IgG2a	Millipore	0.55 mg/ml	MAB3864	1:500 to 1:1000*	Human, equine, porcine, mice, cattle, opossum, dog
Rabbit anti-human myeloperoxidase	Rabbit/IgG	Dako/Agilent	3.2 mg/ml	A0398	1:300	Human, porcine, equine, cattle, mice
Rabbit anti-neutrophil elastase	Rabbit/IgG	Calbiochem/Millipore	5.1 mg/ml	# 481001	1:300	Human
Isotype	**Origin**	**Company**	**Concentration**	**Order number**	**Dilution**	
IgG2a murine myeloma	Murine myeloma	Sigma Aldrich	0.2 mg/ml	M5409-1MG	1:182 to 1:364**	
IgG from rabbit serum	Normal rabbit serum	Sigma Aldrich	10 mg/ml	I5006	1:943	
IgG from rabbit serum	Normal rabbit serum	Sigma Aldrich	10 mg/ml	I5006	1:588	
Secondary antibody	**Host/isotype**	**Company**	**Concentration**	**Order number**	**Dilution**	**Species tested**
Alexa Fluor 488, goat anti-rabbit IgG (H+L)	Goat/IgG	Thermo Scientific	2 mg/ml	A11008	1:1000	Rabbit
Alexa Fluor 633, goat anti-rabbit IgG (H+L)	Goat/IgG	Thermo Scientific	2 mg/ml	A21070	1:1000	Rabbit
DyLight 488, goat anti-mouse IgG IgG (H+L)	Goat/IgG	Thermo Scientific	1 mg/ml	35503	1:500	Mouse
DyLight 633, goat anti-mouse IgG (H+L)	Goat/IgG	Thermo Scientific	1 mg/ml	35512	1:500	Mouse

*Concentrations need to be adjusted to samples derived from different animal species
**Concentration needs to be adjusted to primary antibody

2.4 NET Marker Analysis

2.4.1 Reagents

1. 10 U/ml micrococcal nuclease.
2. EDTA solution (pH 8).
3. Liquid nitrogen.
4. Quant-iT PicoGreen DNA analysis kit.
5. Protein analysis: MPO ELISA, cathelicidin ELISA, neutrophil elastase ELISA.

2.5 Ex Vivo Detection of NET Markers in Body Fluids

2.5.1 Instruments and Reagents

1. Agarose gel electrophoresis (with documentation system).
2. Calf thymus DNA.
3. Agarose.
4. TRIS–borate–EDTA (TBE) buffer.
5. DNA loading dye and DNA ladder (1 kb).

3 Methods

3.1 Isolation of Granulocytes (Human, Porcine, Equine)

1. Use sterile pipette tips for all steps and work under sterile conditions near a flame or in a sterile hood. Bring all solutions to room temperature, except for the cooled H_2O and NaCl solutions.
2. Make a blood smear for Diff Quick/HAEMA staining from the fresh blood (*see* **Note 6**).

3.1.1 Isolation of Human Granulocytes

1. Fresh venous blood from a healthy donor should be collected slowly in Li-Heparin (*see* **Note 1**) tubes and transported as fast as possible without cooling to the laboratory.
2. Pipet 20 ml of PolymorphPrep into a 50 ml polypropylene tube and gently layer 20 ml of blood onto the PolymorphPrep without mixing (*see* **Note 7**).
3. Centrifuge with swing out-rotor at 470 × *g* for 30 min at 20 °C without brake at deceleration (acceleration always on maximum).
4. Remove plasma with a sterile plastic Pasteur pipette. Transfer granulocyte layer with a fresh sterile plastic Pasteur pipette into a new 50 ml polypropylene tube. Fill the tube up to 50 ml with 1× endotoxin-free PBS.
5. Centrifuge at 470 × *g* for 10 min at room temperature with brake at deceleration.
6. Remove supernatant and collect granulocytes in a 15 ml polypropylene tube. Resuspend neutrophils in 5 ml of sterile endotoxin-free H_2O for 15 s.
7. Immediately fill tube with 1× endotoxin-free PBS and centrifuge at 470 × *g* for 10 min at room temperature.

8. Remove supernatant. The pellet should be white. If the pellet is still red, repeat **steps 5** and **6** one more time.
9. Remove supernatant and resuspend pellet in 1 ml cold RPMI medium. Cells can be used now in the NET induction assay (*see* **Note 8**).
10. Count granulocytes with a Neubauer counting chamber and adjust the cell number for the experiment (*see* **Note 9**).

3.1.2 Isolation of Porcine and Equine Neutrophils

1. Fresh venous blood from a healthy donor should be collected slowly in Li-Heparin tube (*see* **Note 1**) and transported as fast as possible without cooling to the laboratory.
2. For porcine blood: dilute 13 ml of blood with 13 ml of 1× endotoxin-free PBS in a 50 ml polypropylene tube. For equine blood: dilute 15 ml of blood with 10 ml of 1× endotoxin-free PBS in a 50 ml polypropylene tube.
3. Mix gently without making air bubbles.
4. For porcine blood: pipet 12 ml of Biocoll separating solution into a 50 ml polypropylene tube and gently layer 12 ml of the blood-PBS mixture onto the Biocoll without mixing. For equine blood: pipet 12.5 ml of Biocoll separating solution into a 50 ml polypropylene tube and gently layer 12.5 ml of the blood-PBS mixture onto the Biocoll without mixing.
5. Centrifuge at 400 × *g* for 20 min at 20 °C without brake at deceleration (acceleration always on maximum).
6. Remove and discard the plasma, peripheral blood mononuclear cell (PMBC) layer, and Biocoll with a glass Pasteur pipette attached to a vacuum pump. After the first centrifugation, no pellet will be visible, only a big red sediment. Do not remove the red sediment but try to remove all Biocoll on the wall of the tube.
7. Add 8 ml of sterile ice cold 0.2% NaCl solution to the red sediment and start a clock for 30 s. Close the lid and invert ten times to dissolve the sediment. This lyses the erythrocytes.
8. After 30 s, immediately add 8 ml of sterile ice-cold 1.6% NaCl solution. Mix gently. The lysis is stopped by adding the 1.6% NaCl solution.
9. Centrifuge at 250 × *g* for 7 min (porcine) or 6 min (equine) in a precooled (4 °C) centrifuge with brake at deceleration and acceleration on maximum.
10. Remove and discard the lysed erythrocytes with a glass Pasteur pipette attached to a vacuum pump.
11. Repeat the lysis once, and the pellet should be white. If not, a third lysis run can be performed. Perform only a maximum of three lysis steps.

12. Resuspend the pellet in 1 ml of cold RPMI medium, and the cells can now be used (*see* **Note 8**).
13. Count the granulocytes with a Neubauer counting chamber and adjust the cell number for the experiment (*see* **Note 9**).

3.2 NET Induction

1. Place 8 mm glass coverslips into wells of a 48-well plate using a vacuum pump to aid in placing the coverslips.
2. Coat the coverslips with 60 μl of poly-L-lysine solution using the manufacturer's protocol, with slight modifications. The poly-L-lysine should build a dome and should not touch the border. Coat for 20 min at room temperature. Aspirate the solution and wash the coverslips three times with 1× endotoxin-free PBS to remove unbound poly-L-lysine. Make sure that all coverslips are always covered with liquid but do not swim. After the final washing, leave liquid on the coverslips. Plates can be stored for 1 week at 4 °C if wrapped with Parafilm. Prepare one plate for each time point.
3. Prepare enough cell suspension so that there will be enough cell suspension to add 2×10^5 cells to each well in a 100 μl volume.
4. Prepare enough stimulus or control solutions so that 100 μl can be added per well. The negative control is RPMI. The stimuli are methyl-β-cyclodextrin (porcine and equine) or PMA (human) (*see* **Note 3**). Duplicates are recommended. Prepare one extra stimulus control for isotype control staining (*see* **Note 10**).
5. Aspirate the liquid from each well and seed 2×10^5 cells per well (100 μl). Mix the stock cell suspension gently by pipetting up and down (do this gently each time before pipetting into each well).
6. Add 100 μl of stimulus or control medium to each well for a final volume of 200 μl/well. (In case of infection with bacteria, it might be necessary to centrifuge plates at 370 g for 5 min at room temperature to bring cells and bacteria into contact.)
7. Incubate the cells for NET induction kinetic analysis with four time points (60, 120, 180, and 240 min) at 37 °C and 5% CO_2.
8. Centrifuge plates at 370 × *g* for 5 min at room temperature.
9. Add 16% PFA solution for a final concentration of 4% (*see* **Note 11**), and incubate for 15 min at room temperature.
10. If staining is not immediately done, storage at 4 °C is possible. If storing, wrap the plate with Parafilm to avoid drying of coverslips.

3.3 NET Staining for Immunofluorescence Microscopy

3.3.1 In Vitro Samples

1. Conduct all staining steps inside the wells (*see* **Note 11**).
2. Wash the PFA-fixed samples three times with 200 μl of 1× PBS. Prevent drying out of coverslips and artifacts by pipetting rapidly for all washing steps. Finally aspirate off the PBS.
3. Add 100 μl of 0.5% Triton X-100 solution to the wells for 5 min to permeabilize cells. Aspirate all liquid with a vacuum pump.
4. Add 100 μl of blocking buffer to each well for 20 min at 20 °C. Aspirate all liquid with a vacuum pump.
5. Add 100 μl of the desired primary antibody (*see* Table 2) in blocking buffer and incubate for 1 h at 20 °C:
   (a) For staining NETs, use mouse anti-DNA/histone IgG2a antibody (*see* **Note 12**) and isotype control (IgG2a from murine myeloma).
   (b) For staining MPO, use rabbit anti-human MPO antibody and isotype control (IgG from rabbit serum).
   (c) For staining neutrophil elastase, use rabbit anti-elastase and isotype control (IgG from rabbit serum).
   Note that a combination of a and b as well as a and c is possible.
6. Wash cells three times with 200 μl of 1× PBS and finally aspirate off the solution.
7. Add 100 μl the relevant secondary antibody (Table 2) in blocking buffer and incubate for 1 h at room temperature in the dark.
   (a) For goat primary antibodies, use DyLight 488 goat anti-mouse IgG or DyLight 633 goat anti-mouse IgG as secondary antibodies.
   (b) For rabbit primary antibodies, use Alexa 633 goat anti-rabbit IgG or Alexa 488 goat anti-rabbit IgG as secondary antibodies.
   A combination is possible and depending if NETs should be red or green labeled.
8. Wash cells three times with 200 μl of 1× PBS and finally aspirate off the liquid.
9. Wash cells one time with 200 μl of distilled H_2O and aspirate.
10. Stain cells with 100 μl of aqueous Hoechst 33342 (1:1000 dilution) for 10 min in the dark at room temperature (*see* **Note 4**).
11. Wash the cells three times with 200 μl of distilled H_2O. Do not aspirate off the liquid.

12. Carefully take the coverslips out of the wells using a round curved cannula and forceps. Avoid scratching and breaking of the coverslips.
13. Embed the coverslips in 3 μl of ProLong Gold on glass slides. Place the coverslips with cells facing down and dry overnight at 4 °C (horizontal position). Five to six 8 mm coverslips can fit on one slide.
14. Surround the coverslips with nail polish on the next day to avoid drying out of the embedded sample.
15. Store samples at 4 °C in the dark until microscopy analysis.

3.3.2 Immunofluorescence Analysis of In Vivo Samples

1. Coat wells with 50 μl of poly-L-lysine solution for 20 min at room temperature. Aspirate the solution and wash the wells three times with 100 μl of 1× endotoxin-free PBS to remove unbound poly-L-lysine. After the final washing, leave liquid in the well. Plates can be stored for 1 week at 4 °C if the plate is wrapped with Parafilm.
2. Collect body fluid and cool immediately (*see* **Note 13**).
3. Count cells in the body fluid with a Neubauer counting chamber or Fuchs-Rosentahl chamber (in case of little amount of cells) and adjust cells to 2×10^5 cells/100 μl. If dilution is needed, use HBSS or RPMI.
4. Pipet 50 μl of body fluid into each well.
5. Centrifuge at 370 × *g* for 5 min at room temperature.
6. Add 16% PFA solution for a final concentration of 4%. Incubate for 15 min at room temperature.
7. If staining is not done immediately, storage at 4 °C is possible. For storage, wrap plate with Parafilm to avoid drying of coverslips.
8. Staining can be performed as described above (Subheading 3.3.2) with the following changes: for washing steps, use only 100 μl per well, and for staining steps, use only 50 μl per well. Do not stain with Hoechst. Embed samples with 10 μl per well of Prolong Gold with DAPI. It is recommended that microscopy analysis is conducted within 5 days after staining to avoid drying. Because of possible high background signals in body liquids, an isotype control is necessary.

3.4 NET Marker Analysis

3.4.1 Sample Preparation (See Fig. 2 for a Schematic Overview of Sample Preparation)

1. Conduct a NET induction assay, as described above, but without glass coverslips inside the wells, or use samples that you want to analyze (e.g., 3D cell culture samples or in vivo samples) (*see* **Note 14**).
2. For plate samples, transfer well contents into 1.5 ml tubes.
3. Centrifuge at 370 × *g* for 10 min.

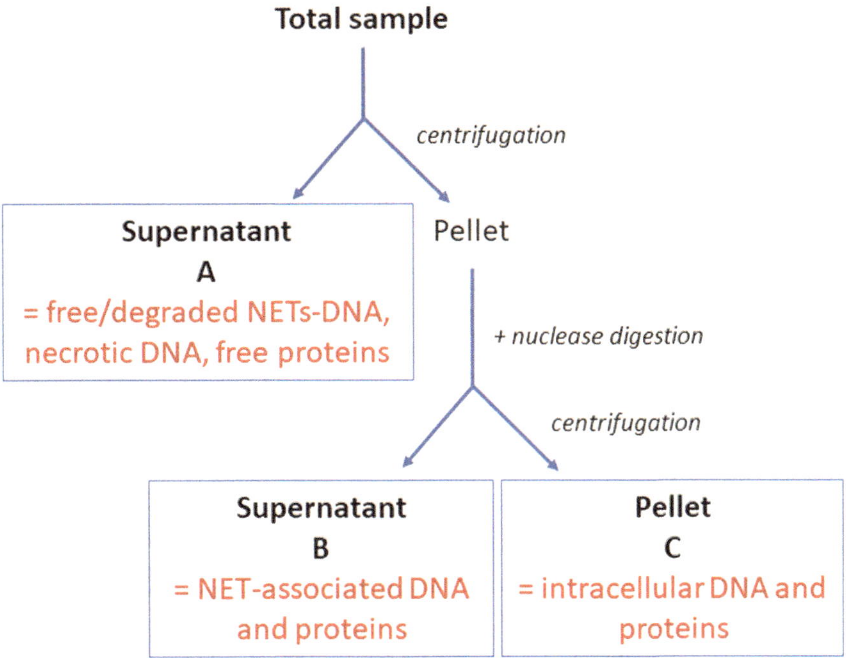

Fig. 2 Schematic for NET marker analysis

4. Collect supernatant (designated as sample A = free NET DNA, necrotic DNA, and free proteins).
5. Resuspend pellet in 125 μl HBSS (designated as sample B).
6. Add 0.5 U/ml micrococcal nuclease to sample B and incubate for 10 min at 37 °C and 5% CO_2.
7. Add 5 mM EDTA to stop the reaction.
8. Centrifuge sample B at 370 × *g* for 5 min.
9. Transfer supernatant (supernatant of sample B = NET-associated DNA and proteins).
10. Add 150 μl of HBSS to the pellet of sample B (designated as sample C = intracellular DNA and proteins).
11. Freeze all samples immediately in liquid N_2. Store samples at −80 °C until further analysis. Samples can be aliquoted for several subsequent analyses to avoid repeated freeze–thaw cycles.

3.4.2 DNA-Detection with Quant-iT PicoGreen

1. Make a standard dilution series (0, 0.001, 0.01, 0.1, and 1 μg/ml) of DNA provided in the PicoGreen kit or use calf thymus DNA instead. Prepare everything in duplicates and include the dilution series on each plate.
2. Thaw the samples and pipet 50 μl of sample or standard per well in a 96 well plate.

Table 3
Commercial ELISA assays used for quantification of NET markers

Name	Company	Order number	Species tested	Comment
LL-37 ELISA kit	Hycult Biotechnology	HK321-02	Human	
Human Deoxyribonuclease I ELISA Kit	MyBioSource Inc.	MBS729766-96	Human	48 well or 96 well
Horse Cathelicidin Antimicrobial Peptide ELISA kit	MyBioSource Inc.	MBS046008-96	Equine	48 well or 96 well
PR-39 ELISA kit	Antibody Research Corporation	811030	Porcine	
PR-39 ELISA	MyBioSource Inc.	MBS288141-96	Porcine	48 well or 96 well
Porcine IL-17 ELISA	Abcam	ab193732	Porcine	
Pig PMAP-36 ELISA kit (competitive EIA)	LifeSpan BioSciences, Inc.	LS-F13412-1	Porcine	
Porcine Antibacterial peptide PMAP-37	LifeScience Market	ELI-37324p	Porcine	
Pig DNase I ELISA kit	Aviva Systems Biology	OKEH03902	Porcine	

3. Dilute PicoGreen 1:200 in TBE buffer (protect from light).
4. Mix the samples 1:2 with diluted PicoGreen solution directly in the 96 well plate (50 μl of sample + 50 μl of PicoGreen for a final volume of 100 μl). Note that you can use up to a total maximum volume of 200 μl per well; however, use the same volume for all samples in one experiment. PicoGreen can be pipetted with a multichannel pipettor into the plate to avoid a much longer incubation for the first pipetted samples compared to the last pipetted samples.
5. Incubate the plate for 5 min at room temperature in the dark.
6. Measure the plate in the microplate reader with the following parameters: Top optics, 485/535 filters, 25 flashes per well, optimized gain.
7. Calculate the amount of DNA in the samples based on the standard curve.

3.4.3 Protein Detection

Several commercial ELISAs are available for the detection of NET-bound and NET-associated proteins (*see* Table 3). Cross-reactivity is possible for granule proteins from different species (e.g., MPO and neutrophil elastase). For animal species-specific antimicrobial peptides, specific ELISAs are needed. All ELISAs can be conducted as described in the user's manuals.

3.5 Ex Vivo Detection of NET Markers in Body Fluids

3.5.1 Detection of NET Markers

The detection of NET markers can be conducted in ex vivo samples, as described in Subheading 3.4 for in vitro samples. The samples should be used fresh or stored at −80 °C. Be aware that neutrophils are destroyed by freezing. Therefore, if a separation between intracellular and extracellularly released components is required, centrifugation (370 × *g* for 5 min) followed by separation of the cells and supernatant is needed *prior to freezing the samples.*

3.5.2 Detection DNase Activity

Since DNase of the host is a marker for NET regulation in the host, a DNase activity test can be conducted and combined with a DNase ELISA (*see* Table 3).

1. As a negative control for DNase activity, use a DNase free medium (e.g., PBS). As a positive control for DNase activity, supplement the DNase free medium (e.g., PBS) with 500 mU micrococcal nuclease.
2. For each sample, pipette 0.5 μg of calf thymus DNA into a 0.5 ml tube. Add 50 μl of negative or positive control or sample. Make sure that everything is mixed together (do not vortex).
3. Close the tube lids and incubate at 37 °C for 1–24 h (depending on DNase activity of your sample). Note that addition of DNase buffer (e.g., 3 mM $CaCl_2$, 3 mM $MgCl_2$, and 300 mM TRIS, pH 7.4) can help for detection of DNase activity. This depends on the sample composition (pH, ions, etc.) and the DNase present in the sample. Different DNases have different pH optima and ion concentration effects. Therefore, different test conditions should be used. Furthermore, time-kinetics can be performed.
4. Prepare 1% agarose gel with DNA markers (e.g., RotiSafe, Gelstain ready-to-use gels).
5. Load the same volume (minimum 20 μl) of each sample mixed with DNA loading dye. Include a 1 kb DNA ladder.
6. Run the gel (100 V, 15–30 min) and visualize the gel with an imager and compare DNA bands in the negative and positive controls with your samples of interest.

4 Notes

1. The anticoagulant used for harvest of fresh blood may highly influence purity and activity of isolated blood cells. For more information, see an example paper from us, which demonstrates that density gradient centrifugation of K_3EDTA blood resulted in higher purity of bovine granulocytes compared to lithium heparin blood [26]. However, EDTA might hamper the assays when studying interaction of bacteria with

neutrophils or NETs, since EDTA has antibacterial activity. Therefore, it is important to carefully select the neutrophil isolation protocol for specific purposes.

2. Contaminations by LPS or other bacterial products can easily lead to prestimulation of cells during isolation methods. A negative control is always needed to control background NET formation and to exclude artifacts. In addition, too harsh mixing of cells might result in artifacts and background NET formation independent of a stimulus.
3. Positive controls for NET induction: several NET inducers are known. Depending on the animal species, they induce NETs to different levels. In human granulocytes, PMA is an efficient NET inducer. In bovine and equine granulocytes, methyl-β-cyclodextrin works efficiently. For PMA storage at −80 °C, a stock solution of 1.6 mM in DMSO is recommended. Avoid freeze–thaw cycles (max. three times) and do not use longer than 6 months after storage. Always prepare fresh working dilutions of methyl-β-cyclodextrin and PMA.
4. Instead of staining DNA with Hoechst 33342, embedding in Prolong Gold with DAPI (4′,6-diamidino-2-phenylindole) is possible. In this case, only three times 1× PBS washing is needed after the staining step with the secondary antibody, followed by embedding in 3 μl Prolong Gold with DAPI.
5. Glass bottom plates: They are available in a 96-well plate format and a 10 well plate format (same size as in the 96-well plate). They are also available in other formats; depending on the cell number in the in vivo sample other formats could be useful. Instead of a glass bottom plate, plastic plates with poly-L-lysine glass coverslips could be used as well.
6. Diff Quick/HAEMA staining of blood smears is recommended to have the possibility to check later if the blood composition was in the standard range. The method is quick and cheap, and stained slides can be stored at room temperature.
7. Layering the gradient: Use a 10 ml sterile serological pipette with pipette boy set on minimum engine speed and the operating mode "release with free outlet" (near flame or under the sterile bench). Place the tube at an angle of 45° and let the blood-PBS mixture slowly run down at the border.
8. Storage of cells: As cells may start to degranulate and become apoptotic, NETotic, or necrotic, immediate use is recommended to avoid nonspecific background stimulation or spontaneous NET release.
9. Mix carefully for counting the cells. Air bubbles and turbulence must be prevented to avoid prestimulation of cells. Trypan blue

stains only dead cells, because of the membrane permeability. Therefore, the counting is an important quality control of isolated neutrophils. A quality standard of maximum 5% dead cells is recommended.

10. Detection of nonspecific binding of antibodies: Control staining using isotype control antibodies in combination with the secondary antibody is always needed, since all kinds of primary or secondary antibodies might lead to false positive staining. Since some bacteria efficiently bind various immunoglobulins, special controls and/or blocking buffer are needed when working with infected cell material [27].
11. Avoid directly pipetting of solutions onto the cell layer on glass coverslips. Pipet PFA and washing solutions during all washing steps at the side of the wells.
12. Concentrations change with antibody batches. Proof and adjust isotype if needed.
13. In vivo samples: An immediate fixation after collection is highly recommended. Processes in cells do not stop outside of the body, and spontaneous necrosis may occur. Therefore, the amount of artifacts increase over time, and false positive NET signals could be possible.
14. One limitation of the PicoGreen assay is that all DNA is detected. Therefore, pathogen DNA can give background signals. Furthermore, some media and stimuli can result in nonspecific background signals. Thus, proper controls must be included (e.g., media alone, pathogens alone, stimuli alone). In addition, negative and positive controls as described in the NET induction assay must be included.

References

1. Brinkmann V, Reichard U, Goosmann C et al (2004) Neutrophil extracellular traps kill bacteria. Science 303:1532–1535
2. Chow OA, von Köckritz-Blickwede M, Bright AT et al (2010) Statins enhance formation of phagocyte extracellular traps. Cell Host Microbe 8:445–454
3. Pilsczek FH, Salina D, Poon KK et al (2010) A novel mechanism of rapid nuclear neutrophil extracellular trap formation in response to *Staphylococcus aureus*. J Immunol 185:7413–7425
4. Yipp BG, Petri B, Salina D et al (2012) Infection-induced NETosis is a dynamic process involving neutrophil multitasking in vivo. Nat Med 18:1386–1393
5. Mitsios A, Arampatzioglou A, Arelaki S et al (2017) NETopathies? Unraveling the dark side of old diseases through neutrophils. Front Immunol 7:678
6. von Köckritz-Blickwede M, Blodkamp S, Nizet V (2016) Interaction of bacterial exotoxins with neutrophil extracellular traps: impact for the infected host. Front Microbiol 7:402
7. De Buhr N, von Köckritz-blickwede M (2016) How neutrophil extracellular traps become visible. J Immunol Res 2016:4604713
8. Fuchs TA, Abed U, Goosmann C et al (2007) Novel cell death program leads to neutrophil extracellular traps. J Cell Biol 176:231–241
9. Neumann A, Völlger L, Berends ETM et al (2014) Novel role of the antimicrobial peptide LL-37 in the protection of neutrophil extracellular traps against degradation by bacterial nucleases. J Innate Immun 6:860–868

10. Mohanty T, Sørensen OE, Nordenfelt P (2018) NETQUANT: automated quantification of neutrophil extracellular traps. Front Immunol 8:1999
11. Rebernick R, Fahmy L, Glover C (2018) DNA area and NETosis analysis (DANA): a high-throughput method to quantify neutrophil extracellular traps in fluorescent microscope images. Biol Proced Online 20:7
12. Ginley BG, Emmons T, Lutnick B et al (2017) Computational detection and quantification of human and mouse neutrophil extracellular traps in flow cytometry and confocal microscopy. Sci Rep 7:17755
13. Lee KH, Cavanaugh L, Leung H et al (2018) Quantification of NETs-associated markers by flow cytometry and serum assays in patients with thrombosis and sepsis. Int J Lab Hematol 40:392–399
14. Proust A, Lévesque JC, Barat C et al (2018) A new tool for detection of extracellular traps. Methods Appl Fluoresc 6:037002
15. Kenny EF, Herzig A, Krüger R et al (2017) Diverse stimuli engage different neutrophil extracellular trap pathways. Elife 6:e24437
16. Gupta AK, Giaglis S, Hasler P et al (2014) Efficient neutrophil extracellular trap induction requires mobilization of both intracellular and extracellular calcium pools and is modulated by cyclosporine A. PLoS One 9:e97088
17. Li P, Li M, Lindberg MR et al (2010) PAD4 is essential for antibacterial innate immunity mediated by neutrophil extracellular traps. J Exp Med 207:1853–1862
18. Wang Y, Li M, Stadler S et al (2009) Histone hypercitrullination mediates chromatin decondensation and neutrophil extracellular trap formation. J Cell Biol 184:205–213
19. Leppkes M, Maueröder C, Hirth S et al (2016) Externalized decondensed neutrophil chromatin occludes pancreatic ducts and drives pancreatitis. Nat Commun 7:10973
20. Altrichter J, Zedler S, Kraft R et al (2010) Neutrophil-derived circulating free DNA (cf-DNA/NETs), a potential prognostic marker for mortality in patients with severe burn injury. Eur J Trauma Emerg Surg 36:551–557
21. Margraf S, Lögters T, Reipen J et al (2008) Neutrophil-derived circulating free DNA (CF-DNA/NETs): a potential prognostic marker for posttraumatic development of inflammatory second hit and sepsis. Shock 30:352–358
22. Megens RT, Vijayan S, Lievens D et al (2012) Presence of luminal neutrophil extracellular traps in atherosclerosis. Thromb Haemost 107:597–598
23. Lin AM, Rubin CJ, Khandpur R et al (2011) Mast cells and neutrophils release IL-17 through extracellular trap formation in psoriasis. J Immunol 187:490–500
24. de Buhr N, Reuner F, Neumann A et al (2017) Neutrophil extracellular trap formation in the Streptococcus suis-infected cerebrospinal fluid compartment. Cell Microbiol 19:1–16
25. Hakkim A, Fürnrohr BG, Amann K et al (2010) Impairment of neutrophil extracellular trap degradation is associated with lupus nephritis. Proc Natl Acad Sci U S A 107:9813–9818
26. Baien SH, Langer MN, Heppelmann M et al (2018) Comparison between K3EDTA and lithium heparin as anticoagulant to isolate bovine granulocytes from blood. Front Immunol 9:1570
27. Nordenfelt P, Björck L (2013) IgG-binding bacterial proteins and pathogenesis. Future Microbiol 8:299–301

Chapter 26

Imaging of Neutrophils and Neutrophil Extracellular Traps (NETs) with Intravital (In Vivo) Microscopy

Iwona Cichon, Michal Santocki, Weronika Ortmann, and Elzbieta Kolaczkowska

Abstract

As we have learned during recent years, neutrophils are not just simple foot soldiers of the innate immune system with a restricted set of pro-inflammatory functions, and instead, they perform sophisticated functions (some of them only recently discovered) both in innate and adaptive immune responses. Neutrophil behavior and functioning should best be studied in situ, at locations where they are executed in a living organism, especially considering that neutrophils are mobile cells, performing their functions in distal body sites and various organs. For this herein we describe an approach to detect neutrophil presence/behavior in various organs (skin, muscle, liver) of alive mice, that is, intravital imaging/microscopy. We describe all surgeries required prior to imaging and share our methods of detection of neutrophils and neutrophil extracellular traps (NETs).

Key words Intravital microscopy, In vivo microscopy, Imaging, Live cell imaging, Microsurgeries, Neutrophils, Neutrophil extracellular traps, NETs, Mice

1 Introduction

Intravital or in vivo microscopy (IVM) is a technique that allows for imaging of dynamic or time-dependent processes occurring in live animals in real time [1]. The advantage of this method over other in vivo studies, and even more so over studies on isolated primary cells (ex vivo) or immortalized cell lines (in vitro), is an ability to capture cells/tissues/processes in situ, at the sites were they actually act or occur. The classical microscopy calls for fixed (often stained) speciment on which various cell types can be observed but it does not allow verification if the cells interact with each other or just colocalize. With intravital microscopy, the interactions can be captured and furthermore, their dynamics, timing and

Iwona Cichon and Michal Santocki contributed equally to this work.

Mark T. Quinn and Frank R. DeLeo (eds.), *Neutrophil: Methods and Protocols*, Methods in Molecular Biology, vol. 2087, https://doi.org/10.1007/978-1-0716-0154-9_26, © Springer Science+Business Media, LLC, part of Springer Nature 2020

consequences can be tracked over time. To give some examples, it was this technique that allowed to describe biphasic neutrophil swarming to the site of injury [2], detect a change of monocyte phenotype from pro- to anti-inflammatory in damaged, healing tissue in real time [3], and show inefficiency of DNase for neutrophil extracellular trap (NET) disassembly in vivo [4]. Overall, IVM functions as a very efficient technique to detect various behaviors of neutrophils, one of the most important first line defenders of innate immunity [1].

Although advanced confocal microscopes are usually used for intravital imaging, some less complex tissues (e.g., cremaster muscle) can also be imaged with rather simple fluorescent microscopes. An advantage of spinning-disk confocal microscopes is their capacity to acquire images with high speed which is important considering imaged rodents are alive. Otherwise, due to animal respiration (the movement of its diaphragm) and/or movement of mobile cells, slowly acquired images might be out of focus. For deep into tissues (300–500 μm) observations, two-photon microscopy should be used as it secures deep penetration with high resolution. As in any fluorescent/confocal microscopy imaging, various cells or structures are labeled with either antibodies conjugated with fluorochromes or specific fluorescent dyes are used for visualization. Alternatively, reporter mice are used.

Historically, the IVM technique is not as novel as one might think. Its roots originate in the nineteenth century, when using very simple light microscopes, the vasculature within the tongue of a live frog was imaged [5]. By the end of the twentieth century, in an era of fluorescent and confocal microscopy, the technique was substantially enhanced embracing not only new microscopic technologies (multiphoton or resonance, to name a few) but also the development of new surgical approaches expanded the repertoire of tissues and organs that can be imaged. It is now possible to visualize not only thin and translucent tissues such as the cremaster muscle or mesentery but also much more complex tissues such as skin. In regard to organs, in addition to the commonly visualized liver, kidney, spleen or brain, recent advancements now also allow us to image dynamic and moving tissues such as the intestine, lung, and heart.

Herein we describe protocols for imaging of neutrophils in vasculature of a mouse ear (skin), cremaster muscle (representing a classical vascular bed) and liver (its unique sinusoids), as well as proceeding surgeries. Accordingly, each paragraph is structured in such a way that it begins with cannulation of a jugular vein, then it is followed by preparation of a given organ – skin, cremaster, or liver, and at last information on imaging of neutrophils and NETs (common for either of the organs) is provided. Although we focus on imaging of neutrophils, other leukocytes can be visualized with the same approaches when the detecting antibodies or reporter mice are accordingly modified.

2 Materials

The described procedures are designed for mice. All animal studies must be approved by the Local/Institutional Animal Care and Use Committee(s). Except for the cremaster muscle preparations that can be performed only on male mice (preferable at least 8–10 weeks old), mouse gender is irrelevant. Imaging of animals younger than 5–6 weeks old is difficult due to their size (especially a diameter of the jugular vein is small in young mice) and also it is more challenging in old mice (6–8 or more months old) due to fat deposition around the veins and in the abdomen.

2.1 Jugular Vein Cannulation

For anesthesia, prepare a mixture of 10 mg/kg xylazine and 200 mg/kg ketamine.

For jugular vein cannulation prepare:

1. Surgical board: Plexiglas board, approx. 20 × 14 cm in size.
2. Polyethylene tubing (0.011″ i.d. × 0.024″ o.d.).
3. 1 mL insulin syringe.
4. Two 30 G × 1/2″ needles.
5. 100 U/mL heparin solution.
6. Removable Scotch tape.
7. Scissors with a sharp tip, scissors with a ball tip, two blunt ended forceps with a curved tip, curved sharp ended forceps.
8. Cotton swabs.
9. Mineral oil.
10. Three pieces of silk suture (approx. 11 cm).

2.2 Ear Skin Preparation (for Imaging of the Vasculature of the Skin)

Prepare:

1. Custom made Plexiglas board with taller edges (Fig. 1a, 4).
2. Coverslips (24 × 24 mm).
3. Stack of microscope slides taped together (8 microscope slides 76 × 26 mm, 1 mm thick, taped together with a tape) (Fig. 1b).
4. 20 mL syringe with 25 G needle filled with saline.
5. Removable Scotch tape.
6. Cotton swabs.
7. Hair removal cream (e.g., Veet).
8. Phosphate buffered saline (PBS).
9. Diamond knife.
10. Blunt ended forceps with a curved tip (0.5 mm).
11. Dust-free paper.
12. 70% EtOH.

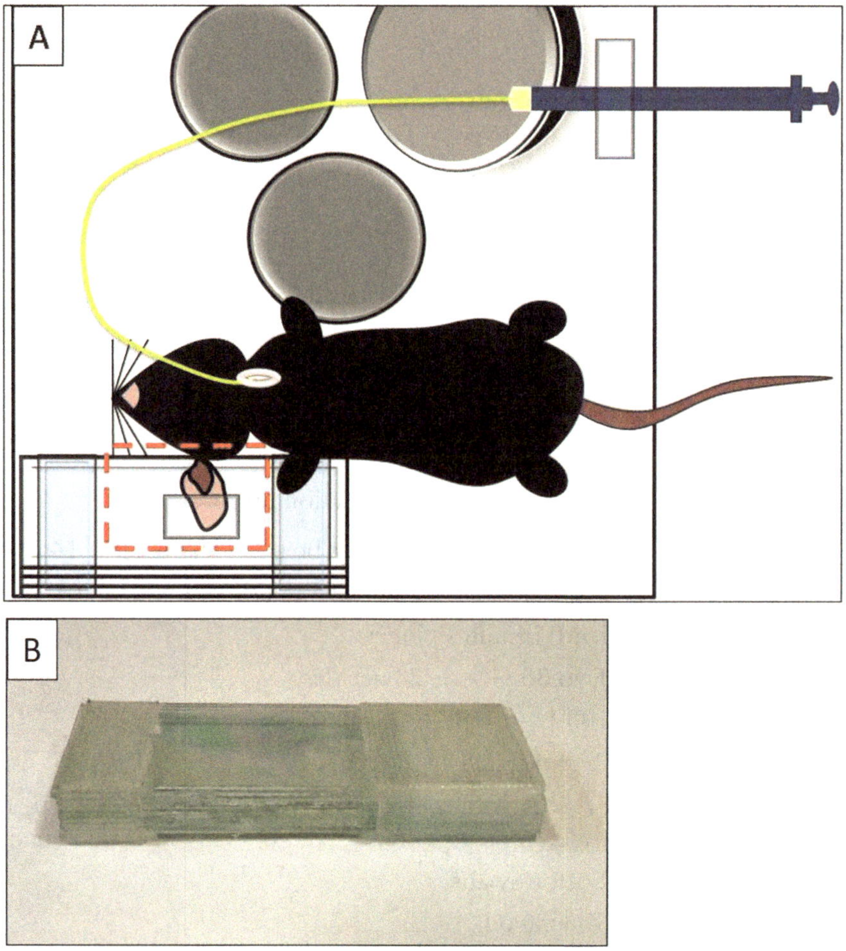

Fig. 1 Ear preparation for in vivo imaging of the skin vasculature in mice. For imaging, a coverslip should be placed on the ventral side of the ear (dotted square, red) (**a**). The ear should be positioned on eight microscopic slides taped together (**b**)

2.3 Cremaster Muscle Preparation (for Imaging of the Classical Vascular Bed)

Prepare:

1. Custom made Plexiglas board with rectangular hole in the upper part between two handles (Fig. 2).
2. Coverslips, one bigger (24 × 60 mm) corresponding to the Plexiglas hole size (Fig. 2b) and smaller ones (24 × 24 mm).
3. Removable Scotch tape.
4. Crystal Scotch tape.
5. 2.5 cm surgical tape (e.g., 3 M Transpore™).
6. 20 mL syringe with 21 G needle filled with saline.
7. Three long pieces of silk sutures (~15 cm long).

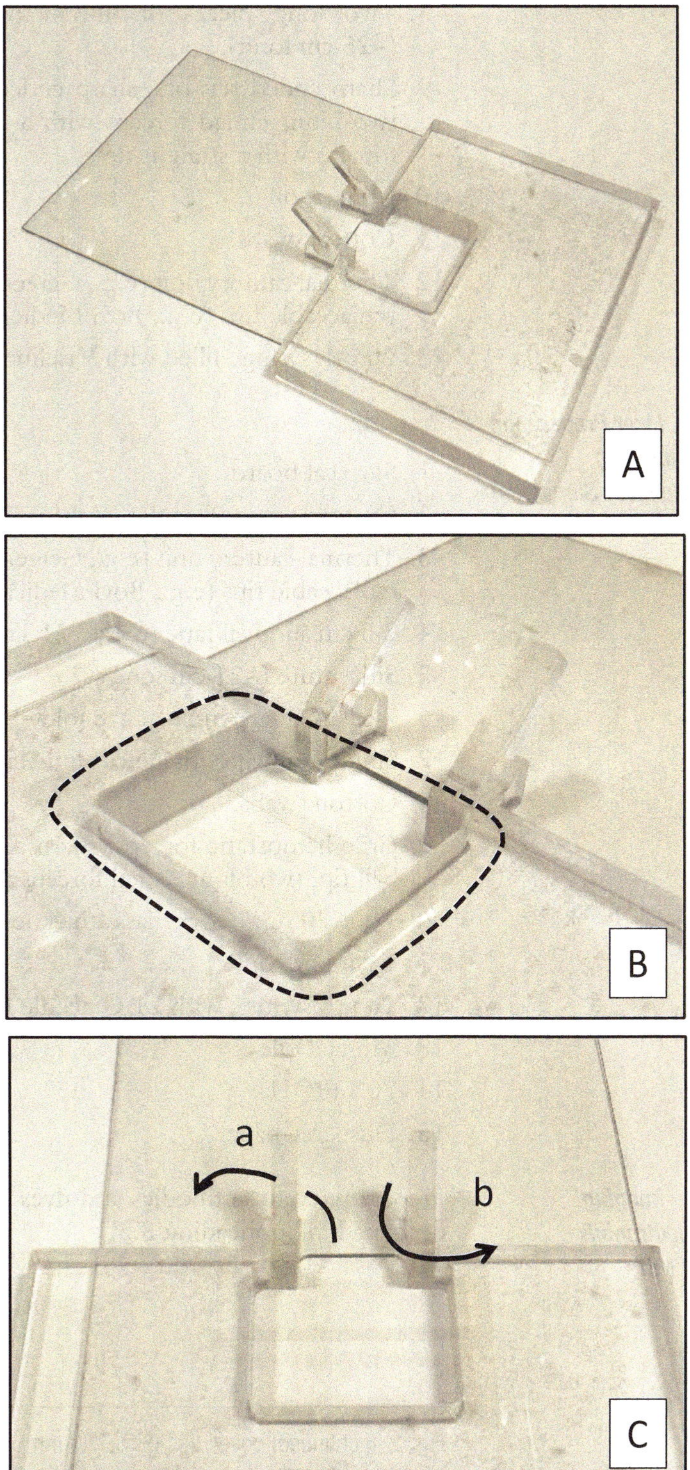

Fig. 2 A board designated for in vivo imaging of the cremaster muscle in mice. The imaging board is a custom-made Plexiglas panel (**a**). Two overlapping

8. Two long pieces of ribbon or any other material tape (~25 cm long).
9. Sharp ended scissors, sharp ended forceps with a curved tip, two blunt ended forceps with a straight tip, two self-closing forceps with a straight tip.
10. Mineral oil.
11. Cotton swabs.
12. Thermal cautery unit (e.g., Geiger Medical Technologies) with replaceable tips (e.g., Bovi Medical Corporation).
13. 20 mL syringe filled with Vaseline.

2.4 Liver Preparation (for Imaging of the Liver Sinusoids)

Prepare:

1. Surgical board.
2. Custom made Plexiglas board with taller edges (Fig. 4).
3. Thermal cautery unit (e.g., Geiger Medical Technologies) with replaceable tips (e.g., Bovi Medical Corporation).
4. 2.5 cm surgical tape (e.g., 3 M Transpore™).
5. Silk suture (~21 cm long).
6. 7.5 × 7.5 cm single gauze folded in half.
7. Single Kimwipes tissue cut in half.
8. Cotton swabs.
9. Crile hemostatic forceps, sharp ended scissors, scissors with a ball tip, two blunt ended forceps with a curved tip.
10. 20 × 20 mm cover glass (thickness 0.13–0.17 mm).
11. Diamond knife.
12. 10 mL syringe with 21 G needle filled with saline.
13. Mineral oil.
14. 70% EtOH.
15. Dust-free paper.

2.5 Imaging of Neutrophils and NETs

Prepare relevant antibodies and dyes and/or use reporter mice as described in Subheading 3.5.

Fig. 2 (continued) coverslips (24 × 60 mm) should be attached to the imaging board between the two handles (marked with a dotted line) (**b**). Orientation of mouse hind limbs during in vivo imaging of the cremaster muscle (**c**): the right hind limb goes under the handle on the left-hand side (**a**), and the left hind limb goes above the handle on the right-hand side (**b**)

3 Methods

3.1 Jugular Vein Cannulation

3.1.1 Anesthesia

Anesthetize a mouse by intraperitoneal injection with the xylazine–ketamine mixture. 10–15 min after injection, check if mouse is in deep sleep by pinching the footpad with forceps. If no limb withdrawal reflexes are observed, proceed to cannulation.

3.1.2 Cannulation

1. The cannula consists of a 22 cm long piece of polyethylene tubing (0.011 × 0.024 in.). Fill a 1 mL syringe with heparin solution and attach a 30 G × 1/2″ needle. Slide the tubing onto the needle.
2. Once the cannula is attached to the syringe, fill tubing with heparin containing saline. Avoid air bubbles in the catheter. Cut the loose end of the cannula to form a semisharp ending that will facilitate its insertion into the vein.
3. Using a surgical tape, immobilize a mouse to the Plexiglas board with its abdomen up (by taping mouse limbs to the board). Turn the board in such a way that the head of the mouse is pointing at you. Then hook the short silk into the front upper teeth and carefully stretch the neck and fix the head in this position by taping the suture to the board.
4. Start surgery on the right jugular vein that is now located on your right-hand side (if cannulation of the right vein is unsuccessful, a subsequent cannulation of the opposite vein is still possible). Skin between the ears and the right-hand side of the neck of the mouse needs to be covered with mineral oil (to control the hair); shaving is optional but not required.
5. A longitudinal incision of about 10–15 mm along the neck up to the clavicle should be made on the right-hand side of the neck. First make a small incision with a sharp scissors and then continue using scissors with a sharp-ball tip. Once the skin is open, the jugular vein should be partially visible under a thin layer of muscles.
6. Using two pairs of forceps with sharp tips, rather than scissors, expose a 4–8 mm section of the vein by separating the skin from the viscera. Carefully remove the connective tissue and fat (in older mice) surrounding the jugular vein. Take time to clean the vein as this will have a significant impact on successful cannulation.
7. Place a loose tie of a 4–0 silk suture (~8–10 cm long) on both cranial and caudal ends of the vessel to maximize the exposure of the vessel. Fix the ends of the silk with tape to the surgical board.
8. Bend the tip of a 30 G × 1/2″ needle with your scissors so that the tip has a 45°. Use this needle to make an incision/opening

by punching the vein with the needle tip (insert the needle 2–3 mm into the vein). Immediately insert the cannula (4–6 mm) into the vein lumen and then use the ligatures at the cranial and caudal ends to secure the catheter to the vessel. Verify if the cannulation was successful by gently pulling the plunger. The syringe should fill with blood that returns to vasculature when the plunger is pushed down. The cannula allows for intravenous application of additional anesthesia and for injection of antibodies and dyes. Remove the silk hooked into the teeth. Subsequently prepare organ/tissue to be imaged.

3.2 Ear Skin Preparation (for Imaging of the Vasculature of the Skin)

1. Place the stack of microscopic slides (Fig. 1b) taped together on the Plexiglas board dedicated for imaging (Fig. 1a) in such a way that it would tightly cling to the tall edge of the Plexiglas board, preventing the slides from sliding.
2. Depilate the mouse ear skin by applying a small amount of hair removal cream with a cotton swab and evenly covering the whole ear on both dorsal and ventral sides (*see* **Notes 1** and **2**). Wait 5 min and then check if it is possible to remove hair with a cotton swab soaked with PBS. If not, wait another minute. Next, remove the hair and excess hair removal cream from the ear with a cotton swab soaked with PBS and then wash the ear with PBS. Gently wipe dry the ear with a paper towel.
3. With one hand, grab the excess of the skin on the mouse abdomen, and with the other hand grab the syringe used for the cannulation of jugular vein and move the mouse from the Plexiglas board used for cannulation to the Plexiglas board dedicated for microscopic imaging (Fig. 1a). Place the mouse on its back in such a way that the ear on the opposite side of the cannulated vein is placed on the stack of microscopic slides (if the right vein was cannulated, the left ear will be imaged and vice versa), ventral side up (Fig. 1a) (*see* **Note 3**).
4. Using a cotton swab soaked with PBS, adjust the ear position (if needed) so the ear lays flat and even. Fix the syringe with attached tubing with the removable tape to the imaging board on the opposite side than the imaged ear, so that it will not move around (Fig. 1a).
5. Using the glass cutting diamond knife, cut the coverslip to the desired size (*see* **Note 4**). Use blunt ended forceps with a curved tip to place the coverslip on the ventral side of the ear and immediately apply the saline underneath the coverslip (*see* **Note 5**). The ear skin is ready for imaging (Fig. 1a) (*see* **Note 6**).

3.3 Cremaster Muscle Preparation (for Imaging of the Classical Vascular Bed)

1. Prepare the board dedicated for in vivo imaging of the cremaster muscle (Fig. 2) (*see* **Note 7**). Attach the larger coverslip (24 × 60 mm) to the Plexiglas board with the crystal tape so that the hole in the board is completely covered (Fig. 2b) (*see* **Note 8**).
2. With one hand, grab the excess of skin on the mouse abdomen, and with the other hand grab the syringe to which cannula is attached and inserted into the jugular vein. Move the mouse from the Plexiglas board used for cannulation to the Plexiglas board dedicated for microscopic imaging of the cremaster muscle (Fig. 2). Place the mouse on its back, with hind limbs toward you. Try to slide in the mouse as far as it is possible toward the coverslip taped to the Plexiglas board.
3. Fold the mouse tail back so that it goes under the body of the animal and use the crystal tape to attach it to the Plexiglas board so that it does not interfere with imaging.
4. Using a piece of ribbon, tie each of the mouse hind limbs to a separate handle, each on the opposite side of the coverslip that is attached to the Plexiglas (Fig. 2c) (*see* **Note 9**). Make a loop with a single piece of ribbon and place it over the mouse left hind limb (the leg on your right-hand side). Tie a knot over the mouse ankle and place the leg above the handle on your right-hand side (Fig. 2c). Slightly stretch the ribbon and attach it to the board with the surgical tape.
5. Proceed the same way with the mouse right hind limb (the leg on your left-hand side). Pay attention that this time after tying the knot over the mouse ankle, the leg is going to be placed under the handle on your left-hand side (Fig. 2c). After doing so, slightly stretch the ribbon and attach it to the board with the surgical tape.
6. Using a cotton swab soaked with mineral oil, cover the surface of mouse scrotum (visibly darker than the rest of mouse skin) with mineral oil.
7. Using blunt ended forceps with a straight tip, grab the tip of the scrotum, gently stretch and cut a small piece, a bit further from the tip of the forceps.
8. Holding the mouse scrotum with one blunt ended forceps with a straight tip, place the other blunt ended forceps inside the scrotum and pull out a piece of tissue.
9. Gently try to expose one of the testis (a light creamy tissue with visible red vessels, surrounded by a jelly-like connective tissue) (Fig. 3a) (*see* **Note 10**). Attach the self-closing forceps with a straight tip on the end of the testis and stretch it gently. Place the self-closing forceps holding the tip of the testes on the Plexiglas board. Rinse the testis with saline.

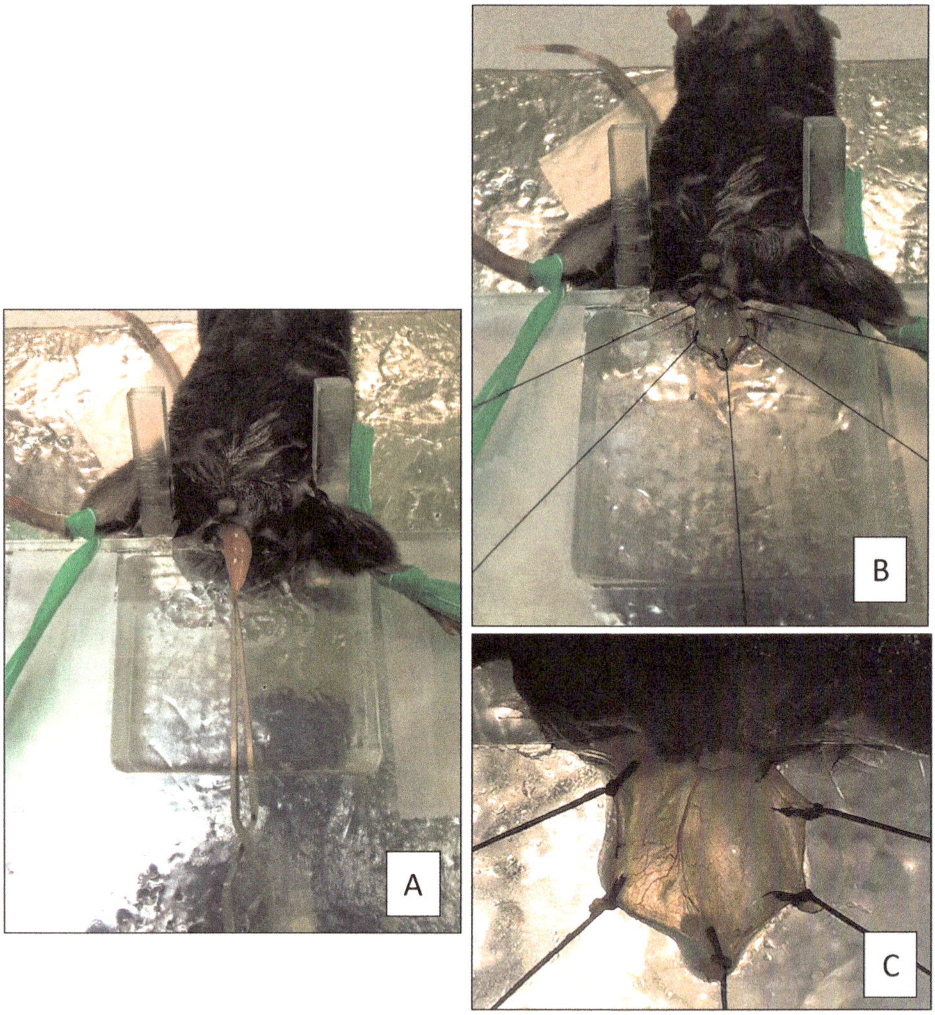

Fig. 3 Exemplary preparation of the cremaster muscle for in vivo imaging. An exposed mouse testis surrounded by a jelly-like connective tissue (**a**). The cremaster muscle stretched with silk sutures (**b**); a close-up on the cremaster muscle vasculature (**c**)

10. Gently clean the exposed testis from the remaining connective tissue using blunt ended forceps with a straight tip.
11. Using two blunt ended forceps with a straight tip create a loop of silk suture around the end of the forceps. First, drag the end of the silk suture below the self-closing forceps, create a loop, initially against the forceps and then slowly move it down to the tip of the forceps. Close the loop on the small piece of testis tissue (slightly below the end of self-closing forceps) by creating a single knot.
12. Open the self-closing forceps and release the testis tissue. Gently pull one of the silk suture's end and fix it to the Plexiglas board, cut the second remaining end next to the knot.

13. Rinse the testis with saline. Using the thermal cautery, make a hole at the beginning of the testis tissue, slightly above the knot. Be careful not to burn through the tissue.
14. Insert the sharp ended forceps with a curved tip inside the testis through the hole made previously between the two layers. Insert the forceps clenched (until you reach the opposite side of the testis), then slightly open the forceps so that the membrane surrounding the testis (the cremaster muscle) will spread on the forceps. Using thermal cautery, cut the tissue spread on the forceps with one continuous move, starting next to the knot and going up toward the scrotum. Rinse thoroughly with saline.
15. Using the self-closing forceps with a straight tip, grab the edge of previously cut cremaster muscle tissue on your right-hand side. Gently stretch the tissue and place the self-closing forceps holding the edge of the cremaster muscle on the Plexiglas board (*see* **Note 11**).
16. Using two blunt ended forceps with a straight tip create a loop of silk suture around the end of the forceps. First, drag the end of the silk suture below the self-closing forceps, create a loop, initially against the forceps and then slowly move it down to the tip of the forceps. Close the loop on the small piece of cremaster tissue (slightly below the end of self-closing forceps) by creating a single knot.
17. Open the self-closing forceps and release the cremaster tissue. Gently pull one of the silk suture's end and fix it to the Plexiglas board and cut the second remaining end next to the knot.
18. Repeat the same procedure on the cremaster muscle tissue on your left-hand side.
19. At this point, the cremaster muscle should be stretched into a characteristic almost rhombus shape (Fig. 3b, c). Rinse the cremaster thoroughly with saline (*see* **Note 12**). If necessary, additional sutures can be used to further stretch the cremaster (*see* **Notes 13** and **14**).
20. Using blunt ended forceps with a straight tip, grab the exposed epididymis (visible as a light creamy folded tissue) and lift it up. Using thermal cautery cut thin ligament of the connective tissue underneath the testis (it attaches the whole testis to the cremaster muscle) taking care not to burn the ductus deferens or the testis itself (*see* **Note 15**).
21. Using the blunt ended forceps with a straight tip, place the testis and epididymis back in the scrotum by gently pushing it all back inside. Rinse the exposed cremaster muscle thoroughly with saline (*see* **Note 16**).

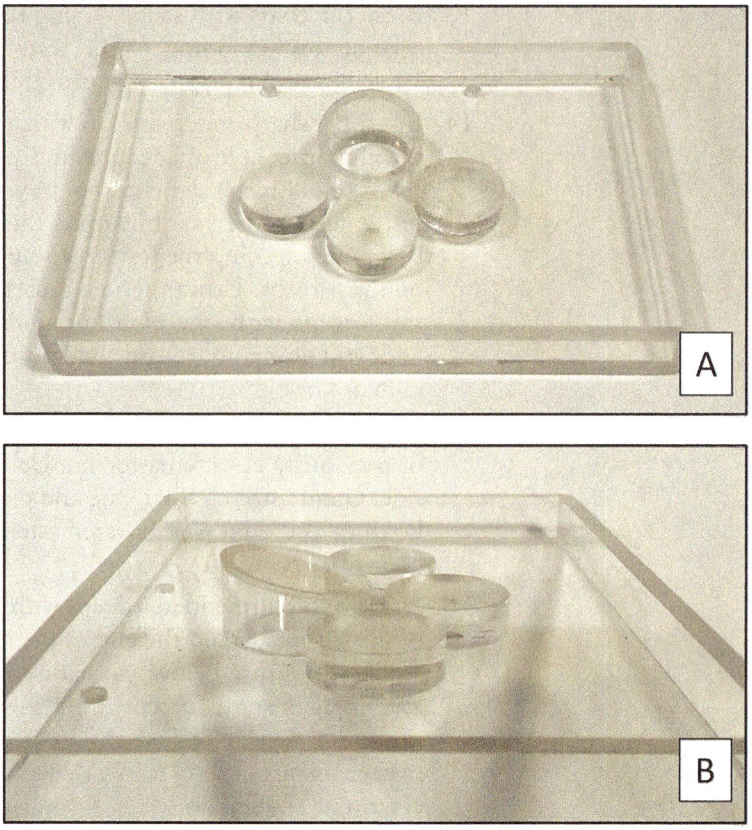

Fig. 4 A board designated for in vivo imaging of the liver. A custom made Plexiglas imaging board with three identical cylinders (on which the mouse body should be placed) and an additional angled cylinder (on which the liver lobe should be placed) (**a**). A close-up from a side on the cylinders (**b**)

22. Take a coverslip (24 × 24 mm) between your thumb and index finger and apply two lines of Vaseline, one for each of the two opposite sides of the coverslip (*see* **Note 17**).
23. Using forceps, place the coverslip over the exposed cremaster muscle, Vaseline side down, and gently push it down so it sticks at both sides. Vaseline lines should be on the left and right side of the cremaster muscle.
24. Apply the saline underneath the coverslip, creating a wet chamber. The cremaster muscle is ready for imaging (*see* **Note 12**).

3.4 Liver Preparation (for Imaging of the Liver Sinusoids)

1. Prepare the board dedicated for in vivo imaging of the liver (Fig. 4). Imaging board is a custom-made Plexiglas board with taller edges and four cylinders, three identical with 30 mm diameters and one angled with a 38 mm diameter (Fig. 4). Stick surgical tape to the board to connect 3 smaller cylinders by creating a cross (Fig. 5a). Take a half of a Kimwipes tissue

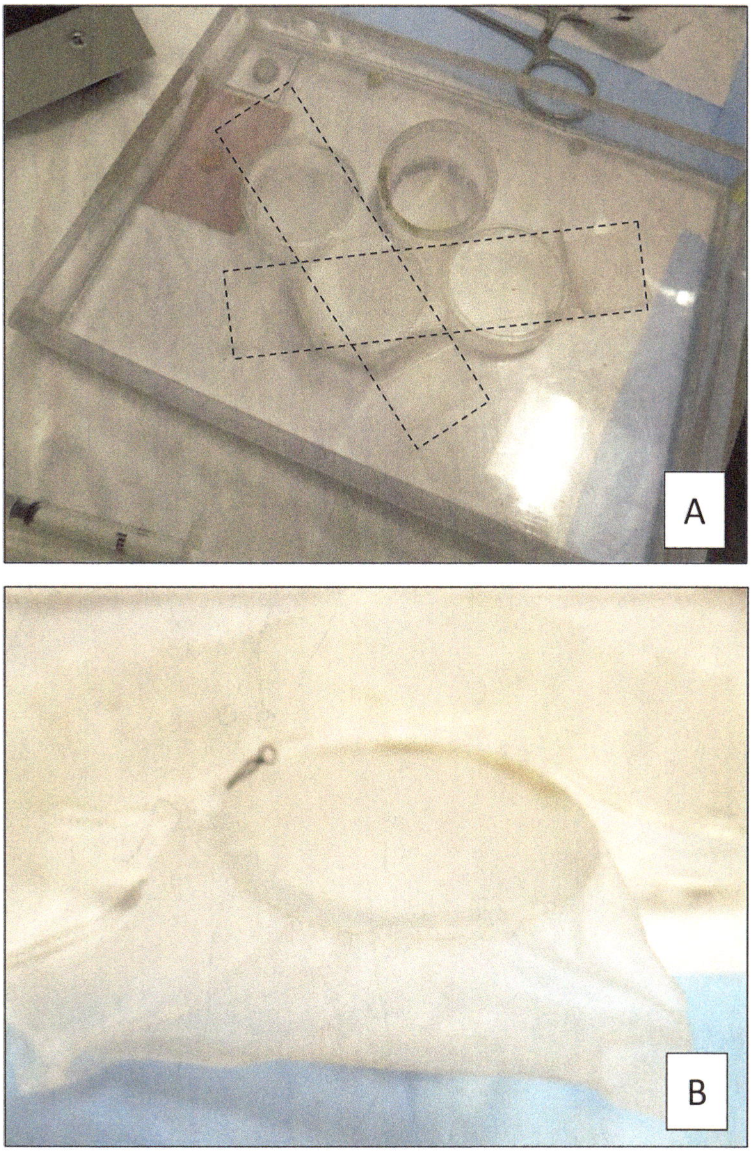

Fig. 5 Preparation of a board for in vivo imaging of the liver. A surgical tape should be placed on three identical (smaller) cylinders (dotted line) to create a cross-like shape (**a**). A close-up on the angled (larger) cylinder. It should be covered by a half of a Kimwipes tissue soaked with saline, and fitted closely on the cylinder (**b**)

and place it on the top of the big cylinder, rinse it with saline and fit it closely around the cylinder (Fig. 5b) (*see* **Note 18**). Place the single gauze halfway through the board (on its right-hand side) and rinse it with saline.

2. After successful cannulation, turn the surgical board so that the tail of the mouse is now facing you and cover the skin of the abdomen using a cotton swab soaked with mineral oil (*see* **Note 19**).

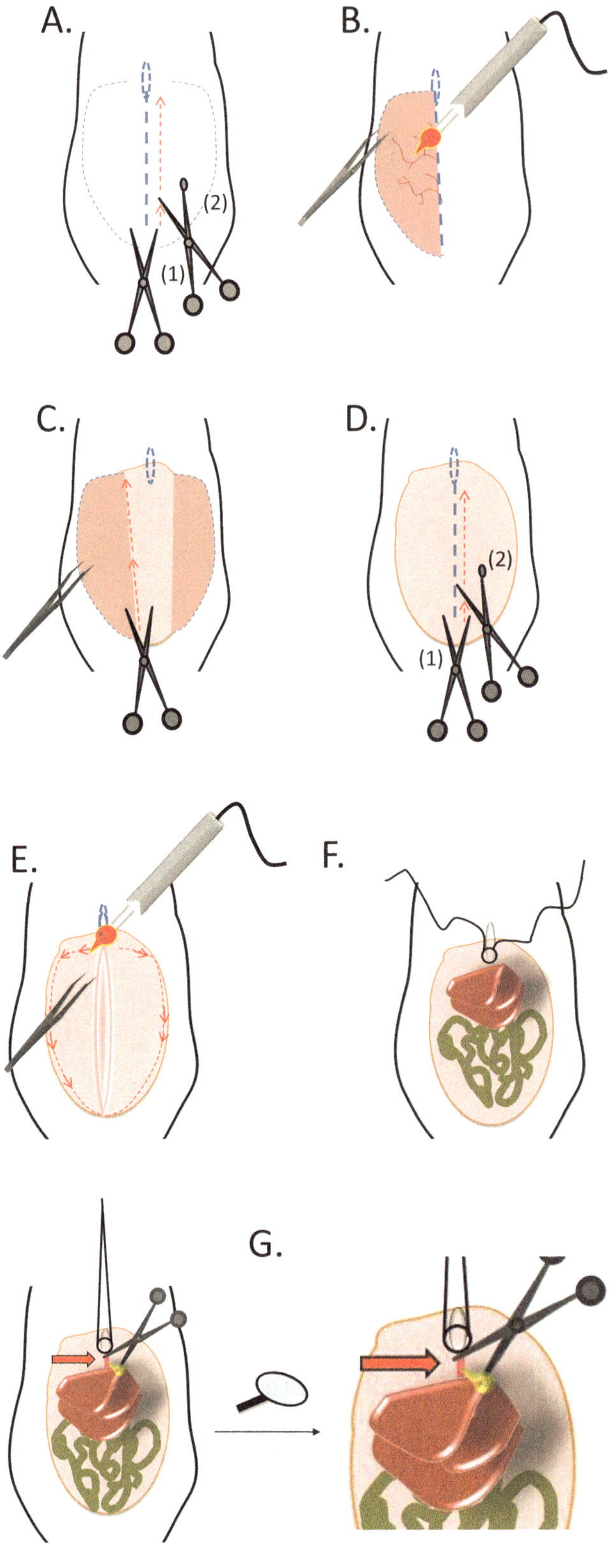

Fig. 6 Main steps of mouse liver preparation for in vivo imaging. (**a**) Open the skin on the abdomen starting from the lower part of the abdomen and continue in

3. Grab the skin on the mouse abdomen. Using blunt ended forceps and fine scissors with a sharp tip make a vertical midline incision starting near the lower abdominal quadrants, and then continue the incision with ball tip scissors until the white sternum appears (Fig. 6a).
4. Using blunt ended forceps, grab the skin on the left-hand side of the incision. Using Crile hemostatic forceps, separate the skin from the peritoneal membrane by clenching and spreading hemostatic forceps between the skin and the peritoneal membrane several times (make so called "pockets") (*see* **Note 20**). Then turn the surgical board so that the head of the mouse is now facing you and proceed the same way with the right side of the mouse body.
5. Return the surgical board to the previous position. Grab the skin on the left-hand side of the incision with forceps (Fig. 6b), revealing the inner, vascularized wall of the skin. Using thermal cautery, coagulate the blood vessels of the skin (Figs. 6b, 7a) (*see* **Notes 21** and **22**). Proceed the same way with the right body side after turning the board as described in Subheading 4.
6. On the outside of the cauterization line, cut off the skin using sharp ended scissors and remove the skin flaps from both sides (in the direction from the lower part of the abdomen to the sternum, Fig. 6c) (*see* **Note 23**).
7. At this point, clean and sterilize all instruments with 70% EtOH to avoid transferring of mouse hair on the peritoneal membrane and internal organs, especially the liver.
8. Using blunt ended forceps, grab the peritoneal membrane in the midline of the lower part of the abdomen and make a small incision at first with sharp ended scissors and then continue with ball tip scissors along the linea alba of the abdomen until xiphoid cartilage appears (Fig. 6d).
9. Using a thermal cautery, remove the peritoneal membrane starting from the xiphoid cartilage along the left rib line to

Fig. 6 (continued) the direction of the sternum. At first use sharp scissors (1), and then follow with ball tip scissors (2). (**b**) Coagulate the blood vessels on the internal side of the skin with a thermal cautery. While thermally closing the vessels hold the skin with forceps. (**c**) Cut off the skin using sharp scissors (hold it with forceps during cutting). (**d**) Make an incision in the peritoneal membrane with sharp scissors (1), then switch to ball tip scissors (2) and continue with the cut through the midline and up to xiphoid cartilage. (**e**) Cut off the peritoneal membrane with the thermal cautery along the costal margin, through the midaxillary line, down to the lower quadrants of the abdomen. (**f**) Make a loop in the center of the silk suture and tie it twice around the xiphoid cartilage. (**g**) Cut off the ligament (pink, red arrow) of the liver using sharp scissors

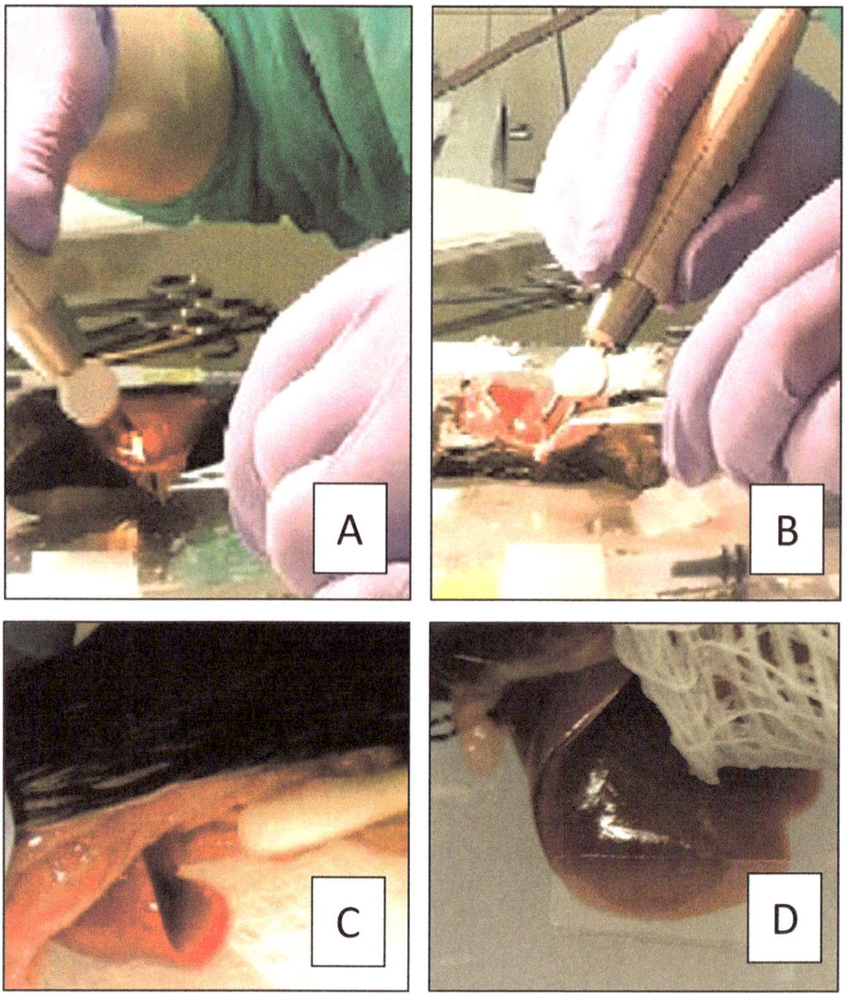

Fig. 7 Representative images from a surgery preparing the mouse for in vivo imaging of the liver sinusoids. (**a**) Coagulation of the mouse skin blood vessels with a thermal cautery. (**b**) Abscission of the peritoneal membrane with the cautery. (**c**) Separation of the left liver lobe from the intestines using a cotton swab tip soaked with saline. The liver should form a bell-like shape. (**d**) The left lobe of the liver prepared for imaging. A small piece of a coverslip should be placed on the tip of the liver. The coverslip should be placed in a horizontal position to the surface of the liver

the lower abdomen. Proceed in the same way on the right-hand side (the cut must be continuous on each side) (Figs. 6e, 7b).

10. Take a long silk suture (~20 cm) and make a loop in the middle, then tie the silk twice around the xiphoid cartilage (Fig. 6f).
11. Gently lift the chest by pulling up the silk so that you can see the ligament that connects the gall bladder with the diaphragm and transect it with sharp scissors (*see* **Note 24**). Carefully release the silk suture (Fig. 6g).

12. Now move the animal to the prepared imaging board (Fig. 5) (*see* **Note 25**). Detach the tape from limbs that kept the mouse immobilized. Turn the plate perpendicular to you, so that the mouse head is now on your left-hand side. With your right hand, grasp paws, and with your left hand grab the cannula and move the animal to the imaging board – carefully lay down the mouse on the right-hand side of the body (in the lateral position) on three identical smaller cylinders and in such a way that the mouse liver is parallel to the angled cylinder (where the liver lobe will be situated). Stretch and fasten the silk suture tied around the xiphoid cartilage and fix it to the imaging board with a tape (parallel to the upper body).
13. Using a cotton swab soaked with saline, gently move the intestines away from the abdomen, separating them from the liver (Fig. 7c). Now, with a cotton swab tip soaked with saline "guide the stomach" to push the liver out (the ventral side of the left lobe of the liver must be gently flipped onto the angled cylinder by the stomach). If accurately positioned the liver lobe should form a bell-like shape (Fig. 7c) (*caution*: never touch the liver with either surgical tools or the cotton tip). After you flipped the liver, move the stomach gently back toward the intestines. Now wrap moistened gauze around stomach and intestines. This will help to avoid contact and pressure on the liver by internal organs, which would increase liver "movement" and could also disturb blood flow in the sinusoids (*see* **Note 26**).
14. Cut a small piece ($\sim 2.5 \times 1$ cm) from a larger coverslip using a diamond knife and carefully place it on the tip of the liver lobe with forceps (*see* **Notes 27** and **28**). Use caution in placing the coverslip in the horizontal position) (Fig. 7d).
15. Fill the space underneath the coverslip with saline to create a wet chamber (*see* **Note 29**). Now the liver is ready for imaging (*see* **Note 30**). Check vitals every 10–15 min, and regularly refill the chamber with saline to keep it moist during the whole procedure (*see* **Notes 31** and **32**).

3.5 Imaging of Neutrophils and NETs

3.5.1 Neutrophil Labeling for the IVM Technique

The jugular vein cannulation and the preparation of an organ of interest for imaging with IVM are a prologue to visualization of target cells/tissues/organs. IVM represents an approach that allows for tracking and visualizing the fate of the cells of interest, helps to understand dynamic processes, cell behavior and cellular interactions occurring in a living organism. In order to visualize cells and biological structures in fluorescent/confocal microscopes, fluorescent labeling is required. This is usually achieved by application of specific (monoclonal) antibodies directed against specific surface antigens on target cells (*see* **Notes 33–35**). To detect murine neutrophils, antibodies against a highly selective marker, such as Ly6G, are commonly used [6]. Despite an early report [7], intravenous application of the anti-Ly6G antibodies does not deplete neutrophils [8]. In the past, anti-Gr-1 antibodies were also frequently used; however, they recognize two epitopes: Ly6G and Ly6C. As the latter marker is also expressed on dendritic cells and subpopulations of lymphocytes and monocytes/macrophages [9], its application is currently considered incorrect and nonselective. For IVM purposes, antibody labeling of neutrophils should be administrated intravenously, and it is recommended to apply antibodies just after the successful cannulation (*see* Subheading 3.1) and before surgery of the organ to be imaged (*see* **Note 36**). In this way, there is time for the antibodies to bind to target cells. However, the antibodies can also be administrated later during real-time imaging. Alternatively, and depending on the setting of a particular experiment, the antibodies can be administered via the tail vein, even several hours prior to the surgery (i.e., cannulation) (*see* **Notes 37**).

Another possibility for labeling neutrophils is to first isolate neutrophils from mice, stain them ex vivo, and then reinject them into animals. In this approach, isolated and unlabeled cells from donor mice are exogenously labeled with nontoxic cell tracker dyes and reinjected into recipient mice. For example, Lämmermann et al. [2] used two cell trackers dyes, one red (CMTPX) and one green (CMFDA). Neutrophils were isolated from two littermates, one with a knockout phenotype and the other one – wild type. Since each neutrophil population was stained with different color, when they were reinjected together into a recipient mouse (wild-type) differences in behavior of the two populations could be imaged with IVM. The use of cell tracker dyes enables the labeling of cell membrane, cytoplasm, and nucleus, and the dye is visible even after several cell divisions (this does not apply to neutrophils, which do not divide after maturation, but might be useful when working, for example, with activated lymphocytes).

Another possibility to visualize neutrophils is the use of reporter mice. They are engineered in such a way that they express a fluorescent protein under a control of a gene promoter; a gene which is uniquely expressed by neutrophils. The exemplary fluorescent proteins are green fluorescence protein (GFP) or red fluorescent protein from *Discosoma* coral (DsRed). The advantage of reporter mice is the constitutive and highly specific expression of the fluorescent protein, so there is no need to use antibodies, and most importantly, the cells will be fluorescent even when localized outside of vasculature. The latter comment refers to the fact that antibodies can only reach cells present in blood. Unfortunately, there are not many reporter mice available for neutrophils. The most common are LysM-eGFP mice, which express the GFP protein under the control of the endogenous lysozyme M promoter [10]. Lysozyme, however, is expressed not only by neutrophils but also monocytes/macrophages. It is possible to distinguish the two cell populations through their respective bright and dim expression of the fluorescent protein and the high motility of neutrophils. However, assignment of cells to either population is somewhat subjective. Nowadays the most specific to neutrophils are *Catchup* mice. They were constructed by modulating the neutrophil-specific locus Ly6G with a knock-in allele expressing Cre recombinase and the fluorescent protein tdTomato [11]. The high brightness and photostability of tdTomato fluorescent protein seems to be a powerful tool for neutrophil tracking [11] and also a useful application for deep-tissue imaging [12].

Another important tool to use in IVM studies of neutrophils are photoactivatable or photoconvertible reporter mice. The use of photoactivatable fluorescent proteins (PA-FPs) and photoconvertible fluorescent proteins (PC-FPs) that mark cells of interest allows observations of the migration and fate of individual cells/subpopulations (neutrophils) in and between tissues after phototransform processes (which occurs in one area/location of interest). There are several types of PA-FPs, namely PA-FPs fluorescence green (e.g., PA-GFP) or red (PA-mRFP1 or PA-mCherry1), and PC-FPs (e.g., Kaede), which can be converted from one fluorescence emission to another (e.g., green to red photoconversion in response of activating light). Wang et al. [13] developed Ly6G-PA-GFP mice and showed their application for IVM. In this elegant study, the authors showed how mobile neutrophils can be tracked when translocating between distant organs (a set of neutrophils was photoconverted in one location and then the cells were detected in another organ by their new color).

3.5.2 Visualization of Neutrophil Extracellular Traps (NETs) with IVM

IVM allows visualization not only of various cell types but also structures present/formed in the imaged tissue or organ. In fact, any structure that can be specifically targeted with a fluorescent dye or antibody can be visualized with IVM. Our group images neutrophil extracellular traps (NETs) cast by neutrophils into the vasculature. NETs are composed of extracellular DNA (extDNA) that forms their backbone, and it is decorated with numerous antimicrobial proteins and enzymes, such as nuclear proteins, histones and also granular proteins, including neutrophil elastase (NE) and myeloperoxidase (MPO) [14]. NETs are involved in various pathological conditions, in which they play a significant role [1, 4, 14]; therefore, it is not surprising that application of IVM for NET imaging was established shortly after their discovery. However, there are several obstacles for NET imaging. First, at least 2–3 components of NETs have to be detected at the same time, and the signal of each of them has to colocalize with the others in order to claim that these are indeed NETs. This is because numerous components of NETs can be released independently of NETs (e.g., during neutrophil degranulation). Therefore, simultaneous colocalization of NET components is principal for these studies. As extDNA functions as a scaffold for the anchored proteins, it is important to visualize. For detection of extDNA, cell impermeable SYTOX Green is most frequently used [4]. Alternatively, SYTOX Orange can be applied, but it can penetrate live cells. Additionally, some protein components of NETs should be detected, and preferentially of various origin (i.e., of nuclear and granular origin). Our group stains for histones (e.g., H2A.X or H3; preferably their citrullinated forms) and NE, respectively [4]. As specificity is always a high demand, to confirm the presence of NETs, one can administer either intravenous DNase I to remove/degrade extDNA or heparin, which dissolves the NET scaffold [15].

4 Notes

1. Hair removal is an important step in the preparation process, as it minimizes the autofluorescence of hair follicles during imaging and assures smooth surface for the coverslip to attach.
2. Avoid grabbing and manipulating the ear with your fingers because even the slightest applied pressure on the ear skin can cause activation of the immune cells in the ear tissue.
3. When adjusting the ear position on the stack of microscope slides, do not be afraid to lift up the mouse head and move it gently above or below the imaging table to spread it flat and even.
4. Before placing the coverslip on the ear, clean it from the dust with 70% EtOH and wipe it dry with the dust-free paper. Any remaining dust particles and/or paper threads may exhibit some autofluorescence potentially interfering with the desired fluorescence signal.
5. When applying saline between coverslip and the ear, be careful not to use too much saline so the coverslip will not detach and float around.
6. During the whole time of imaging, make sure that the ear does not dry out as it may cause damage to the skin and create artifacts. Every now and then, apply fresh portion of saline underneath the coverslip.
7. Make sure to use a male mouse for this procedure.
8. If you do not have one coverslip big enough to cover the hole in the Plexiglas, use two smaller coverslips but be sure they overlap, and firmly secure them to the Plexiglas.
9. Avoid silky ribbons or tapes for hind limb attachment, as they have a tendency to untie prior to or during the imaging, causing mouse limbs to detach from the handles.

10. If you cannot find the testis right away, try to bring it outside by gently pressing on the lower abdomen toward the scrotum.
11. Be careful not to stretch the cremaster tissue too much at any point to avoid tearing it apart. After the sutures are secured in one place, it is possible to pull them to stretch the tissue later on.
12. During the initial prep and while imaging, make sure that the cremaster muscle does not dry out. Every now and then, apply fresh portion of saline on the tissue (during the prep) and underneath the coverslip (during imaging).
13. If the cremaster tissue is thick and it does not lie flat, do not be afraid to fix two extra knots on each side of the cremaster muscle as it might help to flatten it.
14. If the silk suture knot slips out of the tissue, calmly fix it by repeating the procedure with the self-closing forceps with a straight tip and create a new knot. Return the tissue to the previous position.
15. While using the thermal cautery, be sure that the tip of the cauterizer is well warmed up as you want to cut and coagulate at the same time. If at any moment during the prep bleeding occurs, use the cauterizer to stop it by simply touching the bleeding point for a second with the tip of the cauterizer.
16. If the testis and epididymis will not easily go back inside the scrotum, use some extra saline rinse and slide it inside.
17. Before applying Vaseline on the coverslip and placing it on the cremaster muscle, clean the coverslip from the dust with 70% EtOH and wipe it dry with the dust-free paper. Any remaining dust particles and/or paper threads may exhibit some auto-fluorescence, potentially interfering with the desired fluorescence signal.
18. When placing a Kimwipes tissue on the angled cylinder where the liver is placed during imaging, avoid air bubbles and folds.
19. Instead of using mineral oil, the mouse can also be shaved to remove abdominal hair. Do not apply too much mineral oil, otherwise it can detach tapes that hold mouse limbs.
20. If during cutting the skin or the peritoneal membrane any bleeding occurs, use thermal cautery to close the vessels immediately, alternatively you can use cotton swab to stop the bleeding. When separating the skin from the peritoneum, you can additionally use the cauterizer to cut off thin connective tissue.
21. During the coagulation of skin blood vessels, first close the largest vessel from which smaller ones branch out. This facilitates an effective and faster closure of all vessels.

22. While using the thermal cautery, especially for the peritoneal membrane abscission, be sure that the tip is well warmed up.
23. After opening the abdomen apply saline on the peritoneal cavity to keep the internal organs moist.
24. When cutting the ligament, liver lobes adhere to the diaphragm and you may not see the ligament clearly, to overcome this apply some saline between the diaphragm and the liver.
25. Before moving the animal to the imaging board rinse more saline onto gauze that was placed on the plexiglas board during the prep, to make sure it is well soaked. While carrying the mouse to the imaging board be careful not to touch the liver with a silk suture.
26. While adjusting the liver position on the imaging board, do not be afraid to lift up the mouse to situate the liver lobe in the middle of the cylinder; however, always remember to loosen a silk suture before lifting the mouse.
27. The size of the piece of the coverslip depends on the liver lobe size, if it is bigger cut off larger piece to adjust it on the surface of the liver in such a way that it adheres to the liver surface evenly.
28. Before placing the piece of a coverslip on the liver lobe make sure it is clean. Remove any dust with 70% EtOH and a dust-free towel.
29. Do not apply too much saline underneath the coverslip so the liver will not detach and float down.
30. Make sure that the chest of a mouse does not touch the objective during imaging.
31. If mouse breathing disturbs the stable image acquisition take off the coverslip and stretch the silk suture that it is tied around the xyphoid cartilage to move the liver away from the rest of the body.
32. If during operation or image acquisition additional anesthesia is required (the mouse displays symptoms of consciousness) inject anesthetics by cannula or alternatively you can rinse the intestines with the mixture of anesthetics (higher doses of agents are required to induce anesthesia when delivered on the intestines).
33. Antibodies, isotype controls and other dyes should be suspended in sterile saline. To compare obtained results, it is critical to use the same concentration of antibodies and other dyes, and moreover, maintain the same conditions and settings during imaging of both, control and experimental groups.
34. Before intravenous application, desired antibodies should be freshly prepared (they are usually used in concentrations

ranging from 0.5 to 5 μg per mouse), kept on ice, and protected from the light. Depending on the design of the experiment, prepared antibodies should be administered 15–20 min before imaging, for their precise distribution and labeling (the most common approach). The antibodies can be administered at the cannulation step (or before it via the tail vein) or as needed during imaging. In most real-time imaging studies, several antibodies are used simultaneously. In this case, they can be mixed together and injected at the same time. It is of high importance to exclude nonspecific or Fc-receptor–mediated binding by testing all antibodies along with their fluorescent isotype controls.

35. If possible, purchase monoclonal and not polyclonal antibodies, as they recognize only one epitope on the antigen. For our studies, we have used standardized monoclonal antibodies. Neutrophils are labeled with anti-Ly6G antibodies (clone 1A8), and cremaster muscle and skin vasculature are labeled with anti-CD31/PECAM antibodies. To label protein NET components, we use anti-mouse citrullinated histones (H2/H3/H4) and anti-mouse neutrophil elastase (clone M18) antibodies. The backbone of NETs (extDNA) is stained with SYTOX Green. If fluorescently labeled antibodies are unavailable, it is necessary to self-conjugate purified antibodies with protein labeling fluorescent kits, according to the manufacturer's instructions.
36. The administration of antibodies or other dyes during imaging requires a well-functioning cannula at all stages of mouse preparation and imaging.
37. After imaging is finished the mouse should be euthanized with an overdose of the anesthetic (100 μL) followed by cervical dislocation.

Acknowledgments

This work was supported by National Science Center (NCN, Poland) grant K/PBO/000669 from National Science Center, Poland (NCN) to EK.

References

1. Kolaczkowska E, Kubes P (2013) Neutrophil recruitment and function in health and inflammation. Nat Rev Immunol 13:159–175
2. Lammermann T, Afonso PV, Angermann BR et al (2013) Neutrophil swarms require LTB4 and integrins at sites of cell death in vivo. Nature 498:371–375
3. Dal-Secco D, Wang J, Zeng Z et al (2015) A dynamic spectrum of monocytes arising from the in situ reprogramming of CCR2+ monocytes at a site of sterile injury. J Exp Med 212:447–456
4. Kolaczkowska E, Jenne CN, Surewaard BG et al (2015) Molecular mechanisms of NET formation and degradation revealed by intravital imaging in the liver vasculature. Nat Commun 6:6673
5. Hwa C, Aird WC (2007) The history of the capillary wall: doctors, discoveries, and debates. Am J Physiol Heart Circ Physiol 293: H2667–H2679
6. Wojtasiak M, Pickett DL, Tate MD et al (2010) Depletion of gr-1+, but not Ly6G+, immune cells exacerbates virus replication and disease in an intranasal model of herpes simplex virus type 1 infection. J Gen Virol 91:2158–2166
7. Wang JX, Bair AM, King SL et al (2012) Ly6G ligation blocks recruitment of neutrophils via a beta2-integrin-dependent mechanism. Blood 120:1489–1498
8. Yipp BG, Kubes P (2013) Antibodies against neutrophil LY6G do not inhibit leukocyte recruitment in mice in vivo. Blood 121:241–242
9. Hestdal K, Ruscetti FW, Ihle JN et al (1991) Characterization and regulation of RB6-8C5 antigen expression on murine bone marrow cells. J Immunol 147:22–28
10. Faust N, Varas F, Kelly LM et al (2000) Insertion of enhanced green fluorescent protein into the lysozyme gene creates mice with green fluorescent granulocytes and macrophages. Blood 96:719–726
11. Hasenberg A, Hasenberg M, Mann L et al (2015) Catchup: a mouse model for imaging-based tracking and modulation of neutrophil granulocytes. Nat Methods 12:445–452
12. Deliolanis NC, Kasmieh R, Wurdinger T et al (2008) Performance of the red-shifted fluorescent proteins in deep-tissue molecular imaging applications. J Biomed Opt 13:044008
13. Wang J, Hossain M, Thanabalasuriar A et al (2017) Visualizing the function and fate of neutrophils in sterile injury and repair. Science 358:111–116
14. Brinkmann V, Reichard U, Goosmann C et al (2004) Neutrophil extracellular traps kill bacteria. Science 303:1532–1535
15. Fuchs TA, Brill A, Duerschmied D et al (2010) Extracellular DNA traps promote thrombosis. Proc Natl Acad Sci U S A 107:15880–15885

INDEX

Mark T. Quinn and Frank R. DeLeo (eds.), *Neutrophil: Methods and Protocols*, Methods in Molecular Biology, vol. 2087, https://doi.org/10.1007/978-1-0716-0154-9,

9781071601532